ROUTLEDGE HANDBOOK OF WATER LAW AND POLICY

Water plays a key role in addressing the most pressing global challenges of our time, including climate change adaptation, food and energy security, environmental sustainability and the promotion of peace and stability. This comprehensive handbook explores the pivotal place of law and policy in efforts to ensure that water enables positive responses to these challenges and provides a basis for sound governance.

The book reveals that significant progress has been made in recent decades to strengthen the governance of water resource management at different scales, including helping to address international and sub-national conflicts over transboundary water resources. It demonstrates that 'effective' laws and policies are fundamental drivers for the safe, equitable and sustainable utilization of water. However, it is also shown that what might constitute an effective law or policy related to water resources management is still hotly debated. As such, the handbook provides an important and definitive reference text for all studying water governance and management.

Alistair Rieu-Clarke is Chair in Law at the Law School, University of Northumbria, UK.

Andrew Allan is Senior Lecturer in National Water Law at the Centre for Water Law, Policy & Science, University of Dundee, UK.

Sarah Hendry is Senior Lecturer in Water Law at the Centre for Water Law, Policy & Science, University of Dundee, UK.

ROUTLEDGE HANDBOOK OF WATER LAW AND POLICY

Edited by Alistair Rieu-Clarke,
Andrew Allan and Sarah Hendry

Routledge
Taylor & Francis Group

LONDON AND NEW YORK

earthscan
from Routledge

First published 2017
by Routledge

2 Park Square, Milton Park, Abingdon, Oxfordshire OX14 4RN
52 Vanderbilt Avenue, New York, NY 10017

Routledge is an imprint of the Taylor & Francis Group, an informa business

First issued in paperback 2019

British Library Cataloguing-in-Publication Data
A catalogue record for this book is available from the British Library

Library of Congress Cataloging in Publication Data
Names: Rieu-Clarke, Alistair, editor. | Allan, Andrew, editor. | Hendry, Sarah, 1963- editor. Title: Routledge
handbook of water law and policy / Edited by Alistair Rieu-Clarke, Andrew Allan and Sarah Hendry. Other titles:
Handbook of water law and policy Description: New York: Routledge, 2016. | Includes bibliographical references
and index. Identifiers: LCCN 2016028575 | ISBN 9781138121201 (hbk) | ISBN 9781315651132 (ebk) Subjects:
LCSH: Water--Law and legislation. | Water-supply--Law and legislation. | Water rights. Classification: LCC K3496.
R68 2016 | DDC 346.04/691--dc23LC record available at https://lccn.loc.gov/2016028575

ISBN: 978-1-138-12120-1 (hbk)
ISBN: 978-0-367-23106-4 (pbk)

Typeset in Bembo
by Cenveo Publisher Services

CONTENTS

LIST OF ILLUSTRATIONS

Figures

Table

NOTES ON EDITORS AND CONTRIBUTORS

Editors

Andrew Allan is a Senior Lecturer at the Centre for Water Law, Policy and Science at the University of Dundee. His research focuses on national legal frameworks that affect water resource management, and he has a particular focus on ensuring their effectiveness in developing countries. He holds an LLM with distinction from the University of Dundee and an LLB from the University of Edinburgh.

Sarah Hendry is a Lecturer in Law at the Centre for Water Law, Policy and Science, University of Dundee. She has a Bachelor of Laws (Honours) and is a Doctor of Philosophy. Her research and teaching is in comparative national water law and the regulation and governance of water services, with especial interest in frameworks for water law reform.

Alistair Rieu-Clarke is Chair in Law at the Law School, University of Northumbria, UK. He holds a Bachelor of Laws (Honours), a Masters in Natural Resources Law and Policy (with distinction) and is a Doctor of Philosophy. He has almost 20 years' experience in research, teaching, training and consultancy activities focused on deepening knowledge and understanding of how international law can contribute to transboundary water cooperation.

Contributors

Natalia Aguilar Porras is a lawyer graduated from the University of Costa Rica (2006) and incorporated to the national bar in 2007. She has worked for the government of Costa Rica in environmentally related issues for eight years. She has conducted postgraduate studies in Environmental Law in the University of Costa Rica, and is currently undertaking Master studies in Environmental Governance in the University of Freiburg, Germany.

Aziza Akhmouch manages the OECD Water Governance Programme, which helps governments design and implement better water policies for better lives. The programme produces evidence-based economic analysis, benchmarks, peer-reviews and policy dialogues to build consensus and tailor recommendations. She is the author of several publications on water governance and the founder of the OECD Water Governance Initiative, an international multi-stakeholder network gathering twice a year in a Policy Forum. She developed and led the negotiation process towards the OECD Principles on Water Governance adopted by Ministers in June 2015. She holds a PhD in Geopolitics and a MS in International Business.

Delphine Clavreul is a Policy Analyst at the Water Governance Programme of the OECD. Her field of expertise covers a range of governance topics including multi-level governance, stakeholder engagement and water integrity. She contributes to the coordination of the OECD Water Governance Initiative, an international multi-stakeholder network sharing good practices in support of better water governance. She has contributed to several OECD water governance (country and cross-country) reports. Delphine holds an MS in Geopolitics.

Ana María Daza-Clark has been a Teaching Fellow at the University of Edinburgh since 2013, teaching a number of courses within the Department of International Law such as WTO law, international investment law, international commercial arbitration and international law. Ana María's research interests cover the functioning of the international investment arbitration regime and its interaction with other areas of law such as water governance and management. She acts as consultant and counsel for AACNI (an international law firm) and is the editor of the online newsletter *Arbitration Watch*. For several years Ana María worked as Legal Officer and Legal Director at the Public Utility Regulatory System (SIRESE) in Bolivia, advising on appeals against the decision of utility regulators. She holds a PhD from the Centre for Water Law, Policy and Science at the University of Dundee and has specialized in Law and Economics and International and EU Law at the universities of Utrecht and Maastricht in the Netherlands.

Gabriel E. Eckstein is Professor of Law at Texas A&M University, where he focuses on water, natural resources, and environmental law and policy issues at the local, national and international levels. He also serves on the Graduate Faculties of the university's Water Management and Hydrological Science programme and its Energy Institute. Professor Eckstein has served as an expert adviser and consultant for various UN agencies, non-governmental organizations, and other groups on US and international water and environmental issues. He directs the consultancy International H$_2$O Solutions, LLC (www.IH2OS.com), as well as the internet portal International Water Law Project (www.InternationalWaterLaw.org).

Rebecca L. Farnum is a 2012 EPA Marshall Scholar researching environmental peacebuilding as a doctoral candidate at King's College London. She is a Visiting Fellow of the University of East Anglia's Water Security Research Centre. Farnum earned her LLM in International Law from the University of Edinburgh in 2014 and an MSc in Water Security and International Development from the University of East Anglia in 2013. Her research interests include environmental education and activism, the intersection of sustainability and justice, and the geopolitics of Western relations with the Middle East and North Africa.

Richard Franceys (PhD) is a consultant, recently Senior Lecturer in Water Supply and Sanitation Management at Cranfield University, previously at UNESCO-IHE and WEDC, Loughborough. His work focuses upon the management and regulation of extending services to all, investigating aspects of these issues with over 100 utilities, large and small, in over 50 countries. Recent work includes the Community Water Plus investigation in 17 states in India for Australian Aid, advising the four-country WASHCost programme and leading the ten-country Economic Regulation for the Poor study for the Department for International Development, along with consultancy assignments for the World Bank, the World Health Organization, the European Investment Bank and Water and Sanitation for the Urban Poor. Until recently he was a longstanding regional consumer advocate of the Consumer Council for Water in England.

Michael Hantke-Domas (PhD) is the Chief Justice of the Third Environment Court of Chile. He holds a law degree (Andrés Bello National University, Chile) and a PhD (University of East Anglia, UK). He has experience in water and environmental law, governance, public policy, competition and economic regulation, resulting from more than 19 years of involvement in these subjects as judge, academic, regulator, United Nations staff, international consultant and lawyer.

Stephanie Hawkins is a PhD Researcher in Law at the University of Strathclyde's Centre for Environmental Law and Governance. Her research focuses on transboundary groundwater governance, with attention to concepts of power and justice in law and institutions. She is an active researcher member of the London Water Research Group, and is a member of the International Water Resources Association, the International Association for Water Law, the Socio-Legal Studies Association and the Society of Legal Scholars.

William (Bill) Howarth is Professor of Environmental Law in Kent Law School at the University of Kent and Coordinator of Masters Programmes in Environmental Law. He is a past editor of the *Journal of Water Law*; author, co-author or editor of 14 books on different aspects of the law relating to the aquatic environment; and author of over 100 reports, monographs and academic journal articles on diverse aspects of environmental law. He is Honorary Legal Adviser to the Institute of Fisheries Management, a member of the Committee of Fish Legal, the Regional Representative for the International Association for Water Law.

Paul Hutchings (PhD) is a Lecturer at the Cranfield Water Science Institute, where he leads the Community Water and Sanitation MSc programme. His recent research has focused on institutional models for supporting community-managed rural water services in India. His first degree was in Geography at the University of Liverpool, where he also studied for a Masters in Globalisation and Development Studies. Before joining Cranfield, he worked as a researcher at the Chartered Management Institute and spent time in India at the microfinance institution Samhita Community Development Services.

Andrea Keessen is assistant professor at the Institute for Constitutional and Administrative Law and Legal Theory at Utrecht University (UU) and a member of the Utrecht Centre for Water, Oceans and Sustainability Law (UCWOSL). She holds a PhD from Utrecht University in European administrative law. Her research interest is in water law and environmental law from a legal and an empirical perspective. She frequently participates in interdisciplinary research, e.g. in the consortium Governance of Adaptation to Climate Change of the Dutch research programme Knowledge for Climate. See for further information and publications: www. uu.nl/a.m.keessen.

Michael Kidd (B Com LLB LLM PhD Natal) is Professor of Law at the University of KwaZulu-Natal in South Africa. He is an experienced lecturer and researcher in environmental law, water law and administrative law and has published widely in these fields. He is active in the International Union for Conservation of Nature (IUCN) Academy of Environmental Law and has been a board member representing Africa. He is the lead editor of the *South African Journal of Environmental Law and Policy*.

Barbara van Koppen (PhD) is Principal Researcher on Poverty, Gender and Water at the International Water Management Institute (IWMI), Southern Africa Regional Program. She leads multi-country (action-) research in Africa and Asia on water policy and law reform and community-driven water service delivery, also from a human rights perspective. She is lead-author and -editor of five books and (co-) author of over 140 international peer-reviewed publications. She engages in various capacities in national and international policy dialogue. Before joining IWMI, she was Assistant Professor of Gender and Irrigation at Wageningen University and Research Centre, the Netherlands, where she also obtained her PhD.

Christina Leb (PhD) works as Senior Water Resources Specialist at the World Bank and is an associate member of the Platform for International Water Law, housed at the University of Geneva. She holds a doctorate in international law from the University of Geneva and a MA in international relations from the School of Advanced International Studies (SAIS) of Johns Hopkins University, MD. The views presented in the contribution to this book are those of the author alone and in no way reflect the position of the World Bank, its Board of Executive Directors or any of its member countries.

Stephen C. McCaffrey is Distinguished Professor of Law at the University of the Pacific, McGeorge School of Law in Sacramento, CA. He served two terms on the International Law Commission, chaired that body for one of its sessions, and was the Commission's special rapporteur on international watercourses. Professor McCaffrey is a member of the Implementation Committee of the United Nations Economic Commission for Europe (UNECE) Water Convention. He has served as counsel to states in cases before the International Court of Justice and the Permanent Court of Arbitration and has published widely in the field.

Owen McIntyre is a Professor and the Director of Research at the School of Law, University College Cork, Ireland. His principal area of research is environmental and natural resources law, with a particular focus on international and comparative water law. He serves as Chair of the IUCN World Commission on Environmental Law's Specialist Group on Water and Wetlands, as a member of the Project Complaints Mechanism of the European Bank for Reconstruction and Development (EBRD) and a member of the Scientific Committee of the European Environment Agency.

Bjørn-Oliver Magsig (PhD) is a Lecturer in Law at University College Cork, Ireland, where he focuses on international environmental law, water diplomacy and the links between natural resources, international security and equity. Until recently, he worked as a Research Fellow at the Helmholtz Centre for Environmental Research, Leipzig, Germany, leading various interdisciplinary projects revolving around the socio-legal challenges of water scarcity. Bjørn-Oliver has extensive project experience in international law governing transboundary natural resources, serves on the Managing Board of the European Environmental Law Forum (EELF) and is a member of the IUCN World Commission on Environmental Law.

Paula Pacheco Mollinedo is an engineer. She has an MSc in Water Resources, with extensive experience and research skills working in a recognized institution in Bolivia. Currently, Paula is the regional director of the institution Agua Sustentable, a non-profit organization that has been pursuing the objective of contributing to the sustainable management of water and environment at national and international levels for 12 years. As a project manager, she is responsible for leading projects related to climate change adaptation and transboundary water resources, among others.

Laura C. Mulherin is a JD Candidate, May 2017, at the George Washington (GW) University Law School, and has a BA in Public and Urban Affairs, *Magna Cum Laude*,Virginia Polytechnic Institute and State University. Laura serves as a Notes and Projects Editor on *The George Washington Law Review* and as a Writing Fellow for the GW Law Legal Research and Writing Program. Laura is also a member of the GW Environmental Law Association and the GW Law Moot Court Board.

Dr Marian J. Neal (Patrick) is the Transboundary Programme Manager for the Stockholm International Water Institute. She has an interdisciplinary research background in wetland/landscape ecology, water governance, transboundary water management and social and environmental justice. She has worked in the fields of water allocation decision making, complex social-ecological systems and examined how the issue of scale impacts on the management of shared water resources. Dr Patrick has over 15 years of experience in transboundary water management and is currently the manager of the UNESCO Cat II Centre on Water Cooperation in Sweden.

Joshua Newton is an independent consultant and an expert in global political processes and governance related to water, with specific experience working with the Ministerial Processes of several World Water Fora, the Budapest Water Summit and the United Nations' 2030 Agenda on Sustainable Development. Joshua currently manages the Global Water Partnership's Sustainable Development Goals Preparedness Facility and is coordinating the Indus Forum on behalf of the World Bank. Joshua holds a PhD in International Relations from the Fletcher School of Law and Diplomacy at Tufts University.

Nataliya Nikiforova has been working at the UNECE as Environmental Affairs Officer since 2010. She is currently supporting the work under the Protocol on Water and Health to the Convention on the Protection of Transboundary Watercourses and International Lakes (UNECE Water Convention). Prior to her current position, she has been working on the issues related to international water law, particularly in the field of transboundary waters management, and servicing the Implementation Committee of the UNECE Water Convention.

LeRoy C. (Lee) Paddock is Associate Dean for Environmental Law Studies at the George Washington University Law School. His work focuses on environmental compliance and enforcement, environmental governance with particular emphasis on integrating the regulatory system with economic and values-based drivers and governance in the context of emerging technologies, environmental justice, public participation and energy efficiency. Lee holds a BA from the University of Michigan and a JD with High Honors from the University of Iowa. He clerked for Judge Donald Lay of the US Eighth Circuit Court of Appeals.

Marleen van Rijswick is Professor of European and Dutch Water Law at Utrecht University School of Law and she leads the Utrecht Centre for Water, Oceans and Sustainability Law.

Juan Carlos Sánchez Ramírez is a Costa Rican international environmental lawyer. As a Legal Officer of the International Union for the Conservation of Nature, he is part of multidisciplinary teams implementing transboundary water governance and law projects in different regions of the world. He is also a PhD candidate of the University of Dundee and a Hydro Nation Scholar. His current research aims at developing a transboundary water governance framework for improving existing or developing new agreements. By identifying the factors that enable best environment protection, his PhD project intends to serve as guidance for legal arrangements which are better suited for conserving freshwater ecosystems.

Komlan Sangbana is a Research Fellow at the Platform for International Water Law, University of Geneva Law School, and Scientific Collaborator at the Institute for Environmental Sciences, University of Geneva. Dr Sangbana holds a Diplôme d'étude approfondie in Public Law from the University of Lomé, Togo, and a PhD in Public International Law from the University of Geneva. He has been a visiting scholar at the Edinburgh University Law School. From 2012–14, he worked as a researcher for the Swiss National Science Foundation on a project entitled Non-State Actors and the Management of International Freshwater Resources. Dr Sangbana also served as a consultant for the Secretariat of UNECE Water Convention, WATERLEX and Green Cross International. His recent publications include *Public Participation and Water Resources Management: Where Do We Stand in International Law?* (UNESCO, 2015) (co-editor with Mara Tignino).

Melissa K. Scanlan is the Associate Dean of the Environmental Law Program, Professor of Law, Director of the Environmental Law Center and Co-Founder-Director of the New Economy Law Center at Vermont Law School. Prior to joining Vermont Law, she was the University of Wisconsin Law School's Water Law and Policy Scholar and a lead consultant involved in launching the Center for Water Policy at the University of Wisconsin-Milwaukee's School of Freshwater Sciences. She founded and directed the Midwest Environmental Advocates. She earned her law degree and Master of Science in Environmental Science, Policy and Management from the University of California-Berkeley.

Susanne Schmeier (PhD) is currently the Coordinator for Transboundary Water Management at the Deutsche Gesellschaft für internationale Zusammenarbeit (GIZ). Prior to this, she has worked as adviser to the Mekong River Commission (MRC), at the World Bank and with a number of other international and regional organizations and river basin organizations on water resources management. She is also a fellow at the Earth System Governance Project and contributes to a number of research projects. She has published extensively on water management topics, with a particular focus on the legal and institutional dimensions of transboundary water management. Dr Susanne Schmeier holds a PhD in transboundary water management.

Anna Schulz works for the International Institute for Sustainable Development (IISD) Reporting Services as a team leader for *Earth Negotiations Bulletin*, where she specializes in coverage of international climate change and water-related negotiations. She teaches global environmental politics at Boston College. She has also consulted for the UNFCCC Secretariat. She holds a PhD and a MALD from the Fletcher School, Tufts University, MA, in international environmental policy, and negotiation and conflict resolution, and an LLM in international and comparative water law from the University of Dundee. Her dissertation traced the origins of transboundary water law from ancient Mesopotamia to present.

Mia Tamarin is an independent legal scholar exploring the politics of international law, the Israeli–Palestinian conflict and hydro-hegemony. Tamarin is a 2006 United World College Scholar who earned her MA in International Law from the School of Oriental and African Studies in 2014. She is an affiliate member of the London Water Research Group and a founding member of the London School of Psychoanalysis of the Forums of the Lacanian Field. She currently works as a legal consultant in the private sector and serves as Company Secretary of the UK Friends of Abraham's Path.

A. Dan Tarlock is Distinguished Professor of Law at the Chicago-Kent College of Law and holds an AB and LLB from Stanford University. He has written and consulted extensively in water law. His treatise, *Law of Water Rights and Resources*, has been frequently cited by US courts. In 1998, he was the chief report writer for the Western Water Policy Review Advisory Commission report, *Water in the West*, which was one of the first major federal publications to examine the relationship between climate change, urban growth and water use. In 2014, he was appointed to the Technical Committee of the Stockholm-based Global Water Partnership.

Peter S. Wenz is a University Scholar of the University of Illinois, Emeritus Professor of Philosophy at the University of Illinois at Springfield and, most recently, a Visiting Fellow at the University of Canterbury in Christchurch, New Zealand. He has over 40 published articles and book chapters in fields that include environmental ethics, environmental justice, political philosophy and medical ethics. His eight single-authored books are in the same fields, except for his most recent work, *Functional Inefficiency: The Unexpected Benefits of Wasting Time and Money* (Prometheus 2015), which concerns economics and business.

Inga T. Winkler is a lecturer at the Institute for the Study of Human Rights at Columbia University. Previously, she has been in residence as a visiting scholar the Center for Human Rights and Global Justice at NYU, at Stellenbosch (South Africa) and Berkeley. From 2009 to 2014 she was the Legal Adviser to the UN Special Rapporteur on the Human Rights to Water and Sanitation. Inga has also worked as a consultant for various international organizations and NGOs including the European Parliament, the UN Water Supply and Sanitation Collaborative Council, and the Global Initiative for Economic, Social and Cultural Rights. Inga holds a German law degree and a doctorate in public international law.

INTRODUCTION

Andrew Allan, Sarah Hendry and Alistair Rieu-Clarke

'Water is the basis of all things' (Thales 640BC). It is a cross-cutting issue, playing a key role in addressing the global challenges of our time – including climate change, food and energy security, environmental sustainability and the promotion of peace and stability. One of the defining characteristics of the debate regarding water in recent years has been the extent to which it increasingly recognises the many different policy areas to which it is relevant. This includes flood management, economic development and livelihoods, and human health. This creates serious issues with respect to legal and policy frameworks as they struggle to accommodate all of these interlinked perspectives.

The linkages between water and climate change are aptly summed up in a World Bank Report, which observes that, 'the impacts of climate change will be channelled primarily through the water cycle, with consequences that could be large and uneven across the globe' (World Bank Group 2016). Such impacts include greater variability in precipitation, river runoff and water availability, glacier melt, a decrease in agricultural productivity and efficiency of existing water infrastructure, and deteriorating water quality (Bates *et al.* 2008). The poor are most vulnerable to climatic change. Water's centrality to climate change requires a response that ensures that water is also central to adaptation efforts. Adaptive governance mechanisms, while often lacking in many parts of the world, have the potential to enhance resilience to climate change impacts.

Water shortages, as well as degrading water quality, already have a major impact on food security. The World Water Development Report estimates that, as a result of climate change, population growth and unsustainable practices the world is facing a 40 per cent shortfall in water supply by the year 2030 (UNWWDR 2015). Agriculture accounts for 70 per cent of all water withdrawn. With the world population set to grow from 6.9 billion in 2010 to 8.3 billion in 2020, and dietary shifts leading to more water-intensive food production, there is an ever increasing need to put in place legal frameworks that are capable of reconciling competing interests over this finite resource in a way that is both transparent and flexible.

The need to account for the linkages between energy and water are also increasingly being recognised (UNWWDR 2014). The growing demand for energy generation, including from hydroelectric, nuclear and thermal sources, will utilise ever more significant amounts of water. Economic growth is set to double energy consumption in the next 40 years, with Africa's electricity generation estimated to be seven times what it is today (World Energy Council, 2010). This increase will place significant pressure on already constrained water resources which, in turn, may jeopardise reliable and sustainable energy supplies (Rodriguez *et al.* 2013). Again, ensuring that water is available to support energy security demands effective law and policy frameworks, at multiple levels, in order to reconcile potentially competing interests in an equitable and transparent manner. Many large hydropower projects, for instance, take place in rivers shared by two or more states. This raises political challenges, and demands a stable legal framework that allows both upstream and downstream states to share data and information, communicate effectively, and come up with equitable solutions for the governance of their shared resources.

One of the most pressing needs is to reconcile cross-sectoral interests in a way that protects aquatic ecosystems, as these underpin the sustainability of all other uses of water. Adopted in 2005, the Millennium Ecosystem Assessment recognised that humans have changed ecosystems more rapidly and extensively than in any comparable period. This, in turn, 'has resulted in substantial and largely irreversible loss in the diversity of life on Earth' (MEA 2005). Anthropogenic changes have also led to a loss in the services that such ecosystems provide to populations across the world. Freshwater ecosystem services are central to human well-being, thus providing provisioning services (drinking water), regulating services (climate and food regulation), supporting services (soil formation and nutrient cycling) and cultural services (spirituality, aesthetics, education and recreation) (Mayers *et al.* 2009). Mayers *et al.* argue that governance is key to managing water ecosystem services, and also, '*the core challenge… in sustaining ecosystem services and the needs of the poor in all regions*' (2009, emphasis added).

Finally, the linkages between water and security and peace are well recognised. While wars between states solely over water appear unlikely, growing water shortages have been seen as a key factor in exacerbating tension between different users at a local, provincial, national and transboundary level (UN News Centre 2008). Such disputes may range from conflicts between farmers and herders, to tension between states sharing transboundary rivers, lakes and aquifers. Law plays an important role in both putting appropriate mechanisms in place by which to avoid conflicts and, where such conflicts do arise, providing the means by which they can be resolved in a peaceful manner. Such an objective is clear in Article 1 of the UN Charter, which calls upon states to maintain international peace and security through effective collective action.

The challenges faced in the governance of water have not been ignored by the international community. While it is arguable that there is no such thing as a global agenda, in water or any other policy area, it is nonetheless possible to trace the development of a set of policies over the last 30 years, in the fields of both water resources and water services (see Newton, Chapter 26, this volume). These are not always cohesive: especially at a global level, there are competing priorities, although it is arguably the failure to provide basic services for so many people that has kept water at the policy forefront. Some progress has been made. Since 1990, an estimated 2.6 billion people have gained access to an improved drinking water source and 2.1 billion people have gained access to an improved form of sanitation. However, in 2015, 663 million people still lacked improved drinking water sources and 2.4 billion people still lacked access to an improved sanitation facility (UNICEF and WHO 2015). Multiplying the complexity of policy and legal responses by linking water to a multitude of different considerations makes implementation and coherent responses at national levels difficult, and resolution of these problems will be a key factor in the success of the sustainable development goals (SDGs).

While it is feasible to trace international cooperation over water back to the nineteenth century (see Newton, Chapter 26, this volume) and treaties relating to transboundary waters much further (see McCaffrey, Chapter 15, this volume), Agenda 21 (UN 1992) stands out as a particularly significant milestone in the evolution of policy related to water. As a global plan of action, endorsed by the UN and its Member States and building upon the 1977 Mar del Plata Action

Plan, Agenda 21 devoted a chapter to freshwater resources, whereby states committed to take action in areas including water resource management, water allocation, water quality and the supply of water services. In each set of actions, there was recognition of the need for reform of the legislative and regulatory environment.

Agenda 21 was also preceded by the Dublin International Conference on Water and the Environment, which resulted in the 'Dublin Statement'. This set out four principles: that freshwater is a finite and vulnerable resource; that its development and management should be based on a participatory approach; that women play a central part in water management; and that water has an economic value in all its competing uses and should be recognised as an economic good. These were subsequently reformulated into three principles by the World Bank (2004): the ecological principle (river basin management, environmental protection and managing land and water together); the institutional principle (subsidiarity and the inclusion of all stakeholders); and the instrument principle (a scarce resource requires incentives and economic instruments to manage effectively).

As regards the last, it is important to note the recognition in the Dublin sub-text that, first, there is a basic right of access to water. Otherwise, the 'special nature' of water risks disappearing in a purely economic analysis of service provision and cost recovery, at the expense not just of the basic human needs of those who cannot pay, but also of ecological needs, and of what might best be described as the spiritual aspects of water. This special nature is reflected in the preamble to the European Community's Water Framework Directive (WFD): '[w]ater is not a commercial product like any other, but, rather, a heritage which must be protected, defended and treated as such' (European Parliament and Council 2000).

The debate around the Dublin Principles has been dominated by principle four and the market-oriented approach; this has fostered deep polarity in the debate around water services (see Hantke-Domas, Chapter 5, this volume).

The Johannesburg Summit on Sustainable Development, ten years after Rio, took forward the global sustainable development agenda with the emphasis on delivery rather than new policies (UN 2002; and see Schulz, Chapter 27, this volume). There was a specific requirement for all signatories to produce integrated water resource management (IWRM) plans and water efficiency plans at all levels by 2005. There was also provision for better water pollution control, recognising the benefits for public and ecosystem health. Efficient use and better mechanisms for access and allocation were called for, and water and sanitation issues were still a priority. Despite the rather disappointing response to this initiative, all of these policy areas are relevant to chapters of this book; all the policy documents above make mention of stable and transparent regulation as one tool for better management.

At the start of the twenty first century, the broad policy objectives received new focus with the production of the millennium development goals (MDGs) (UN General Assembly 2000). Goals in relation to water included halving the proportion of people without safe drinking water and (after Johannesburg) access to basic sanitation, by 2015. Water was recognised as a cross-cutting issue, relevant to all the MDGs. In 2015, the drinking water target was achieved at a global level, but not that for sanitation, with the biggest deficits in sub-Saharan Africa, southern and eastern Asia (UNICEF and WHO 2015). The international community moved towards a post-2015 agenda at the 'Rio +20' Summit in 2012.

The 'Outcomes' document (UN 2012) reaffirms many existing high-level commitments, including sustainable development and poverty eradication, and emphasises good governance, and human rights, including to water and sanitation. In the few paragraphs on water, there is commitment to the progressive realisation of these rights, as well as the role of ecosystems, the need to manage water pollution and treat wastewater (see Howarth, Chapter 6; Hendry, Chapter

7; and Paddock and Mulherin, Chapter 13); the management of floods (Allan, Chapter 10) and drought (Tarlock, Chapter 11) and the use of non-conventional water sources (all this volume).

The international community has now moved to the creation of new SDGs to take us to 2030 (UN General Assembly 2015) (and see Schulz, Chapter 27, this volume). In preparation for a water goal, there were three thematic sub-groups: on water, sanitation and hygiene; water resources management; and wastewater and water quality. Their report (UN-Water *et al.* 2012) stressed the need to move away from narrow goals and silos, build collaboration and recognise that water will continue to cut across all development and poverty-alleviation activities. It proposed ambitious goals and targets, including universal access to basic services, but, further, that a rights-based approach to water needs to move beyond a narrow perception of water and sanitation and recognise policy interlinkages, especially with food, as well as the inter-generational principle of sustainable development. The relationship between water and other critical sectors – for example, the so called 'water–food–energy nexus' and the multiple impacts of climate change – is identified. So too is the need to address 'water for nature', to ensure the continuation of the services that ecosystems provide. In the last decade, the debate around ecosystems, an ecosystems approach and ecosystem services, has informed thinking in many disciplines, including law – and, within that, water law – nationally and internationally. On wastewater and water quality, there is recognition that a combination of urbanisation and population growth means we are all downstream users now. The report urges the collection and treatment of wastewater; as with solid waste, there is a critical need to begin to see this as a valuable resource base and to overcome some of the taboos and negative perceptions, which move wastewater and sanitation too far down the policy agenda. As might be expected, there is a recognition of a growing debate around water security (itself a term with many meanings, Magsig, Chapter 28, this volume); the need for governments to work with many stakeholders; the need for capacity development; and, of course, the need for finance.

The SDGs themselves include Goal 6 on water; but water is also relevant to many of the other goals, including poverty alleviation.

In terms of scope, the *Routledge Handbook of Water Law and Policy* has sought to pick up on the main established and emerging issues related to water law and policy. The *Handbook* is structured in three parts. The first part is primarily dedicated to a comparative analysis of different national water law and policies across a range of sub-topics. In recognition of the fact that around 42 per cent of the world's population live in river basins and aquifer systems shared between states, the second section of the *Handbook* is dedicated to transboundary water law. The final section of the *Handbook* is dedicated to a series of chapters that address cutting-edge topics that are likely to influence the future development of both national and transboundary water law and policy. Finally, the *Handbook* concludes by extrapolating some of the key insights that have been offered by each of the contributors, and suggesting future topics that might be developed.

The *Routledge Handbook of Water Law and Policy* aims to provide comprehensive coverage of a topic that has grown in significance and stature over the last few decades, and will no doubt continue to grow as the challenges relating to access to water become ever more pressing. In so doing, the *Handbook*, like water, cuts across many sectors, spatial scales and geographic regions. The contributors also reflect a diverse group of lawyers and policy experts, scholars and practitioners, from different parts of the world. The intention in pulling together these contributions is to provide the reader – be they a student, researcher or practitioner 'in the field' – with a reference book that covers key sub-topics of water law and policy. The *Handbook* should therefore be of benefit to those delivering or participating in courses related to water law and policy, as well as for practitioners that are working on the design and implementation of water law and policy across a wide range of settings. For the research community, the *Handbook* will guide researchers as to the current state of the art with in the field of water law and policy. Additionally, the *Handbook* draws upon the collective knowledge and experience of its contributors to suggest ways in which water law and policy is evolving, gaps in our current knowledge and understanding, and suggestions on future research topics.

References

Bates, B.C., Kundzewicz, Z.W., Wu, S. and Palutikof, J.P. (eds) (2008) *Climate Change and Water: Intergovernmental Panel on Climate Change (IPCC) Technical Paper VI*. Geneva: IPPC Secretariat.

European Parliament and Council (2000) Framework Directive on Water Policy 2000/60/EC.

Mayers, J., Batchelor, C., Bond, I., Hope, R., Morrison, E. and Wheeler, B. (2009) *Water Ecosystem Services and Poverty Under Climate Change*. London: International Institute for Environment and Development.

MEA (2005) *Ecosystems and Human Well-being*. Washington, DC: Island Press.

Rodriguez, D.J., Delgado, A., De Laquil, P. and Sohns, A. (2013) *Thirsty Energy*. Washington, DC: World Bank.

Thales (640BC) see *Ancient Greek Philosophy* [online]. Available from: http://www.iep.utm.edu/g/greekphi.htm

UN General Assembly (2000) Millennium Declaration, A/RES/55/2.

UN General Assembly (2015) Transforming Our World: the 2030 Agenda for Sustainable Development, A/RES/70/1.

UNICEF and WHO (2015) 25 Years Progress on Sanitation and Drinking Water – 2015 Update and MDG Assessment [online]. Available from: http://data.unicef.org/corecode/uploads/document6/uploaded_pdfs/corecode/Progress-on-Sanitation-and-Drinking-Water_234.pdf.

UN (United Nations) (1992) Agenda 21: An Agenda for the 21st Century, A/Conf. 152/126.

UN (2002) Report of the World Summit on Sustainable Development Incorporating the Johannesburg Declaration and Plan of Implementation (Johannesburg), A/Conf.199/20.

UN (2012) Report of the UN Conference on Sustainable Development (Rio+20) 'The Future We Want' A/CONF.216/L.1.

UN News Centre (2008) Ban Ki-moon Warns that Water Shortages are Increasingly Driving Conflicts [online]. Available from: http://www.un.org/apps/news/story.asp?NewsID=25527#.V1KGILSwq6w

UN-Water, UNDESA and UNICEF (2012) *The Post 2015 Water Thematic Consultation Report* [online]. Available from: http://www.unwater.org/downloads/Final9Aug2013_WATER_THEMATIC_CONSULTATION_REPORT.pdf [last accessed 31 May 2016].

UNWWDR (2014) *Water and Energy*. Paris: UNESCO.

UNWWDR (2015) *Water for a Sustainable World*. Paris: UNESCO.

World Bank (2004) *Water Resources Sector Strategy: Strategic Directions for World Bank Engagement*. Washington, DC: World Bank.

World Bank Group (2016) *High and Dry – Climate Change, Water, and the Economy*. Washington, DC: World Bank.

World Energy Council (2010) *2010 Survey of Energy Sources*. London: World Energy Council.

PART 1

Comparative national water law and policy

1

CUSTOMARY WATER RIGHTS AND LEGAL PLURALISM

Barbara van Koppen

Introduction

Customary (or 'local') water law governs the water investments and uses by the majority of citizens in low- and middle-income countries. These are the rural and peri-urban small-scale water users with diversified agriculture-based livelihoods. They depend in many ways on water. Since time immemorial, they have invested in water infrastructure and institutional arrangements to adjust to the vagaries of climate and use water throughout the seasons for their multiple domestic and productive water needs. Yet, hardly any low- or middle-income country recognizes customary water arrangements in their statutory water laws. Bolivia is a rare exception to a rule of formal denial, forced conversion and erosion of customary law by statutory law (Roth *et al.* 2005; Boelens *et al.* 2007).

Paradoxically, the opposite holds in high-income countries: wherever customary water law occurs, this is recognized as functional, accepted and nurtured. An example is the Water Tribunal of Valencia in Spain, which was constituted more than 1,000 years ago. The same holds at much larger scales, as Bruns (2005) writes:

> Acceptance of traditional water rights, even when these have not been formally registered, has been a key principle underlying river management in Japan. Existing users were not forced to register merely to defend their access. Instead, the law established the principle that they have legal standing to protect their interests when necessary. The River Laws of 1896 and 1964 provided a formal basis in state law, through which agencies and courts could take account of existing rights. The principle of being 'deemed to have obtained permission' illustrates one way of reducing conflicts between state and local law without forcing local rules to explicitly conform to the criteria and formulations of state law.

This chapter aims to unravel this paradox, with a focus on permit systems. Permit systems (or license or concession systems; all terms refer to the same tool) are widely promoted across the globe as the supposedly most effective form of statutory law amidst plural water laws. While already almost universally adopted across low- and middle-income countries in Latin America and Africa, more Asian states and countries have also started adopting permit systems as a strongly promoted core ingredient of integrated water resource management (IWRM) (Van Koppen *et al.* 2014b).

In disentangling this paradox, we acknowledge legal pluralism as the co-existence of several, often overlapping normative frameworks according to which people interact with each other with regard to a bundle of rights and obligations vis-à-vis water resources (Von Benda-Beckmann *et al.* 1998). Society's sources of legitimacy of these frameworks vary, and reflect power relations (Boelens and Dávila 1998). Accordingly, water laws include customary (or 'local' or 'community-based'), religious, statutory and so-called 'project law'. The latter are the – often implicit – bundle of rights and obligations that public infrastructure projects create (Meinzen-Dick and Nkonya 2007).

The various regimes can blend. For example, customary and religious law can overlap. 'Customary law' can also refer to a formal codification of local arrangements as perceived by officials. Codification risks 'freezing' vital elements of local arrangements (Boelens *et al.* 2005). In order to avoid confusion about such blends, many scholars prefer the terms 'living customary law' or 'local law' (Hellum *et al.* 2015). Moreover, unlike the terms 'customary' or 'traditional', the term 'local law' captures the highly dynamic nature of small-scale water users' livelihoods and investments and community-based institutional arrangements. This chapter will refer to 'local' law. Blending also occurs as so-called 'forum shopping', in which people combine elements of each framework to best suit their interests. For example, women may refer to their constitutional rights to contest exclusion from prevailing male-dominated customary arrangements. Blending is contested when one normative framework, including claims to ownership, is imposed, even with violence, at the expense of the other. Thus, small- or large-scale users with a formal permit can claim lawful access to water in areas with customary arrangements and use physical force to actually prevent local contesters from accessing the same sources (Van Eeden 2014).

Hodgson (2016, n.p.) proposes a useful distinction between formal and informal water rights regimes, or, as he calls it, formal and informal 'water tenure'. He defines water tenure as 'the relationships, whether legally or customarily defined between people, as individuals or groups, with respect to water resources'. He proposes an exhaustive identification of various typologies of 'water tenure' that, moreover, explicitly seeks the inputs of many more disciplines than lawyers alone. His typology of tenure arrangements includes:

- those that are defined by formal law ('traditional' formal water rights, 'modern' formal water rights, regulatory licensing, agency control, water supply contracts, commonhold water tenure, investment contracts, *de minimis* uses, exempt commercial uses and reserves/minimum flow requirements), as well as
- those that are not defined by formal law (customary or local law water tenure, water tenure under religious law, informal tenure, assumed and impossible tenure and unrecognized tenure).

Governance of water tenure, and also water tenure itself, are elements of water governance as a whole. Unlike Hodgson, we propose to explicitly include human rights to water as one of the various formal legal frameworks that govern people's mutual relationships with regard to water. The links between human rights framework and water governance are increasingly invoked (Hellum *et al.* 2015).

The analysis by Shah (2007) is particularly relevant for tracing why local water law is the dominant law for the majority, but denied in permit systems in low- and middle-income countries (although cherished for a minority in high-income countries' water laws). He makes the more encompassing distinction between formal and informal institutional arrangements within low-, middle- and high-income economies at large, and water economies in particular. As shown in Figure 1.1, the nature of the water economies and their institutional arrangements are inextricably linked to economic development. In low-income countries, most water users are primary

	Stage I: Completely Informal	Stage II: Largely Informal	Stage III: Formalizing	Stage IV: Highly Formal Water Industry
% of water users in the formal sector	<5%	5–35%	35–75%	75–95%
Examples	Sub-Saharan Africa	India, Pakistan, Bangladesh	Mexico, Thailand, Turkey, Eastern China	USA, Canada, Western Europe, Australia
Dominant mode of water service provision	Self-supply and informal mutual-help community institutions	Partial public provisioning but self-supply dominates	Private-public provisioning, attempts to improve service and manage the resource	Rise of modern water industry; high intermediation; self-supply disappears
Human, technical and financial resources used by water sector; % of total water use self-supplied; Rural population as % of total; Cost of domestic water as % of per caput income; Cost of water service provision	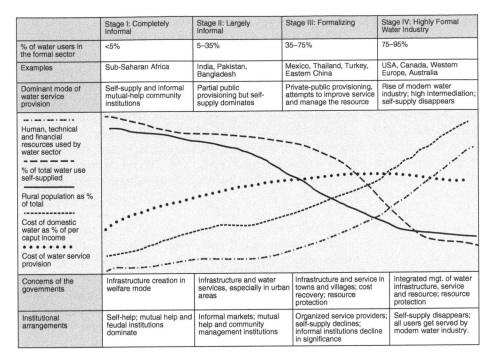			
Concerns of the governments	Infrastructure creation in welfare mode	Infrastructure and water services, especially in urban areas	Infrastructure and service in towns and villages; cost recovery; resource protection	Integrated mgt. of water infrastructure, service and resource; resource protection
Institutional arrangements	Self-help; mutual help and feudal institutions dominate	Informal markets; mutual help and community management institutions	Organized service providers; self-supply declines; informal institutions decline in significance	Self-supply disappears; all users get served by modern water industry.

Figure 1.1 Community-based water law and water resources management reform in developing countries
Source: Shah (2007)

water users directly taking water from the source and investing in small-scale water infrastructure for self-supply. Local institutional arrangements are embedded in the prevailing feudal or tribal environments. In this context, the preoccupation of the state and development partners used to be, and should continue to be, support to infrastructure development in a welfare mode.

In contrast, in high-income formalized economies, water provision is largely organized by state-regulated parastatals and companies as a business. In urban areas, and even in rural areas, most water users are known, well organized and able to pay for the (often partially subsidized) water services. Only a few per cent of the population are farming and irrigating, and these are often well organized, self-regulating and reachable by the water administration.

Shah's analysis points at the inevitable mismatch when supposedly generic, blanket water governance measures borrowed from high-income countries with formalized economies are imposed on low- and middle-income countries with informal water economies (with local law). The following section will briefly sketch these informal water economies. The third section will trace the very different histories and current implications of permit systems that are promoted as a blanket legal tool, to illustrate how permit systems are a clear example of precisely that risk. Human rights frameworks will appear a useful reference to analyze these implications. The conclusions and recommendations will present an alternative formal perspective, based on human rights, on water tenure in low- and middle-income.

Local legal frameworks for the management of water resources

Informal local water tenure in informal water economies can be characterized by ten features that are commonly reported in the literature, in particular from Nepal (Ostrom 1992; Yoder and

Martin 1998;Von Benda-Beckmann and Von Benda-Beckmann 2000), Latin America (Boelens and Dávila 1998; Boelens *et al.* 2005; Roth *et al.* 2005; Boelens *et al.* 2007); India (Shah 2007) and Sub-Saharan Africa (Ramazotti 1996; Sokile 2006; Derman *et al.* 2007;Van Koppen *et al.* 2007; Komakech 2013).

The first feature was mentioned above. While livelihoods diversify with migration and urbanization, large proportions of the rural and even peri-urban population continue to depend to a significant extent on water-dependent *agriculture-based livelihoods* and agriculture's forward and backward linkages. They need water for a range of uses: drinking, other domestic uses, livestock, fishing or aquaculture, cropping, horticulture, crafts, small-scale enterprise and cultural uses. Informal arrangements prevail: in rural Sub-Saharan Africa, 90 per cent of the territory is governed under customary arrangements for land tenure and related water and other natural resources (Hodgson 2004).

Second, people eke out an agriculture-based living in often harsh conditions of highly variable and unpredictable availability of water resources. Their age-old coping strategies to buffer against this variability are based on *directly using and reusing multiple water sources, or protecting against too much water*, in the local water cycles: precipitation, run-off, ponds, wetlands and groundwater. In their coping strategies, people move to the water resources or make the water move to them.

A third feature is that people make rules. When moving to the water resources – for example, rivers – water for domestic uses is taken upstream of places for livestock watering. On a much larger scale, settlement patterns follow seasonal flooding. Similarly, people move to and manage wetlands for cropping, grazing and fisheries.

When people move the multiple water sources to themselves for self-supply or to protect and drain too much water, *infrastructure* is key. As individual or collective primary water takers, they invest in infrastructure, building weirs and diverting rivers and run-off; storing water; retaining or draining soil moisture; using the fertility that floods bring to temporarily inundated and vacated land; and digging groundwater wells and recharging those, for household uses, livestock and irrigation. As a rule, people design *multi-purpose* infrastructure for as many uses as possible to meet the fullest range of their needs. Single-use designed infrastructure is the exception – for example, in the case of groundwater wells in distant fields. This differs from the planners and engineers in either the irrigation or water, sanitation and hygiene sectors, who design infrastructure for the single uses of their sectoral mandates. Not surprisingly, users immediately transform these single-use-designed schemes into *de facto* multiple-use schemes (Van Koppen *et al.* 2014a).

Fourth, these water arrangements are highly *dynamic*. The recent availability of plastics, pvc pipes, motorized pumps and rural electrification has considerably accelerated these informal investments. Within only a few decades, informal groundwater irrigation has outstripped formal surface irrigation in, for example, India. Growing population pressure on smaller plots or new market opportunities in urbanizing economies are other drivers for these investments. In urbanizing areas, specialized informal water service providers fill the void of unmet water needs, particularly in poorer neighbourhoods. Investments are positively and negatively affected by the broader rapidly changing rural landscapes of remittances, rural transport and energy infrastructure, communication technologies and rapidly increasing numbers of young people, but also forced displacements and conflicts. With the predicted greater variability and more extreme events under climate change, these age-old and ever-changing responses to nature's and man-made water resource fluctuations will become even more important.

A fifth feature of local water law is the important link between water entitlements and investments. Individuals and groups who invest in the infrastructure tend to have strongest claims to the water conveyed. This process of 'hydraulic property rights creation' (Coward 1986) is continuously confirmed through contributions to infrastructure maintenance. Except for the small quantities of water to 'quench one's thirst', investors may exclude others (which may not always be feasible). Inheritance, land transfers, scheme extensions or new schemes upstream or downstream incrementally alter the precise claims.

Sixth, within communal schemes, a wide range of *water sharing arrangements* has been documented. Transparent member organizations with a relatively accountable leadership set and implement these rules. A specialized and modestly compensated water guard often implements the rules set. An illustration of these rules from an irrigation association in Nepal is that the owner of a plot that is to be irrigated by the water guard is not allowed to be present during that period, as that would inevitably lead to negotiations, if not intimidation or bribery of the water guard.

Water sharing arrangements also exist between schemes, sometimes over long river stretches. At agreed times, upstream users close their intakes, leaving water for downstream users. Relatives living hundreds of kilometres upstream of major flood plains might inform their relatives in the temporary settlements in the flood plains about rising river levels and flooding of the plain a few days later. Sharing arrangements are flexible principles that can be adjusted to any specific situation. In case of water scarcity, a commonly found rule is that everybody should first receive an equal minimum. Only once those needs are equitably met, can others – for example, larger landowners – claim more water. While access to riparian strips, or land with springs, naturally favours the land title holders, ways of access often remain open to community members. Nobody can be denied from using water for their livelihoods (Derman *et al.* 2007). No one claims exclusive ownership of naturally available water resources, as 'water is given by god'.

The seventh feature is the consideration to *conserve* water and other natural resources upon which livelihoods so critically depend. Villagers set rules and organize to prevent pollution – for example, by removing animal cadavers from streams. An even more proactive example is the spontaneous groundwater recharge movement in Gujarat. Here, religious leaders, wealthier outmigrants and villagers collaborated to start area-wide recharging of groundwater wells, well before the government had acted with regard to groundwater overdraft (Shah 2007).

The eighth feature underpins the above: efforts to avoid conflict and maintain precious social relations in a bottom-up manner. Immediate conflict resolution between the parties is promoted – for example, by stating that if two plots holders cannot solve their conflicts, both will lose the plot, as noted in wetlands in Burkina Faso (Van Koppen 2000). When this is insufficient, nested *conflict resolution* mechanisms are often in place. Mediation takes place at step-wise higher and more authoritative levels of judgment, as embedded in community structures. Instead of the 'winner takes all' justice of western courts, the emphasis is on the creation of consensus and repair of relationships through compensation.

In contrast to these eight (and other) positive features, the embeddedness of local water arrangements in wider socio-economic settings implies that feudal *hierarchies* are reproduced as well. This is a ninth feature. Men and male elite leaders tend to dominate hydraulic property rights creation, including the precious resource of technical skills. In areas where women have weaker land rights, women are even more strongly excluded from the planning and design of land-bound infrastructure (Von Benda-Beckmann and Von Benda-Beckmann 2000). Women may also be excluded from the right to contribute labour for maintenance as confirmation of their water rights, and, instead, be summoned to hire male labourers. Changing those norms, even where (young) men massively out-migrate, appears a slow process. Unfortunately, while public agencies' stated goals are to redress inequalities, their ample opportunities to realize this are not always used – quite the contrary.

The brief sketch of local water law in this section is not meant to romanticize local water law in any way. The last and pervasive feature is that rural and peri-urban agriculture-based livelihoods remain vulnerable and fragile, and extreme poverty remains widespread. This justifies why *support* for infrastructure development by governments, NGOs and international development agencies (or 'public' support) is needed to improve welfare. Unfortunately, as discussed next, water laws and policies, in particular permit systems – that work well in formalized high-income countries but are implemented as blanket measure across the world – create even more problems.

Relationship with formalizing statutory law: permit systems

History of permit systems in Europe

The following look at the history of permit systems is based on Caponera (1992) and elaborates typologies that Hodgson proposes. It clarifies, first, that permit systems are rooted in Roman water law since 500BC – the world's oldest surviving water law. (From that perspective, the name 'modern' water law seems questionable.) Second, this historical perspective sheds light on the two sides of the same coin intrinsic to current permit systems: on the one hand, the dimension of regulation, as in regulatory licences, in which a state administration legitimately regulates pollution, over-use, inundation risks, etc., in the public interest; on the other hand, there are fundamental ownership questions in state-backed legal 'tenure security' dimensions of so-called 'modern' law. Permit systems followed entirely different pathways in Europe and in its colonies. Opposite dimensions dominated and caused the paradox mentioned.

When Roman water law emerged around 500BC in the small area around Rome, it made the famous distinction between public and private waters, while vesting the authority to regulate public waters in the common interest in 'the collective'. With the growing expansion of the Roman Empire up to 500AD throughout large parts of today's Europe, the 'collective' gradually became the Roman emperor. Further, dispossession of any prior and future claims to water resources by conquered tribes became the outright aim simply by declaring those waters as being 'public waters', so under the authority of the same emperor. In Europe, after the fall of the Roman Empire, the emperor's and his lower vassals' expropriation weakened in favour of customary rights. With the French Revolution, France and civil law countries of most of continental Europe revived Roman law. Accordingly, the Napoleonic Code of 1804, classified water into private waters (located below, along or on privately owned land) and public waters (which were confined only to 'navigable' or 'floatable' waters) requiring a permit for rights of use (with related water rates). Obviously, the new emancipating social middle classes negotiated that most waters were classified as private waters.

In the same era, British common law countries developed the riparian doctrine, which does not recognize any form of ownership, not even by the Crown. The very specific general good of water is for common use (*res comunis omnium*). Riparian landowners got equal (riparian) use rights over water for free utilization, without the need for administrative intervention to it. Riparians have strong rights vis-à-vis newcomers beyond the riparian strips, who had to negotiate hard for their entrance. Water courts were established to settle the inevitable disputes. In many countries following the common law system, laws, ordinances, regulations or other legal enactments were issued for administering or regulating specific subjects related to water. They sprung from needs arising from local conditions.

During the next one and a half centuries, Europe's economies and water economies gradually formalized. States invested heavily in public infrastructure to catalyze the growth of the water economies. Gradually, public agencies, parastatals, public–private partnerships, hydro-power plants, municipal and industrial water service providers and private companies established effective technical and institutional control over the nation's water resources. Almost all former primary water takers became secondary users and clients of water service providers or members of irrigation groups and water users' associations. In France, the definition of public waters only slightly expanded in 1910, when other waters could be included, if acquired by the state for the purpose of public works. Pollution issues became more important. The 'environment' emerged as a new water user in its own right. Earlier investments in, for example, large dams had to be undone for recreational purposes. Stronger state control was needed and fully accepted for such regulatory roles. More uniformity, also eliminating overlap, facilitated an effective regulatory state. Only in 1964 did the French laws include more waters in the public domain, such as those necessary for domestic water supply, navigation, agricultural and industrial production. Also, the law no longer speaks of private waters but of 'non-domanial waters' – but they still cannot be revoked without compensation. A new criterion of 'public interest' was introduced, which further limited the sector of privately owned waters. In sum, in the civil law countries it was only very gradually that 'the increasing intervention of the administration and the introduction of the water use permit system' rendered 'obsolete the former subdivision between public and private water ownership and all water utilizations are being submitted increasingly to regulatory control' (Caponera 1992, p. 79).

In England and Wales, the common law riparian system also only changed with the Water Resources Law of 1963, when general licensing for the abstraction of water was imposed by statute. An authority became responsible for authorizing water abstractions above certain thresholds. Nevertheless, many features of riparianism were preserved. Moreover, under other legislation, licences carefully considered prior water uses: no licence for the abstraction of surface or underground water could be granted without prior public notice being given, so that persons whose rights or interests were affected were able to file their objections. The state is explicitly not seen as owner but 'has the power to control water utilisations' (Caponera 1992, p. 114).

History of permit systems in Europe's colonies

A completely different history of formal water law is found in Europe's colonies. There, Europeans introduced Roman water law with the primary aim of outright dispossession of conquered tribes from prior and future claims to their water resources. The Papal Bull in 1493 encouraged the expropriation of water resources in most parts of Latin America by Spanish and Portuguese colonizers through permit systems. In this Bull, Pope Alexander VI:

> gave the catholic kings all newly discovered lands, including waters. Water use became the object of special king's permits (Mercedes) granted by the Spanish government authorities for certain purposes, such as domestic drinking needs and irrigation. Such permits could be revoked.... The violation of permit requirements could be punished with a fine.
>
> *Caponera 1992, p. 49*

In Sub-Saharan Africa, civil law colonizers classified waters, as in France, as public or private, with public waters only being those 'navigable or floatable'. Public waters were vested in the colonial governors. However, as Caponera describes:

> Later, due to climatic circumstances, i.e., of the fact that most African streams are seasonal and therefore non-navigable during certain periods of the year with the consequence that very little is left to the public domain, the distinction between navigable and

non-navigable waters disappeared and, generally, all waters were placed in the public domain. Under this regime, every use of public water is subject to the obtention of an administrative authorization, permit or concession.

Caponera 1992, p. 99

British rulers brought complex and fragmented forms of common law, often related to specific water uses. In South Africa, the riparian rights system was adopted in 1912. Groundwater became private water. This law grafted the expropriation of water resources on the land grab through the well-vested British title deed system. Africans were dispossessed from water tied to 91 per cent of their land resources lost under territorial segregation, while the declaration of homelands as state land rendered the dispossession complete. At independence in 1994, hardly any black or coloured person was formally entitled to water. In other countries like Rhodesia, Kenya and Tanzania, the British quickly turned to colonial ownership and permit systems as well. Initially, customary law was recognized in the sense that only the colonizers were obliged, or entitled, to apply for a permit, often under a veneer of sophisticated regulatory duties. The complete dispossession came with the formal imposition of this dual obligation and entitlement to all citizens, except for those exempted, ironically in the name of 'equality'.

In Tanzania, for example, under German colonization, registration to vest rights was only open to foreign settlers. Water uses 'under native law and custom' were recognized but with secondary status. Natives could only participate in decision-making through 'duly authorized representatives' or 'in addition to the District Commissioner'. Customary law was tolerated, but only 'where it did not conflict with the interests of the colonial state'. The British took Tanzania over from Germany after the First World War. The later Water Ordinance of 1948, chapter 257, stipulated: 'the entire property in water within the Territory is hereby vested in the Governor, in trust for His Majesty as Administering Authority for Tanganyika'. It is only in the Water Ordinance of 1959 that the option of registration (not of a permit, or 'water right' as it was called) was extended to 'native' water users, but not obligatory. Urban water supply and water use for mining operations were regulated separately.

At independence, both in Latin America and Sub-Saharan Africa, ownership of the nation's water resources shifted from the colonial powers to the new independent states. Then, the right to apply for a permit became an obligation for all citizens. Only certain small- and micro-scale uses were exempted. Thus, Tanzania's new government under Nyerere declared that 'all water in Tanganyika is vested in the United Republic' under the Water Utilization (Control and Regulation) Act 1974, section 8. This Ordinance (section 14) rendered registration obligatory for all who 'divert, dam, store, abstract and use' water. From then onwards, only such registered water uses were considered to be lawful. In Tanzania and elsewhere, this formally finished the unfinished business of colonial dispossession of customary law.

For decades, these laws remained dormant, while the state focused on infrastructure development in a welfare mode. However, under the banner of IWRM and neo-liberal notions of 'water as an economic good' since the 1990s, key financers and donors from high-income countries started encouraging and paying governments to revise the colonial laws for a 'modern' outlook (Van Koppen *et al.* 2014b). The early water reforms in countries like Chile and Mexico are also variations on the theme of this legacy.

Across Africa, but also in, for example, Mexico, the major revision of the laws concerned the tying of fees to permits under the aggressive promotion of the notion that 'water is a scarce economic good'. Legal systems were perverted into fiscal systems. Moreover, the laws now had to be *implemented*, including among the majorities of rural and peri-urban small-scale water users governed under local law. Expectedly, the early experiences of implementation of permit systems highlight complete failures in terms of, first, serving as a legitimate, regulatory tool in the way high-income countries had developed permit systems; second, in terms of establishing property rights regimes in line with the constitutions and human rights. As a result, instead of supporting the countries' majority of small-scale water users, the latter are currently the losers. The following findings illustrate issues at stake.

The state perspective: weakening regulation

State regulation is widely seen as a legitimate task for a water administration. However, the newly promulgated water laws fail to enable effective state regulation because, first of all, they appear *impossible to implement*, given the large numbers of small-scale water users obliged to apply for a permit – the majority of whom are administration-illiterate, if not totally illiterate, without internet and bank accounts and remote, requiring high transport costs. Only fractions of eligible permit holders are reached.

For example, in Ghana, the Water Resources Commission Act of 1996 led to the setting up of a Water Use Permits Data Base, which is essentially a volume-based billing tool. It can generate invoices and record payments of fees for water use. Twelve years later, in 2008, 154 formal large-scale users had been entered as 'permit holders'. They included municipal water service providers, mines, a few large-scale irrigators and industrial and food processers. In that year, a total of US$180,000 was collected from these 154 permit holders. This is 40 per cent of the amount invoiced. There is another data base for the registration of the millions of other water users but this has hardly any entries. The legal status of these customary water uses remains contested terrain (Ampomah and Adjei 2009).

Moreover, the efforts to reach these small-scale users are also disproportionate compared to the volumes regulated as a result of the immense inequalities in water use. In Tanzania, 3,680 water use, effluent discharge and drilling permits had been allocated across the entire country by 2014, after almost 20 years of intensified efforts to implement them. However, the *Water Sector Status Report* complained that no donor was found to be interested in financing the implementation of permits (MoWI 2010). Inequalities are immense. In Tanzania's Wami Ruvu basin, Sumuni (2015) found that 960 permits had been allocated. The 30 largest permits used 89 per cent of the total volume of water allocated. The other 930 permit holders only used 11 per cent of total water volumes permitted.

Even in South Africa, with a rural population of about 20 million and with its long-standing and relatively well-resourced water administration, only 70,000 water uses have been registered (out of which only part is licensed). Moreover, among those registered, the registration of just the 10 per cent largest users already accounts for 70–90 per cent of the volume of water registered. Obviously, the full legal procedure of licensing is much more resource-intensive, and efforts to reach the tens of thousands of water users of the other 10–30 per cent even more disproportionate. These logistic burdens of trying to reach many small-scale users negatively affect the factual enforcement of any regulatory conditions. Indeed, the question is whether any regulatory purpose is achieved at all by nation-wide application of permits.

The major logistic burdens render permit systems counter-productive taxation systems. In Sub-Saharan Africa, both bureaucrats and water users commonly refer to the new permit systems as *taxation* systems. The new basin organizations are supposed to function on the basis of these taxes. Fiscal measures for under-sourced states can be legitimate. However, the state's costs for both legal procedures and revenue collection among many small-scale water users are, obviously, higher than any net revenue raised. Moreover, the experience in Mexico shows that revenue collection worked much better from 1989–92, when it was a well-defined and well-targeted fiscal measure. Once it became part and parcel of the revived nation-wide permits (concessions) introduced in 1992, revenue collection declined (Garduno 2005). Revenue collection now even risks being jeopardized by the major implementation problems and fraud of the concessions system. It is noted that Mexico exempts all agricultural water uses from the obligation to apply for a concession.

Ironically, permit systems are not only logistically impossible to implement, but their design also intrinsically hampers *regulation of over-use or pollution*. This is because permits of any duration become the only way to use water lawfully, so attempts to regulate are always accompanied by granting much-desired, state-backed water tenure security. This tying of regulatory measures to tenure security enables the tail (an entitled user) to wag the dog (the regulator). The more powerful and administration proficient benefit most from this. They can most easily obtain these first-class entitlements. They find their ways to the water bureaucracies much more easily than small-scale water users. Even without considering the likely bribery for permits, large-scale users pay the highest volume-related taxes that the state regulators in basin organizations need for their salaries. Regulating those who provide one's bread and butter erodes regulatory power by design. Further, permit holders can demand compensation and, in the case of markets of tradable permits, obtain entitlements almost free on a first come, first served basis. They can resell later for good profits. This form of speculation occurred with powerful foreign and other large-scale investors in Chile's water markets. In response, the rules were changed with respect to speculation (Bauer 2004). In sum, the state has to choose between either strengthening a water user's water tenure security, also vis-à-vis the state itself, or regulating, so necessarily curtailing that user; the two cannot go together.

It is obvious that regulatory administrations need, as an absolute minimum, to be *informed* about water uses. Yet, the maintenance of a cadastre is already quite a logistic burden; many permit registers have shown to even default in that, as a high proportion of permits appeared outdated and 'not operational any more'. If being informed about water use at a more central level is the key purpose, there are many much cheaper methods by which to do that, certainly with modern mapping and remote sensing methods.

These experiences raise the question afresh: how can under-resourced states that face a minority of powerful water users and a large majority of small-scale users who are almost unreachable, cost-effectively implement water regulation, and in which public interest? Before answering this question, weaknesses in providing formal water tenure security for informal water users are discussed.

Informal users' perspectives: further marginalization in water entitlements

From the perspectives of the majority of rural and peri-urban water users, the renewed efforts to *implement* permit systems are, essentially, the belated confrontation of the vesting of ownership in colonial settlers and the silent transfer of those claims to the independent state. Communities widely contest the state's ownership claims, also as grounds for taxation, as long as there is no clear state service delivered in return. This negatively affects the state's legitimacy.

A common assumption is that local water laws are not eroded, but that local laws just need to be converted into permit systems, within a 'granted' period of a couple of years. However, conversion is impossible. As mentioned above, even just a codification risks freezing living local law. Moreover, if conversion to permit systems were possible at all, it would marginalize small-scale water users in at least three other ways, intrinsic to permits.

First, state capacity to inform and to respond to permit applications is simply lacking, especially for the more remote and smaller-scale users, who are less administration proficient. Yet, these citizens are still obliged to comply with certain rules; failure to do so may lead to imprisonment. South Africa's *National Water Resources Strategy* formally acknowledges that 'current licensing processes are often costly, very lengthy, bureaucratic and inaccessible to many South Africans' (DWA 2013). Any water administrator would orally admit the same. Moreover, as permits are typically written in the name of the male adult of the household, women are even more disadvantaged.

Second, some water laws impose certain conditions for those who can apply for a permit, such as land ownership, as in Kenya (Mumma 2007), or expensive measurement obligations, as in Zimbabwe (Derman *et al.* 2007). These conditions exclude those who cannot meet those requirements, again typically the smaller-scale customary water users.

Last, but not least, permit systems arbitrarily relegate some citizens to a status of exempted users. These are the smallest and micro-users – for example, those who irrigate less than 1 hectare or only use manual water abstraction devices. As long as permits are linked to more water tenure security, being exempted erodes any tenure security vis-à-vis water users with a permit (Hodgson 2004). These smallest-scale and micro-users disproportionately include the poor and poorest, especially women.

Thus, the second question arises: how should governments of low- and middle-income countries avoid offering the opportunity to protect the water tenure security for some (those who are already wealthier), and structurally and categorically not for others (the poor, typically governing water under local law)? This question is even more pertinent for countries that have adopted constitutions and have ratified human rights that commit to: equal and fair treatment; gender equality; the right to food and an adequate standard of living (so to water uses that contribute to realizing those rights); procedural rights; or historical justice and protection of indigenous people's rights.

Possible answers and conclusions

The chapter disentangled the paradox of the precarious situation of local law in low- and middle-income countries as opposed to high-income countries, where permit systems have become the dominant form of statutory law but nurture the small pockets where customary laws prevail. The history of permit systems showed how the entitlement (ownership and dispossession) and regulatory dimensions of permit systems evolved in completely different directions. This raised two new questions about the role of the state vis-à-vis both dimensions, with three possible answers.

First, Europe's former colonies face a unique form of state ownership of water resources and permits as entitlements. This colonial legacy raises unprecedented questions and opportunities to address tenure security. Currently, the powerful minority benefit from the entitlement dimensions of permits. Instead, formal water entitlements need to be conceptualized afresh. As formal owner of water resources, the state has the power to redefine entitlements. An obvious first step is recognition and protection of local law (as widely adopted in the case of customary land tenure). This would also contribute to realizing constitutional and human rights commitments that can well be reflected in local law as well (Derman *et al.* 2007): every living being has the right to water volumes needed for basic domestic *and* productive uses to realize rights to food and adequate standard of living. Only once that minimum for all is realized can specific individuals obtain larger shares. In South Africa, this option is discussed (but not adopted) in terms of *priority* general authorizations for all. The minister can issue this with a stroke of the pen (Van Koppen and Schreiner 2014).

Water uses are expanding, which affects prior uses. In mediating these changing entitlements, the state would focus in particular on new investments by large-scale and high-impact newcomers. Current and future local uses are to be formally protected through a bottom-up process, from local government up to central and transboundary agencies. Through negotiations, impacts on available water resources can be minimized – for example, by large-scale users' storage development. Anyone negatively affected by newcomers needs to be compensated.

Thus, instead of allowing large-scale and administration-proficient users to speculate and obtain tradable permits, these entitlements dimensions of current permits of the powerful minority should also be removed. This also serves the following possible answer.

Second, as regulator in the public interest, it is counter-productive to give entitlements as a condition to observe regulations. The state's task in high-, middle- and low-income countries alike is more feasible, obviously, if all permits become, in Hodgson's typology, regulatory licences. This is also the nature of permits in Europe. For legitimate targeting and enforcement of regulation, only those who can logistically be reached should be obliged to apply for such licences (or waste water discharge permits, for that matter). Moreover, specific regulatory goals, such as information or taxation, may well be achieved much more effectively through entirely separate measures to regulatory permits. Above all, regulation in the public interest would cherish whatever local arrangements are in place for millions of water users, which already function well and are equitable and sustainable. States should identify, broadly record, understand, encourage and build upon those arrangements.

Third, these and other solutions to reform permit systems would enable the state and partners to revert to the agenda that has already been snowed under by the legalistic debates: the agenda of developing infrastructure in a welfare mode, taking the myriad local investments and communities' own priorities as starting point.

References

Ampomah, B. and Adjei, B. (2009) Statutory and customary water rights governing the development and management of water infrastructure and technologies in the Volta basin [unpublished project report]. Accra: Water Resources Commission, International Water Management Institute, Water Research Institute, and Challenge Program on Water and Food.

Bauer, C. (2004) *Siren Song: Chilean Water Law as a Model for International Reform.* Washington, DC: Resources for the Future.

Boelens, R. and Dávila, G. (1998) *Searching for Equity: Conceptions of Justice and Equity in Peasant Irrigation.* Assen, Netherlands: Van Gorcum.

Boelens, R., Bustamante, R. and de Vos, H. (2007) Legal pluralism and the politics of inclusion. In B. van Koppen, M. Giordano and J. Butterworth (eds), *Community-based Water Law and Water Resources Management Reform in Developing Countries.* Oxford: CABI, chapter 6.

Boelens, R., Gentes, I., Guevara, A. and Arteaga, P. (2005) Special law: recognition and denial of diversity in Andean water control. In D. Roth, R. Boelens and M. Zwarteveen (eds), *Liquid Relations: Contested Water Rights and Legal Complexity.* New Brunswick, NJ: Rutgers University Press.

Bruns, B. (2005) Routes to water rights. In D. Roth, R. Boelens and M. Zwarteveen (eds), *Liquid Relations: Contested Water Rights and Legal Complexity.* New Brunswick, NJ: Rutgers University Press.

Caponera, D.A. (1992) *Principles of Water Law and Administration: National and International.* Rotterdam: Balkema.

Coward, W. E. Jr (1986) State and locality in Asian irrigation development: the property factor. In K.C. Nobe and R.K. Shanpath (eds), Irrigation management in developing countries: current issues and approaches. Proceedings of an Invited Seminar Series sponsored by the International School for Agricultural and Resource Development (ISARD). *Studies in Water and Policy Management.* No. 8. Boulder and London: Westview Press.

Derman, B., Hellum, A., Manzungu, E., Sithole, P. and Machiridza, R. (2007) Intersections of law, human rights and water management in Zimbabwe: implications for rural livelihoods. In B. van Koppen, M. Giordano and J. Butterworth (eds), *Community-based Water Law and Water Resources Management Reform in Developing Countries.* Oxford: CABI.

DWA (Department of Water Affairs), Republic of South Africa (2013) *National Water Resource Strategy*, 2nd edn. Water for an Equitable and Sustainable Future. Department of Water Affairs, Pretoria, South Africa.

Garduno, H. (2005) Lessons from implementing water rights in Mexico. In B. Bruns, C. Ringler and R. Meinzen-Dick (eds), *Water Rights Reform: Lessons for Institutional Design*. Washington, DC: International Food Policy Research Institute.

Hellum, A., Kameri-Mbote, P. and van Koppen, B. (2015) The human right to water and sanitation in a legal pluralist landscape: perspectives of southern and eastern African women. In A. Hellum, P. Kameri-Mbote and B. van Koppen (eds), *Water is Life: Women's Human Rights in National and Local Water Governance in Southern and Eastern Africa*. Harare: Weaver Press, chapter 1.

Hodgson, S. (2004) Land and water – the rights interface. FAO Legislative Study 84. Rome: Food and Agricultural Organization of the United Nations.

Hodgson, S. (2016) Exploring the concept of water tenure. Land and water discussion paper 10. Rome: Food and Agriculture Organization of the United Nations.

Komakech, C.H. (2013) Emergence and evolution of endogenous water institutions in an African river basin: local water governance and state intervention in the Pangani River basin, Tanzania. PhD thesis. Delft UNESCO-IHE and International Water Management Institute.

Meinzen-Dick, R. and Nkonya, L. (2007) Understanding legal pluralism in water and land rights: lessons from Africa and Asia. In B. van Koppen, M. Giordano and J. Butterworth (eds), *Community-based Water Law and Water Resources Management Reform in Developing Countries*. Oxford: CABI.

MoWI (Ministry of Water and Irrigation). United Republic of Tanzania (2010) *Water Sector Status Report 2010*. Dar-es-Salaam: Ministry of Water and Irrigation United Republic of Tanzania.

Mumma, A. (2007) Kenya's new water law: an analysis of the implications of Kenya's Water Act, 2002 for the Rural Poor. In B. van Koppen, M. Giordano and J. Butterworth (eds), *Community-based Water Law and Water Resources Management Reform in Developing Countries*. Oxford: CABI.

Ostrom, E. (1992) *Crafting Institutions for Self-governing Irrigation Systems*. San Francisco: ICS Press.

Ramazzotti, M. (1996) *Readings in African Customary Water Law*, FAO Legislative Study 58 Rome: FAO.

Roth, D., Boelens, R. and Zwarteveen, M. (eds) (2005) *Liquid Relations: Contested Water Rights and Legal Complexity*. New Brunswick, NJ: Rutgers University Press.

Shah, T. (2007) Issues in reforming informal water economies of low-income countries: examples from India and elsewhere. In B. van Koppen, M. Giordano and J. Butterworth (eds), *Community-based Water Law and Water Resources Management Reform in Developing Countries*. Oxford: CABI.

Sokile, C. (2006) Analysis of institutional frameworks for local water governance in the Upper Ruaha catchment. Unpublished PhD thesis. Dar-es-Salaam: University of Dar-es-Salaam.

Sumuni, P. M. (2015) Influence of institutional set-up on performance of traditional irrigation schemes, a case study of Nyandira Ward, Mvomero District Tanzania. A dissertation submitted in partial fulfilment of the requirements for the degree of Masters of Science in Irrigation Engineering and Management of Sokoine University of Agriculture, Morogoro, Tanzania.

van Eeden, A. (2014) *Whose Waters: Large-Scale Agricultural Development in the Wami-Ruvu River Basin*. 60 Credit. Ås, Norway Norwegian University of Life Science, Department of International Environmental and Development Studies.

van Koppen, B. (2000) Gendered water and land rights in construction: rice valley improvement in Burkina Faso. In B. Bruns and R. Meinzen-Dick (eds), *Negotiating Water Rights*. New Delhi: Sage.

van Koppen, B. and Schreiner, B. (2014) Priority general authorizations in rights-based water use authorization in South Africa. In M. J. Neal (Patrick), A. Lukasiewicz and G.J. Syme (eds), *Supplemental Issue Why Justice Matters in Water Governance*. Water Policy 16(S2), pp. 59–77. London: IWA. DOI: 10.2166/wp.2014.110.

van Koppen, B., Giordano, M. and Butterworth, J. (eds) (2007) *Community-based Water Law and Water Resources Management Reform in Developing Countries*. Oxford: CABI.

van Koppen, B., Smits, S., Rumbaitis del Rio, C. and Thomas, J. (2014a) *Upscaling Multiple-Use Water Services: Accountability in the Water Sector*. London: Practical Action, IWMI/WLE – International Water and Sanitation Centre IRC – Rockefeller Foundation.

van Koppen, B., van der Zaag, P., Manzungu, E. and Tapela, B. (2014b) Roman water law in rural Africa: the unfinished business of colonial dispossession. *Water International* 39(1).

Von Benda-Beckmann, F. and von Benda-Beckmann, K. (2000) Gender and the multiple contingencies of water rights in Nepal. In R. Pradhan, F. von Benda-Beckmann and K. von Benda-Beckmann (eds),

Water, Land, and Law: Changing Rights to Land and Water in Nepal. Proceedings of a workshop held in Kathmandu. March 1998. Kathmandu: FREEDEAL, Wageningen Agricultural University, Erasmus University Rotterdam, pp. 17–38.

Von Benda-Beckmann, F., von Benda-Beckmann, K. and Spiertz, J. (1998) Equity and legal pluralism: taking customary law into account in natural resource policies. In R. Boelens and G. Davis (eds), *Searching for Equity: Conceptions of Justice and Equity in Peasant Irrigation.* Assen, Netherlands: Van Gorcum, pp. 57–69.

Yoder, R. and Martin, E. (1998) Water rights and equity issues: a case from Nepal. In R. Boelens and G. Dávila (eds), *Searching for Equity: Conceptions of Justice and Equity in Peasant Irrigation.* Assen, Netherlands: Van Gorcum.

2

A COMPARATIVE ANALYSIS OF THE PUBLIC TRUST DOCTRINE FOR MANAGING WATER IN THE UNITED STATES AND INDIA

Melissa K. Scanlan

Introduction

The starting point for understanding water law and policy is that water is a public commons held in trust; a corollary to this is that private rights are usufructuary and secondary to public rights (*Glass v Goeckel* 2005, 65, 68; *RW Docks & Slips v State* 2001, 788 (citing *State v Bleck* 1983) 498). These may seem like remarkable propositions in a world where privatization is increasing. Yet, this chapter, which anchors the study of water law in the public trust doctrine, will explore how this functions in two countries: the United States and India.

I compare the public trust doctrine in the United States and India because no two countries have a more developed public trust jurisprudence. Both countries, having a common ancestry as British colonies, have their legal roots in English common law. Both countries are constitutional democracies, both use a common law system that relies on courts to develop the law based on precedent and both have national supreme courts whose decisions are binding on all lower courts, as well as state-level high or supreme courts (Rajamani and Ghosh 2012, p. 139). At the same time, these countries face very different social and economic pressures on their water resources. Yet in both the courts have applied the public trust doctrine to resolve conflicts over scarce water resources.

The public trust doctrine encompasses the concepts that the sovereign holds shared water resources in trust for the public's use and enjoyment. Like private trusts, with the public trust there is an identified trustee (the government), beneficiaries (every member of the public) and trust property.

Traditionally, public trust property in England was limited to tidal waters. In the United States, at the time of the original formation of the states, the trust property was described as tidal waters, as well as submerged lands below those waters. Then the court expanded it to navigable freshwaters and non-navigable tidal waters (The Daniel Ball 1871, 563–64 (navigable freshwaters); *Phillips Petroleum Co v Mississippi* 1988, 483 (non-navigable tidal waters)). Courts have articulated the trust property boundaries over time to reflect resources of importance to the public. It has been described as including beaches – both below the ordinary high-water mark or tideline and above, groundwater, wetlands and non-navigable tributaries (*Illinois Central*

Railroad v Illinois 1892, 452–3 (navigable waters and submerged lands); *City of Berkeley v Superior Court* 1980, 362 (tidelands); *Nat'l Audubon Soc'y v Superior Court* 1983, 727–8 (non-navigable tributaries); *Glass v Goeckel* 2005, 65, 68 (beaches below ordinary high-water mark); *Matthews v Bay Head Improvement Ass'n* 1984, 363–6 (dry sand beaches above high tide); In *re Water Use Permit Applications* 2000, 467) (groundwater); *Lake Beulah Mgmt Dist v State Dep't of Natural Res* 2011, 84–5, 88 (groundwater hydrologically connected to navigable water); *Just v Marinette Cnty* 1972, 769) (wetlands)). Although this book focuses on water law, it is notable that in both the United States and India the doctrine has been applied beyond these water boundaries to parks, air and all public natural resources (*Gould v Greylock Reservation Comm'n* 1966, 114 (state park); *Friends of Van Cortlandt Park v New York* 2001, 1053 (park); Hawaii Const 1959, art 11, sec 1 (all public natural resources); Texas Const 1876, art 16, sec 59 (all natural resources); *Bonser-Lain v Te. Comm'n on Envtl. Quality* 2012 (air and atmosphere); *MC Mehta v Kamal Nath* 1997, 21).

In the United States, the doctrine is a matter of state law, which makes it highly fragmented with variations in its scope and vitality across the 50 states (*PPL Montana LLC v Montana* 2012, 1235). However, some argue for a federal public trust doctrine, and there is a live dispute in the federal courts about whether the US government is a trustee over waters for which it alone has sovereignty, such as the territorial seas of its coastline (Blumm and Wood 2013, ch 11; *Kelsey Cascade Rose Juliana v United States* 2016, 18). A meticulous and exhaustive analysis of those differences is beyond the needs of this chapter. Here, you will find general principles distilled from prominent law reviews and cases. One primary law review that orients this chapter is Professor Joseph Sax's seminal 1970 article (Sax 1970). Although there have been hundreds of subsequent law reviews on the subject, this one merits focus here because it has been used repeatedly in cases developing the doctrine in the United States and India (Klass 2006, 707, n. 38; *MC Mehta v Kamal Nath* 1997, 388; *MI Builders Private Ltd v Radhey Shyam Sahu*, 1999, 518; *Intellectuals Forum, Tirupathi v State of AP & Ors* 2006, 18).

Over 40 years ago, Professor Sax wrote that the public trust doctrine would be a 'useful tool of general application for citizens seeking to develop a comprehensive legal approach to resource management problems' (Sax 1970, p. 474). In order to serve this function, Professor Sax thought that the doctrine needed to articulate a legal right in the public, be enforceable against the government and advance modern environmental quality concerns (ibid.). In the countries that have robust public trust doctrines, courts enforce these elements.

I have woven these aspects together with additional case law and scholars' subsequent thinking to present four general principles that are important to understanding the public trust doctrine. These are that the public trust doctrine: 1) guards against privatizing important public water resources or commons; 2) prohibits the full privatization of certain water resources; 3) places a continuing duty on the trustees to supervise and protect the trust; and 4) encourages long-term stewardship that is aligned with sustainable development and protective of intergenerational equity. In this chapter, these four principles of the public trust doctrine will be used as an organizing framework to analyze the public trust doctrine in the United States and India.

United States

State governments, acting as trustees on behalf of the public, protect public rights, which, at the time of the founding of the United States, were understood to involve commerce, navigation, fishing and passage (walking). A majority of states have now interpreted public rights to include all forms of water-oriented recreation, and some states protect ecological integrity and water quality.

The public trust doctrine guards against privatizing important public resources or commons (Sax 1970, pp. 491–5)

The public trust doctrine contributes to water management by more clearly articulating how powers are separated between the sovereign trustee(s) and the judiciary. Professor Sax argued that judges should take a more exacting review of a water trustee's decisions when a case involves: 1) diffuse public uses threatened by resource alienation below market value; 2) giving private interests the power to make public resource decisions; or 3) reallocating public uses to private uses (Blumm and Guthrie 2012, pp. 753–4 (citing Sax 1970, pp. 561–3)). These situations call for scepticism and less deferential judicial review of the actions of the state trustee (which can be either a legislature or an administrative agency).

In the United States, this judicial scepticism has manifested itself into something akin to a substantive rule of statutory interpretation that looks for a clear statement from the legislature. Professor Sax identified a judicial presumption that the trustee cannot privatize public resources without a clear statutory directive, and analyzed cases in Massachusetts, Wisconsin and California showing this in operation (Sax 1970, pp. 491–545).

In the lodestar United States case, *Illinois Central*, the Supreme Court upheld the Illinois legislature's rescission of a grant to a private railroad corporation of the bed of Lake Michigan in the Chicago harbour (*Illinois Central* 1892, 452). The Court described Illinois' title to the land under the harbour's water as one held in trust for the people, and ruled that attempting to transfer the entire property to the railroad was inconsistent with the state's duty as a trustee (*Illinois Central* 1892, 452–3). This case has been interpreted over time not to prevent all privatization of public trust resources, but as a restriction on the trustee to not 'substantially impair' the public interest in the remaining trust resources (*Illinois Central* 1892, 452). Thus, when there appears to be a privatization of public resources, there is a judicial presumption that the state cannot privatize public resources without a clear statutory directive and a legislative finding that the action would not 'substantially impair' the remaining public trust resources (Sax 1970, pp. 494–6; *Gould v Greylock Reservation Comm'n* 1966, 121–4; Araiza 2012).

This operates like other substantive canons of statutory interpretation that require a clear statement from the legislature, such as the constitutional avoidance doctrine: courts narrowly construe a statute in order to avoid constitutional doubts unless the legislation clearly states otherwise. Such a rule of interpretation furthers democracy by ensuring elected representatives in the state legislature have publicly considered and expressed their intent to authorize a privatization of trust resources (Sax 1970, pp. 498–9). This moves the locus of decision-making when there is a privatization underway away from administrative agencies and back to the democratically elected legislature.

The public trust doctrine prohibits the full privatization of certain water resources

Even with a clear statement from the legislature, however, some public trust resources cannot be fully privatized. Some states have articulated a perpetual public trust easement – for instance, in beaches and submerged lands under navigable or tidal waters (*Glass v Goeckel* 2005, 62; *City of Berkeley v Superior Court* 1980; *RW Docks & Slips v State* 2001, 788). Some courts have interpreted this public trust easement to allow the public the right to walk on beaches, even where a private party holds a deed to the property (*Glass v Goeckel* 2005, 62). Other courts have interpreted this as a public trust servitude to prevent a private property owner of tidelands from destroying the tidelands (*Marks v Whitney* 1971, 381). These limitations on private property use are consistent with the English common law origins of the public trust doctrine, which recognized an overlap of public and private rights in the same property (*jus privatum* and *jus publicum*) and an inability to alienate or extinguish the public attribute when the sovereign conveyed private title (Hale 1888, 22; Scanlan 2013, pp. 327–8, nn. 260–1).

The import is that even if the legislature clearly states its intent to extinguish the *jus publicum* in this public trust property, it is beyond the power of the state to do so. Cases from Michigan, Wisconsin and Arizona exemplify this concept. The supreme courts of Wisconsin and Michigan have emphasized that, on navigable waters, public rights limit private title – a concept that is 'vital' to public trust law (*Glass v Goeckel* 2005, 65, 68; *RW Docks & Slips v State* 2001, 788 (citing *State v Bleck* 1983, 498)). Although the state may 'convey lakefront property to private parties, it necessarily conveys such property subject to the public trust' (*Glass v Goeckel* 2005, 65; *Marks v Whitney* 1971, 382 (citing *Yates v Milwaukee* 1870, 504); *Ill Steel Co v Bilot* 1901, 856; *Diana Shooting Club v Husting* 1914, 819 (citing *Willow River Club v Wade* 1898)).

The Michigan Supreme Court based its holding in *Glass v Goeckel*, which protected the public's right to walk Great Lakes beaches, on a view of the public trust doctrine as a limitation on state sovereignty: in other words, because the state cannot abdicate its trustee responsibilities to protect public rights in the Great Lakes and its beaches up to the ordinary high-water mark – even if the state had issued patents to private parties that extended below the high-water mark – it could not have conveyed away the public trust easement (*Glass v Goeckel* 2005, 62, 65). The court declared that 'the sovereign must preserve and protect navigable waters for its people' (*Glass v Goeckel* 2005, 63).

In *Arizona Center for Law in the Public Interest v Hassell* (1991), the Arizona Court of Appeals addressed whether legislation substantially relinquishing the state's interests in riverbed lands violated the public trust doctrine. '[T]he court stated it would scrutinize closely any state action that might violate the public trust doctrine and, applying such scrutiny, found the legislation violated the state's public trust principles' (Klass 2006, 732 (citing Ariz Ctr For Law in the Pub Interest v Hassell 1991, 166–73)).

The Arizona Legislature attempted again in 1995 to diminish the public trust by enacting a statute to declare that '[t]he public trust is not an element of a water right' and cannot be part of adjudicating private water rights (*San Carlos Apache Tribe v Superior Court ex rel Maricopa* 1999, 199 (citing *Ariz Rev Stat Ann* 2003, sec 45-263(B)). The Arizona Supreme Court invalidated the provision in *San Carlos Apache Tribe v Superior Court ex rel Maricopa*. The court held that Arizona's public trust doctrine provides a constitutional limitation on legislative power to give away resources held in trust for the public (*San Carlos Apache Tribe* 1999, 199).

The public trust doctrine places a continuing duty on the trustees to supervise and protect the trust (Nat'l Audubon Soc'y v Superior Court 1983)

The government has a duty and a right to revise previously issued water use permits to private parties. This allows for adaptive management to be carried out over time by revising permits to balance and respond to new information.

In California's *Mono Lake* decision (*Nat'l Audubon Soc'y v Superior Court* 1983), the state supreme court applied the doctrine to take a fresh look at a 40-year-old decision to issue private water rights to the Los Angeles Department of Water and Power to divert water from non-navigable tributaries of Mono Lake (*Nat'l Audubon Soc'y v Superior Court* 1983, 728). The court's holding has been regarded as significant for both expanding the scope of the public trust doctrine to non-navigable tributaries and clarifying the contingent nature of private rights in water such that the state has the power to reconsider a water rights allocation decision at any time, based on a consideration of the effect on the public trust.

The public trust doctrine does not function as a trump card, but as a requirement that the state must weigh the impacts on the public trust against the private water rights, and must accommodate public rights as far as feasible (*Nat'l Audubon Soc'y v Superior Court* 1983, 728).

The public trust doctrine encourages long-term stewardship aligned with sustainable development and is protective of intergenerational equity

The public trust doctrine not only addresses what natural resources should be managed by the trustee, but also how they should be managed. Principally, resources in the public trust should be managed in order to protect intergenerational equity and developed economically in a way that stewards and preserves natural resources. In short, the trustees should manage resources in ways that do not deplete the trust.

An example of intergenerational equity arises in the Hawaii Supreme Court's public trust decision in 2004, in which it considered whether a reservation of water for instream use was protected by the public trust, and whether the state water commission sufficiently considered this when it granted a water permit. The court held that the commission failed in its duty, in part, because 'a reservation of water constitutes a public trust purpose with respect to the state's continuing trust obligation' to ensure water resources for present and future generations (In re *Wai'ola O Moloka'i Inc* 2004, 694).

Similarly, in *Brooks v Wright*, the Alaska Supreme Court contrasted private trusts from the public trust in natural resources, emphasizing intergenerational equity as a distinguishing feature (1999, 1032–4). The court explained that private trusts typically require the trustee to maximize economic yield, but the public trust, based on Alaska's constitution, 'requires that natural resources be managed for the benefit of all people under the assumption that both development and preservation may be necessary to provide for future generations, and that income generation is not the sole purpose of the trust relationship' (*Brooks v Wright* 1999, 1032).

India

Like the United States, India draws its public trust doctrine from Roman and English common law. In defining the doctrine, India borrows heavily from US law and Professor Sax's 1970 article (*MC Mehta v Kamal Nath* 1997). However, India also grounds its public trust doctrine in ecological or natural law and in its national constitution. India contributes further by applying sustainable development concepts, such as the 'precautionary principle' and 'polluter pays' to public trust cases (*MC Mehta v Kamal Nath* 1997, 22–3). Citing a prior court decision and the Brundtland Report, the Indian Supreme Court described the polluter pays and precautionary principles as shifting to the polluter the burden of going forward with a project and of restoring damaged areas. 'It is thus settled by this Court that one who pollutes the environmental must pay to reverse the damage caused by his acts' (*MC Mehta v Kamal Nath* 1997, at 22–3).

In its first public trust doctrine case, the Indian Supreme Court in *MC Mehta v Kamal Nath* expanded the scope of the doctrine beyond its traditional grounding in tidal and navigable waters to 'all ecosystems operating in our natural resources' (1997, 21). The Supreme Court summarized the Indian public trust doctrine as follows:

[our legal system] includes the public trust doctrine as part of its jurisprudence. The state is the trustee of all natural resources which are by nature meant for public use and enjoyment. Public at large is the beneficiary of the sea-shore, running waters, airs, forests and ecologically fragile lands. The state as a trustee is under a legal duty to protect the natural resources. These resources meant for public use cannot be converted into private ownership.

Ass'n for Env't Prot v Kerala 2013, 356–7 (quoting
MC Mehta v Kamal Nath 1997, 21)

The court in *Kamal Nath* based this all-inclusive natural resources scope of the trust on several findings. First, the court noted that environmental science provides us with certain laws of nature that cannot be changed by legislative fiat, and that an 'understanding of the laws of nature must therefore inform all of our social institutions' (*MC Mehta v Kamal Nath* 1997, 15–16 (quoting Hunter 1988, 314–16)). Second, the public trust doctrine rests on the principle that the:

air, sea, waters and the forests have such a great importance to the people as a whole that it would be wholly unjustified to make them a subject of private ownership. . . . [B]eing a gift of nature, they should be made freely available to everyone irrespective of the status in life.

MC Mehta v Kamal Nath 1997, 17

Third, it interpreted US public trust cases to include 'the protection of ecological values' as a purpose of the trust and concluded that ecological concepts are a 'relevant factor to determine which lands, waters or air are protected by the public trust doctrine' (*MC Mehta v Kamal Nath* 1997, 21).

The Indian public trust doctrine has a natural law foundation. In *Fomento Resorts & Hotels v Minguel Martins* the Supreme Court explained that:

The Indian society has, since time immemorial, been conscious of the necessity of protecting environment and ecology. The main moto of social life has been 'to live in harmony with nature'. Sages and Saints of India lived in forests. Their preachings contained in Vedas, Upanishadas, Smritis etc. are ample evidence of the society's respect for plants, trees, earth, sky, air, water and every form of life. It was regarded as a sacred duty of every one to protect them. In those days, people worshipped trees, rivers and sea, which were treated as belonging to all living creatures. The children were educated by their parents and grandparents about the necessity of keeping the environment clean and protecting earth, rivers, sea, forests, trees, flora fauna and every species of life.

Fomento Resorts & Hotels v Minguel Martins 2009, 30

In its most recent public trust case, the Indian Supreme Court asserted that the '[m]ajority of people still consider it as their sacred duty to protect the plants, trees, rivers, wells, etc., because it is believed that they belong to all living creatures' (*Ass'n for Env't Prot v Kerala* 2013, 353–4).

The Indian Supreme Court has also grounded the public trust doctrine in several provisions of its Constitution: Articles 21, 48A, and 51A. In *Fomento Resorts & Hotels v Minguel Martins*, the Court identified amendments created in 1976: Article 48A in Part IV and Article 51A in Part IVA. Article 48A imposes a 'duty. . . on the State to endeavour to protect and improve the environment and safeguard forests and wild life of the country' (*Fomento Resorts & Hotels v Minguel Martins* 2009, 30; *Ass'n for Env't Prot v Kerala* 2013, 356). Article 51A makes it a 'duty of every citizen of India to protect and improve the natural environment including forests, lakes, rivers and wild life and to have compassion for living creatures' (*Fomento Resorts & Hotels v Minguel Martins* 2009, 30; *Ass'n for Env't Prot v Kerala* 2013, 356).

In a case about a private mall development in a public park, the Indian Supreme Court grounded the public trust doctrine in Article 21 of the Indian Constitution, which declares that '[n]o person shall be deprived of his life or personal liberty, except according to procedure established by law' (*MI Builders Private Ltd v Radhey Shyam Sahu* 1999, 466). The Supreme Court applied this to water in *Intellectuals Forum* (2006, 20) and the High Court of Kerala also interpreted this article to apply to water and more:

> The right to life is much more than the right to animal existence and its attributes are many fold, as life itself. A prioritization of human needs and a new value system has been recognized in these areas. The right to sweet water, and the right to free air, are attributes of the right to life, for these are the basic elements which sustain life itself.
>
> *Attakoya Thangal v Union of India 1990, 583*

Thus, the Indian public trust doctrine is based on natural law 'from time immemorial', constitutional law, English common law and United States cases such as *Illinois Central* and *Mono Lake*, which the Indian Supreme Court appears to have incorporated into its jurisprudence.

Compared to the United States, with cases dating back to the founding of the country, the body of Indian law is more recent, with its first public trust case coming from the Indian Supreme Court in 1997 (*MC Mehta v Kamal Nath*). However, during this comparatively brief history, India has articulated a doctrine whose scope applies to all natural resources, and the courts have had the opportunity to apply it to a wide variety of public resources: riverbed alteration, a state-owned forest (*MC Mehta v Kamal Nath* 1997), a public park (*MI Builders Private Ltd v Radhey Shyam Sahu* 1999, 466), groundwater (*Perumatty Grama Panchayat v State* 2004, para 34 (reversed on appeal by *Hindustan Coca-Cola Beverages v Perumatty Grama Panchayat* 2005), pending final determination in the Supreme Court of India); *West Bengal v Kesoram Indus Ltd* 2004, 104), a beach access path 200 metres from the high-tide line (*Fomento Resorts & Hotels v Minguel Martins* 2009, 11), natural gas (*Reliance Natural Res Ltd v Reliance Indus Ltd* 2010, para 97–8) and 2G wireless wavelengths (*Ctr For Pub Interest Litigation v Union of India* 2012, 161, 246). The water resource cases will be described more fully below along the lines of the four public trust principles: *MC Mehta v Kamal Nath; Fomento Resorts & Hotels v Minguel Martins; Intellectuals Forum, Tirupathi v State of AP & Ors.; Ass'n for Env't Prot v Kerala;* and *Perumatty Grama Panchayat v State.*

The public trust doctrine guards against privatizing important public resources or commons

In its first application of the public trust doctrine, the Indian Supreme Court considered the legality of a classic example of a government agency official favouring a narrow private interest and attempting to privatize riverbed, floodplain and forest resources that had been a public commons. The Court considered whether Span Resort, which developed a tourist resort in Kullu-Manali valley, could dredge, blast and otherwise block the flow of the Beas river and create a new channel to divert the river to at least 1 kilometre down stream, as well as build in a state forest. The resort development was described as a project aimed to carry out 'Kamal Nath's dream of having a house on the bank of the Beas in the shadow of the snow-capped Zanskar ranges' (*MC Mehta v Kamal Nath* 1997, 1–4, 7–8, 13–14). Kamal Nath, the Minister of Environment and Forests who approved the project after it was initiated without permission, was found to have members of the Nath family with business interests in the company (*MC Mehta v Kamal Nath* 1997, 1–4).

The Court ultimately held that the Span Resort project violated the public trust doctrine, voided the prior approvals and leases and ordered restoration of the resources. Given that the leases to the company for 'purely commercial purposes' involved a 'large area of the bank of the River Beas which is part of a protected forest,' 'ecologically fragile land' and 'part of the riverbed', the court had no hesitation in holding that the Himachal Pradesh Government committed a patent breach of the public trust held by the State Government (*MC Mehta v Kamal Nath* 1997, 22).

In reaching this holding, the court in *Kamal Nath* used a separation of powers analysis that puts the onus on the elected lawmaking body to provide a clear statement of intent when public natural resources are leased to narrower commercial interests.

> The resolution of this conflict in any given case is for the legislature and not the courts. If there is a law made by Parliament or the State Legislatures the courts can serve as an instrument of determining legislative intent in the exercise of its powers of judicial review under the Constitution. But in the absence of any legislation, the executive acting under the doctrine of public trust cannot abdicate the natural resources and convert them into private ownership, or for commercial use.
>
> *1997, 21*

In *Kamal Nath* and subsequent cases, the Indian Supreme Court favourably quoted Professor Sax to describe its role in public trust cases:

> When a State holds a resource. . . for the free use of the general public, a court will look with considerable scepticism upon any governmental conduct which is calculated either to relocate that resource to more restricted uses or to subject public uses to the self-interest of private parties.
>
> *1997, 17; MI Builders Private Ltd v Radhey Shyam Sahu 1999, 518;*
> *Intellectuals Forum, Tirupathi v State of AP & Ors 2006, 18*

Building on this foundation, the Indian Supreme Court in *Intellectuals Forum* explained that the state trustee's duty to 'protect the people's common heritage' requires a 'more demanding' level of Court scrutiny of government actions than required for 'the government's general obligation to act for the public benefit' (Intellectuals Forum 2006, 18).

In *Ass'n for Env't Prot v Kerala*, the Supreme Court was asked to review the state of Kerala's approval of the Department of Tourism's development of a restaurant in Manalppuram Park on 'reclaimed land' that had previously formed part of the Periyar river (2013, 362–3). The Court described the appeal as 'illustrative of the continuing endeavour of the people of the State [of Kerala] to ensure that their rivers are protected from all kinds of man made pollutions and/or other devastations' (*Ass'n for Env't Prot v Kerala* 2013, 361).

After favourably citing the public trust decisions of the Indian Supreme Court and scrutinizing the facts of the controversy, the Court refused to defer to the state's approval of the tourism development project. The Court found the state had approved the project without conducting a legally required environmental review, analogous to the National Environmental Policy Act in the United States. Thus, the Court held the state violated the environmental review law that India established to implement Article 48A of the Indian Constitution. Such a violation, according to the Court, undermined the 'fundamental right to life guaranteed to the people of the area under Article 21 of the Constitution' (*Ass'n for Env't Prot v Kerala* 2013, 368–9). The Court ordered the respondents to demolish the restaurant (*Ass'n for Env't Prot v Kerala* 2013, 369).

The public trust doctrine prohibits the full privatization of certain water resources

In *Kamal Nath* the Supreme Court summed up India's public trust doctrine as an expansive doctrine that prevents private ownership of all water (and all natural resources) (1997, 21). The Court applied the public trust doctrine to invalidate the approvals and leases the government had given to a private resort granting use and alterations to a riverbed, floodplain and forest (Kamal Nath 1997, 22, 24).

In *Fomento Resorts & Hotels v Minguel Martins,* the Indian Supreme Court considered whether the development of a private resort to attract international tourists to a new beachside hotel, yoga centre, health club and water sports facilities could destroy a traditional footpath that provided access to a public beach (*Fomento Resorts & Hotels v Minguel Martins* 2009, 35). The property used for this development included a traditional public walkway about 200 metres from the high-tide line that was used 'by the members of the public including the fisher folk' to access the sea 'from time immemorial, without objection whosoever, openly, peacefully and continuously and as a matter of right' (*Fomento Resorts & Hotels v Minguel Martins* 2009, 23).

The Court found that public access existed prior to the acquisition of the property for the resort development, and despite an agreement between one of the parties and the President of India and that the developer constructed an alternative access by constructing a road, parking area and public footpath, the court held the developer cannot deprive 'members of the public of their age old right to go to the beach' through the original access way (*Fomento Resorts & Hotels v Minguel Martins* 2009, 28). The holding relied on the public trust doctrine:

> This doctrine puts an implicit embargo on the right of the State to transfer public properties to private party if such transfer affects public interest, mandates affirmative State action for effective management of natural resources and empowers the citizens to question ineffective management thereof.
>
> *Fomento Resorts & Hotels v Minguel Martins 2009, 28*

The Court went on to elaborate that the state holds in trust, 'even if private interests are involved', a wide variety of 'renewable and non-renewable resources, associated uses, ecological values or objects in which the public has a special interest (i.e. public lands, waters, etc.)' (*Fomento Resorts & Hotels v Minguel Martins* 2009, 28–9).

The Court said it was reiterating law from prior decisions of the Indian court that the state trustee:

> cannot transfer public trust properties to a private party, if such a transfer interferes with the right of the public and the Court can invoke the public trust doctrine and take affirmative action for protecting the right of people to have access to light, air and water and also for protecting rivers, sea, [water] tanks, trees, forests and associated natural eco-systems.
>
> *Fomento Resorts & Hotels v Minguel Martins 2009, 32*

Applying this formulation of the public trust doctrine, the Court held that the private developer must maintain a longstanding public access and road to the beach (*Fomento Resorts & Hotels v Minguel*

Martins 2009, 32). Since the developer had destroyed the access, the Court ordered the developer to restore access and demolish the extension of the hotel that prevented the public from using its traditional footpath to reach the beach (*Fomento Resorts & Hotels v Minguel Martins* 2009, 39).

The public trust doctrine places a continuing duty on the trustees to supervise and protect the trust

The Indian corollary to the *Mono Lake* case is the Kerala High Court's decision in *Perumatty Grama Panchayat v State* (2004), in which the court used the public trust doctrine to impose restrictions on groundwater pumping for commercial purposes. In *Mono Lake* a state supreme court similarly relied on the common law public trust doctrine to justify restrictions on long-standing private rights to divert and use water out of the basin of origin.

The Indian case centres on Coca-Cola, which obtained a licence from Perumatty Grama Panchayat (the traditional local government) for the Hindustan Coca-Cola Beverages Bottling Plant in Plachimada, and started production of non-alcoholic beverages in 2000 (*Perumatty Grama Panchayat v State* 2004, para 2.). Coca-Cola's product was dependent on a single natural resource, water, pumped from the groundwater of the area (*Perumatty Grama Panchayat v State* 2004, para 2). In the production of 1 litre of beverage, they used 3.75 litres of water (*Perumatty Grama Panchayat v State* 2004, para 9). 'The estimated water utilization figures for production year 2002 were 1,41,015 cubic metres (m^3) out of which 37,604 m^3 will be exported outside the plant in the form of beverage' (*Perumatty Grama Panchayat v State* 2004, para 9).

Similar to the law of capture used by Texas in the United States, and based on English common law, the accepted practice had been that a landowner was not liable for injury and was 'free to extract any amount of ground water which is available under' the land it owns (*Perumatty Grama Panchayat v State* 2004, para 13). Thus, at the time Coca-Cola started operating there was 'no law governing the control or use of ground water' (*Perumatty Grama Panchayat v State* 2004, para 13).

In 2003, after widespread citizen protests, the Panchayat attempted to not renew the licence on the grounds that it was no longer in the 'public interest' because 'excessive exploitation of ground water by the Coca-Cola Company in Plachimada is causing acute drinking water scarcity in Perumatty Panchayat and nearby places' (*Perumatty Grama Panchayat v State* 2004, para 2). A higher level of government ordered the Panchayat to continue the licence pending an expert study of groundwater, and the High Court of Kerala was called on to determine whether Coca-Cola's extraction of groundwater was legal (*Perumatty Grama Panchayat v State* 2004, para 13).

Describing groundwater as 'a nectar, sustaining life on earth', the court established that 'ground water is a national wealth and it belongs to the entire society' (*Perumatty Grama Panchayat v State* 2004, para 13). The court found that the prior cases allowing unlimited extraction of groundwater can no longer apply due to the 'sophisticated methods used for extraction like bore-wells, heavy duty pumps' and the 'emerging environmental jurisprudence' based on Article 21 of the Indian Constitution, Principle 2 of the Stockholm Declaration of 1972, and the public trust doctrine (*Perumatty Grama Panchayat v State* 2004, para 13 (quoting *MC Mehta v Kamal Nath* 1997, para 24)).

After quoting extensively from *Kamal Nath* to describe India's public trust doctrine, the court 'safely concluded that the underground water belongs to the public' (*Perumatty Grama Panchayat v State* 2004, para 13). The state 'trustee' has a duty to protect groundwater against excessive exploitation and trustee inaction is 'tantamount to infringement of the right to life of the people guaranteed under Article 21 of the Constitution of India' (*Perumatty Grama Panchayat v State* 2004, para 13). The court reasoned that because the 'right to clean air and unpolluted water' are part of the right to life under Article 21, the government is 'bound to protect ground water from excessive exploitation' (*Perumatty Grama Panchayat v State* 2004, para 13). In other words, contrary to the rule of capture, groundwater does not belong to the overlying land-owner, and is to be managed by the state trustee for the benefit of all.

The court then applied the public trust doctrine in a way that favours domestic and agricultural use of groundwater over commercial use that exports water from its basin of origin.

> Normally, every land owner can draw a reasonable amount of water, which is necessary for his domestic use and also to meet the agricultural requirements. . . [By contrast, Coca-Cola extracted] 510 kilolitres of water. . . per day, converted into products and transported away, breaking the natural water cycle.
>
> *Perumatty Grama Panchayat v State 2004, para 13*

The court opined that no 'great knowledge of Science of Ecology is necessary' to infer that this excessive extraction will create an ecological imbalance (*Perumatty Grama Panchayat v State* 2004, para 13). The court further described a tragedy of the commons when it reasoned that if Coca-Cola can extract and export groundwater with no controls or liability, then every landowner in the area has a similar potential to extract and export huge quantities of water, ultimately creating a desert (*Perumatty Grama Panchayat v State* 2004, para 13). The court held that the groundwater extraction by Coca-Cola was 'illegal' because they had 'no legal right to extract this much of national wealth' (*Perumatty Grama Panchayat v State* 2004, para 13). It is beyond the power of the state trustee 'to allow a private party to extract such a huge quantity of ground water, which is a property, held by it in trust' (*Perumatty Grama Panchayat v State* 2004, para 13).

The court ordered that the company should be treated like every other landowner and permitted to draw an amount of groundwater 'equivalent to the water normally used for irrigating the crops' on the amount of land owned (in this situation, 34 acres) and for domestic purposes (*Perumatty Grama Panchayat v State* 2004, para 15, para 4 of Order). This case was reviewed again by a panel of judges for the High Court of Kerala and the panel held that 'ordinarily a person has right to draw water, in reasonable limits, without waiting for permission from the Panchayat and the Government' (*Hindustan Coca-Cola Beverages v Perumatty Grama Panchayat* 2005, para 43). This decision has been appealed and is awaiting an opinion by the Supreme Court of India. However, in another decision by the Indian Supreme Court, involving mining, the Supreme Court favourably discussed the earlier case of *Perumatty Grama Panchayat v State* (2004) 1 KLT 731 (Kerala HC) and described the groundwater and public trust law in India as follows: 'Deep underground water belongs to the State in the sense that doctrine of public trust extends thereto.' A landowner has the right to use groundwater only to the extent that it does not affect other's rights and only to further the purpose for which the land is held (*West Bengal v Kesoram Indus Ltd,* AIR 2005, 104). In summary, with many legal and political challenges surrounding the Coca-Cola facility in Plachimada, it has remained closed since 2004, the year the High Court of Kerala ruled in this case based on the public trust doctrine (GlobalResearch 2010).

The public trust doctrine encourages long-term stewardship aligned with sustainable development and protective of intergenerational equity

In *Fomento Resorts & Hotels v Minguel Martins,* India's Supreme Court addressed the duty the public trust doctrine imposes on resource managers, water rights holders and all members of the public that is focused on the long-term and intergenerational equity. Each is called on to focus on sustainable management of shared resources and each has a separately articulated 'duty' or 'obligation' to 'not impair' public trust resources or values (*Fomento Resorts & Hotels v Minguel Martins* 2009, 29). As to general members of the public who use water (air or land) and associated natural ecosystems, the Court described an 'obligation to secure for the rest of us the right to live or otherwise use that same resource or property for the long term and enjoyment by future generations' (*Fomento Resorts & Hotels v Minguel Martins* 2009, 29). As to the 'water right holder', the Court described their 'obligation to use such resources in a manner as not to impair or diminish the people's rights and the people's long term interest in that property or resource, including down-slope lands, waters and resources' (*Fomento Resorts & Hotels v Minguel Martins* 2009, 29).

In summary, the Court sees the public trust doctrine as 'a tool for exerting long-established public rights over short-term public rights and private gain' (*Fomento Resorts & Hotels v Minguel Martins* 2009, 29). 'We reiterate that natural resources including forests, water bodies, rivers, seashores, etc. are held by the State as a trustee on behalf of the people and especially the future generations' (*Ass'n for Env't Prot v Kerala* 2013, 359–61 (quoting *Fomento Resorts & Hotels v Minguel Martins* 2009)).

Drawing on Professor Sax's work, the Supreme Court in *Intellectuals Forum* identified three types of restrictions the public trust doctrine places on government action: 1) the property subject to the trust must not only be used for a public purpose, but it must be held available for use by the general public; 2) the property may not be sold, even for fair cash equivalent; and 3) the property must be maintained for particular types of use; either traditional uses, or some uses particular to that form of resources (*Intellectuals Forum* 2006, 18).

The Court then reviewed a government housing development project on government-owned land that was found to be a 'systematic destruction of percolation, irrigation and drinking water tanks in Tirupati', a place that was also undergoing rapid population expansion that exceeded the existing infrastructure. The Court found 'overwhelming evidence' that the tanks had existed since 1500AD and had been used for irrigation, for lakes, and were 'furthering percolation to improve the ground water table, thus serving the needs of the people in and around these tanks' (*Intellectuals Forum* 2006, 21). The Court noted that the tanks are 'communal property and the State authorities are trustees to hold and manage such properties for the benefit of the community and they cannot be allowed to commit any act or omission which will infringe the right of the Community and alienate the property to any other person or body' (*Intellectuals Forum* 2006, 21). Due to government housing developments to accommodate rapid population growth, much of the natural resources of the lakes (tanks) had been lost and considered irreparable (*Intellectuals Forum* 2006, 21).

Applying principles of sustainable development, including intergenerational equity, the Court was called to balance the 'basic human need' for shelter in the context of poverty and urbanization against the state trustee's duty to protect and improve the water supply (*Intellectuals Forum* 2006, 18–19; *AP Pollution Control Board v Prof MV Nayudu* 1999). The Court in *Intellectuals Forum* understood that it needed to adjudicate whether 'economic growth can supercede' environmental protection for the good of current and future generations (*Intellectuals Forum* 2006, 19–20). It stated that '[M]erely asserting an intention for development will not be enough to sanction the destruction of local ecological resources' (*Intellectuals Forum* 2006, 16–17). The Court held that the development project violated the first and third aspects of the public trust doctrine, above (*Intellectuals Forum* 2006, 18–19). The Court found further support for its holding in principles of sustainable development, including intergenerational equity, and Articles 21, 48A and 51A of the Indian Constitution (*Intellectuals Forum* 2006, 19–20).

The Court ordered no further construction of housing in Peruru and Avilala tanks and corrective measures for recharging the water supply (*Intellectuals Forum* 2006, 18–19). It struck the balance of sustainable development by refraining from ordering demolition of the housing developments that had already impaired the water supply of the area; but instead ordering proactively there be no further construction, that each existing house install rooftop rain harvesting systems, and that no new groundwater wells be allowed, among other restoration and water recharge activities (*Intellectuals Forum* 2006, 22–3).

Conclusion

The public trust doctrine has survived and evolved over the millennia to adapt to changing societies and their water management challenges. This chapter has shown how the doctrine is applicable in vastly different circumstances, by comparing the use of the doctrine in the United States and India. Whether contesting access to the beach for recreation in the United States or protecting scarce drinking water resources for a rapidly growing and urbanizing population in India, the courts have used the public trust doctrine to assert governmental protections of shared resources. It has proven to be a doctrine that courts utilize to guard against privatizations, impose a continuing duty to sustainably manage the commons and protect long-term intergenerational equity.

While development of the doctrine in the United States has focused on state property title and perpetual easements when states have granted private titles, the Indian doctrine is far more ecologically grounded and expansive – including all of the country's natural resources. Although the Indian courts were inspired by the US courts, they have gone further in protecting public trust rights by finding the national constitution is a basis for such protections. Indian courts have based public trust protections on several provisions, including Article 21's protections of life and liberty.

Perhaps the development of the doctrine between the two countries is now at a point where learning will be reciprocal and the US will adopt the Indian approach to find public trust rights are protected by the US Constitution's safeguards of life and liberty (US Const. 5th Amendment). The current controversy where this could happen will come from the US District Court of Oregon, where Magistrate Judge Thomas Coffin issued an order that recommended denial of motions to dismiss a suit brought by a group of young people who alleged excessive carbon emissions are threatening their future (*Kelsey Cascade Rose Juliana v United States* 2016, 24).

Unlike prior US public trust cases addressed in this chapter, which were based on state law, here the plaintiffs argued the federal government of the United States, in its sovereign role, is a trustee of the territorial ocean waters and atmosphere of the nation. They further argued the government's actions and inactions regarding climate change are contrary to the public trust doctrine and violate the Due Process Clause of the US Constitution, which protects 'fundamental rights' (*Kelsey Cascade Rose Juliana v United States* 2016, 17, 20).

> At this stage of the proceedings, the court cannot say that the public trust doctrine does not provide at least some substantive due process protections for some plaintiffs within the navigable water areas of Oregon. Accordingly, the court should not dismiss any claims under the public trust doctrine to that extent.
>
> *Kelsey Cascade Rose Juliana v United States 2016, 23*

Plaintiffs are a group of youth who allege the effects of climate change will harm them and future generations to a greater extent than others. The Magistrate articulated the role the judiciary can play to break logjams of this nature.

The intractability of the debates before Congress and state legislatures and the alleged valuing of short term economic interest despite the cost to human life, necessitates a need for the courts to evaluate the constitutional parameters of the action or inaction taken by the government. This is especially true when such harms have an alleged disparate impact on a discrete class of society.

Kelsey Cascade Rose Juliana v United States 2016, 8

Since the public trust doctrine has a history of being engaged to protect intergenerational equity, this case holds the potential for a broad court order to address climate change.

References

Araiza, W.D. (2012) The public trust doctrine as an interpretive canon. *University of California at Davis Law Review* 45(3), pp. 693–740.

Ariz Ctr for Law in the Pub Interest v Hassell [1991] 837 P 2d 158.

Ass'n for Env't Prot v Kerala [2013] 7 SCR 352.

Attakoya Thangal v Union of India [1990] AIR 580.

Blumm, M. and Guthrie, R. (2012) Internationalizing the public trust doctrine: natural law and constitutional and statutory approaches to fulfilling the Saxion vision. *University of California at Davis Law Review* 45(3), pp. 741–808.

Blumm, M.C. and Wood, M.C. (2013) *The Public Trust Doctrine in Environmental and Natural Resources Law.* Durham: Carolina Academic Press.

Bonser-Lain v Tex Comm'n on Envtl Quality [2012] No D-1-GN-11-002194.

Brooks v Wright [1999] 971 P 2d 1025.

City of Berkeley v Superior Court [1980] 606 P 2d 362.

GlobalResearch (2010) *Coca-Cola Causes Serious Depletion of Water Resources in India* [online]. Available from: http://www.globalresearch.ca/coca-cola-causes-serious-depletion-of-water-resources-in-india/18305 [accessed 6 April 2016].

Constitution of India 1950, art. 21, s. 48A, 51A.

Constitution of the State of Hawaii 1959, art. 11, s. 1.

Constitution of the State of Texas 1876, art. 16, s. 59.

Ctr for Pub Interest Litigation v Union of India [2012] 3 SCR 147.

Diana Shooting Club v Husting [1914] 145 NW 816.

Fomento Resorts & Hotels v Minguel Martins [2009] INSC 100.

Friends of Van Cortlandt Park v New York [2001] 750 NE 2d 1050.

Glass v Goeckel [2005] 703 NW 2d 58.

Gould v Greylock Reservation Comm'n [1966] 215 NE 2d 114.

Hale, M. (1888) *A Treatise de Jure Maris et brachiorum ejusdem.*

Hindustan Coca-Cola Beverages v Perumatty Grama Panchayat [2005] 2 KLT 554.

Hunter, D.B. (1988) An ecological perspective on property: a call for judicial protection of the public's interest in environmentally critical resources. *Harv Envtl. L. Rev*, 12, p. 311.

Illinois Central Railroad v Illinois [1892] 146 US 387.

Illinois Steel Co v Bilot [1901] 84 NW 855.

In re Wai'ola O Moloka'i Inc [2004] 83 P 3d 664.

In re Water Use Permit Applications (Waiahole Ditch I) [2000] 9 P 3d 409.

Intellectuals Forum, Tirupathi v State of AP & Ors [2006] 3 SCC 549.

Just v Marinette Cnty [1972] 201 NW 2d 761.

Kelsey Cascade Rose Juliana v United States [2016] 6:15-cv-01517-TC.

Klass, A.B. (2006) Modern public trust principles: recognizing rights and integrating standards. *Notre Dame Law Review* 82(2), pp. 699–754.

Lake Beulah Mgmt Dist v State Dep't of Natural Res [2011] 799 NW 2d 73.

Marks v Whitney [1971] 491 P 2d 374.

Matthews v Bay Head Improvement Ass'n [1984] 471 A 2d 355.

MC Mehta v Kamal Nath [1997] 1 SCC 388.

MI Builders Private Ltd v Radhey Shyam Sahu [1999] 6 SCC 464.

Nat'l Audubon Soc'y v Superior Court [1983] 658 P 2d 709.

Perumatty Grama Panchayat v State [2004] 1 KLT 731.

Phillips Petroleum Co v Mississippi [1988] 484 US 469.

PPL Montana LLC v Montana [2012] 132 S Ct 1215.

Rajamani, L. and Ghosh, S. (2012) India. In R. Lord, S. Goldberg, L. Rajamani and J. Brunnée (eds), *Climate Change Liability: Transnational Law and Practice*. Cambridge: Cambridge University Press, chapter 7.

Reliance Natural Res Ltd v Reliance Indus Ltd [2010] INSC 374.

RW Docks & Slips v State [2001] 628 NW 2d 781.

San Carlos Apache Tribe v Superior Court ex rel Maricopa [1999] 972 P 2d 179.

Sax, J.L. (1970) The public trust doctrine in natural resources law: effective judicial intervention. *Michigan Law Review* 68(3), pp. 471–566.

Scanlan, M.K. (2013) Shifting sands: a meta-theory for public access and private property along the coast. *South Carolina Law Review* 65(2), pp. 295–372.

The Daniel Ball [1871] 77 US 557.

West Bengal v Kesoram Indus Ltd [2004] AIR 2005 SC 1646.

3

WATER RIGHTS AND PERMITTING

A South African approach

Michael Kidd

Introduction

At the time of South Africa's transition to democracy in 1994, access to natural resources was heavily skewed in favour of the minority white population. This was particularly so in the case of water, the right to use being predominantly held by white riparian landowners. The common law which provided for riparian rights would obviously not be an appropriate tool to provide for redress in access to water. Consequently, the law was developed not only to manage and conserve water in a water-scarce country, but also to provide for increased equity in access. This would be done by placing the water within the control of the national government, which would then allocate water rights by an administrative decision-making process – that is, permitting.

This chapter examines the permitting system in South Africa (water use licensing) within the context of South African water law's approach to integrated water resource management. In many respects, it is a story of what ought to be rather than what is, and, consequently, the chapter will highlight the several shortcomings that may be experienced in respect of allocating rights by a permitting system.

The chapter will briefly outline the history of South African water law (the common law), which will demonstrate the need for a different approach to riparian rights when considered in the context of social transformation. This will be followed by an explanation of the current South African law relating to water rights. Analysis of the implementation of the law will then reveal its strengths and weaknesses. The South African example is informative in that its water laws are often seen as innovative. This chapter will indicate that the attractiveness of the laws on paper do not necessarily translate into good practice. One of the purposes of this piece is to identify the shortcomings and, where possible, means of addressing these. South Africa's experience with a water use licensing-based system in order to achieve integrated water resources management (IWRM) (with a particular transformative slant) is undoubtedly a useful example (positive and negative) of an alternative approach for any other jurisdiction where priority-based (or similar) water access regimes are not promoting desirable social and environmental objectives.

The South African common law

European settlers in South Africa first followed Roman-Dutch law, which regarded the state as *dominus fluminis* (master of the rivers), with rights being allocated by the state. By the end of the

1800s, however, as a result of judicial decisions such as *Hough v van der Merwe* (1874) and *Van Heerden v Wiese* (1880) (that appeared to have been influenced by the law of the United States: Milton 1995), 'the doctrine of riparian rights had become firmly established throughout South Africa' (De Wet 1959, p. 33). Despite misgivings about the suitability of such a system in a water-scarce country, the law was not reformed and when there was legislative intervention, the legislation mostly confirmed the common law position (see Kidd 2009, pp. 90–1).

By the time of the enactment of the 1956 Water Act,[1] the legislature saw fit to introduce more state control of water in order to provide for mining and secondary industries, but the control mechanisms in the Act were not much used and the widespread reality on the ground was that riparian rights still held sway: shortly after the Act's promulgation, De Wet observed that the Act was a 'half-hearted attempt to restore to the community the rights lost by a process of judicial legislation, and the doctrine of "riparian rights" is by no means dead' (De Wet 1959, p. 35).

Consequently, at the time of the transition to democracy in 1994, most rights to water were enjoyed by riparian landowners, overwhelmingly white farmers. On the other side of the coin, millions of black people had little or no access to safe water (DWAF 1997). The Constitution of the Republic of South Africa, 1996, provides in s 27 for the right to water as follows:

> (1) Everyone has the right to have access to (a.)…, (b.) sufficient food and water;…
> (2) The state must take reasonable legislative and other measures, within its available resources, to achieve the progressive realization of each of these rights.

In the light of this right and other rights, including the right to an environment not harmful to health or well-being, equality, dignity, right to life, property and administrative justice, the government decided on four key principles that would underpin the new South African water law: first, the government would be the custodian of national water resources in order to manage effectively a critical strategic resource; second, there must be equitable access to water by all people in the country; third, the hydrological cycle is a single system and the water needs of the environment are crucial for the healthy operation of that cycle; and fourth, the rights of neighbouring countries to shared water resources would be recognized (WLRP 1996).

The National Water Act 36 of 1998

Section 3 of the National Water Act 1998 (the Act)[2] provides:

> As the public trustee of the nation's water resources the national government, acting through the Minister, must ensure that water is protected, used, developed, conserved, managed and controlled in a sustainable and equitable manner, for the benefit of all persons and in accordance with its constitutional mandate.

This 'public trustee' role is to be exercised primarily by means of water use licensing. The 1997 White Paper that preceded the Act recognized that government would have to take into account an estimated 40,000 permits, allocations, or scheduling provisions in terms of the 1956 Water Act; 800 water court orders covering water use on 30,000 properties; and 5 million boreholes (DWAF 1997). This meant that it would be all but impossible to replace overnight a system of riparian rights with a system of allocations based on administrative decision-making, and this is the reason (possibly influenced also by concerns about what the effect would be on the economy, particularly commercial agriculture – see Morvik 2014) that the decision was taken to provide for a transitionary period during which existing water rights would remain legally recognized.

The water rights allocation process operates within an elaborate water management statutory system that is based on balancing the principles mentioned above. Space does not permit detailed explanation and discussion of the Act as a whole, but the overall statutory objective is as follows. Ultimately, the national department responsible for water affairs is responsible for all decision-making, but the Act envisages the creation of catchment management agencies (CMAs) (initially 19, now nine), which will be responsible for establishing their own catchment management strategies, all of which must be consistent with the National Water Resource Strategy (NWRS) (see Chapter 2 of the Act). The CMAs may be assigned powers (including licensing powers) under the Act, but at the time of writing this has not yet been done – all decisions are still made by the national department.[3]

Whereas the 2003 NWRS expressly recognized that the management strategy for South Africa's water would be one of integrated water resources management (DWAF 2003), this has been somewhat refined in the 2013 NWRS, which is based on a philosophy of 'developmental water management',

> which can be considered part of IWRM principles in practice and which takes, as a central premise, the fact that water plays a critical role in equitable social and economic development, and that the developmental state has a critical role in ensuring that this takes place.
>
> *DWA 2013, p. 14*

In other words, developmental water management can be interpreted as IWRM with a special emphasis on equitable allocation of water resources. Equity entails three dimensions: equity in access to water services (including basic human needs); equity in access to water resources (for productive purposes); and equity in access to the benefits of water resource use ('water must be allocated so that it brings maximum benefit to all, whether directly or indirectly') (ibid., p. 45).

Any authority responsible for any decision-making in relation to water is required to give effect to the NWRS when exercising any power or performing any duty in terms of this Act (s 7), so the NWRS provides the ultimate strategic framework within which more specific decisions, outlined below, must be taken.

What the Act refers to as the 'protection' of water resources, aiming at water quality imperatives, requires the Minister to establish a system of classification of water resources (s 12), followed by the actual classification of water resources (s 13), so that water resources are then categorized into different classes (on the basis, essentially, of the water quality of, or the extent of human impact upon, the resource). The Minister in this process is to determine, for all or part of every significant water resource,[4] resource quality objectives. This process is to be followed, as soon as possible thereafter, by determination of the Reserve (s 16), which is a central concept in the Act that has led to its reputation as especially innovative legislation. The Reserve is the quantity and quality of water required to satisfy basic human needs and to protect aquatic ecosystems in order to secure ecologically sustainable development and use of the relevant water resource (s 1). These decisions (determining class and Reserve) are prerequisites for the water licensing process, but are complex, resource-intensive and time-consuming processes, as evidenced by the fact that the classification system was published only in 2010[5] and the first set of proposed classifications (in respect of only six catchments) published in 2014.[6]

Alongside the Reserve, probably the other most important concept in the Act is that of 'water use'. This notion includes not only extraction of water from a water resource, but also a number of actions that can be described broadly as any activity that has an impact on a water resource, including diverting a stream and putting substances into a watercourse (see s 21). The importance of this concept is that a person may only carry out a water use for which he has a licence in terms of the Act, unless any one of three exceptions pertains. The first is if the water use is permissible under Schedule 1 of the Act, which contains what can be described as water uses that would have a minimal impact, such as 'reasonable domestic use in that person's household, directly from any water resource to which that person has lawful access' and watering of animals that are not being intensively farmed. Schedule 1 would also cover the emission of sewage and industrial effluent into municipal sewers, not because this would have a minimal effect but because these activities are regulated by the operators of the relevant water treatment works, usually by means of municipal by-laws (see Kidd 2011, p. 169). The second exception is if the water use is permissible as a continuation of an existing lawful use, defined, in essence, as any lawful use during a period two years prior to the commencement of the Act (1 October 1998). This means that all of the riparian rights holders (mostly white farmers, as pointed out above) would have been existing lawful users on this date. The third exception is if the water use is permissible in terms of a general authorization issued under s 39. General authorizations are aimed at reducing the administrative burden whereby common water use practices, generally not that large in impact, are authorized in relation to any person who carries on that activity. Those who are using water under general authorizations are required to comply with the requirements set out in the general authorizations, usually including self-monitoring and reporting requirements. Several general authorizations have been published and are in operation (see Kidd 2011, p. 78).

The objectives of the National Water Act will be difficult to achieve if most water is still used on the basis of existing lawful use, so this was intended to be temporary, even if the reality is that in 2015 there is still widespread water usage under this exception. The legal mechanism in the Act that is designed to replace the existing lawful use is compulsory licensing. Section 43 provides:

If it is desirable that water use in respect of one or more water resources within a specific geographic area be licensed –

(a) to achieve a fair allocation of water from a water resource…
   (i) which is under water stress; or
   (ii) when it is necessary to review prevailing water use to achieve equity in allocations;
(b) to promote beneficial use of water in the public interest;
(c) to facilitate efficient management of the water resource; or
(d) to protect water resource quality,

the responsible authority may issue a notice requiring persons to apply for licences for one or more types of water use contemplated in section 21.

The initiation of such a process would require all potential users of water, including existing rights holders – whether under licence or on the basis of existing lawful use, to apply for a licence. The decision (the outcome of a process described in Chapter 4 Part 8 of the Act) will be based on balancing the demand, determined by adding up the total amount for which applications have been made, and the amount of water available for a particular geographic area, most often a particular catchment. The amount of water available is calculated by determining the total amount of water, and subtracting the amount of water that is required for the Reserve and international obligations. The remainder can then be allocated to the applicants. This process requires classification of water resources and determination of the Reserve, so can only be set into motion once the necessary science underpinning the decision has been finalized (s 15 requires the responsible authority, when exercising any power or performing any duty in terms of the Act, to give effect to any determination of a class of water resource and the resource quality objectives and any requirements for complying with the resource quality objectives; and s 18 has a similar requirement in relation to giving effect to the Reserve). This is one of the reasons that the conversion of the old order to a system of universally licensed water use has been so slow (at the time of writing, only three compulsory licensing processes, together covering a very small area in the country, have been finalized).

In addition to the determination of how much water is available for allocation, there are several other factors that have to be taken into account. In any licensing process (not just compulsory licensing), the Act sets out the factors that must be taken into account in reaching a decision. Section 27 reads:

> In issuing a general authorisation or licence a responsible authority must take into account all relevant factors, including –
>
> (a) existing lawful water uses;
> (b) the need to redress the results of past racial and gender discrimination;
> (c) efficient and beneficial use of water in the public interest;
> (d) the socio-economic impact –
>   (i)   of the water use or uses if authorised; or
>   (ii) of the failure to authorise the water use or uses;
> (e) any catchment management strategy applicable to the relevant water resource;
> (f) the likely effect of the water use to be authorised on the water resource and on other water users;
> (g) the class and the resource quality objectives of the water resource;
> (h) investments already made and to be made by the water user in respect of the water use in question;
> (i) the strategic importance of the water use to be authorised;
> (j) the quality of water in the water resource which may be required for the Reserve and for meeting international obligations; and
> (k) the probable duration of any undertaking for which a water use is to be authorised.

It ought to be immediately apparent that any decision on a water use licence application is not straightforward and requires a substantial amount of data, only some of which is the outcome of water classification and determination of the Reserve. The list in s 27 is also open-ended, so not necessarily confined to the factors explicitly mentioned. The 'equity' requirement in subsection (b) is particularly important, and the Department has run into some controversy by insisting, in some decisions, that this factor is the overriding factor, to the extent that the licence was refused in some cases where the decision would not further the equity objective. In other words, the decision was made on the basis that subsection (b) was a trump.

Some of these decisions were taken on review and the courts, correctly, decided that the Department was wrong to take this approach. In *Guguletto Family Trust v Chief Director, Water Use, Department of Water Affairs and Forestry* (2011), the court held that the Department was 'wrong in law' to regard the equity factor as the overriding factor, because the correct approach was that all relevant factors, including the 11 mentioned in section 27, should be taken into account and then balanced 'without attaching due weight to any one'. The equity issue and the relevant judicial authority are discussed further below.

Further aspects of the licensing regime to note are that the licence must be reviewed at intervals not longer than five years (s 28(1)(f)) and that conditions may be attached to all licences and general authorizations, relating, inter alia, to the protection of water resources, the stream flow regime and other water users (existing and potential); water management practices; return flow and discharge or disposal of waste (including permissible effluent standards) and so on.

Given the scope of the meaning of 'water use', there will be a large variety of licence applications for consideration and decision. These will range from a relatively straightforward application for extraction of water in a catchment where there is a large excess of available water and all the relevant classifications have already been done, to an application by a mining facility that involves extraction of water, storage of wastewater, deviation of streams and disposal of effluent.

Having considered the licensing regime under the National Water Act, it is now possible to consider the benefits and shortcomings of such a regime within the overall context of the pursuit of IWRM.

Benefits of water use licensing

If one considers IWRM in the classical sense of a 'process which promotes the coordinated development and management of water, land and related resources, in order to maximise the resultant economic and social welfare in an equitable manner without compromising the sustainability of vital ecosystems' (Global Water Partnership 2000, p. 22), one consequently envisages water management as having three simultaneous objectives: economic efficiency, equity and environmental sustainability (Van Koppen and Schreiner 2014a, p. 543). It is unrealistic to expect historical, racially skewed patterns of rights holding to be able to fulfil these goals, for obvious reasons. Similarly, relying on the market, particularly where one starts with a position of inequality, will not serve these purposes either; as Bauer observed in the case of Chile, the market cannot adequately provide for equity and environmental sustainability (Bauer 2004). South Africa, along with other countries, has decided that the pursuit of IWRM requires allocation to be carried out by administrative decision-making through licensing (Van Koppen *et al.* 2014). This makes sense from at least an intuitive perspective, since there are several factors that have to be balanced in order to provide for optimal utilization of the resource, whether the utilization is extractive or an activity that has a detrimental impact on water, such as effluent disposal. The relevant factors may often not be pulling in the same direction; so, for instance, a decision that makes the most sense from an economic efficiency perspective may be detrimental to improved equity.

Providing for a state decision-maker to balance these factors in order to make a decision has the benefit (at least in theory) of prioritizing policy objectives, aligning policy objectives (for instance, decisions relating to water and to agricultural practices), and allowing for public participation. Thus not only the applicant for water rights, but other interested and affected parties (including other users and government agencies) may have input into the decision.

A further, very important, theoretical benefit is that a licensing system may provide for both consistency, in the case of like cases being treated alike, and yet avoid a 'one size fits all' approach that may arise if allocation decisions are governed by only legislative rules. This is illustrated by the difference between uniform effluent standards – one size fits all – which may be set by legislation and to which all users would have to conform, and a receiving water quality objectives approach, where decisions may be taken in relation to a particular user and that user's impact on the particular water resource.

In South Africa, however, the theory is often not translated into reality and the reasons for this may be found among the several drawbacks to licensing, which are discussed below. Some of these are universal while others are, if not unique to, more likely to be found in the South African context.

Drawbacks of water use licensing

Administrative burdens/inefficiency

Since the decision-making for a water use licence in South Africa is complex, and involves processing a considerable amount of information, it is not surprising that there are often considerable delays in making decisions. In July 2015, in response to a parliamentary question, the Minister of Water Affairs indicated that there were a total of 1,530 licence applications still waiting decision, more than half of which pertained to applications made before 2014 – 384 were from 2011 and before (PMG 2015). One of the contributing causes of the administrative inefficiency is shortage of staff and failure to fill vacant posts (Funke *et al.* 2007, p. 1244).

In addition to the administrative burden on the state, a permitting system has been observed to have a negative impact on poorer water users, especially women, who are faced with disproportionate transaction costs compared with other applicants who have better access to legal and technical expertise (Van Koppen *et al.* 2014, pp. 54–5).

Weaker legal status of exempted uses

The main reason why certain water uses are exempted from the requirement of a licence is administrative convenience. While the holders of existing lawful use rights are regarded as having recognized legal rights (on the basis of pre-1994 law), the status of Schedule 1 users may be precarious. As Hodgson observes, such a *de minimis* right is a 'curious type of residuary "right"' (Hodgson 2004, p. 19): 'While they may be economically important to those who rely on them, it is hard to see how they provide much in the way of security' (ibid.). Possible problems in enforcing such rights are likely to have more of a negative effect on pastoralists and the poor (Van Koppen and Schreiner 2014b, p. 5). It is not, however, administratively feasible to consider licensing such uses. A possible solution to this problem is placing such uses under general authorizations (ibid.), but that will entail amending the Schedule and will add to existing administrative burdens.

Recognition of customary water uses

In South Africa, the recognition of existing lawful use had the result that limited customary law rights to water maintain legal recognition under the National Water Act, a situation that Van Koppen *et al.* (2014, p. 53) observe is not replicated in any other African country. In other countries (not just in Africa) administrative licensing systems have completely replaced customary, often communal, water use rights and this has often had negative consequences in such communities (Van Koppen *et al.* 2014).

Equity failures

While one of the main imperatives of South Africa's post-1994 water law has been redressing past inequity in access to water, there has been little progress in this regard in the nearly 20 years of the Act's existence. The 2008 Water Allocation Reform Strategy (DWAF 2008) stipulated targets of national water rights allocation of 60 per cent in black hands and 50 per cent to women by 2024. Yet only 1,518 of the 4,284 licences issued between 1998 and 2012 for new water rights were allocated to historically disadvantaged individuals and the amount allocated was only 1.6 per cent of the total water allocated in this period (Van Koppen and Schreiner

2014a, p. 553). As pointed out above, decisions regarding the equity factor in s 27 as an overriding factor in licensing applications have been set aside on review by the courts in the *Guguletto* case (supra) and *Goede Wellington Boerdery (Pty) Ltd v Makhanya NO* (2011). I have argued elsewhere that the courts were correct in holding that the equity consideration is not overriding in the sense that it acts as a trump (and, consequently, if absent the application must inevitably fail), but equity does have a heightened role to play in the overall scheme of the Act (Kidd 2012). Nevertheless, there are apparently moves to amend the Act to make it clear that equity plays a priority role (Van Koppen and Schreiner 2014a, p. 555). Whether this involves merely giving it priority or whether it will act as a trump remains to be seen.

Will such an amendment help to meet the equity backlog? It is submitted that the approach of the Department in this regard is misguided. The push for equity is ultimately based on the Constitutional right to equality, which involves not only non-discrimination but also active efforts at redress (affirmative action, in the widest sense of the term). Thus it is insufficient if the Department merely approaches licensing in a way that does not discriminate between appli-cants; it needs to do more by actively encouraging access by black people to water rights, particu-larly for productive purposes. This objective is not necessarily furthered by refusing an application for a water licence from an applicant who is not black (which, essentially, was the effect of the decisions mentioned above). What would the consequences of such a policy be in a catchment where there are no black users?

Other problems with attempting to achieve a rigid target of black and women rights holders by a particular date is that it does not take account of changes in land ownership or control. While water rights are not legally 'attached' to land (as was the case in the riparian system), it will only be once blacks or women have control of land on which productive use of water can be carried out (probably primarily in the agricultural context), that they would be applying for water rights. It would be an exercise in futility allocating water to persons who were unable to utilize it because they did not have access to land on which to do so. This highlights the impor-tance of land reform and land restitution, which is happening at a far slower rate than initially intended. But even if the speed of land reform were such that it matched the equity targets for water, there are other flaws.

First, the targets suggest that people will be effectively forced into agriculture, despite the fact that the overall population impetus is away from rural areas and farming (SALGA 2014).[7] Second, many applicants for water use licences are juristic persons, which are without race or gender. How do these affect the equity targets? It may be argued that the targets for increasing black rights could be met by using the Broad-Based Black Economic Empowerment criteria to ascribe race (in effect) to juristic persons, but there is no equivalent legal sleight-of-hand that can be used to meet the gender objectives. Finally, rights can be transferred in terms of the Act to suc-cessors-in-title (s 51), subject only to informing the relevant authority. So the state may exercise control over the racial and gender profile of water rights holders in the licensing process, but it has no control at all over the effective transfer of the licence when property changes hands.

Furthering equity is essential, but the Department of Water Affairs is doing itself no favours by having such rigid equity targets. They cannot be reached without a radical improvement in land reform initiatives, over which the Department has no control. A far more nuanced approach would be preferable and more capable of achievement.

Alignment of policies

As argued in the preceding section, there is a critical disjuncture between efforts at transforming the water sector in relation to equity, and the pace of the land reform process. This will have the effect of making impossible the achievement of the equity targets under the National Water Act. This is but one example of the absence of policy alignment in South Africa. The Act requires the licensing authority to consider, inter alia, the efficient and beneficial use of water in the public interest and the strategic importance of the water use to be authorized (s 27). This is a difficult task when government policies are contradictory. For example, the National Development Plan calls for the existing 1.5 million hectares of land under irrigation to be expanded by another 500,000 hectares, 'through the better use of existing water resources and developing new water schemes' (National Planning Commission 2011, p. 197), and yet the existing water resources in the country are all but fully committed. A second example is the lack of articulation in certain geographical areas between industrial development and such development's impact on water resources (Funke *et al.* 2007, p. 1244).

Corruption

South Africa suffers considerably from corruption, which is not a unique phenomenon in developing (and some other) countries. Administrative decision-making is particularly susceptible to corruption, which is why licensing has been avoided in some jurisdictions (Van Koppen *et al.* 2014, p. 55) and one of the reasons why Bronstein (2002, pp. 477–8) argued against water use licensing in favour of a market approach in South Africa. Corruption can, however, be kept in check by insistence on procedural fairness (particularly in respect of interested and affected parties), which is a requirement of South African administrative law.[8]

Economic inefficiency

The ubiquity of neoliberal economic drivers in recent decades (Dellapenna 2008) has resulted in numerous calls for the market to be used to regulate water management, and in particular in South Africa (for example, Tewari and Oumar 2013; Niewoudt 2000; Bronstein 2002). It is argued that an administrative permitting system is insufficiently flexible to allow for market forces to operate in such a way to generate water pricing that ensures the optimal valued use of water (Niewoudt 2000, p. 33). Others, however, argue that the market is inappropriate for managing water resources, and certainly is not conducive to achievement of the social and environmental objectives of IWRM (Dellapenna 2008; Bauer 2004). Irrespective of the theoretical benefits of the market, a prerequisite for its use is an initially fair and equitable allocation of water use rights (Tewari and Oumar 2013, p. 13), which was (and still is) manifestly not the case in South Africa. (There is limited scope within South African law for trade of water rights, but it falls far short of what could be regarded as a market-based system. The predominant paradigm is administrative allocation, since trade – if carried out in terms of the law – is still subject to administrative authorization (s 25 of the Act).)

Long-term investment

Water use licences in South Africa must be reviewed at least every five years under the Act (s 28(1)(f)) and they have a maximum licence period of 40 years (s 28(1)(e)). Moreover, any person exercising water use rights under licence or existing lawful use may become subject to a compulsory licensing process during which rights to water are reduced. Because of the consequent long-term lack of security of rights to water, this may be a disincentive to long-term investment in agricultural property and water infrastructure (Tewari and Oumar 2013, pp. 7–8).

Compliance, monitoring and enforcement

It need hardly be said that the effectiveness of any law depends ultimately on effective compliance and enforcement. This observation is valid for any legal duty, whether based on statutory or licence requirements. In licensing, however, the decision-maker has more opportunity to assess the specific compliance and monitoring needs of the licence under scrutiny, as opposed to a statutory duty which is 'one size fits all'. It has been observed in the South African context that the licensing of a water use is often, in effect, seen as an end to the regulatory process in itself because the licence holder's compliance with the conditions of the licence are not monitored. Particularly in the mining sector, where mines were permitted under the 1956 Act to engage in certain polluting activities that are now proscribed in terms of licence conditions issued under the 1998 Act, mines simply continue to behave as they always have in the absence of compliance monitoring (Van der Walt 2015). This is evidence that supports a widely held perception (probably based on good cause) that the promotion of mining is regarded by government as significantly more important than water conservation (see, for example, Kings 2015). While ineffective compliance monitoring is not just a problem relating to a licensing process, issuing licences with elaborate conditions in the absence of effective compliance monitoring means that the conditions are not worth the paper on which they are written.

Problems associated with the 'water allocation discourse'

When the state acts as trustee of water resources, as in South Africa, and is responsible for licensing water use, it then plays the role of deciding how to license water use: 'to formulate principles and guidelines and make judgments on who should have water and on what basis', which are ultimately policy decisions (Morvik 2014, p. 189). As Morvik (ibid.) argues:

> Policymakers, as the agents of discourse production, will always highlight some features and downplay others; they will frame issues in particular ways, creating categories and labels that open up some avenues for action, whilst closing down others. The notion of 'allocation discourses' refers to the ways in which categories of users are portrayed and particular arguments marshalled to make the case for conditions and allocation mechanisms that favour this set of users rather than that.

In the South African context, the prevalent 'allocation discourse' has served, Morvik argues, to privilege existing lawful users (particularly mines), which are regarded as economically beneficial water users, while downplaying the (often huge) negative impacts they have on water resources and consequently on other water users. Historically disadvantaged individuals (primarily black people, who are to be the beneficiaries of improved equity in access to water rights) are, despite increased commitment to policy initiatives aimed at equity, nevertheless still disadvantaged by a perceived emphasis on their being able to use the reallocated resources efficiently and effectively.

This is, perhaps, the social-science perspective of the fact that administrative lawyers would call the water use licensing process in South Africa a polycentric process – one that is 'many-centred', and a decision made in respect of which 'has multifarious consequences, meaning that polycentric issues cannot be decided in isolation from other issues' (Hoexter 2012, p. 86). This does have the potential to make the licensing process somewhat opaque, which conflicts with the general administrative law requirement that the applicant ought to be made aware of the policy requirements that apply to her application in advance (in South Africa, authority for this is

Tseleng v Chairman, Unemployment Insurance Board (1995)). The problem with this, as Morvik is suggesting, is that the policy considerations are not necessarily all conscious or explicit. The result, as Bronstein (2002, p. 483) observes, is that polycentric water licensing decisions 'cannot be unravelled by judicial review' and, consequently, that the outcomes of the 'highly politicized' decisions are 'inherently uncertain and thus predictability, consistency and other "rule of law" values are eroded'.

Conclusion: balancing positive and negative

The analysis above suggests that the licensing process is beset by so many problems that it may be worthwhile investigating an alternative approach. Bronstein's observation (2002, p. 483) is that it is not correct to view discretionary licensing as 'never an acceptable form of regulation but [its costs] can only be offset by clearly defined benefits'. In concluding this chapter, it is necessary to consider what the benefits of licensing are and whether the costs identified above can be reduced.

The first point to bear in mind is that, in a water-scarce country, management of water is important in order to ensure that there is sufficient water for use in the public interest. This must be balanced with the peculiarly South African imperative of removal of discriminatory patterns of water use rights holdings. Is there a realistic alternative to licensing in order to pursue IWRM in South Africa?

The only possible alternatives are reliance on the market or retention of the riparian rights approach. Neither would be acceptable options for regulation of the South African water sector. Whatever the other shortcomings of the market model may be (see Dellapenna 2008; Bauer 2004), the critical impediment of the market in the South African context is that there was a highly inequitable allocation of the resource to start with and, 17 years after the National Water Act came into operation, the inequity has not yet been significantly addressed. Retention of riparian rights would suffer from the same problem. The objectives of IWRM and, in particular, achievement of equity in South Africa can only realistically be achieved through state control, which entails administrative decision-making by the state in relation to allocation of water.

Several of the problems revolving around administrative inefficiency could conceivably be alleviated considerably by devolution of decision-making power to the CMAs (Chapter 7 of the Act). There have been enormous delays in the establishment and commencement of full functioning of the CMAs, the reasons for which are beyond the scope of this chapter, but until now the Minister has not devolved power to even the CMAs that have been established. Devolution of such power, when it happens, will undoubtedly contribute to better administrative efficiency.

It has also been suggested that greater use can be made of general authorizations, which operate as a 'blanket licence' for specified activities. More widespread use of general authorizations may well alleviate administrative burden, while also allowing for legally better-recognized use rights in some circumstances (Van Koppen and Schreiner 2014b). Whichever options are followed in order to improve the efficiency of the system, conditions of licences or general authorizations still need effective enforcement, which adds another challenge to the mix.

The historical legacy that South Africa had to confront in relation to access to water meant that neither the market nor common law approach of riparian rights were appropriate, arguably leaving administrative licensing as the only realistic approach, despite its several flaws. There are means (current and potential) of addressing some of these flaws at least. The achievement of IWRM through balancing all water needs – basic human needs, productive use requirements and environmental sustainability – can only be achieved through effective data collection, appropriate interpretation of the data, efficient and equitable water allocation and effective enforcement. This clearly needs adequate human resources (which has a financial cost) and this is one of the main problems faced in South Africa. This is probably the most important lesson that can be learned for other jurisdictions that use, or are contemplating using, administrative licensing approaches to water management – it is resource intensive.

The bottom line is that the South African government, as trustee of the nation's water resources, must show better political will to ensure that water receives the management it warrants in a water-scarce country. Too often, water is seen as a secondary priority (behind mining, for instance) and this is often a significant cause underlying some of the shortcomings in the licensing and management process.

Notes

1 Water Act 54 of 1956.
2 National Water Act 36 of 1998.
3 Several of the catchment management agencies have not yet been established as of the end of 2015.
4 'Water resource' is defined in s 1 of the Act as including 'a watercourse, surface water, estuary, or aquifer', and 'watercourse' defined as '(a) a river or spring; (b) a natural channel in which water flows regularly or intermittently; (c) a wetland, lake or dam into which, or from which, water flows; and (d) any collection of water which the Minister may, by notice in the *Gazette*, declare to be a watercourse, and a reference to a watercourse includes, where relevant, its bed and banks'.
5 Regulations for the Establishment of a Water Resource Classification System GN R 810, *Government Gazette* 33541 (17 September 2010).
6 GNs 818-823 in *Government Gazette* 37999 of (19 September 2014). As of April 2016, these have not yet been made final.
7 Productive use is not confined to agriculture – it could also be mining and industry, but it is unlikely that the targets specified by the Department will be met without a substantial increase in black, and black women, farmers.
8 See s 3 of the Promotion of Administrative Justice Act 3 of 2000 and s 33 of the Constitution of the Republic of South Africa, 1998.

References

Bauer, C.J. (2004) *Siren Song: Chilean Water Law as a Model for International Reform*. Washington, DC: Resources for the Future.
Bronstein, V. (2002) Drowning in the hole of the doughnut: regulatory overbreadth, discretionary licensing and the rule of law. *South African Law Journal* 119, pp. 469–83.
De Wet, J.C. (1959) One hundred years of water law. *Acta Juridica, 1959*, pp. 31–5.
Dellapenna, J.W. (2008) Climate disruption, the Washington Consensus, and water law reform. *Temple Law Review* 81, pp. 383–432.
DWAF (Department of Water Affairs and Forestry) (1997) *White Paper on a National Water Policy for South Africa*. Pretoria: South African Government.
DWAF (2003) *National Water Resources Strategy*. Pretoria: South African Government.
DWAF (2008) *Water Allocation Reform Strategy*. Pretoria: South African Government.
DWA (Department of Water Affairs) (2013) *National Water Resources Strategy*, 2nd edn. Pretoria: South African Government.
Funke, N., Oelofse, S.H.H., Hattingh, J., Ashton, P.J. and Turton, A.R. (2007) IWRM in developing countries: lessons from the Mhlatuze catchment in South Africa. *Physics and Chemistry of the Earth* 32, pp. 1237–45.
Global Water Partnership (2000) *Integrated Water Resources Management*. Stockholm: Global Water Partnership.
Goede Wellington Boerdery (Pty) Ltd v Makhanya NO (2011) Unreported case 56628/2010 (GNP). Available from: http://www.saflii.org/za/cases/ZAGPPHC/2011/141.pdf (accessed 25 April 2016).
Guguletto Family Trust v Chief Director, Water Use, Department of Water Affairs and Forestry (2011) Unreported case A556/10 (GNP) Available from: http://cer.org.za/virtual-library/judgments/high-courts/guguletto-family-trust-v-chief-director-water-use-department-of-water-affairs-forestry-and-another (accessed 8 November 2016).
Hodgson, S. (2004) *Land and Water – The Rights Interface*. FAO Legislative Study 84. Rome: Food and Agricultural Organization of the United Nations.

Hoexter, C. (2012) *Administrative Law in South Africa*, 2nd edn. Cape Town: Juta.

Hough v van der Merwe (1874) Buch 148.

Kidd, M. (2009) South Africa: the development of water law. In J. Dellapenna and J. Gupta (eds), *The Evolution of the Law and Politics of Water*. New York: Springer.

Kidd, M. (2011) *Environmental Law*, 2nd edn. Cape Town: Juta.

Kidd, M. (2012) Fairness floating down the stream? The Water Tribunal and administrative justice. *South African Journal of Environmental Law and Policy* 19, pp. 25–50.

Kings, S. (2015) Water be damned, the mines are what count. *Mail & Guardian*. 20 November. Available from: http://mg.co.za/article/2015-11-19-water-be-damned-the-mines-are-what-count (accessed 30 November 2015).

Milton, J.R.L. (1995) The history of water law 1652–1912, in Land and Agriculture Policy Centre, Submission to the Department of Water Affairs and Forestry (unpublished).

Morvik, S. (2014) A fair share? Perceptions of justice in South Africa's water allocation reform policy. *Geoforum* 54, pp. 187–95.

National Planning Commission (2011) *National Development Plan: Vision for 2030*. Pretoria: South African Government.

Niewoudt, W.L. (2000) Water market institutions in Colorado with possible lessons for South Africa. *Water SA* 26(1), pp. 27–34.

PMG (Parliamentary Monitoring Group) (2015) Questions to the Minister of Water and Sanitation (14 July). Available from: https://pmg.org.za/committee-question/183/ (accessed 30 November 2015).

SALGA (South African Local Government Association) (2014) *Draft Integrated Urban Development Framework*. Pretoria: South African Government.

Tewari, D.D. and Oumar, S. (2013) Is the water permit system a panacea or a bed of inefficiency? The case of South Africa. *Water Policy* 15(4), p. 570.

Tseleng v Chairman, Unemployment Insurance Board 1995 (3) SA 162 (T).

Van Heerden v Wiese (1880) Buch AC 5.

Van der Walt, L. (2015) Environmental auditor, personal communication.

Van Koppen, B. and Schreiner, B. (2014a) Moving beyond integrated water resource management: developmental water management in South Africa. *International Journal of Water Resources Development* 30(3), pp. 543–58.

Van Koppen, B. and Schreiner, B. (2014b) Priority general authorisations in rights-based water use authorisation in South Africa. *Water Policy* 16(S2), pp. 59–77.

Van Koppen, B., van der Zaag, P., Manzungu E. and Tapela B. (2014) Roman water law in rural Africa: the unfinished business of colonial dispossession. *Water International* 39(1), pp. 49–62.

WLRP (Water Law Review Panel) 1996. *Report to the Minister of Water Affairs and Forestry of the Water Law Review Panel: Fundamental Principles and Objectives for a New Water Law in South Africa*. Pretoria: South African Government.

4

THE EU APPROACH FOR INTEGRATED WATER RESOURCE MANAGEMENT

Transposing the EU Water Framework Directive within a national context – key insights from experience

Marleen van Rijswick and Andrea Keessen

Introduction

This chapter discusses the approach of the EU to IWRM by means of the Water Framework Directive (WFD) (European Parliament and Council 2000). The WFD came into force in the year 2000. Its key distinguishing feature is the focus on the protection of ecosystems instead of a sole focus on pollution control. It can be seen as an example of a successful legal implementation of IWRM for four reasons. First, it takes a river basin approach to enable water management across administrative borders. Second, through the combination of chemical and ecological goals it integrates the environmental regulation of industrial, household and agricultural water pollution. Third, the WFD emphasizes sustainable and equitable water use by requiring balanced groundwater use, minimum environmental flow of surface water, prescribing economic analysis of water uses and cost recovery for water services, including environmental and resource costs. Fourth, it leaves room for economic and social concerns by containing exemptions for not achieving 'good ecological status' (see further below) by 2015 and by containing ambitious public participation provisions to encourage and enable citizens to be involved in water management. It should be noted, however, that one depends on the Member States for the implementation of the WFD and access to justice.

The WFD introduced a procedural approach to establishing regulation and introduced new concepts and legal instruments, which caused uncertainties as to interpretation and enforceability, resulting in large differences between Member States (Keessen *et al.* 2011). In the last 15 years, many discussions have taken place on the correct interpretation of provisions, concepts and obligations that follow from the Directive. Recent case law sheds more light on the effectiveness of the approach. The WFD is a huge step forward in IWRM in Europe, but should be further developed to combat the important challenges of our time: climate change resulting in water scarcity and floods; the relationship with granting access to new products and substances on the market; and the development of a distribution mechanism to regulate the amount of water use and pollution across transnational river basins. As agricultural pollution and water use is one of the most important threats, this needs to be addressed – not only in the field of water law and policy, but also in the field of agricultural policy. Future developments should focus more on the human right to water and sanitation, improving societal resilience and combining mitigation and adaptation strategies in the field of climate change.[1] We will, therefore, take a closer look at the EU approach to IWRM, as laid down in the WFD, analyze its main advantages and shortcomings and search for inspiration for other countries that need to improve their IWRM.

We will discuss specific topics in EU water law, starting with the river basin approach, followed by the environmental goals. Then the position of water law within the broader field of environmental law will be described, as many threats for achievement of a good water status also require action in other policy fields. In the following sections, special attention will be given to the adaptive approach of the WFD and the new way of regulation within EU law. Special attention will be paid to public participation and the increasing role of economic instruments as characteristics of the new regulatory approach. We will discuss the new design of water legislation within the EU, giving special attention to the tension between flexibility and effectiveness. While the WFD takes water management in the EU one step further by its holistic and adaptive approach, we see that Member States are also struggling with the new EU approach for IWRM. The environmental goals are taken as an illustration to show this struggle and explain it. This chapter concludes with lessons learned, and questions if these lessons might be valuable for other river basins districts or countries.

A river basin approach in EU water law

The WFD is a major step forward in IWRM. Its innovative and adaptive approach, which pays specific attention to the protection of aquatic ecosystems, and the great emphasis on involvement of the public in all stages of the policy process, were a source of inspiration for many countries across the world (Dai 2015). Using the river basin scale for water management follows the UN/ECE Water Convention (Helsinki Convention 1992) and is logical from a water management perspective. Member States have to identify river basins and assign them to individual river basin districts (art. 3 WFD). Each Member State has to ensure appropriate administrative arrangements, which include the identification of appropriate competent authorities, both on the national and the international level, since many river basins in Europe are transboundary.

However, the discretion in the set-up of the authorities is such that national water authorities differ considerably in their tasks, legal status and competences. These institutional differences between Member States sharing the same river basin hinder the transboundary management of river basins (Van Rijswick *et al.* 2010).

Consequently, despite the river basin approach, river basins remain managed within national borders (European Commission 2007a). Currently, the most significant and widespread pressures are diffuse pollution (in particular by agricultural sources), physical degradation of water ecosystems and, particularly in Southern Europe, the overexploitation of water. Many low-status water bodies are located in densely populated areas and regions of intensive, often unsustainable, water use. Unfortunately, the Commission and EEA reports reveal little progress (European Commission 2015; EEA 2015).

Environmental goals and standard setting

The WFD adopts a multilevel approach to goal and standard setting. Most of the implementing regulation is established by the lower levels in the hierarchy – that is, at the national, regional and (sub-) river basin levels. At all levels, the river basin approach has to be embedded in the existing regulatory framework, with a patchwork of water-related Directives and relevant Directives in other fields of environmental law. Protecting ecosystems asks for differentiated quality standards instead of the uniform standards prevalent in most EU environmental Directives. The WFD provides for flexibility to enable the development of a substantive water policy at the appropriate level, because geophysical circumstances and the effects of climate change differ per region (Van Holten and Van Rijswick 2014; Raadgever *et al.* 2011; cf. Keessen *et al.* 2011). This is reflected in the open and flexible goals of the EU regulatory water framework. The WFD environmental goal for surface waters is to achieve 'good chemical and ecological status' in 2015, while ground-waters should then be in a good chemical and quantitative status. The chemical goals are set at the EU level for the most common dangerous substances, and at the national level for less dangerous substances or for those substances that are only problematic in some Member States. The ecological goals for surface waters, which include minimum flow levels, are established at the (sub-) river basin level and then compared using an intercalibration method to calibrate standards among similar water types. This makes the status of the water bodies classifiable and thus comparable at the EU level. The WFD prescribes good quantitative status for groundwater as the achievement of a balance between abstraction and recharge.

The benchmark against which the achievement of the 'good ecological status' goal is to be tested is the best status achievable, 'high' status. This is defined as the biological, chemical and morphological conditions associated with no or very low human pressure (Howarth 2006). These reference conditions have to be set at the (sub-) river basin level based on expert advice from ecologists (Dieperink *et al.* 2012; Raadgever *et al.* 2011). 'Good status' is achieved if only a slight deviation from high status is present. Ecologists have, however, criticized the benchmark as being unrealistic (e.g. Paganelli *et al.* 2011).

Many Member States had severe problems in dealing with the elaboration of the ecological standards of the WFD (Keessen *et al.* 2010). The social context plays an important role in determining how these high ecological objectives can be compatible with human impacts (Moss 2008; Bijker *et al.* 2009; Dieperink *et al.* 2012). While the WFD integrates the measures to be taken under other water Directives (e.g. the Nitrates Directive, European Council 1991), it allows the Member States to invoke justifications for not attaining the good status.

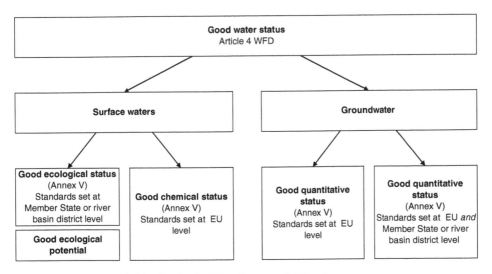

Figure 4.1   Environmental objectives in the Water Framework Directive

These justifications have to be included in the river basin management plans and are subject to disclosure and public participation obligations. Moreover, the WFD also gives the European Commission a role. It may only accept a justification in situations that fit the conditions of the four exceptions mentioned in the WFD. These exceptions are 1) new sustainable developments, 2) *force majeure*, 3) a postponement of the deadline for meeting the objectives and 4) a lowering of the objectives. A proportionality test and an assessment of costs and benefits are among the conditions that apply for their use. The main conditions are that all affordable and practically feasible measures must have been taken before a justification can be invoked (Van Kempen 2012, 2014a). However, in the period 2009–15, many Member States only applied existing and feasible measures, instead of designing cost-effective and appropriate measures to achieve the WFD objectives (European Commission 2015).

Although the WFD aims to be the framework for all water management in the EU, its main focus is on water quality issues. It is complemented by the Floods Directive (FD) (European Parliament and Council 2007), which addresses the risk of floods. The FD also sets an 'open goal' – reducing the adverse effects of flooding – and allows the Member States to set 'appropriate objectives' for the management of flood risks in their river basin districts to achieve this goal (art. 7 FD). The WFD is also complemented by a Water Scarcity and Drought Strategy (WSDS) (European Commission 2007b), which is a non-binding policy instrument. Its goal is to mitigate water scarcity and droughts, and it offers a set of policy options to be implemented under the WFD.

Water in the broader context of environmental protection

The achievement of good status of all European waters implies actions outside the realm of water law. The ecological WFD goals can create synergy where water bodies overlap with natural areas

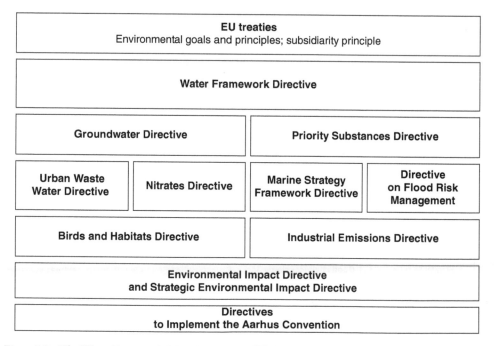

Figure 4.2 The Water Framework Directive as part of EU environmental law

protected by the EU Habitat and Bird Directive regimes. For instance, in the Netherlands, the implementation of the WFD led to the creation of many eco-friendly riverbanks. Its integrative impetus on land use, and, in particular, on urban and agricultural activities, is less obvious. It is crucial to overcome the land–water boundary, as over-abstraction, pollution and floods are often caused by land use decisions that do not take the effects of these activities on water into account.

The main accomplishment of the WFD is the integration within the field of water. Prior to the WFD, European water legislation mainly contained environmental quality standards and emission limit values which each Member State could implement separately in the various policy domains. The river basin approach prompted the creation of new institutional arrangements with competent authorities that operate at a (sub-) river basin scale instead of the traditional Member States, regions, provinces and municipalities. Moreover, in international river basins, the Member States have to cooperate to achieve shared goal setting, planning and risk assessments. However, the Member States are only obliged to discuss their plans and measures in international river basin committee meetings to try to achieve a coordinated overarching management plan and programme of measures. The available instruments to realize this cooperation are traditional international treaties between riparian states (Hey and Van Rijswick 2011). In practice, we see a strong use of informal cooperation for smaller watercourses.

Adaptive water management

The approach in the WFD is based on an adaptive six-year policy cycle (see Figure 4.3 for details). In sum: an analysis of human and natural impacts; followed by setting goals and standards at EU and national levels; proceeding with the plans and programmes to meet the goals; and accompanied by an obligation to develop a sophisticated monitoring network; leading to revised plans and programmes for the next planning period. This approach should be able to improve Europe's river basins and aquatic ecosystems in the period between the years 2000 and 2027.

Enabling adaptation presumes, first, assessments and monitoring to obtain reliable and up to date information for decision-making in a changing environment (Baaner 2011b). The WFD obliges the Member States to make, for each river basin district, an analysis of its characteristics,

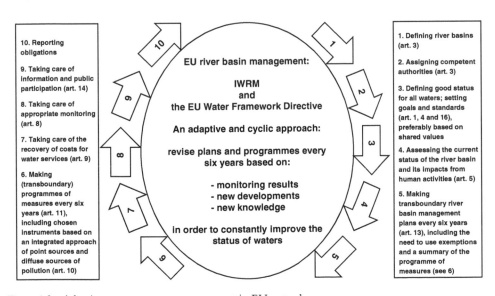

Figure 4.3 Adaptive water resource management in EU water law

the impacts of human activities on the status of surface waters and groundwater, and an economic analysis of water use (art. 5 WFD; Brouwer *et al.* 2005). Having this information available makes it not only possible to set adequate goals and standards but also to take adequate, proportionate and fair measures to improve and protect the status of water bodies (Howarth 2009b). The monitoring programmes established under the WFD should address water scarcity and drought risks as well.

Adaptability also refers to changing the course of action on the basis of the information acquired through assessments and monitoring. The WFD provides for a mandatory cyclical planning process, based on the assessments and monitoring results, to protect and improve the ecological, chemical and quantitative status of river basins. The WSDS recommends that water scarcity be taken into account in the context of WFD assessments and monitoring (Beijen *et al.* 2014). The plans have to be made and reviewed every six years. Public participation in this process should ensure that local input is taken into account. The plan has to contain a programme of measures to reach good status. From 2015, the timescale for the creation of flood risk management plans and programmes of measures will be synchronized with the WFD timescales for plans and programmes of measures to further realize integrated water management (art. 9 WFD). This will make it easier to achieve synergies – for instance, by taking measures to boost ecosystem storage capacity for water.

The first generation of plans and programmes dated from 2009 and expired in 2015. In the meantime, new plans and programmes have been drafted to cover the next planning period, which lasts from 2015 until 2021. This six-year planning cycle facilitates learning and enables adaptive water management. Unfortunately, it also enables the Member States to postpone the realization of the WFD goals and objectives at least until the third planning cycle has ended in 2027 (Howarth 2009a; Keessen *et al.* 2008). It is not fully clear to what extent plans and programmes of measures may or must be changed during the planning period. It appears – as mentioned above – from the text of the WFD that changes must be made if monitoring reveals that the goals as set out in the plan will not be met without additional measures. Yet the Member States may, in these circumstances, also be entitled to invoke a WFD exemption. In case of an extreme event, the WFD allows for temporary derogations. The exemption of 'new sustainable development' can be used to justify adaptation to climate change measures such as new flood defence infrastructure or the diversion of a river (*Nomarchiaki* 2012; *Bund für Umwelt und Naturschutz Deutschland eV versus Bundesrepublik Deutschland* 2015).

Unfortunately, it is therefore possible that the advantages of the flexible and adaptable European river basin approach may, at the same time, seriously hamper the effectiveness of the legal framework (Green *et al.* 2013). The vast differences between Member States constitutes another problem – in particular, for transboundary river basins. Taking regional, climatological and societal differences into account in decision-making appears to enable less-legitimate differences to creep in – for instance, concerning the level of ambition to protect waters and the choices to prioritize and facilitate the water needs of the various users. Interviews with civil servants involved in the implementation of the WFD and questionnaires completed by legal experts from various Member States revealed that Member States adopt different approaches (Keessen *et al.* 2010). There is a tension between the adaptability and the effectiveness of water management with regard to the achievement of environmental goals.

Public participation and access to justice

In line with the UN/ECE Aarhus Convention (1998), EU water law contains mandatory disclosure provisions and encourages public participation (European Community 2002). Article 14

WFD obliges the Member States to inform and consult the public when defining goals, making plans and adopting measures. This requires transparency and a clear explanation of the proposed measures (European Commission 2007a). These requirements also apply to drought risk management plans and measures if a Member State has integrated drought risk management in its WFD plans and programmes of measures. The WSDS does not require public participation in the price setting of drinking water and sanitation services. Yet this aspect deserves scrutiny, particularly in countries that suffer from water scarcity and droughts (Quesada 2011; OECD 2011). Disclosure and participation requirements similar to those of the WFD are provided by the FD. Within the ambit of flood risk management, the FD promotes openness by the mandatory use of flood hazard maps, showing areas prone to flooding, and flood risk maps, showing the potential adverse consequences of floods.

Common goals and necessary measures may be hard to define and to realize when so many, often opposing, interests are at stake (De Smedt and Van Rijswick 2015). If disputes arise, the *Janecek* case reveals that when Directives take a programmatic approach like the WFD and the FD, citizens can only enforce their right that the Member States establish plans and programmes of measures (*Janecek v Freistaat Bayern* 2008; Backes and Van Rijswick 2013). They cannot enforce compliance with the specific measures in a plan or programme. Citizens depend on national law for access to the courts as long as the Directive on Access to Justice in Environmental Matters is pending. However, the European Court of Justice (ECJ) has held that the Member States have to interpret their laws in the light of the Aarhus Convention (*Lesoochranárske zoskupeni* 2011). Apparently, and in accordance with international obligations, the ECJ feels entitled to control the faithful implementation of the Aarhus Convention in the Member States because of the Europeanization of environmental law (Jans 2011). After all, a lack of access to justice undermines the position of citizens in water management and diminishes their role in the implementation of water management measures.

Economic instruments

The WFD, the FD and the WSDS promote the balancing of the different interests through financial instruments (Lindhout 2012). The WFD requires economic analysis of all water uses and cost recovery, based on the principles that the polluter pays and the user pays (Howarth 2009a). The WSDS recommends the introduction of the user pays principle for all essential uses to encourage efficient water use (European Commission 2007b, p. 3). The FD does not refer to cost recovery or measures, but it prescribes an economic perspective in the mandatory assessment of the adverse consequences of future flooding. The implementation of cost recovery measures differs wildly among the Member States (Aragao 2013; Reese 2013; Lindhout 2013). Often they are not used beyond paying for drinking water supply and wastewater treatment (European Commission 2007b, p. 3). It appears from the ECJ case *European Commission v Federal Republic of Germany* (11 September 2014) that Member States have a large amount of discretion in this respect, as the court considers that cost recovery is only one of the instruments to achieve the goals of the WFD and that both use and scope are discretionary (Gawel 2014) – although, at the same time, it was held that all water uses are covered by 'water services'.

Environmental goals and emerging case law

The most important provision of the WFD – Article 4 with its environmental objectives – remained unclear for more than 15 years even though it is crucial for a proper understanding of the WFD system. In the *Weser* case (*Bund für Umwelt und Naturschutz Deutschland eV versus*

Bundesrepublik Deutschland, 1 July 2015), the ECJ finally shed light on the legal meaning of not only the environmental obligations, but also the system of the WFD as a whole. The case focused on two important questions regarding the requirements that follow from the environmental objectives of the WFD. The first theme concerns the relationship between water quality norms and plans, programmes, projects and authorizations. Are these only goals for and relating to management planning or are they, more or less, binding for concrete authorizations of projects and permitting decisions? This question is important because, if the environmental quality standards only relate to policy and plans, the flexibility and policy discretion would by far overrule the effectiveness of the Directive.

The second theme concerns the 'stand still principle', which is referred to as 'no deterioration' in the WFD. What does 'no deterioration' mean, exactly, and which exceptions are possible? In the following section, we discuss the environmental goals and their place within the system of the WFD as a whole more in depth.

Within the EU, the views on the legal scope and meaning of environmental quality objectives are very diverse. According to German and Dutch authorities, the environmental objectives of Article 4 (1) only relate to river basin management plans and the Member States' programmes of measures for water. Therefore, they are not applicable as testing criteria for the approval of individual projects. Other countries, like the UK and Poland, take a stricter view and think that the standards of Article 4 (1) WFD also have to be applied to decisions on concrete, individual projects (Conclusion AG 23 October 2014, C-461/13, point 33). In earlier comparative research on the implementation of the WFD in the Netherlands, Luxembourg, Belgium, Germany, France, England and Wales, Denmark, Italy, Romania, Spain and Portugal, similar disagreements on the meaning of Article 4 (1) WFD were found (Keessen *et al.* 2010). Although not all implementation legislation came into force at the moment of the research (2009), it became clear that Members States had different opinions on: 1) whether the environmental objectives of Article 4 WFD should be regarded as obligatory for results or best practices and 2) whether 'no deterioration' refers to all deterioration, or only applies when a water body shifts to a lower status class. Furthermore, there are large differences between 3) the selection criteria for water bodies that have been designated as artificial or heavily modified (not good ecological status), with less stringent ecological potential required, and 4) the way external integration with other policy fields – agriculture, spatial planning, nature conservation – has been organized.

The ECJ is quite clear on the questions related to the no-deterioration principle and the relationship between environmental requirements and policy instruments in the programme of measures in the *Weser* case. The Court held that both the prohibition of deterioration of the quality of a water body and the obligation to ensure a good water status by the year 2015 are binding requirements and not simply goals for management plans. Hence, they also have to be applied on request for permits to allow a certain use of water. Furthermore, these standards must be applied strictly and do not allow flexibility or derogations other than those mentioned in the Directive itself.

Also, the prohibition on allowing deterioration not only applies to the management plans and to the programmes of measures determined on the basis of these management plans, but also to the measures necessary to implement these programmes – for example, permitting decisions and project authorizations. The Court decided that the main environmental obligations of Article 4 (1) are to be seen as binding criteria for the application of all other instruments prescribed in the Directive, a clarification of the Directive that surely will increase its effectiveness. Furthermore, the Court stressed that permitting new projects may result in a failure to comply with the main objectives and concluded that the prohibition of deterioration is very strict and absolute in time. The prohibition of deterioration is a binding testing criterion for all permits that may have an influence on the quality status of surface waters.

The second topic discussed by the Court concerned the content of the prohibition to allow deterioration of surface waters. A main issue discussed over the last decade in European water law and policy has been whether the WFD would prohibit *any* worsening of the quality of water bodies at all, or only prohibit changes that would lead to a lower-status class (Backes and Van Rijswick 2015; Faßbender 2013; Keessen *et al.* 2010). In Germany, scholars did not agree on an interpretation. Most authorities – also in the Netherlands – applied a less stringent approach – that is, the change of class as a whole. In France, this interpretation, whereby a water body only deteriorates if it passes to a lower water class, was even codified (Article 212–13 Code de l' Environment). However, in recent years, German courts started to follow the stricter approach. In its answer to this European disharmony, the Court concluded that deterioration means any worsening, even within the same class, and that the effectiveness of the Directive would be hampered seriously if one could only speak of deterioration when the whole water body dropped a status class – which would seriously jeopardize the achievement of the final objectives of the Directive and does not fit in the textual and systematic analysis of the Directive.

Lessons learned for the EU and beyond

In what way might the EU approach in the Water Framework Directive inspire other countries or river basin authorities? In other words, what are the advantages of the WFD, and what elements ask for special attention in case other countries want to learn from the EU approach?

One important feature of the WFD is the emphasis on ecosystem protection, with special but difficult-to-implement requirements for ecological objectives. Moreover, the holistic and adaptive approach and the systematic and cyclic planning and programming, backed up and informed by monitoring obligations and results, are a huge step forward. The requirements for public participation enable citizens to be involved in water management. In theory, the relationship with adjacent policy fields has been addressed, although, in practice, this system could be much improved because the main threats to the achievement of good status for waters and sustainable and equitable water management cannot be sufficiently addressed and regulated through a reform of water legislation alone. These threats need to be tackled under the regulatory frameworks of other policy fields (Van Rijswick 2016).

The European Commission's 2015 assessment report reveals that if the WFD acts only as an integration tool for the existing water legislation – with only existing measures applied to improve water quality – the ambitious goals of the WFD will not be achieved in good time (European Commission 2015). The Commission also suggests that exemptions are applied too widely and without appropriate justification. These remarks should be taken into account if other countries want to learn from the WFD. Too many exemptions might hamper the effectiveness of the regulatory framework, certainly in a transboundary river basin (Van Kempen 2012). Bringing the social and the ecological dimensions of water management in balance also seems advisable and requires action beyond the field of water management. The European Commission proposes further integration of water objectives into other policy areas, such as the Common Agricultural Policy (CAP) and the European Structural and Investment Funds, and the wise use of economic instruments and incentives (European Commission 2015). Since the WFD does not explicitly address water scarcity, and as over-abstraction increasingly threatens European river basins, more action on water quantity is needed. This might be the same in many other river basins across the world. The European Commission suggests that unaffected Member States should take preventive measures, while others should review and enforce permits (ibid.). Actually, the EU could go much further and promote the development of a distribution mechanism to be used in the various river basins (Van Kempen 2014b; Van Rijswick 2008). Another tool could be learning from best practices in the various Member States by investigating what constitute constraining and enabling factors for these best practices.

More research also needs to be done on the quest for more and more policy discretion with regard to goal achievement and problem solving. The main challenge will be how to take the inspiring idea of holistic, adaptive and sustainable water management one step further to an effective legal framework aimed at achievement of the ambitious but necessary goals of the WFD in order to solve water and climate change-related problems. Another question is to what extent the EU model has specific features unique to Europe and to what extent it might serve as a model elsewhere (see Dai 2015 for a comparison with China).

It is clear that adopting a new approach raises new questions and uncertainties. These problems can be overcome afterwards, but could better be addressed before implementing a new approach. One of the issues that hampered smooth implementation in the EU was the lack of clarity of many obligations following from the Directive. This lack of clear concepts and obligations should be avoided if possible, although this could make the legislative process more difficult. The ruling of the ECJ in the *Weser* case shows that the EU regulatory system should be considered as an holistic system and not as a bundle of individual requirements. It makes clear that the different provisions cannot be understood properly in isolation but should be interpreted within the whole system of the WFD. All provisions are interconnected and build up a modern system of water law that tries to combine a need for flexibility, policy discretion and subsidiarity, on the one hand, and an ongoing improvement and effective protection of Europe's waters, on the other hand. Copying only elements of this approach may therefore cause unexpected and unwanted consequences. By emphasizing that the environmental goals are strictly binding obligations, the Court chose the best way to combine the sometimes conflicting needs of flexibility, policy discretion and ambitious water policies necessary to fulfil the needs of current and future generations. Flexibility remains an important advantage of the WFD, and Member States can partly set their own goals, although after inter-calibration. They can focus on the main water problems in their country and choose those measures that are most effective and fit well in the national context. Transboundary cooperation and management is explicitly addressed and Member States are free to choose the way they want to shape their cooperation. If there are serious reasons for postposing or lowering goals, the Directive offers numerous possibilities, although under strict conditions.

However, effectiveness in achieving the goals is just as important. The environmental goals and obligations are more binding than many Member States expected or hoped. The fact that the WFD takes a cyclic approach will lead to changes in regulatory approaches in some of the Member States. It will lead, for example, to discussions on what kind of programmatic approach (Groothuijse and Uylenburg 2014) could be designed to meet the requirements of the Directive. In this regard, it is important to take a different approach for those water bodies that are already at good status and those water bodies that do not meet the standards yet. In this respect, especially, Article 11(5) WFD is an important provision. It makes clear that, as long as the standards are not met, a link between quality standards, authorizations for projects and concrete decisions should be stronger than in those cases where the status of a water body is already good. Even when all measures are combined within a programme of measures, individual authorizations should be revised if the requirements of Article 4 are not met. Therefore, each programmatic approach should take into account specific circumstances of the water body regarding scale, natural characteristics and human impacts. It will make a difference whether the quality status of the large transboundary River Rhine or Danube has to be improved or whether it concerns an isolated lake or small water body. Most Member States will prefer to develop legislation applicable to the whole country or to large river basin districts. Because the WFD puts a strong emphasis on the goal to be achieved and requires that the causes of water pollution or ecological degradation are tackled at the source of the problem, the Member States have less flexibility in their programme of measures. It is recommended that a formal link be established between environmental objectives and water quality standards, and the authorization of projects or concrete licenses. The same goes for a necessary link between water quality objectives and decisions taken in other policy fields like urban planning, nature conservation, environment, agriculture or infrastructure.

Another issue that should be addressed when developing a regulatory framework for IWRM is whether, and if so when, economic considerations can be taken into account. Should this be done in the phase of goal setting, standard setting, planning, making programmes of measures, or only when one wants to rely on an exemption? In the EU, it took a court judgment from the ECJ to make clear that only economic considerations should be borne in mind when a Member State relies on an exemption as provided for by the Directive.

Future developments of EU water management should focus more on the human right to water and sanitation, as this is lagging behind many other countries in the world (Van Rijswick 2012), improving societal resilience[2] (Walker *et al.* 2002; Van Rijswick and Keessen 2016) and combining mitigation and adaptation strategies in the field of climate change.

Notes

1 This chapter builds on Van Rijswick and Keessen (2016); Van Rijswick (2016); Backes and Van Rijswick (2015); Green *et al.* (2013); Van Rijswick and Havekes (2012) and Keessen and Van Rijswick (2012).
2 Resilience refers to the capacity of a social-ecological system to absorb disturbance and reorganize while undergoing change so as to still retain essentially the same function, structure, identity and feedbacks.

References

Aarhus Convention (1998) Convention on Access to Information, Public Participation in Decision-making and Access to Justice in Environmental Matters (Aarhus Convention), done at Aarhus, Denmark, 25 June, entered into force 2001, http://www.unece.org/fileadmin/DAM/env/pp/documents/cep43e.pdf

Aragao, A. (2013) Water pricing and cost recovery in water services in Portugal. *Journal for European Environmental and Planning Law* 10(4), pp. 333–54.

Baaner, L. (2011a) Programmes of measures under the Water Framework Directive: a comparative study. *Nordic Environmental Law Journal* 1, pp. 31–52.

Baaner, L. (2011b) The programme of measures of the Water Framework Directive: more than just a formal compliance tool. *Journal for European Environmental and Planning Law* 8(1), pp. 82–100.

Backes, Ch. and Van Rijswick, H.F.M.W (2013) Effective environmental protection: towards a better understanding of environmental quality standards in environmental legislation. In L. Gipperth and C. Zetterberg, *Miljörättsliga perspektiv och tankevändor, Vänbok till Jan Darpö & Gabriel Michanek*. Uppsala: Iustus Förlag AB, ISBN 9789176788424, pp. 19–50.

Backes, C. and Van Rijswick, H.F.M.W. (2015) Ground breaking landmark case on environmental quality standards?: the consequences of the CJEU 'Weser-judgment' (C-461/13) for water policy and law and quality standards in EU environmental law. *Journal for European Environmental and Planning Law* 12, pp. 363–77.

Beijen B.A., Van Rijswick, H.F.M.W. and Anker, H.T. (2014) The importance of monitoring for the effectiveness of environmental directives: a comparison of monitoring obligations in European environmental directives. *Utrecht Law Review* 10(2), May. https://www.utrechtlawreview.org/articles/abstract/10.18352/ulr.273/, URN:NBN:NL:UI:10-1-115821, pp. 126–35.

Bijker, W.E., Bal, R. and Hendriks, R. (2009) *The Paradox of Scientific Authority: The Role of Scientific Advice in Democracies*. Cambridge, MA: MIT Press.

Brouwer, R., Schenau, S. and van der Veeren, R.J.H.M (2005) Integrated river basin accounting in the Netherlands and the European Water Framework Directive. *Statistical Journal of the United Nations Economic Commission for Europe* (22)2, pp. 111–31.

Bund für Umwelt und Naturschutz Deutschland eV versus Bundesrepublik Deutschland (Weser) (2015) ECJ, C-461/13.

Craig, R.K. (2010) Adapting water law to public necessity: reframing climate change adaptation as emergency response and preparedness. *Vermont Journal of Environmental Law* 11, p. 709.

Dai, L. (2015) China's water resources law in transition. Diss. Utrecht University.

De Smedt, P. (2010) Water-related tools for climate change adaptation in the Flemish region: the art of linking water objectives to spatial planning. *Journal for European Environmental and Planning Law* 7(3), pp. 287–301.

De Smedt, P. and van Rijswick, H.F.M.W. (2015) Nature conservation and water management: one battle? In C.H. Born, A. Cliquet, H. Schoukens, D. Misonne and G.Van Hoorick (eds), *The Habitats Directive in its EU Environmental Law Context, European Nature's Best Hope?* Abingdon: Routledge-Earthscan.

Dieperink, C., Raadgever, G.T., Driessen, P.P.J., Smit, A.A.H. and van Rijswick, H.F.M.W. (2012) Ecological ambitions and complications in the regional implementation of the Water Framework Directive in the Netherlands. *Water Policy* 14(1), pp.160–73.

EEA (2015) The European environment – state and outlook 2015: synthesis report. European Environment Agency, Copenhagen.

European Commission (2007a) *Communication from the Commission to the European Parliament and the Council, Towards sustainable water management in the European Union, First stage in the implementation of the Water Framework Directive 2000/60/EC*, COM (2007) 128 final.

European Commission (2007b), Communication 'Addressing the challenge of water scarcity and droughts', COM (2007) 414.

European Commission (2009) *Report from the Commission to the European Parliament and the Council in accordance with article 18.3 of the Water Framework Directive 2000/60 on programmes for monitoring of water status*, COM (2009) 156 final.

European Commission (2015) *Communication from the Commission to the European Parliament and the Council, the Water Framework Directive and the Floods Directive: actions towards the 'good status' of EU water and to reduce flood risks*, COM(2015) 120 final. Available from: http://ec.europa.eu/environment/water/water-framework/impl_reports.htm [accessed 17 August 2016].

European Commission v Federal Republic of Germany (2014) ECJ, C-525/12 ecli:eu:C:2014:2202.

European Community (2002) *Guidance on Public Participation in Relation to the Water Framework Directive*. Luxembourg: Office for Official Publications of the European Communities.

European Council (1991) *Nitrates Directive* 1991/676/EEC.

European Parliament (2008) *Water Scarcity and Drought Strategy: Addressing the Challenge of Water Scarcity and Droughts in the European Union*. European Parliament resolution of 9 October (2008/2074(INI)).

European Parliament and Council (2000) *Water Framework Directive* 2000/60/EC.

European Parliament and Council (2007) *Floods Directive* 2007/60/EC.

Faßbender, K. (2013) *Zur aktuellen Diskussion um das Verschlechterungsverbot der Wasserrahmenrichtlinie*, EurUP 2013, p. 70.

Gawel, E. (2014) Cost recovery for 'water services'/critical review of the EU Court of Justice conclusions of Advocate General Jääskinen in case C-525/12. *Europäische Wasserrahmenrichtlinie*, 19 August, 2.

Ginzky, H. (2006) Exemptions from statutory water management objectives: requirements, spheres of responsibility, unresolved implementation issues. *Journal for European Environmental and Planning Law* 3(2), pp. 117–31.

Green, O., Garmestani, A., van Rijswick, H. and Keessen, A. (2013) EU water governance: striking the right balance between regulatory flexibility and enforcement? *Ecology and Society* 18(2), p. 10. http://dx.doi.org/10.5751/ES-05357-180210 [online]. http://www.ecologyandsociety.org/vol18/iss2/art10

Groothuijse, F. and Uylenburg, R. (2014) Everything according to plan? Achieving environmental quality standards by a programmatic approach. In Marjan Peeters and Rosa Uylemburg (eds), *EU Environmental Legislation, Legal Practice on Regulatory Strategies*. Cheltenham: Edward Elgar, pp. 116–45.

Guidance on Public Participation in Relation to the Water Framework Directive. Luxembourg: Office for Official Publications of the European Communities.

Hegger, D., Driessen, P., Dieprink, C., Wiering, W., Raadgever, T. and van Rijswick, M. (2014) Assessing stability and dynamics in flood risk governance: an empirically illustrated research approach. *Water Resources Management* 28(12), pp. 4127–42, DOI 10.1007/s11269-014-0732-x.

Helsinki Convention (1992) Convention on the Protection and Use of Transboundary Watercourses and International Lakes (Helsinki Convention), Helsinki, entered into force, 1996. http://www.unece.org/fileadmin/DAM/env/water/pdf/watercon.pdf

Herman, C. (2010) Will the Floods Directive keep our feet dry? Policies and regulations in the Flemish region and Scotland. *The Journal of Water Law* 21, pp. 156ff.

Hey, E. and van Rijswick, H.F.M.W. (2011) Transnational water management. In O. Jansen and B. Schöndorf-Haubold (eds), *The European Composite Administration.* Antwerp and Cambridge: Intersentia, pp. 227–49.

Hofmann, E. (ed.) *Wasserrecht in Europa.* Baden Baden: NOMOS.

Howarth, W. (2006) The progression towards ecological quality standards. *Journal of Environmental Law* 18(1), pp. 3–35.

Howarth, W. (2007) The European Community approach to flood defence. *Journal of Water Law* 18, pp. 115ff.

Howarth, W. (2009a) Cost recovery for water services and the polluter pays principle. ERA Forum.

Howarth, W. (2009b) Aspirations and realities under the Water Framework Directive: proceduralisation, participation and practicalities. *Journal of Environmental Law* 21(3), pp. 391–417.

Janecek v Freistaat Bayern (2008) ECJ, C-237/07 [2008] ECR I-2607.

Jans, J. (2011) Who is the referee? Access to justice in a globalized legal order. A case analysis of ECJ judgment C-240/09 Lesoochranárske zoskupenie [2011] ECR I-0000 of 8 March. *Review of European Administrative Law* 1, pp. 85–97.

Josefsson, H. (2015) Good ecological status, advancing the ecology of law. Diss. Uppsala Universitet.

Keessen, A.M. and van Rijswick, H.F.M.W. (2012) Adaptation to climate change in European Water Law and Policy, *Utrecht Law Review*, November, pp. 38–50, www.utrechtlawreview.org

Keessen, A.M., van Kempen, J.J.H. and van Rijswick, H.F.M.W. (2008) Transboundary river basin management in Europe: legal instruments to comply with European water management obligations in case of transboundary water pollution and floods. *Utrecht Law Review* 4(3), pp. 35–56.

Keessen, A., Van Kempen, J.J.H., Van Rijswick, H.F.M.W., Robbe, J. and Backes, C.W. (eds) (2010) European river basin districts: are they swimming in the same implementation pool?. *Journal of Environmental Law* 22(2), pp. 197–222.

Keessen, A., Runhaar, H., Schoumans, O. and Zwart, K. (2011) The need for flexibility and differentiation in the protection of vulnerable areas in EU environmental law: the implementation of the Nitrates Directive in the Netherlands. *Journal for European Environmental and Planning Law* 8(2), pp. 141–64.

Lesoochranárske zoskupenie VLK v Ministerstvo životného prostredia Slovenskej republiky (2011) ECJ, C-240/09 [2011] ECR I-0000.

Lindhout, P.E. (2012) A wider notion of the scope of water services in EU water law: boosting payment for water-related ecosystem services to ensure sustainable water management?. *Utrecht Law Review* 8(3).

Lindhout, P.E. (2013) Application of the cost recovery principle on water services in the Netherlands. *Journal for European Environmental and Planning Law* 10(4), pp. 309–32.

Moss, D. (2008) The Water Framework Directive: total environment or political compromise?, *Science of the Total Environment* 400, pp. 32–41.

Nomarchiaki Aftodioikisi Aitoloakarnanias v Ipourgos. Perivallontos, Khorotaxias kai Dimosion Ergon (2012) ECJ, C43/10 [2012] ECR I-0000.

OECD (2011) *Water Governance in OECD Countries: A Multilevel Approach.* OECD Studies on Water. Paris: OECD.

Paganelli, D., Forni, G., Marchini, A., Mazziotti, C. and Occhipinti-Ambrogi, A. (2011) Critical appraisal on the identification of reference conditions for the evaluation of ecological quality status along the Emilia-Romagna coast (Italy) using M-AMBI. *Marine Pollution Bulletin* 8, pp. 1725–35.

Quesada, M. (2011) *Water and Sanitation Services in Europe, Do Legal Frameworks Provide for Good Governance.* UNESCO: Centre for Water Law, Policy and Science, University of Dundee.

Raadgever, G.T., Dieperink, C., Driessen, P.P.J., Smit, A.A.H. and Van Rijswick, H.F.M.W. (2011) Uncertainty management strategies: lessons from the regional implementation of the Water Framework Directive in the Netherlands. *Environmental Science and Policy* 14(1), January.

Reese, M. (2013) Application of the cost recovery principle on water services in Germany. *Journal for European Environmental and Planning Law*, pp. 355–77.

Van Holten, S. and Van Rijswick, M. (2014) The consequences of a governance approach in European environmental Directives for flexibility, effectiveness and legitimacy. In M. Peeters and R. Uylenburg (eds), *EU Environmental Legislation: Legal Perspectives on Regulatory Strategies.* Cheltenham: Edward Elgar, pp. 13–47.

Van Kempen, J.J.H. (2012) Countering the obscurity of obligations in European environmental law: an analysis of Article 4 of the Water Framework Directive. *Journal of Environmental Law* 24(3), pp. 499–533. DOI: 10.1093/jel/eqs020.

Van Kempen, J.J.H. (2014a) Obligations of the Water Framework Directive: dealing with problems of interpretation. In M. Peeters and R. Uylenburg (eds), *EU Environmental Legislation: Legal Perspectives on Regulatory Strategies*. Cheltenham: Edward Elgar, pp. 146–72.

Van Kempen, J.J.H. (2014b) Good river basin management in Europe: distribution and protection. *Water Governance*. 4(3), pp. 31–6.

Van Rijswick, H.F.M.W. (2008) *Moving Water and the Law: On the Distribution of Water Rights and Water Duties within River Basins in European and Dutch Water Law*. Inaugural address, Utrecht University. Groningen: Europa Law.

Van Rijswick, H.F.M.W. (2012) Searching for the right to water in the legislation and case law of the European Union. In H. Smets (ed.), *The Implementation of the Right to Safe Drinking Water and Sanitation in Europe/Le droit à l'eau potable et à assainissement, sa mise en oeuvre en Europe*. Report for the 6th Forum Mondial de l'Eau, Marseille. Paris : L'Académie de l'Eau, pp. 87–113.

Van Rijswick, H.F.M.W. (2015) Mechanisms for water allocation and water rights in Europe and the Netherlands: lessons from a general public law perspective. *Journal of Water Law*, April (24), pp. 141–8.

Van Rijswick, H.F.M.W. (2016) Trans-jurisdictional water governance in the European Union. In J. Gray, C. Holley and R. Rayfuse (eds), *Transnational Water Governance*. Abingdon: Routledge, pp. 62–78.

Van Rijswick, H.F.M.W. and Havekes, H.J.M. (2012) *European and Dutch Water Law*. Groningen: Europa Law.

Van Rijswick H.F.M.W. and Keessen, A. (2016) Integrated water law and climate change: a EU perspective. In D. Farber and M. Peeters (eds), *Encyclopedia of Environmental Law: Climate Change Law*. Cheltenham: Edward Elgar.

Van Rijswick H.F.M.W., Gilissen, H.K. and van Kempen J.J.H. (2010) The need for international and regional transboundary cooperation in European river basin management as a result of new approaches in EC water law. *ERA Forum* 11(1), pp. 129–57.

Walker, B. *et al.* (2002) Resilience management in social-ecological systems: a working hypothesis for a participatory approach. *Conservation Ecology* 6(1), p. 14.

5

WATER MARKETS

Michael Hantke-Domas[1]

Introduction

Water provision is part of civilization. The Mesopotamian, Egyptian, Indian, Chinese, Khmer, Incan, Aztec and other empires developed around water sources. The human settlement meant that water needed to be controlled, made accessible and distributed equitably among uses and users (Caponera 1992).

Water uses may vary across societies but always share some similarities. Water is needed for human consumption, for health, for food, for the environment, for industry, for energy, for tourism, for religious motives and for cultural reasons. Its physical characteristics make it difficult and expensive to transport and to accumulate, depending on geography and climate. Furthermore, water is part of a natural system – the water cycle (Smith 2009).

As stakes are high on water use, water is usually a contentious matter for different groups, as each of them want to benefit from water in different ways. Governments are involved to different degrees and depths in water allocation (Dinar *et al.* 1997). In democratic societies, the realm in charge of solving divergent opinions on matters of public interest is the public space (i.e. government and parliament). On that account, water may be allocated for agrarian reforms, income redistribution, security, settlement in remote regions, protection of the environment, industrial development, or protection of industries, to name a few (ibid.).

As Dinar *et al.* (ibid.) suggest, there are roughly four main schemes of water allocation. The first one is *control by governments*[2] as 1) water is difficult to handle as a tradeable good; 2) it is perceived as a public good; and 3) water services and major irrigation infrastructure require massive investments. It follows that the only institution able to leverage such funding is a government.[3] The second scheme is *marginal cost pricing*,[4] by which the final user is charged with all the costs of receiving water. It has to include, among other things, externality costs, scarcity value, investments and distribution costs. The third one is *user-based allocation*,[5] where water users cluster into local groups to attain different objectives, such as irrigation, domestic water supply, or animal watering. The fourth scheme is *water markets*,[6] which I will define later. This scheme is not fixed, hence, several combinations of different schemes will be found around the world.

Public interest decisions influence public policy and carve water law. However, water law in every country has particularities that influence public policy.

Despite water law having evolved into sophisticated frameworks, it is not possible to satisfy every interest in water, and broad concerns remain about access (e.g. use), allocation (e.g. right to use) and management (e.g. quality and pollution control, control and protection of water-works, environmental protection and watershed management). The rules laid down by law are inserted in a social space where other institutions steer the satisfaction of multiple interests (Sharma 2012). Consequently, the role of law in water allocation and management is to set the permanent rules by which a society will satisfy interests, inspired by certain normative judge-ments[7] about how to do it.

Current water law tends to address those broad concerns by adopting different public policy strategies. The most widely proclaimed to be adopted being the integrated water resources man-agement (IWRM). In general, IWRM advocates ideas such as: water resources need to fulfil human needs and ecosystem protection; enhancing private and institutional investment; incor-porating participation in water management; balancing interests ranging from general welfare to environment protection; informing decisions from an economic, social and environmental per-spective; regulating monopolies for raw water and water services; and introducing economic tools (Global Water Partnership 2000).

Another strategy is the application of the neoclassical economic idea that markets allocate efficiently scarce resources. As water is limited, markets may allocate it to benefit everyone by maximizing general welfare. This simple, but strong statement has vast implications for access, allocation and management of water. Examples of this approach, which is much less adopted than IWRM – although they can co-exist – vary from countries where markets are largely free of regulation to operate and deliver the promise of maximizing general welfare (e.g. Chile) to countries that confine markets to the exchange of the remaining water after the environment and human consumption have been served (e.g. Australia).[8]

The concept of a water market can be used to define legislation that has adopted neoclassical economics ideas, or to define concrete geographical areas where such laws are applied. In both cases, water law blends with competition law, as both aim to generate and protect conditions for competition to bring allocative efficiency. The study of water markets needs to take into account the study of particular markets subject to competition law, and I will argue along these lines.

Indeed, competition law aims at preventing the exercise or the gain of 'market power', or the ability of a firm to raise prices to earn economic rents (Gavil *et al.* 2002). The importance of competition law in water markets stems from the fact that unregulated or even regulated markets may not be efficient if firms have market power. In this chapter, I will argue that water markets are prone to fail and competition law should be imposed to correct their inefficiencies.

Water markets in economic theory

Fresh water is in high demand around the world. Lack of plentiful water obliges societies to decide how to allocate it, among competing needs to be fulfilled. A market is an option for such allocation, as it provides signals (i.e. prices) to people and institutions by which 'to reconcile deci-sions about consumption and production' (Begg *et al.* 2003, p. 4).

Theoretically, markets maximize welfare[9] by allocating scarce resources to their most efficient use. Despite there being multiple ways of allocating resources that achieve general welfare, usually there will be one result that can be regarded as superior to others according to welfare economics[10] (Perman *et al.* 1999). The prevalent criterion, inspired by utilitarian moral philosophy,[11] stems from the idea that social welfare 'consists of some weighted average of the total utility level enjoyed by all individuals in the society' (ibid., p. 5). Economists have come up with the concept of economic efficiency (or allocative efficiency or Pareto optimality).

A distribution (or one result) is efficient as long as the interacting parties in a transaction become better off. Another efficient allocation arises from a transaction in which one party is better off and the other keeps her original position before entering the transaction. A third possibility, called Kaldor–Hicks optimality, derives from a transaction where one party is better off enough to compensate the other party that was made worse off. In other words, 'benefits from the reallocation must outweigh all costs' (Colby and Bush 1987, p. 238).

In order to attain economic efficiency, a market must be perfectly competitive. Conditions for such outcome are 1) all firms sell the same product (homogeneous product); 2) buyers and sellers have the same information regarding price and quality (perfect information); 3) buyers and sellers cannot determine the price at which the product is sold, because the market defines it (price taking); 4) buyers and sellers do not bear costs in their transactions (no transaction cost); 5) all production costs are endured by producers (no externalities); 6) firms can enter and leave markets without assuming special expenses (free entry and exit); 7) buyers can buy small parts of outputs, and firms can produce them (perfect divisibility of output) (Carlton and Perloff 2000).

Conditions for perfect competition are rare in the real world, however they provide for ideal parameters to assess real markets. Hence, it might be possible to determine how deviated a market is from economic efficiency (ibid.).

Water enters into the picture of markets as an asset that produces services (e.g. consumption, irrigation, energy generation, environmental, aesthetic, cultural, leisure, tourism and transport). As an asset, water will be an input (i.e. natural resource) for producing goods and services.[12]

Water markets have been defined as transactions where the value of water is different from the value of land and improvements, where buyers and sellers act voluntarily, and prices are negotiable between buyers and sellers (Colby and Bush 1987, p. 1). In a strict sense, a water market can be defined as the set of transactions over defined units of raw water in bulk, where buyers and sellers have complete information about price and quality, at prices defined by markets (and not the parties), without imposing cost on non-parties, and at no costs either to enter, to participate, or to exit.[13]

Besides economic efficiency, water markets are advocated for the allocation of water to the most valuable uses, access to water, internalization in prices of scarcity and easy exit of markets (Sharma 2012). Rosegrant and Binswanger (1994) have suggested that water markets can improve not only efficiency but also equity and sustainability of water use in developing countries. It is equally possible for a water market to dissipate if existent, economic rents[14] are extracted from users by public officials supervising allocation (Easter and Huang 2014).

Water markets may adopt different structures, such as agricultural markets (i.e. water transfer to high-value crops), inter-basin markets (i.e. that ability to move bulk water between watersheds) and inter-sector markets (i.e. the exchange of water between different uses, such as energy, industry, irrigation, environment). Other markets may involve just groundwater or water supply options contracts – the reservation of a use for future specified conditions (Berbel *et al.* 2014; Easter and Huang 2014; Maziotis *et al.* 2013; Adler 2008). Of course, other water markets may include part or the entirety of these markets.

The first requirement for a water market is to have a defined unit of raw water in bulk (e.g. rivers, lakes, aquifers, desalinated seawater and others). This is to say that the market has to be conceptualized as a *homogeneous* asset. Across countries, and even inside them, water might be accounted by different methods. One of the most common is volume over time, say litres per second; but other ways might be equally useful, as long as the entire market trades on one definition. In some countries, additional elements should be defined beforehand, such as uses that water might be applied to, or amounts of water allowed being withdrawn in a certain period. Nevertheless this requirement only captures the consumptive aspect, not the real impact of water on the hydrological system, as users upstream and downstream, and ecosystems are not incorporated into the consumption decision of the water – asset – owner.[15]

Complete *information* refers to access to legal, administrative, investment and hydrologic information.[16] Water is a flow resource that varies in amount and quality over time, depending on the geographical area, climate, weather, institutional arrangements, custom and, increasingly, politics. If a buyer wants to acquire water, they need to know beforehand how much water they will be able to withdraw. Usually, there will be regulations about how to extract water from surface and ground sources (development costs). Important information will concern water management to ensure effective use and, more generally, conditions for the use of existing infrastructure such as canals or dams. Other information will relate to transfer approval requirements (e.g. court hearings or hydrologic and environmental studies regarding the impact of transfer) between geographical points within a watershed or to a different basin, or between different uses.[17] Another important consideration will be about the existence of current and potential litigation, as there might be tribal claims, existing claims, or traditional uses, which might impair the normal use of water in a specific area. Governments can decide to build dams or other irrigation infrastructures that may impair or facilitate water use. The quality of water is another piece of central information, depending if it is used for human consumption, irrigation, bathing, tourism, or aesthetic pleasure. Differences in income levels and access to capital are accepted hindrances to obtaining information (Maziotis *et al.* 2013). Finally, different authorities might intervene in water management, such as environmental agencies regarding inflow water, health authorities in respect to human consumption, agricultural agencies regarding the quality of irrigation water, municipalities on human consumption, watershed agencies or user's associations concerning water management, to name a few. This is just a simple enumeration of important information to have before entering into the market to facilitate a complete assessment of the entitlement a water purchase can entail.

Prices should be set by the market. This is of central concern for competition legislation, as we will see later. Suffice to say that prices should be taken from the market, either from spot markets[18] or from other means such as newspaper sections. The more difficult it is to learn about the price, the more suspicion about its interference there will be.

Externalities are present in the whole water cycle (Siebert *et al.* 2009). Externalities in the water cycle may be caused by extraction, harvesting, diversion or storage. One of the critical externalities is the return flow, when water users can utilize all of their extracted water. In this case, rivers and aquifers are exploited unsustainably as no water is left for users downstream and for ecosystems.[19] Other problems may arise from the pollution of returned water, and even from land-use practices that may waterproof areas, changing the amount, quality and speed of water flow. Furthermore, water is an essential element for humans and the environment, spawning a myriad of interrelations. Access to fresh water is a human right;[20] life, health, education, security and equality are at stake (UN Water 2014). Its use for production (e.g. agriculture, industry, energy) creates multiple interrelations on water management, such as the use of water for energy generation may deplete a dam, foregoing its regulation objective for future irrigation. The same is true the other way around. If a water source is mostly used for productive activities, other non-productive uses might not be accounted for, such as environmental sustainability, amenities and tourism.

Additionally, land use, fertilizer use, new technologies, climate change, macroeconomic conditions, regulation and political climate may introduce risks to water markets, as 'Actual and perceived risks are often a barrier to the use of economic instruments' (Kerr 2013, p. ii). The mitigation of these risks can be tackled either by regulation or by internalizing costs in prices. In water, risks may arise from drought or flood, which may (to an unknown degree and extent) damage ecosystems, their functions and services. Furthermore, they may hamper water quality by turbidity, increase in algae, or pollution. Regulations may help mitigate these risks, but they may also create new risks or increase the existing ones by using poorly designed instruments (Kerr 2013). These negative externalities cut across the water cycle and are a source of criticism against water markets (Dellapenna 2000).

From an economic point of view, the existence of externalities show that unintended negative effects on a market are not internalized by prices, in that transactions are not economically efficient and thus diminish welfare. In other words, the transacting parties earn the benefits of water transaction, and losses are borne by third parties or the society. The only possibility for correction is price internalization, but unregulated markets will not achieve this objective (Perman *et al.* 1999, p. 135). Regulation is the best option to internalize costs when markets fail (Baldwin and Cave 1999).

Water markets face multiple *transaction costs*.[21] McCann and Garrick (2014) recognize physical and institutional determinants in such costs, such as technology, environment, governance structures, policy design, property rights definition, historic water use patterns and even path dependency. Furthermore, establishing and registering water rights is expensive, as legislation needs to be put in place to define them, separate them from land, define them with precision (e.g. amount, time, use, area of withdrawal), protection mechanisms and externalities (e.g. environmental and return flow). Additionally, public registration is important in keeping a record of transactions, as it is the existence of a formal place to trade (Maziotis *et al.* 2013; Sharma 2012, p. 172).

Free access to water markets is essential. The highest stake is human consumption, which involves human dignity and life. Some of the proponents of human rights recommend free access to water up to an amount that satisfies basic needs. This idea does not trump the fact that water provision comes at a cost, but can still be free through cross-subsidies. Hodgson (2006) affirms that water rights have nothing to do with the human right to water; however, such distinction is less clear when water is recognized as a resource common to everybody as it erodes the preclusion element (or exclusion in consumption) of a property right.

Besides the complexities that the human right to water imposes on markets, free entry is an issue in these kinds of markets. The problem stems from a lack of information and transaction costs, making it difficult (or costly) for people facing asymmetries of information to learn how to access the water market, bearing in mind the intricacies of it.

With regard to market exit, transaction costs may make it difficult to achieve, as long as administrative procedures can add exit costs (e.g. to sell a water right might be necessary for administrative approvals of transfer and new points of extraction).

Property rights are necessary in order for markets to bring economic efficiency (Perman *et al.* 1999, p. 8). In order to trade, a buyer has to have the right to use an asset. Economists maintain that only well-defined property rights avoid externalities, allowing for economic efficiency (Carlton and Perloff 2000, p. 83). According to Casado-Pérez (2015), governments should clearly define property rights on water, with their consequent apportionment quota – to be respected even during drought, security and tradability.[22] Colby and Bush suggest four requirements necessary for property rights in water markets:

1) Completely specified and enforced so that all individuals know the privileges and restrictions associated with holding a water right and the penalties for their violation;
2) Exclusive so that benefit and costs associated with water use and transfer decisions accrue to the decision makers (buyers, sellers, right holders), not third parties;
3) Comprehensive so that all attributes and uses of water that generate value can be represented by water rights, including water quality, instream flow levels and so on; and

4) Transferable so that water rights holders may transfer rights in response to an attractive offer and water resources can gravitate to their highest value uses.   [*1987, p. 23*]

Governments and parliaments have a large role to play in creating water markets as there are many enabling conditions, such as original allocation of water, or establishing the legal framework for the market to operate, and certainly creating property rights (Colby and Bush 1987; Dinar *et al.* 1997; Sharma 2012; Casado-Pérez 2015).

As most of the requirements for perfect competition are not present or even attainable by water markets,[23] water markets fail. As Colby and Bush (1987) assert, 'Competitive markets have many desirable attributes but the interdependence and public good characteristics associated with water resources imply that a perfectly competitive market is not a feasible water allocation process' (p. 31).

Besides, water can be considered a public good (Dellapenna 2000; Dinar *et al.* 1997; Sharma 2012). Dellapenna (2000) affirms that water 'has long been considered to be the quintessential "public good"' (p. 329). Dellapenna (2000) does not recognize that markets for bulk raw water exist,[24] and this might be an example of the public good nature of water. Despite this, it is possible to exclude someone else from consumption – hence it is not an economic public good. It is difficult to accept that people may not have access to it (Sharma 2012). This last predicament is precisely the source of the public good characteristic of water, as there are cultural – even moral – reasons for not excluding members of society from accessing it[25] (Dellapenna 2000). Libecap (2010) is more sceptical about the use of the argument of public good by special interests 'to weaken property rights and the efficiency benefits they can provide for incentives for wise use, conservation, and exchange' (p. 20). Bannock *et al.* (1998, cited by Sharma 2012) suggest that water has a public and a private good characteristic, which makes water stand in the middle of government management (public goods) and markets (private goods).

Casado-Pérez (2015) recognizes a divide between free market environmentalists,[26] for whom markets should be left out of government interference, and those who oppose markets and their commodification of water.[27] Despite this divide, still a non-market management of water has economic and equity implications. Hence, critical conditions for water markets are transferrable to regulation by control of government, marginal cost pricing or user-based allocation.

Casado-Pérez (ibid.) argues that water markets might fail as the result of the lack of government facilitation of market transactions, thus government failure produces market failures. An alternative answer places failures in social opposition to marketization of water (e.g. anti-privatization, resistance to water trading and private ownership of water rights) (Easter and Huang 2014).

Another explanation might be that markets are difficult to achieve, as water is not only a resource, but also the source of a myriad of interests within a watershed. Failure can also result from reliance on an economic model that, by simplification, is unable or inadequately able to deal with the complexities of water-related problems (Sharma 2012). Most authors recognize the importance of institutions for water markets to succeed, so corrections inspired in institutional economics (such as the case of Sharma 2012) should be regarded in more detail. The role of conflict resolution and the role of courts should be taken into account.[28]

As water markets present several failures, is it possible to keep calling them markets? Professor Joseph Dellapenna (2001) suggests a misunderstanding between markets and administrative use of economic incentives. The former refers to enticing measures to promote efficient interactions in water management, but its aggregation does not transfigure them into a market, which requires strict conditions to be attained. Sharma (2012) takes another stance, by suggesting that water markets are not a solution for all water-related problems, rather they may be a part of the solution. He suggests three modifications to market analysis to enhance markets: 1) integration of institutions (incorporate the institutional variable into the analysis, such as social norms and rules), 2) integration of case-specific rules (e.g. water law) and 3) emphasis on the importance of markets and evolution (markets create knowledge for agents to adapt to competition conditions) for competition and risk control.

There are many suggestions to enhance water markets. The most salient suggestions are those aimed at the correction of failures to obtain a homogeneous product, free entry and exit, perfect information, avoidance of externalities and transaction costs, and much more (Maziotis *et al.* 2013; Sharma 2012). The introduction of fairness in water allocation has been proposed, by recognizing past uses and how to solve the preference between public ownership of water rights and private ownership (Easter and Huang 2014). In the field of transaction costs, if property rights and governance are in place, other issues such as biophysical complexity, cultural values, conflict and lobbying should be considered (McCann and Garrick 2014).[29]

It is important to note that perfect water markets attain only economic efficiency, but not other objectives, such as equity, or fairness. This is an important distinction, as markets should be regarded, when perfect, as the best scheme for allocating scarce resources to the most valuable use, or the use that produces more returns (economic). Markets do not concern themselves with equity, so usually governments intervene to introduce corrections. One example is the human right to water. In this context, different authors have suggested different requirements for a water market to succeed.

The standard response to a market failure is regulation (Baldwin and Cave 1999). The idea is that governments will be able to make corrections that will allow markets to work. The underlying idea is still that water markets are a scheme to pursue.[30] Other responses might suggest avoiding returning to a market solution, or to adopt mixed or completely different schemes.

Summing up, only with certain preconditions and assumptions are markets an efficient way to allocate scarce resources to their most valuable (price) use. In the case of water, markets can produce important economic gains, but the complexities endured by public policy are so many and so difficult to deal with that caution should be taken to promote them. Another layer of issues that have to be borne in mind in the case of existing water markets is how to protect consumers from anti-competitive conduct arising in a market prone to fail, as water markets have been shown to do.

Competition law issues of water markets

Economic efficiency is achieved under perfect competition. The antithesis of this goal is a monopoly, where there is only one seller in the market. This position of absolute dominance allows the seller to charge a price above marginal cost, extracting economic rents directly from users. Elasticity (negative) of water demand by users may indicate that they might be willing to do whatever it takes to acquire water. As the monopolist faces no competition (there is no substitute for water or possibility of market access), they do not have incentives to produce efficiently, generating a social loss (spending more than the minimum), which is paid by the buyer. A similar problem occurs when there is a single buyer and multiple sellers (monopsony) (Gellhorn and Kovacic 1994).

Another possible structure is the oligopoly where there are few sellers and, among them, there is interdependence so that in order to set prices and quantities they will always be attentive to what their competitors do. In practice, none of them reduce prices to compete for a larger market since others will respond in the same manner and further lower prices. Consequently, vendors will focus on coordination and anticipation (ibid.).

Within these forms of industrial organization including monopolistic competition, a series of behaviours that water market agents incur and, *prima facie*, depart from the economic efficiency. There are behaviours that reflect agreements between competitors in the same market. If these behaviours seek to exercise market power, and avoid competition, we face collusive conduct. The strongest manifestation is the cartel (Whish 2003). In a water market, if sellers agree on the price charged to buyers, then there is collusion. Similarly, if the agreement relates to amounts that can be sold, there will be collusion. Another example of the same thing occurs when, in a water stream, two or more sellers have the exclusivity to sell or to lease water in a part thereof; as this gives them an oligopolistic position in the assigned part of the river. There is open collusion when the deal was sealed in writing, and concealed when competitors behave as if an agreement has been made (Gellhorn and Kovacic 1994). Most national legislation protects consumers against anti-competitive behaviours, with different nuances.[31]

Other behaviours that may occur in a market are vertical practices.[32] These occur as products are manufactured at different levels (e.g. raw materials, intermediate and finished product) (ibid.). In this production chain, different conducts contrary to competition (e.g. excess charges, quantity discounts, sales pricing, fixed amounts, or exclusivity clauses) may appear.

For water markets, vertical practices are more difficult, since the natural resource is not transformed. However, it could be that market power is exercised on the sale or refusal to sell, as well as the price, and this may have effects on a parallel market. For example, in agriculture, it may hinder the entry of a new competitor into a watershed by way of a refusal to sell to the next. Without water, the land cannot be cultivated, but also it is not possible to develop mining activities, some manufacturing industries, energy production, or construction.

At this point, a pattern of behaviour contrary to competition in water markets arises in the form of abuse of dominant position, which can manifest in horizontal behaviour whether these are collusive or exclusionary. Barriers to entry to the water market support the dominant position, which may include a water rights holder. We have seen water markets are prone to fail due to transaction costs and lack of homogeneity of the product, which reinforces the dominant position of the incumbent.

In order to determine market power, it is important to define the relevant market, both product and geographical. In my view, this definition is complex. In terms of product, it is necessary to look at the substitute goods for water. Can the water sold by water utilities compete with fizzy drinks? The substitution approach most commonly used is cross-price elasticity. To do this analysis requires a lot of information, which is easy to find in many water markets.

Following clarification of the product in question, it is necessary to define the relevant geographic area for the analysis of competition. Water can be confined in an aquifer, circulate in a river or a spring or a canal, or be impounded; there are multiple possibilities. A river can go from a mountain to the sea, and that river can be a tributary of other rivers, which belong to the same watershed. What is relevant is to determine the geographic area to weigh market power. Thus, a market can be a canal, a section of a river, several sections of a river, the entire course, a combination of a river and an aquifer, cover a watershed, a sub-basin and even several basins.[33]

After such definitions, we must figure out the degree of market concentration. The aim is to understand the market power of water holders at a defined market (product and geographic). There are several methods, none of them exempt from criticism. The traditional is the Herfindahl–Hirschman index (HHI), which corresponds to the sum of the squares of the market share. If a market yields more than 1,000 points, but is lower than 1,800 points, and increases in suspected transaction by less than 100 points indicator, not exceeding 1,800 points, then it is a market with an acceptable level of competence. However, if the index is above 1,800 and the operation helps to increase the rate by more than 50 points, then there is a presumption of illegality that must be studied.

Consequently, the complexity of the analysis of competition, although brief, shows spaces that competition authorities should keep in mind when evaluating whether a water market is competitive or not.

Conclusions

Water markets often fail, because of a myriad of circumstances. Conditions for market existence are difficult to attain, and non-economic stakes may not contribute to solving them. Besides, water markets are complex physical systems that require information usually particular to them, ranging from sub-basins to transboundary watersheds.

Markets can be an effective way of allocating water. If they are perfect, the distribution maximizes welfare. Even when they are not perfect markets may signal or show efficient transactions. Nevertheless, water markets should be regarded in their true perspective in terms of public policy. Markets only provide a good answer about how to maximize welfare, but not other societal goods such as environmental flows (Grafton *et al.* 2011). Even in the case of perfect water markets, they cannot deliver equitable solutions different from utilitarian Pareto optimality, thus, they cannot recognize that water is a human right or that the environment limits water property rights. In the same line, Carl Bauer (2008) when referring to the Chilean case, recalls Daniel Bromley (1989), to recognize the incompatibility of water markets with IWRM and even sustainable development: 'economic efficiency cannot serve as an objective guideline for public policy because it assumes an initial set of institutional arrangements and distribution of wealth, yet dealing with those institutions and that distribution are the essence of public policy' (p. 10).

Focusing on economic efficiency, competitive markets are a desirable scheme for water allocation, so it makes sense to promote them. Notwithstanding, public policy cannot be incomplete. It is not advisable to start with some elements of a market – for example, kicking off with the definition of property rights – and leave for the future conditions of information or transaction costs. The most likely result will be to create privileged water agents (the ones who have the rights) that, in the future, will lobby against market corrections.

Notes

1 I would like to thank Andrei Jouravlev and Carl Bauer, for their thoughtful comments on the draft. Of course, all mistakes remain mine. The competition law part has been already presented at the VII Jornadas Australes de Medio Ambiente y Derecho, 7 and 8 October 2015, Universidad Austral de Chile, Valdivia, Chile.
2 This system is extensively used in Latin America.
3 Today, the leveraging by the private sector for big investments is possible. Sanitation coverage in urban Chile rose from 22.5 per cent to 99.93 per cent, between 2000 and 2014, following a massive privatization and concession process (Superintendencia de Servicios Sanitarios 2015).
4 Examples are found in France for irrigation, water services in Chile and industrial uses in the US, Japan, India, Brazil and Israel.
5 This scheme is found in Bali (Indonesia), India and communal irrigation in Portugal.
6 This scheme is principally found in Chile, the west of America and Australia.
7 For different normative judgements in water management, see Kowarsch (2011), Kowarsch and Schröer (2011), Liu *et al.* (2011), and Priscoli *et al.* (2004).
8 There is a large amount of water markets case studies (Adler 2008; Alegría *et al.* 2002; Bauer 1997, 2008; Becerra 2005; Bjornlund *et al.* 2014; Bretsen and Hill 2008; Brewer *et al.* 2007; Brooks and Harris 2008; Chong and Sunding 2006; Colby 1988; Culp, Glennon, and Libecap 2014; Easter and Huang 2014; Garrick *et al.* 2011; Grafton *et al.* 2011; Hadjigeorgalis 2009; Hernández-Mora and Del Moral 2015; Holden and Thobani 1996; Kiem 2013; Kirby *et al.* 2014; Libecap *et al.* 2010, 2011; Libecap 2010; McMahon and Smith 2011; Qureshi *et al.* 2009; Rosegrant and Binswanger 1994; Waye and Son 2010; Wheeler *et al.* 2013; Zaman *et al.* 2009; Zekri and Easter 2005; Zhang 2007).

9 Welfare is defined as: 'Enjoyment of the necessary resources for a worth-while life. The welfare is the attempt to organize society so that at least the minimum income and public services necessary for this are available to all a country's inhabitants' (Black 1997, p. 502).

10 Welfare economics can be defined as: 'The part of economics concerned with the effects of economic activity on welfare' (Black 1997, p. 503).

11 Utilitarianism can be defined as: 'Utilitarians hold that there is one principle that sums up all our moral duties. The ultimate moral principle is that we should always do whatever will produce the greatest possible balance of happiness over unhappiness for everyone who will be affected by our action' (Rachels 1998, p. 401).

12 This understanding – based on utilitarianism – competes with alternative ethical approaches to water. For example, liberal thinkers have promoted equality of access to water by elaborating the idea of a human right to water. This conceptual ethical clash does not trump the idea of economic efficiency, but its ethical foundation. In general, see Perman *et al.* (1999).

13 See Dinar *et al.* (1997).

14 Rent is 'a payment to the owner of an input beyond the minimum necessary to cause it to be used' (Carlton and Perloff 2000, p. 65).

15 This is an idea shared by Andrei Jouravlev in commenting on this chapter.

16 In general, see Colby and Busch (1987).

17 For a list of information problems, see Colby and Busch (1987) and Easter and Huang (2014).

18 In Chile, there was an attempt to create an electronic water market in the watersheds of River Limarí and River Maipo. One of the encumbrances found was the lack of public information on prices and the poor quality of information produced by the Chilean Water Agency (*Dirección General de Aguas*) (Cristi 2011). It is important to note that spot water markets may reflect prices for temporary exchange of certain quantities between nearby users, and may operate under different rules than markets for water-use rights (Dinar *et al.* 1997).

19 This is an idea shared by Andrei Jouravlev in commenting on this chapter.

20 The United Nations recognizes this human right as necessary for the attainment of other human rights (UN Water 2014). Despite this recognition, still there are some sceptics.

21 On this subject see Colby and Busch (1987), Wang (2012), Garrick *et al.* (2011), Berbel *et al.* (2014), Qureshi *et al.* (2009), Casado-Pérez (2015).

22 Nevertheless, the legal thought has evolved to suggest that property rights are not absolute, particularly from an environmental perspective. Mr Ricardo Lorenzetti (2011), Chief Justice of the Argentinian Supreme Court, recognizes that property rights have an environmental dimension. He suggests that right holders are not isolated but are part of the society, living and limited by the 'natural' duty of solidarity with others, even the unborn (p. 48).

23 See Colby and Bush (1987) and Dellapenna (2000).

24 Dellapenna (2000) believes that water markets are not markets, but only economic administrative incentives.

25 Paradoxically, in Chile water is a common resource belonging to everyone ('*bien nacional de uso público*') according to its Civil Code, but there is a perpetual property right to the use of water when a licence has been issued. Exclusion is achieved by a criminal offence committed by those who withdraw water without a licence, with the exception of water extraction for personal use.

26 Two representatives of this line of thought are Anderson and Leal (2001) and Huffman (1997).

27 Three representative views are Blumm (1997), Radin (1987), Sharma (2012). Casado-Pérez (2015) suggests a review of the debate undertaken by Spaulding III (1997). A recent review of opposing arguments in Alberta (Canada) can be found in Bjornlund *et al.* (2014).

28 There is a traditional Water Tribunal in Valencia, Spain.

29 For a set of transaction costs affecting the design of water markets see McCann and Garrick (2014).

30 One proponent of addressing water market failures via government intervention is Casado-Pérez (2015).

31 In the US, the Sherman Act (1860), the Clayton Act (1914) and the Federal Trade Commission Act (1914) have a longstanding tradition in this field. In Europe, the EC Treaty and national legislation are very strong. For nuances, see Motta (2004).

32 Horizontal restraints are 'concerted acts that restrain competition between firms at the same level of production or distribution'; whereas, vertical restraints are restrictions 'imposed by the seller on the buyer (or vice versa) or on what is called a vertical relationship' (Gellhorn *et al.* 2004, p. 333).

33 For interesting research on market power in water markets, see Ansink and Houba (2012).

References

Adler, Jonathan H. (2008) Warming up to water markets. *Regulation* 31, p. 14.

Alegría, Maria Angélica, Fernando Valdés and Lillo, Adrián (2002) El Mercado de Aguas: Análisis Teórico Y Empírico. *Revista de Derecho Administrativo Económico*, pp. 169–85.

Anderson, Terry L. and Leal, Ronald D. (2001) *Free Market Environmentalism: Revised Edition*. Basingstoke: Palgrave.

Ansink, Erik and Houba, Harold (2012) Market power in water markets. *Journal of Environmental Economics and Management* 64(2), pp. 237–52.

Baldwin, Robert and Cave, Martin (1999) *Understanding Regulation: Theory, Strategy and Practice*, 1st edn. New York: Oxford University Press.

Bauer, Carl J. (1997) Bringing water markets down to earth: the political economy of water rights in Chile, 1976–1995. *World Development* 25(5), pp. 639–56.

Bauer, Carl J. (2008) The experience of Chilean water markets. In Expo Zaragoza. https://goo.gl/iCmXDC

Becerra, Manuel Rodríguez (2005) La Posible Creación de Mercados de Agua Y La Gobernabilidad de Este Recurso En Colombia. *Revista de Ingeniería* 22, pp. 94–102.

Begg, David, Fischer, Stanley and Dornbusch, Rudiger (2003) *Foundations of Economics*, 2nd edn. London: McGraw Hill Education.

Berbel, J., Bouscasse, H., Calatrava, J., Duponteil, A., Giannocarro, G., Garrido, A., Figureau, C. *et al.* (2014) Water markets scenarios for Southern Europe: new solutions for coping with increasing water scarcity and drought risk? Final Project Report. Water Cap and Trade. IWRM-NET Initiative. https://goo.gl/VbXTOJ

Bjornlund, Henning, Zuo, Alec, Wheeler, Sarah and Xu, Wei (2014) Exploring the reluctance to embrace water markets in Alberta, Canada. In William Easter and Qiuqiong Huang (eds), *Water Markets for the 21st Century. What Have We Learned?* Global Issues in Water Policy 11. Dordrecht: Springer, pp. 215–37.

Black, John (1997) Welfare. *Disctionary of Economics*. Oxford Paperback Reference. Oxford: Oxford University Press.

Blumm, Michael C. (1997) The fallacies of free market environmentalism. *Harvard Journal of Law and Public Policy* 15.

Bretsen, Stephen N. and Hill, Peter J. (2008) Water markets as a tragedy of the anticommons. *William and Mary Environmental Law and Policy Review* 33, p. 723.

Brewer, Jedidiah, Glennon, Robert, Ker, Alan and Libecap, Gary D. (2007) Water markets in the West: prices, trading, and contractual forms. National Bureau of Economic Research.

Bromley, Daniel (1989) *Economic Interests and Institutions: The Conceptual Foundations of Public Policy*. New York: Basil Blackwell.

Brooks, Robert and Harris, Edwyna (2008) Efficiency gains from water markets: empirical analysis of Watermove in Australia. *Agricultural Water Management* 95(4), pp. 391–9.

Caponera, Dante (1992) *Principles of Water Law and Administration: National and International*. Rotterdam: Balkema.

Carlton, Dennis W. and Perloff, Jeffrey M. (2000) *Modern Industrial Organization*, 3rd edn. Reading, MA: Addison-Wesley.

Casado-Pérez, Vanessa (2015) Missing water markets: a cautionary tale of governmental failure. *New York University Environmental Law Journal* 23, pp. 157–241.

Chong, Howard and Sunding, David (2006) Water markets and trading. *Annual Review of Environment Resources* 31, pp. 239–64.

Colby, Bonnie G. (1988) Economic impacts of water law–state law and water market development in the Southwest. *Natural Resources Journal* 28, p. 721.

Colby, Bonnie and Bush, David B. (1987) *Water Markets in Theory and Practice: Market Transfers, Water Values, and Public Policy*. Boulder, CO: Westview Press.

Cristi, Oscar (2011) Proyecto: Mercado Electrónico Del Agua (MEDA). Presented at the seminar Modernización del Mercado del Aguas en Chile, Santiago de Chile. http://goo.gl/qOYG57

Culp, Peter W., Glennon, Robert J. and Libecap, Gary (2014) *Shopping for Water: How the Market Can Mitigate Water Shortages in the American West*. Washington, DC: Island Press.

Dellapenna, Joseph W. (2000) The importance of getting names right: the myth of markets for water. *William and Mary Environmental Law and Policy Review* 25, p. 317.

Dinar, Ariel, Rosegrant, Mark W. and Suseela Meinzen-Dick, Ruth (1997) *Water Allocation Mechanisms: Principles and Examples*. 1779. World Bank. http://goo.gl/MTCNL4

Easter, William and Huang, Qiuqiong (2014) Water markets: how do we expand their use? In William Easter and Qiuqiong Huang (eds), *Water Markets for the 21st Century. What Have We Learned?* Global Issues in Water Policy 11. Dordrecht: Springer.

Garrick, Dustin, Whitten, Stuart and Coggan, Anthea (2011) Understanding the evolution and performance of water markets and allocation policy: a transaction cost analysis framework. *Ecological Economics* 88, pp. 195–205.

Gavil, Andrew I., Kovacic, William and Baker, Jonathan B. (2002) *Antitrust Law in Perspective: Cases, Concepts and Problems in Competition Policy.* St Paul, MN: Thomson West.

Gellhorn, Ernest and Kovacic, William (1994) *Antitrust Law and Economics*, 4th edn. St Paul, MN: West Group.

Gellhorn, Ernest, Kovacic, William and Calkins, Stephen (2004) *Antitrust Law and Economics*, 5th edn. St Paul, MN: Thomson West.

Global Water Partnership (2000) Integrated water resources management. TAC Background Papers No. 4. Stockholm: Global Water Partnership.

Grafton, R. Quentin, Libecap, Gary, McGlennon, Samuel, Landry, Clay and O'Brien, Bob (2011) An integrated assessment of water markets: a cross-country comparison. *Review of Environmental Economics and Policy* 5(2), pp. 219–39.

Hadjigeorgalis, Ereney (2009) A place for water markets: performance and challenges. *Applied Economic Perspectives and Policy* 31 (1), pp. 50–67.

Hernández-Mora, Nuria and Del Moral, Leandro (2015) Developing markets for water reallocation: revisiting the experience of Spanish water mercantilización. *Geoforum* 62, pp. 143–55.

Hodgson, Stephen (2006) *Modern Water Rights: Theory and Practice.* 92. Rome: Food and Agriculture Organization.

Holden, Paul and Thobani, Mateen (1996) *Tradable Water Rights: A Property Rights Approach to Resolving Water Shortages and Promoting Investment.* World Bank.

Huffman, James L. (1997) Institutional constraints on transboundary water marketing. In Terry L. Anderson and Peter J. Hill (eds), *Water Marketing: The Next Generation.* Lanham: Rowan & Littlefield.

Kerr, Suzi (2013) Managing risks and tradeoffs using water markets. Motu Working Paper.

Kiem, Anthony S. (2013) Drought and water policy in Australia: challenges for the future illustrated by the issues associated with water trading and climate change adaptation in the Murray–Darling Basin. *Global Environmental Change* 23(6), pp. 1615–26.

Kirby, John M., Connor, Jeffrey, Ahmad, Mobin-ud-Din, Gao, Lei and Mainuddin, Mohammed (2014) Climate change and environmental water reallocation in the Murray–Darling Basin: impacts on flows, diversions and economic returns to irrigation. *Journal of Hydrology* 518, pp. 120–9.

Kowarsch, Martin (2011) Diversity of water ethics: a literature review. Working Paper prepared for the research project Sustainable Water Management in a Globalized World.

Kowarsch, Martin and Schröer, Katharina (2011) What should water ethics be about? The problem with identifying problems. Working Paper prepared for the research project Sustainable Water Management in a Globalized World.

Libecap, Gary D. (2010) Water rights and markets in the US semi arid West: efficiency and equity issues. Working Paper prepared for the research project Sustainable Water Management in a Globalized World.

Libecap, Gary D., Grafton, R. Quentin, Landry, Clay, O'Brien, Robert J. and Edwards, Eric C. (2010) Water scarcity and water markets: a comparison of institutions and practices in the Murray–Darling Basin of Australia and the western US. ICER Working Paper No. 28/2010.

Libecap, Gary D., Grafton, R. Quentin, Edwards, Eric C., O'Brien, Robert J. and Landry, Clay (2011) A comparative assessment of water markets: insights from the Murray-Darling Basin of Australia and the western US. Working Paper prepared for the conference on The Evolution of Property Rights Related to Land and Natural Resources, September.

Liu, Jie, Dorjderem, Amarbayasgalan, Fu, Jinhua, Lei, Xiaohui, Liu, Huajie, Macer, Darryl, Qiao, Qingju, *et al.* (2011) Water ethics and water resource management. Working Group 14 Report 14. Ethics and Climate Change in Asia and the Pacific (ECCAP) Project. Bangkok: UNESCO.

Lorenzetti, Ricardo L. (2011) *Teoría Del Derecho Ambiental.* Colección Internacional. Bogotá: TEMIS.

Maziotis, Alexandros, Calliari, Elisa and Mysiak, Jaroslav (2013) Robust institutions for sustainable water markets: a survey of the literature and the way forward. FEEM Working Paper, 24 June.

McCann, Laura and Garrick, Dustin (2014) Transaction costs and policy design for water markets. In William Easter and Qiuqiong Huang (eds), *Water Markets for the 21st Century: What Have We Learned?.* Global Issues in Water Policy 11. Dordrecht: Springer, pp. 11–32.

McMahon, Tyler G. and Smith, Mark Griffin (2011) The Arkansas Valley super ditch: a local response to 'buy and dry' in Colorado water markets. Available from: http://papers.ssrn.com/sol3/papers.cfm?abstract_id=1922444

Motta, Massimo (2004) *Competition Policy: Theory and Practice*. Cambridge: Cambridge University Press.

Perman, Roger, Ma, Yue, McGilvray, James and Common, Michael (1999) *Natural Resources and Environmental Economics*, 2nd edn. Harlow: Longman.

Priscoli, Jerome Delli, Dooge, James and Llamas, Ramón (2004) Water ethics: overview. Essay 1. Series on Water and Ethics. Paris: UNESCO International Hydrological Programme/World Commission on Ethics of Scientific Knowledge and Technology.

Qureshi, Ejaz, Shi, Tian, Qureshi, Sumaira and Proctor, Wendy (2009) Removing barriers to facilitate efficient water markets in the Murray–Darling Basin of Australia. *Agricultural Water Management* 96, pp. 1641–51.

Rachels, James (1998) Nietzsche and the objectivity of morals. In Scott Arnold, Theodore Benditt and George Graham (eds), *Philosophy Then and Now*. Malden, MA: Blackwell, pp. 385–414.

Radin, Margaret J. (1987) Market-inalienability. *Harvard Law Review* 100.

Rosegrant, Mark W. and Binswanger, Hans (1994) Markets in tradable water rights: potential for efficiency gains in developing country water resource allocation. *World Development* 22(11), pp. 1613–25.

Sharma, Dhruv (2012) A new institutional economics approach to water resource management. University of Sydney. http://goo.gl/wHMYzR

Siebert, Elizabeth, Young, Mike and Young, Doug (2009) *Exposure Draft – Draft Guidelines for Managing Externalities: Restoring the Balance*. Canberra: Department of the Environment, Water, Heritage and the Arts.

Smith, Bryant Walker (2009) Water as a public good: the status of water under the General Agreement on Tariffs and Trade. *Cardozo Journal of International and Comparative Law* 17, pp. 291–314.

Spaulding III, Norman W. (1997) Commodification and its discontents: environmentalism and the promise of market incentives. *Stanford Environmental Law Journal* 16.

Superintendencia de Servicios Sanitarios (2015) Informe de Gestión Del Sector Sanitario 2014. Santiago de Chile: Superintendencia de Servicios Sanitarios.

UN Water (2014) International Decade for Action 'Water for Life' 2005–2015. Focus Areas: The Human Right to Water and Sanitation. *United Nations*. http://www.un.org/waterforlifedecade/human_right_to_water.shtml

Wang, Yahua (2012) A simulation of water markets with transaction costs. *Agricultural Water Management* 103, pp. 54–61.

Waye, Vicki and Son, Christina (2010) Regulating the Australian water market. *Journal of Environmental Law* 22 (3), pp. 431–59.

Wheeler, Sarah, Garrick, Dustin, Loch, Adam and Bjornlund, Henning (2013) Evaluating water market products to acquire water for the environment in Australia. *Land Use Policy* 30(1), pp. 427–36.

Whish, Richard (2003) *Competition Law*, 5th edn. London: LexisNexis Butterworths.

Zaman, A.M., Malano, Hector M and Davidson, Bart (2009) An integrated water trading-allocation model, applied to a water market in Australia. *Agricultural Water Management* 96(1), pp. 149–59.

Zekri, Slim and Easter, William (2005) Estimating the potential gains from water markets: a case study from Tunisia. *Agricultural Water Management* 72(3), pp. 161–75.

Zhang, Junlian (2007) Barriers to water markets in the Heihe River Basin in northwest China. *Agricultural Water Management* 87(1), pp. 32–40.

6

WATER POLLUTION AND WATER QUALITY

Shifting regulatory paradigms

William Howarth

Introduction

Approached from a United Kingdom perspective, and specifically that of England, the following discussion investigates the role of national law in addressing issues of water pollution and water quality. Although an aim is to provide observations on the current state of the law in this field, it is also important to see this as an outcome of an accumulative progression involving a sequence of distinct historical stages that have been supplemented and consolidated over time. Moreover, it is a progression that is ongoing, with contemporary needs to address a continually evolving conception of pollution prompting a rethink of the role of law in this exercise. In this endeavour, England provides a particularly precocious and developed national perspective upon the evolution of water pollution and water quality laws, and serves admirably to illustrate the stages of the progression which provide the main structure for this chapter. Equally, the stages of development in English water pollution and water quality law that are identified may serve as reference points for comparative discussion of other jurisdictions.

The early onset of industrialisation in England, and the appreciation of its demographic, social and public health impacts, provided the context in which laws were first employed, ostensibly, to legislate away the most extreme impacts of the industrialisation upon the aquatic environment. Naively, perhaps, this early legislation seemed to suppose that clear rivers, stillwaters and groundwaters were possible without hindering the progress of industrialisation and without the need for costly and burdensome enforcement responsibilities. Nonetheless, these first crude legislative initiatives have formed the main bedrock for prohibitive measures that continue to the present. The reactive and prohibitive response to industrial pollution in nineteenth-century water pollution legislation is, therefore, taken as *stage one* in the discussion that follows.

Stage two in the process comprises a range of adaptations and extensions of the basic criminalisation of water pollution, which sought to instil a greater degree of relevance, practicality and effectiveness into this branch of water law. Amongst other things, this included refinements of the blunt use of the criminal law to allow licensing of environmentally benign activities. In effect, this involved a realisation that the original industrial-prohibitive model of pollution control was not sufficiently discerning to recognise and evaluate the range of activities that cause damage to the water environment, and was not sufficiently sophisticated to reflect the available legal options. Centrally, perhaps, this second stage is characterised by a shift of attention towards anticipatory controls upon land use (both industrial and non-industrial) as a means towards ensuring satisfactory water quality. Whilst the industrial-prohibitive paradigm was modified, it was not superseded. Stage two marks a shift of attention towards *preventative* regulation of a range of potentially problematic *land use* activities, as a means towards achieving a satisfactory state of the aquatic environment. Accepting some over-simplification, stage two is dubbed the *anticipatory land use control* stage.

Stage three is termed the *strategic water quality management* stage. This is characterised by the recognition that water quality regulation should be regarded as a means towards an end: the realisation of a satisfactory state of the aquatic environment. This 'satisfactory state' may be defined in terms of the needs of a range of users: water for drinking water supply purposes, capable of meeting public health requirements; water that is needed to serve the ecological purpose of supporting different kinds of biodiversity and ecosystems; and/or natural water which is seen as in inherent need of protection, irrespective of its practical use, perhaps for aesthetic or other reasons which recognise the intrinsic or non-utilitarian value of the aquatic environment. The question of what is to count as 'satisfactory' for these different purposes is clearly one of major technical and ethical complexity, not least because waters may have multiple uses, demanding different quality objectives to be achieved for different purposes. The overriding feature, from a legal viewpoint, is that particular legal powers, to regulate activities that impact upon the water environment, must now be used *purposively* in order to achieve the relevant objective. In stage three, particular legal powers which may have previously been used in a reactive way to respond to past pollution incidents, or in an anticipatory way to prevent future incidents, have now gained a strategic purpose in that they are now to be used as a means towards facilitating the end of securing water quality standards.

Stage four has been dubbed the *post-industrial environmental regulation stage*, largely for want of a better name. In part this recognises the decline of heavy industries in the UK and other developed countries, which have historically been seen as the major cause of pollution, and that industrial water pollution has actually served to mask a range of other factors that contribute to the unsatisfactory state of the aquatic environment. In part, these may arise from land use activities where the link between these activities and the impacts on the water environment is unknown or uncertain (termed *diffuse* pollution). These impacts may arise as the cumulative result of consumption of goods and services, and diverse lifestyle choices. This rather vague characterisation recognises that the new wave of pollution challenges reflect a shift of attention from the manufacturing of products to their *use* by consumers. Products containing chemicals, such as pharmaceutical products, for example, may find their way into the aquatic environment by diverse routes, but clearly elude environmental regulation of a traditional kind adopted in respect of industrial pollution (and see Chapter 7, this volume, on Contaminants of emerging concern). The relatively new awareness is that satisfactory water quality can only be achieved by regulating the use of products, throughout their life cycle, and that the regulated patterns of consumption may actually be quite remote from the water environment that they seek to protect. The post-industrial regulatory challenge is that of relating pollution controls to the actions of individual consumers of goods and services, rather than the manufacturing and land use activities that have previously been seen as the centre of legal attention.

The four stages that have been identified, provide a convenient structure for the exposition that follows, but, in reality, legal development may not follow the rigid historical sequence that may seem to be suggested. For quite long periods, the different stages have run concurrently rather than consecutively. The progression is 'accumulative' in the sense that one stage builds upon another rather than replacing it as the successive stages consolidate with one another. Despite this, it is instructive to conceive of the stages as having broad chronology following the sequence described below. Likewise, it is instructive to see contemporary concerns as a consequence of the preceding stages of development of water pollution and quality law. However, before discussing these stages of development in detail, some preliminary observations are needed on the legal meaning of the centrally important terms *water pollution* and *water quality*.

Water pollution and water quality

The first challenge in addressing the law on water pollution is the elusiveness of the concept of pollution itself. What counts as 'pollution' seems to be capable of immense variation for historical, scientific, social, economic and cultural reasons and may be best regarded as an inherently relative concept. Beyond that, the idea seems to be systematically ambiguous. An ordinary English word seems capable of having quite different meanings in different grammatical contexts. These meanings range from: the origin or source (of something harmful); the transference route or causative process of the harm; the particular entry of the harmful thing into an uncontained part of the environment; and the impact and/or kind and extent of the resulting environmental damage. Hence, it is often unclear whether 'pollution' is the offending substance, the offending activity, the resultant damaged state of the environment or something else (Howarth and McGillivray 2001, p. 7).

As a matter of everyday language, the European Environment Agency website (undated) quotes a summary of a definition taken from Wikipedia (undated) that is formulated as follows.

> Pollution is the introduction of contaminants into the natural environment that cause adverse change. Pollution can take the form of chemical substances or energy, such as noise, heat or light. Pollutants, the components of pollution, can be either foreign substances/energies or naturally occurring contaminants. Pollution is often classed as point source or nonpoint source pollution.

Although this general language definition of 'pollution', as a process or mechanism leading to some part of the environment being polluted by substances that may be termed *pollutants*, is broadly helpful for many purposes, it may not be sufficiently precise for legal purposes.

A relatively early attempt to define the key term in national law was as follows:

> 'Pollution of the environment' means the release into any environmental medium from any process, of substances which are capable of causing harm to man or any other living organism supported by the environment. For these purposes 'harm' was defined to mean harm to the health of living organisms or other interference with the ecological systems of which they form a part and, in the case of man, includes offence caused to any of his senses or harm to his property.[1]

The difficulty with this brave attempt to give legal meaning to the concept of pollution is its breadth and circularity. It is quite difficult to see what, if anything, is excluded and the meanings of 'pollution' and 'harm' seem to be mutually defined ('pollution' is something that causes 'harm' and 'harm' is something that is caused by 'pollution').

Comparable difficulties in defining the concept of pollution arise in EU environmental law. Hence, the definition of 'pollution' used in the EU Water Framework Directive (European Parliament and Council 2000) is as follows.

> Pollution means the direct or indirect introduction, as a result of human activity, of substances or heat into the air, water or land which may be harmful to human health

or to the quality of the aquatic ecosystems or terrestrial ecosystems directly depending on aquatic ecosystems, which result in damage to material property, or which impair or interfere with amenities and other legitimate uses of the environment.

Art. 2(33)

Alongside the difficulties that have been noted with the nation law characterisation, this definition emphasises the anthropocentric character of definition: 'pollution' is something caused by humans and resulting in certain kinds of harm to humans. This might or might not correspond to everyday intuitions about changes in environmental quality that might arise from non-human causes, such as natural events which have adverse impacts upon environmental quality. Beyond that, it is not clear why ecological damage, without more, should not be sufficient to constitute pollution.

Potentially a more explicit and legally useful definition of pollution, or the lesser state that is sometimes termed *contamination*, involves a recognition of a cause–effect relationship and the need for a particular mechanism by which this may arise. Hence, pollution has been legally characterised as the state that exists where three elements are found to be present: a source, a pathway and a target. This approach is adopted towards the national regime for remediation of contaminated land,[2] where it is advised that a determination should be made that land is contaminated where three elements are found:

1) a contaminant or pollutant, which has the potential to cause harm or pollution of controlled waters;
2) a receptor or target, consisting of a living organism or group of living organisms, an ecological system, some piece of property of a specified kind or controlled waters; and
3) a pathway, consisting of one or more routes or means by or through which the receptor could be exposed to, or affected by, the contaminant identified either through general scientific knowledge of the nature of the contaminant or the circumstances of the land in question.

Department for Environment, Transport and the Regions (2000) Annex 3 ¶ 11 and 19;
Department for Environment, Food and Rural Affairs (2006a) and Department for
Environment, Food and Rural Affairs (2012)

This characterisation of pollution, as the combination of three distinct elements, seems helpful in emphasising the need for transference of the undesirable thing between points of origin and destination, and the recognition that harm is dependent upon the different sensitivities of the recipient or victim in the pollution process.

However, the potential for extreme variability in each of the three elements is evident. Pollutant sources were historically regarded as industrial activities involving emissions of toxic chemicals, but it has been progressively realised that things originating from agricultural and domestic activities are equally capable of being regarded as pollutants. The concept of a pollutant is far from having any defined boundaries. The recipient of pollution may have been traditionally seen as a human being, particularly, where an adverse public impact upon health was the key concern, but the targets for pollution have been progressively extended to encompass the environmental media (where a reduction in the quality of water, air or land is involved) and the non-human living constituents of the environment media (where impacts on biodiversity and ecosystems are at issue). Finally, the idea of a pathway for contamination seems to raise a spectrum of possibilities: from the direct dumping of a contaminant on the target; the transmission of a pollutant by the environmental media (water, air or land); to the range of diffuse transmission mechanisms where contamination takes place through an uncertain or unknown route.

The three elements that are needed for a finding that pollution exists are almost infinitely variable. Nonetheless, the useful implication of the threefold legal test is that pollution is not present when any one (or more) of the three elements is absent: first, because the pollutant does not have the potential to cause harm; second, because there is no thing that is capable of being harmed; and third that there is no means of transference by which the pollutant is transported to a thing that is harmed. The differing implications of this are important for the discussion of the different stages in development of water pollution and water quality law that follow.

Water quality standards

Environmental quality law is, first, about preventing an unsatisfactory state of the environment and, second, about achieving a satisfactory state. These twin goals are equally applicable to the aquatic environment, where laws can be seen as either negatively or positively formulated environmental imperatives: first, concerned with *preventing* water pollution and, second, directed towards *achieving* a satisfactory state of the aquatic environment. However, there are some major differences between using law in negative (prohibitive) and positive (facilitative) ways.

Whilst early English environmental legislation concerned itself with the negative of preventing pollution, later legislation has increasingly engaged with the positive question of what is to count as a *satisfactory* state of environmental quality and how law might be most effectively used to achieve that state. This satisfactory state of the aquatic environment is one in which relevant water quality standards are met and maintained. For these purposes, a water quality *objective* is a general statement as to the short- or long-term aims for the use of a particular water. The realisation of a particular water quality objective is secured by the achievement of the relevant, technically and numerically expressed, physical, chemical and biological parameters comprising the relevant water quality *standard*. The water quality standard is formulated at such a level as to ensure that the water is of sufficient quality to meet the use identified as its objective (Howarth and McGillivray 2001, p. 28). Characteristically, water quality standards are intended to be achieved within a relatively short duration. In this respect, they contrast with target-based approaches towards environmental management. Hence, the EU environment programme target that the Union should be living with the ecological limits of the planet by 2050 might be seen as an important overall aspiration for EU environmental management, but not as an environmental quality standard (European Commission 2013).

The reluctance to impose statutory standards upon the quality of effluent discharges or to formulate statutory standards for the quality of receiving waters may be seen as a distinctive feature of English water quality law (Richardson *et al.* 1982, ch. 3). Whilst explicit and precise water quality standards were resisted, there was a longstanding practice of using water quality standards of an imprecise kind. Typically standards of this kind would require effluent discharging activities to be conducted 'so as not to become a nuisance' or so as not to be 'injurious to health', with no further indication as to what qualitative standards of water quality are likely to give rise to these.[3]

Although the legal mechanisms to specify water quality standards and to require these to be met by particular waters were eventually provided for in national law, these mechanisms have only been used to realise water quality standards that have been formulated under EU water quality legislation. This illustrates a characteristic feature of English water quality law: its reluctance to adopt statutory water quality standards and to require these to be met. Hence, as will be seen, national legislative inaction has resulted only in the application of standards formulated at an EU level.

Stage one: the prohibitive response to industrial pollution

Stage one in the progression of national water quality law is seen as the prohibitive response to industrial water pollution, which is characterised by the criminalisation of industrial activities that were regarded as excessively damaging to the environment. This response may be seen as reactive in that it was not seen as a means towards securing any overall strategic goal for the quality of the environment, but rather as a recognition that certain activities gave rise to unacceptable polluting consequences.

The first general UK enactment entirely concerned with water pollution was the Rivers Prevention of Pollution Act 1876. The success of this Act in counteracting the industrial water pollution at the time of its enactment may have been limited. Nonetheless, the Act put in place a form of criminal prohibition of water polluting activities that has remained substantially unchanged to the present day.

Summarising greatly, the 1876 Act created an offence concerning water pollution by solid matter where any person 'put or cause to be put or to fall or knowingly permit to be put or to be carried into any stream' any of a range of matter which, 'either singly or in combination with other similar acts of the same or any other person, interfered with its due flow, or polluted its waters' (s.2 1876 Act). A further offence arose where any person caused to fall or flow or to be carried into any stream any solid or liquid sewage matter. The offence was circumscribed by qualifications which served significantly to limit its practical effect. Where any sewage matter entered any stream the person causing or knowingly permitting this did not commit an offence if it could be shown that the 'best practicable and available means' had been used to render the sewage harmless (s.3). The 1876 Act was also concerned with manufacturing and mining pollution in respect of which the basic offence was committed by any person who caused to fall or flow or knowingly permitted to fall or flow, or to be carried into any stream any poisonous, noxious, or polluting liquid proceeding from any factory or manufacturing process (s.4). This offence was again made subject to an exception where the best practicable and reasonably available means was used to render harmless the polluting liquid passing into the stream.

Despite the apparent breadth of the prohibitive measures provided for under the 1876 Act, it was largely ineffective in its objective of preventing the pollution of rivers. Primarily, this was due to the extensive limitations upon enforcement action under the Act. In respect of manufacturing and mining pollution, prosecution proceedings were not to be taken except by a sanitary authority and only then with the consent of the Local Government Board. The Local Government Board, in giving or withholding consent to proceedings, was to have regard to the industrial interests involved in the case and to the circumstances and requirements of the locality. In particular, the Board was not to give its consent to proceedings by a sanitary authority of any district which was the 'seat of any manufacturing industry', unless they were satisfied that means for rendering harmless the poisonous, noxious, or polluting liquids proceeding from the processes were reasonably practicable and available, and that no material injury would be inflicted on the interests of industry by instigating proceedings. In effect, therefore, an Act which purported to outlaw the main forms of river pollution was emasculated by excessive restraints upon enforcement which, in practical terms, rendered the law a dead letter (s.6, and also Turing 1952 and Hammerton 1987).

Moving forward to the present, the central offence concerning water pollution in England is that a person must not cause or knowingly permit a water discharge activity, except under and to the extent authorised by an environmental permit.[4] A 'water discharge activity' for these purposes is (subject to certain exempt activities) defined to include the discharge or entry to inland freshwaters, coastal waters or relevant territorial waters of any poisonous, noxious or polluting matter (Howarth 1993), waste matter, or trade or sewage effluent.[5]

Recognising that styles of legislative drafting have changed over the years, the substance of the present offence can be traced back from the Environmental Permitting Regulations 2010 to the Water Resources Act 1991, the Water Act 1989, the Control of Pollution Act 1974, the Rivers (Prevention of Pollution) Act 1951 and back to the Rivers Pollution Prevention Act 1876. Despite simplification of the wording, the substance of the offence has remained substantially the same. Since 1876, an offence has been provided for where a person 'causes of knowingly permits' the entry of matter that is 'poisonous, noxious or polluting' into rivers and other kinds of watercourse, stillwaters and groundwaters.

In practical legal terms this continuity is particularly relevant because the caselaw under the earlier legislation remains relevant to the interpretation of the present offence. This caselaw involves cases, determined at the highest level, to the effect that 'causing' the entry of polluting matter is a matter of strict liability, which does not require the prosecution to establish intention, recklessness, negligence or any other kind of fault (*Alphacell Ltd. v Woodward* 1972; *Empress Car Company (Abertillery) Ltd. v National Rivers Authority* 1998). By contrast, 'knowingly permitting' involves a failure to prevent an entry where this is accompanied by knowledge of the entry and some degree of control over the circumstances that give rise to the entry (*Environment Agency v Biffa Waste Services* 2006). Although the main criminal offence incorporates strict liability for causing a polluting entry, this needs to be contrasted with the guidance used by the prosecuting authority, which recognises that 'offences that are committed deliberately, recklessly or with gross negligence are more likely to result in prosecution' (Environment Agency 2015, p. 12). Hence the theoretical legal position, that fault is irrelevant to commission of the offence, needs to be contrasted with the practical reality that prosecution is unlikely to be pursued where the pollution incident arises as a result of a genuine accident.

Despite strict liability and the proven durability of the original formulation of the main water pollution offence, it remains as an example of an early, and rather crude, kind of environmental prohibition. There are four notable aspects to its rudimentary character.

First, there is the mismatch between the offence, which requires only the *entry* of polluting matter into relevant waters, and the definition of pollution discussed above which requires *harm* as an essential element of pollution. The peculiarity of the offence is that actual environmental harm is irrelevant to the commission of the offence, it is merely the *entry* of polluting matter that needs to be shown. Clearly, the degree of harm that results will be a relevant factor in the decision to prosecute (Environment Agency 2011) and a relevant factor in relation to sentencing (*R v Milford Haven Port Authority* 2000; Sentencing Council 2014), but environmental harm is not relevant to the commission of the offence as such. This is curious, since it might be thought that an offence concerning water pollution would require *actual* pollution to be shown, but under this formulation of the offence there is no requirement for this.

Second, on the location at which the offence takes place, an offence of causing the entry of polluting matter occurs at the place where the polluting matter enters the relevant water, usually where an effluent outflow pipe transmits the offending matter into the relevant water. Hence, this type of pollution is commonly termed *point-source pollution*. This is important because pollution may enter waters by other means than an effluent outfall, such as where surface water runoff or groundwater percolation carries contaminants into waters without there being any distinct point of entry. In effect, the offence serves to criminalise one kind of pollution whilst not addressing other kinds of diffuse pollution which may be at least as damaging in their environmental impact.

Third, the basic offence of causing the entry of pollution matter was formulated as a blunt, all or nothing, kind of criminal prohibition, which made little provision for exceptions in relation to activities that might be considered *de minimis* or benign so far as the receiving environment is concerned. Clearly, the blanket criminalisation of every kind of industrial emission into the aquatic environment, without regard to harmfulness or consequences, seems extreme. Perhaps the original restrictions upon enforcement were intended as a counterbalance to a criminal offence of extraordinary breadth. It might also be appreciated that the 1876 formulation of the offence was arrived at long before environmental licensing had gained common currency. The idea of using an authorisation granted by a regulatory body, originally termed a *discharge consent*, to allow an emission to take place without this constituting a criminal offence was first provided for in 1951.[6]

Discharge consenting,[7] or environmental permitting as it has become known under the Environmental Permitting Regulations 2010,[8] is an invaluable way of narrowing and refining the scope of a widely formulated blanket criminal offence. Providing that the discharger stays within the limits of the consent that has been given for the discharge, no offence is committed. This allows the regulatory authority extensive discretion to undertake fine tuning in setting the parameters of a permit to take account of the sensitivity of the receiving waters and a range of other factors, including the need to meet relevant EU water quality measures. Although the role of environmental permitting is now seamlessly interwoven with the criminal prohibition on water pollution, the fact remains that the original offence was formulated without this facility being available.

Fourth, and perhaps most important, is the unavoidable fact that the enactment of any criminal offence needs to take account of the capacity for its enforcement. The initial difficulties concerning this have been noted in respect of the enforcement responsibilities and restrictions upon enforcement under the 1876 version of the water pollution offence. The clear inadequacies in enforcement of the original offence had the result that it was largely unenforceable, though, with hindsight, a cynically minded observer might have regarded this as the ulterior intention all along!

Enforcement of environmental law, including the water pollution offence, has been a matter of continuing concern for as long as criminalisation has been a legal response. Impartial enforcement of environmental law requires the establishment of public bodies with a sufficient degree of independence, both from central government and from the operators of the installations that they regulate (see, generally, International Network for Environmental Compliance and Enforcement 2009). Further, environmental law enforcement involves a level of technical expertise beyond that required in most policing contexts. This has a cost in respect of providing the resources needed to employ a sufficient number of adequately qualified and equipped persons. In short, environmental criminal law enforcement is an expensive activity which will only work effectively if the costs involved are fully met. Enacting a water pollution offence without making adequate financial provision for its enforcement is a self-defeating exercise and yet enforcement has remained largely neglected and underfunded for so long as the prohibition has been in existence, and remains precarious (Kaminski 2016; Marshall 2014; Marshall 2013). In short, a criminalisation approach to water pollution is only as effective as the resources directed towards enforcement (see, generally, Department for Environment, Food and Rural Affairs 2006b and WRc 2006).

Stage two: anticipatory land use controls

The characteristic feature of early industrial pollution control is its reactivity to past pollution incidents of a kind usually arising from effluent discharges from manufacturing industries. To some extent, criminal convictions and sanctions may act as deterrents and fulfil a useful retributive role in punishing acts that are regarded as beyond the boundaries of what is environmentally acceptable. However, they do nothing to rectify the environmental harm that has been caused by pollution. Cost recovery powers, to enable regulatory authorities to recover the reasonable expenses of clean-up and restoration activities, are a useful adjunct to a pollution prosecution to ensure that the polluter pays for the economic cost of the incident.[9] Again, these serve to impose a cost upon the polluter, rather than preventing environmental harm arising in the first place.

The second stage in the progression, recognises that, alongside punishing water polluters, the law should be used in a forward-looking way, and directed towards ways of preventing pollution arising in the first place. Insofar as this involves ensuring that industrial and other kinds of effluent are of a sufficiently high standard to prevent harm to the environment, an ongoing preventative role may be seen to operate through the imposition of environmental permitting requirements. Conditions within permits can be formulated to ensure that the quality of effluent is benign to the aquatic environment, perhaps taking into account special uses that may be made of the receiving water, as where water is used for supply purposes or where a watercourse has a particular ecological sensitivity. Certainly, environmental permitting can be seen to have a preventative function. However, beyond permitting, there are a range of other actions, largely involving anticipatory controls upon land use, that regulate activities that might give rise to pollution, regardless of whether any pollution has actually occurred.

A good illustration of the preventative approach is to be seen in powers that are given to the Environment Secretary to make, so called 'precautionary'[10] regulations against pollution under the Water Resources Act 1991. The power allows for regulations to be made prohibiting a person from having custody or control of any poisonous, noxious or polluting matter unless prescribed works, precautions and other steps have been taken for the purpose of preventing or controlling the entry of the matter into any controlled waters.[11] The statutory provision is worded as an enabling power, so that there is no *duty* upon the Environment Secretary to make any particular regulations, but only a *power* do so where this is thought necessary.

The regulatory power has been exercised to make the Water Resources (Control of Pollution) (Silage, Slurry and Agricultural Fuel Oil) (England) Regulations 2010.[12] These Regulations require persons with custody of silage, livestock slurry or certain fuel oil to carry out works and take precautions and other steps for preventing pollution of controlled waters. Similar, preventative regulations have been made in respect of the design of oil storage tanks with a view to minimising the risk of pollution of water by imposing construction and bunding (containment) requirements.[13] The point to be emphasised is that the criminal offences that are provided under these regulations relate to the unlawful storage of polluting matter. It does not need to be shown that any *actual* water pollution has occurred – hence, the new role of the law is in criminalising a *failure to prevent* pollution occurring, rather than causing any entry of polluting matter into water. Under the second stage, *water pollution* law has been transformed and extended into preventative *land use* law.

Whilst the provisions for 'precautionary' regulations may be seen as *substance-specific* in that they allow controls to be imposed upon the storage of potentially polluting substances, alternative provision is made for making preventative regulations that are *location-specific* in their application. Hence, the Environment Secretary is empowered to make regulations to prohibit or restrict the carrying on *in a particular area* of activities which are likely to result in the pollution of waters, through the designation of a water protection zone. Where this is done, a prohibition or restriction may be imposed upon the carrying on in the designated area of activities specified or described in the order.[14] This is a potentially far-reaching power to control a wide range of land use activities within a designated area where these are seen as hazardous with regard to water pollution. Remarkably, the power to create a water protection zone has only been exercised on one occasion, in relation to the designation of the River Dee Catchment Water

Protection Zone.[15] Nonetheless, the value of the water protection zone approach has been reaffirmed by the creation of further designation powers to allow the designation of zones for the purpose of meeting requirements arising under the EU Water Framework Directive; however, it is understood that no further water protection zones have yet been designated under the new powers.[16]

Another context where restrictions upon land use are imposed for the purpose of preventing water pollution is in respect of nitrate contamination of water as a result of applications of manure and chemical fertiliser to agricultural land. The application of nitrate fertilisers enhances crop growth, but excess nitrate washes off agricultural land or percolates through soil to enter groundwater or watercourses, resulting in enrichment, termed *eutrophication*, which may involve dramatic algal growth with damaging effects on ecosystems in the receiving waters (Boyle 2014). Nitrate contamination of waters is seen as a prominent example of diffuse pollution, where the link between the source of pollution and its impacts cannot be clearly determined. The regulation of this kind of pollution at the point where the nitrate enters a watercourse is not an option. Controls must be imposed on upstream land use activities and, particularly, upon the application of nitrate to agricultural land.

The need for measures to protect water quality from agricultural nitrate contamination has been recognised for some time and national measures were first adopted in area-specific powers allowing the relevant minister to designate areas of agricultural land as a nitrate sensitive areas. This enabled farmers within the area to enter into compensation agreements in respect of changes in agricultural practice that involved a reduction in the application of nitrate to land.[17] The initial national approach towards addressing the problem of diffuse pollution by nitrate was termed a *pilot scheme*, in the sense that it was intended as a relatively short-term measure to ascertain the impacts of changes in farming practice upon water quality. As a means of assessing this, the nitrate sensitive area scheme provided evidence of the effectiveness of controlling nitrate application in improving water quality (Ministry of Agriculture, Fisheries and Food 1993; House of Commons Agriculture Committee 1997; MAFF 1993). However, the national pilot scheme, involving only voluntary participation by farmers in return for compensatory payments, was superseded by a more mandatory system of controls needed to implement the EU Agricultural Nitrates Directive (91/676/EC). Amongst other things, this stipulates limits upon quantities of nitrate that may be applied to agricultural land in nitrate vulnerable zones that are nationally designated for the purpose of implementing the Directive.[18]

Whilst further examples of anticipatory land use controls could be given,[19] precautionary regulations, water protection zones and the designation of areas for nitrate control purposes are sufficient to illustrate the essential elements of this second stage. The reactive, prosecution–after–the-event approach that characterised the first legal inroads into industrial pollution control needs to be *supplemented* (rather than *replaced*) by a secondary approach. This anticipatory land use control approach has taken different forms, but they are all directed towards preventing land use activities that involve a pollution hazard, and are applied prospectively – and, in some instances, irrespective of whether any actual water pollution has occurred. Prevention is certainly better than cure in respect of environmental protection and these illustrations of the second stage show that effective protection of the aquatic environment requires control of land use activities at locations that are often quite remote from the waters that are vulnerable to pollution.

Stage three: strategic water quality management

Whilst the reactive-prohibition approach involves outlawing of unacceptable industrial activities and the anticipatory land use control approach involves regulating certain kinds of failure to prevent pollution, neither of these provide any explicit indication of what state of the aquatic environment the law is seeking to achieve. By contrast, the third stage, termed *strategic water quality management*, involves a purposeful or goal-orientated regulatory approach towards water quality. This approach is characterised by the purposive use of law to achieve specified standards for water quality (Howarth 2006). The important difference is that whilst law may be used to prohibit polluting activities (in stage one) and to require measures to prevent pollution (in stage two), these are negatively formulated aims (as things to be avoided). Strategic water quality management involves using law to realise positively formulated objectives for the quality of the aquatic environment.

Because, strategic water quality management is a more sophisticated kind of legal approach, it tends to come later in the historical progression of water quality law which is being considered. In the UK, the link between pollution incidents and chronic pollution had been recognised for some time, but there was a longstanding national resistance towards any kind of binding statutory specification of the environmental standards that must be met by watercourses and other waters.

Eventually, national legal provision for strategic water quality management was provided under the Water Resources Act 1991, through a three-stage mechanism. First, the Environment Secretary is empowered to prescribe a classification system for the quality of watercourses and other waters according to criteria specified in regulations.[20] Although the basis for making water classification systems under this power was open-ended, in practice regulations were normally made to implement EU water quality directives.[21] The sole exception to this was a set of Regulations which prescribed a system for classifying the general ecological quality of inland freshwaters, though these regulations were never brought into effect and have now been repealed.[22]

The second stage in the process involved the specification of water quality objectives, for the purpose of maintaining and improving the quality of particular waters. This was done by the Environment Secretary serving a notice on the Environment Agency specifying a classification under a water quality classification system and a date by which compliance with the relevant water quality objective had to be met for particular waters. Thereafter, the quality of the waters had to be maintained in compliance with the water quality requirements set out in the relevant classification system.[23]

The third step in the process, was the imposition of a general duty to achieve and maintain the waters in accordance with the specified water quality objective. Hence, it is the duty of the Environment Secretary and the Environment Agency to exercise their powers, under the Water Resources Act 1991 and the Environmental Permitting Regulations 2010, so as to ensure, so far as it is practicable, that the water quality objectives specified for any waters are achieved and maintained at all times.[24]

The upshot of this three-stage process is that the national provisions for strategic water quality management allow legal requirements to be put in place so that powers of regulatory bodies must be used for the purpose of ensuring that waters actually meet a satisfactory quality. However, this approach is more of a theoretical than a practical guarantee of satisfactory water quality. In the first place, it requires that the regulatory bodies actually have sufficient stringent legal powers to ensure that the specified standards are actually met and, in the second place, it depends upon sufficiently demanding water quality standards having been put in place by the Environment Secretary. In addition, overall effectiveness depends upon there being sufficient political will to match the desired level of implementation with the human, technical and financial resources that are needed to realise this.

How the environmental duties might work in practice is not spelt out. In the case of England, the powers of the Environment Agency could be used to impose strict environmental permit conditions upon a discharge to the aquatic environment or, in an extreme case, to prohibit any emission whatsoever. The various preventative mechanisms summarised above might also be used to restrict or prevent land use activities of a kind that might cause water quality to fall below its required water quality standard. However, where unsatisfactory water quality arises from causes that cannot be addressed through Environment Agency regulatory powers, it is difficult to see how a statutory water quality standard could be met and maintained. For this reason, perhaps, the duty upon the Agency is to use its legal powers purposively, as a useful recognition of the need for a strategic approach, but this falls some way short of a duty *actually to realise* a statutory water quality standard.

As it turned out, the diverse issues concerning implementation were never tested in practice. The debate about national standards for water quality was largely overtaken by the need to implement water quality standards that derived from various EU directives. The broadly formulated national powers to impose water quality requirements were only ever exercised to meet regionally formulated legal requirements and the powers in relation to a purely national initiative on water quality were never exercised. Although, the provisions for national water quality standards remain on the statute book, the prospect of them being used for any nationally determined purpose seems remote, given the need to secure good status of waters, in accordance with the EU Water Framework Directive (Art. 4). The chemical, physical and ecological elements of the good status requirement should, in principle, be as useful as any water quality objectives that might have been imposed at national level. The end result, therefore, is that the pursuit of water quality standards as a strategic role in water management has become a regional rather than a national endeavour.

Stage four: post-industrial environmental regulation

The previous three sections have shown the regulatory progression with regard to water pollution and water quality, advancing through three stages: from reactive criminalisation of industrial activities, anticipatory land use controls and strategic water quality management. The paradox of this progression is that adopting increasingly sophisticated approaches to aquatic environmental regulation, at each stage, serves to reveal new kinds of regulatory challenge that lie beyond. Tackling point-source pollution served to put into relief the need for controls on land use and diffuse pollution. Preventative measures concerning land use put into prominence the question of what standard of environmental quality is being sought and the need for a strategic approach to achieve this. Each step in the progression has served to change the perception of the environmental regulatory challenge that needs to be confronted.

Similarly with the fourth stage in the progression, post-industrial circumstances, in which the new focus is placed upon activities that are ever more remote from the waters they affect and only register as a relevant environmental concern when their cumulative impact is taken into account. Moreover, the measures needed to achieve satisfactory water quality need to be integrated into a wider spectrum of environmental legislation in which water protection aspects diminish in significance.

With the decline of the most seriously polluting industries, the emerging perception of the environmental challenge shifts towards the ordinary daily-life activities of persons living in developed countries and the cumulative impact of consumption patterns, transportation decisions and a diverse spectrum of lifestyle choices about goods and services. Few, if any, of these things will be seen by the participants to have any great relevance to the state of the environment, even less the aquatic environment, but cumulatively they form the basis for future environmental concerns. In a post-industrial society, the main focus of pollution has shifted towards the impact of consumption of goods and services. Stages one to three in the progression that has been recounted have not ceased to be relevant, but a further post-industrial integrative stage is needed to address the new perception of the water and environmental protection challenge.

The decline of heavy industries, with the greatest pollution potential, is a gradual phenomenon which may be seen as having some broadly beneficial environmental consequences. Hence, the European Environment Agency's, *Europe's Environment: The Fourth Assessment* (European Environment Agency 2007, p. 106) attributed improvements in water quality to a decline in heavy industry during the industrial recession of the 1990s and noted that, since that recession, there has been a shift towards less-polluting industries. Likewise, the post-2008 economic recession has been identified by the Agency as having a beneficial impact on the EU's total greenhouse emissions trends (European Environment Agency 2011). It might well be observed that the decline in heavy industry in the EU and UK may be matched by industrial revolutions in developing countries to which manufacturing has been transferred. The reduction of industrial pollution in UK may have been achieved by the export of pollution to newly industrialised countries, but the point remains that, from a national perspective at least, there has been a significant decline in pollution attributable to industrial activities (Department for Environment, Food and Rural Affairs 2013 and Environment Agency 2013).

The de-industrialisation process has taken place over several decades and the environmental effects of this transition are ongoing, with the full implications remaining to be seen. However, there are some important indicators of the kinds of environmental problem that are likely become most prominent in England (and other post-industrial countries) and the kinds of measure needed to address future environmental quality concerns, particularly as these arise from the consumption of new kinds of consumer products. For further discussion of this, see Chapter 7, this volume, on Contaminants of emerging concern.

To the extent that unsatisfactory water quality can be attributed to the consumption of consumer products, the only feasible way to address this is through the regulation of those products. Insofar as national law is concerned, legal mechanisms by which hazardous substances or articles can be restricted for environmental protection reasons have been in place for some time, allowing prohibition of the import, use, supply or storage of these things.[25] This is a particularly useful power where the substances or articles concerned are not capable of being effectively regulated by other means such as environmental emission controls. This regulatory power is provided for in national law; however, the practical reality is that consumer goods are traded within a single EU market and national law mechanisms infringing upon the free movement of goods are likely to be subject to intense scrutiny by the European Commission.[26] As a consequence, national law provisions are likely to fall foul of free movement problems, even when apparently motivated by genuine environmental concerns (*Commission v Germany* 2003; Zander 2004, Docherty 2004).

Clearly, the answer to the problem of imposing environmental product controls to protect the aquatic environment lies in a pan-EU approach. Essentially, this approach has now been put in place under the 2006 EU REACH Regulation.[27] Despite being a concerned with environmental protection generally, and not specifically concerned with the water environment, arguably, REACH is the most powerful and far-reaching anti-water-pollution measure ever adopted at either the national or EU level.

Accepting that the REACH Regulation is an EU rather than a national measure, it provides a good example of the kind of product or consumption-orientated approach that is needed to address a post-industrial kind of pollution. Possibly, if a sufficiently precautionary approach is taken towards authorisation, it may serve to address recently discovered kinds of water quality concern, such as the serious challenges raised by endocrine-disrupting substances in the aquatic environment.

However, it must be recognised that every piece of environmental legislation has a cost, even if it not entirely clear what that cost is or upon whom it will fall. The shift from environmental regulation of production to regulation of consumption is especially politically sensitive. Past measures to restrict the worst excesses of industrial pollution, or regulate environmentally hazardous land uses, have gained fairly wide public support. Measures which would have the effect of restricting consumer choices of products and services may well meet with public opposition. Not least, this is because they are perceived to be depriving consumers of something that they have been accustomed to having and, in many instances, they cannot understand why this should be an environmental problem. Yet the ban on free plastic bags in shops seems to have been remarkably trouble-free (BBC News 2015a, 2015b). Environmental law develops within the bounds of political feasibility and there are limitations upon what can be accomplished within that constraint. So far as the future progress of post-industrial pollution is concerned, this is sure to be a major consideration.

Concluding observations

This chapter has provided an account of English water pollution and water quality laws which has condensed a massive amount of legal detail into the minimal content that is needed to support a particular characterisation of the way that laws of this kind evolve. This is the contention that water laws follow an accumulative progression which passes through just four key stages of historical development. The aim has been to relate these phases of legal development to evolutions in thinking about the underlying environmental concerns that these laws are seeking to address. The high level of selectivity in the discussion is fully recognised, but it is hoped that this has been sufficient to meet the overall aims of drawing out interactions between legal and environmental thinking. A bolder contention, which takes the discussion beyond the remit of the chapter, is that the stages of development that have been discerned in English law may also be illustrated in other jurisdictions and serve as reference points to assess the level of development of different national laws.

Notes

1 Environmental Protection Act 1990 s.1, and see a similar definition now provided under Reg.2 Environmental Permitting (England and Wales) Regulations 2010 (SI 2010/675).
2 Under Part IIA Environmental Protection Act 1990, as amended, and the Contaminated Land (England) Regulations 2006 (SI 2006/1380).
3 Public Health Act 1875 ss.27 and 91, and Howarth and McGillivray 2001, p. 23.
4 Regs.12 and 38 Environmental Permitting (England and Wales) Regulations 2010 (SI 2010/675).
5 Environmental Permitting Regulations 2010 ¶ 3(1) of Schedule 21.
6 The Rivers (Prevention of Pollution) Act 1951 s.7 first required the consent of the appropriate river board to be obtained for an emission into the aquatic environment, though earlier examples of this approach can be seen in the Public Health (Drainage of Trade Premises) Act 1937.
7 Under the Water Resources Act 1991, a defence to the principal water pollution offences arose where a discharge was made under and in accordance with various kinds of environmental licences and permits, including 'discharge consents' under s.88(1) of the 1991 Act.
8 Essentially the same result is achieved by Reg.12 Environmental Permitting Regulations 2010, where the key offences are operating a regulated facility or causing or knowingly permitting a water discharges activity, *except under and to the extent authorised by an environmental permit*. In effect, authorisation serves as a defence to what would otherwise be a criminal offence.
9 In relation to the recovery of costs for anti-pollution works and operations undertaken by the Environment Agency with regard to water pollution, see s.161 Water Resources Act 1991, as amended by the Water Resources Act 1991 (Amendment) (England and Wales) Regulations 2009 (SI 2009/3014)

and the Environmental Permitting (England and Wales) Regulations 2010 (SI 2010/675). Powers to recover clean-up and restoration costs are also provided under other legislation, not specifically concerned with water pollution.

10 The term 'precautionary' is confusing here because, in EU environmental law, precaution and prevention are regarded as different ideas. See Art. 191(2) Treaty on the Functioning of the European Union setting out 'precaution' and 'prevention' as the basis for distinct environmental policy principles.

11 Water Resources Act 1991 s.92(1)(a).

12 SI 2010/639, replacing SI 1991/324.

13 Under the Control of Pollution (Oil Storage) Regulations 2001 SI 2001/2954.

14 Under the Water Resources Act 1991 s.93(1),(2).

15 See Water Protection Zone (River Dee Catchment) Designation Order 1999 (SI 1999/915) and Water Protection Zone (River Dee Catchment) (Procedural and Other Provisions) Regulations1999 (SI 1999/916).

16 Water Resources Act 1991 (Amendment) (England and Wales) Regulations 2009 SI 2009/3104.

17 Under Water Resources Act 1991 ss.94 and 95.

18 Implemented by Nitrate Pollution Prevention Regulations 2008 (SI 2008/2349) and Department for Environment Food and Rural Affairs (2009).

19 For example, the regulation of land use to protect species and habitats where an area of water is designated as a Site of Special Scientific interest under s.28 Wildlife and Countryside Act 1981, or where land use may be restricted under the Town and Country Planning legislation in order to prevent contamination of water that may be used for water supply purposes.

20 s.82(1) Water Resources Act 1991.

21 For example, the Surface Waters (Dangerous Substances) (Classification) Regulations 1989 (SI 1989/2286, and later regulations) now repealed by Surface Waters and Water Resources (Miscellaneous Revocations) Regulations SI 2015/524, which listed a number of 'dangerous substances' designated under EU Dangerous Substances Directive 2006/11/EU and gives the concentration of each which should not be exceeded in fresh or marine waters on an annual basis. Similarly, the Bathing Waters (Classification) Regulations 1991 (SI 1991/1597) to implement water quality standards established under the EU Bathing Waters Directive 2006/7/EC establish a system of water quality classification for bathing waters. Now revoked by Bathing Water Regulations SI 2013/1675.

22 The Surface Waters (River Ecosystem) (Classification) Regulations 1994 (SI 1994/1057) revoked by Reg.2(d) Surface Waters and Water Resources (Miscellaneous Revocations) Regulations SI 2015/524.

23 Water Resources Act 1991 s.83(1).

24 Water Resources Act 1991 s.84(1), amended by Sch.26(1) ¶ 8(4)(a) Environmental Permitting (England and Wales) Regulations SI 2010/675.

25 Environmental Protection Act 1990 s.140.

26 See Art. 114(4) Treaty on the Functioning of the European Union on the adoption of stricter national provisions for protection of the environment and De Sadeleer (2014).

27 The REACH Regulation, Reg. 1907/2006 concerning the Registration, Evaluation, Authorisation and Restriction of Chemicals, and see European Commission (2001).

References

All web references were last accessed at 8 April 2016.

Alphacell Ltd. v Woodward [1972] 2 All ER 475.

BBC News (2015a) Plastic bag charge: shoppers in England have to pay 5p, 5 October. Available from: http://www.bbc.co.uk/news/business-34438030

BBC News (2015b) Tesco plastic bag use down 80% since 5p charge, 5 December. Available from: http://www.bbc.co.uk/news/business-35013520

Boyle, S. (2014) The case for regulation of agricultural water pollution. *Environmental Law Review* 16(1), pp. 4–20.

Commission v Germany (2003) Case C-512/99 ECR I-845.

De Sadeleer, N. (2014) *EU Environmental Law and the Internal Market.* Oxford: Oxford University Press.

Department for Environment, Food and Rural Affairs (2006a) Circular 01/2006 *Environmental Protection Act 1990: Part 2A Contaminated Land.* Available from: https://www.gov.uk/government/uploads/system/uploads/attachment_data/file/69309/pb12112-circular01-2006-060817.pdf

Department for Environment, Food and Rural Affairs (2006b) *Review of Enforcement in Environmental Regulation.* London: Defra.

Department for Environment, Food and Rural Affairs (2009) *Protecting Our Water, Soil and Air: A Code of Good Agricultural Practice for Farmers, Growers and Land Managers.* Available from: https://www.gov.uk/government/uploads/system/uploads/attachment_data/file/268691/pb13558-cogap-131223.pdf

Department for Environment, Food and Rural Affairs (2012) *Environmental Protection Act 1990: Part IIA Statutory Guidance.* Available from: https://www.gov.uk/government/uploads/system/uploads/attachment_data/file/223705/pb13735cont-land-guidance.pdf

Department for Environment, Food and Rural Affairs (2013) *Environmental Statistics: Key Facts*, January. Available from: http://data.defra.gov.uk/env/doc/Environmental%20Statistics%20key%20facts%202012.pdf

Department for Environment, Transport and the Regions (2000) Circular, 02/2000, *Environmental Protection Act 1990: Part 2A Contaminated Land.* Available from: https://www.gov.uk/government/uploads/system/uploads/attachment_data/file/223705/pb13735cont-land-guidance.pdf

Docherty, M. (2004) Challenging a Commission refusal to allow the introduction of more stringent national measures. *Environmental Law Review* 6, pp. 120–6.

Empress Car Company (Abertillery) Ltd. v National Rivers Authority [1998] Env LR 396.

Environment Agency v Biffa Waste Services [2006] EWHC 1102 Admin.

Environment Agency (2013) *Pollution Incidents Report September 2013.* Available from: https://www.gov.uk/government/uploads/system/uploads/attachment_data/file/292843/LIT_8547_b70a6b.pdf

Environment Agency (2015) *Enforcement and Sanctions: Guidance.* Available from: https://www.gov.uk/government/uploads/system/uploads/attachment_data/file/468315/LIT_5551.pdf

European Commission (2001) *White Paper on Strategy for a New Chemicals Policy* COM (2001) 88. Available from: http://eur-lex.europa.eu/legal-content/EN/TXT/PDF/?uri=CELEX:52001DC0088&from=EN

European Commission (2013) Decision No 1386/2013/EU on a General Union Environment Action Programme to 2020. *Living Well, Within the Limits of Our Planet.*

European Environment Agency (2007) *Europe's Environment: The Fourth Assessment.* Available from: http://www.eea.europa.eu/publications/state_of_environment_report_2007_1

European Environment Agency (undated) website under *Definitions and Sample Images.* Available from: http://glossary.eea.europa.eu//terminology/sitesearch?term=pollution

European Environment Agency (2011) *Greenhouse Gas Emission Trends and Projections in Europe: Tracking Progress Towards Kyoto and 2020 Targets* EEA Report No 4/2011. Available from: http://www.eea.europa.eu/publications/ghg-trends-and-projections-2011

European Parliament and Council (2000) Directive establishing a framework for community action in the field of water policy (the Water Framework Directive) 2000/60/EC.

Hammerton, D. (1987) The impact of environmental legislation. *Water Pollution Control* 86, pp. 333–44.

House of Commons Agriculture Committee (1997) *Environmentally Sensitive Areas and Other Schemes under the Agri-Environment Regulation.* HC Paper 45-I.

Howarth, W. (1993) Poisonous, noxious or polluting: contrasting approaches to environmental regulation. *Modern Law Review* 56(2), pp. 171–87.

Howarth, W. (2006) The progression towards ecological quality standards. *Journal of Environmental Law* 18(1), pp. 3–35.

Howarth, W. and McGillivray, D. (2001) *Water Pollution and Water Quality Law.* Shaws. Available from: http://eur-lex.europa.eu/legal-content/EN/TXT/PDF/?uri=CELEX:32013D1386&from=EN.

International Network for Environmental Compliance and Enforcement (2009) *Principles of Environmental Compliance and Enforcement Handbook.* Available from: http://www.inece.org/principles/PrinciplesHandbook_23sept09.pdf

Kaminski, I. (2016) Unions call for EA staffing levels to be maintained. *ENDS Report* News, 6 January.

MAFF (Ministry of Agriculture, Fisheries and Food) (1993) *Pilot Nitrate Sensitive Areas Scheme: Report on the First Three Years.*

Marshall, A. (2013) Environment Agency to cut 15% of staff within the year. *ENDS Report*, Analysis, 29 October.

Marshall, A. (2014) Environment Agency cuts: surviving the surgeon's knife. *ENDS Report,* Analysis, 3 January.

Richardson, G., Ogus, A. and Burrows, P. (1982) *Policing Pollution: A Study of Regulation and Enforcement.* Oxford: Clarendon.

Sentencing Council (2014) *Environmental Offences Definitive Guideline.* Available from: https://www.sentencingcouncil.org.uk/wp-content/uploads/Final_Environmental_Offences_Definitive_Guideline_web1.pdf

Turing, H.D. (1952) *River Pollution.* London: Edward Arnold.

Wikipedia (undated) Under *Pollution.* Available from: https://en.wikipedia.org/wiki/Pollution

WRc (2006) *The Effectiveness of Enforcement of Environmental Legislation.* Available from: http://citeseerx.ist.psu.edu/viewdoc/download;jsessionid=6786448F5E9765ED5405E26EE181E285?doi=10.1.1.103.9053&rep=rep1&type=pdf

Zander, J. (2004) The 'Green Guarantee' in the EC Treaty: two recent cases. *Journal of Environmental Law* 16(1), pp. 65–79.

7

CONTAMINANTS OF EMERGING CONCERN

Sarah Hendry

Introduction

This chapter will examine the policy and law relevant to the management of what are often described as 'emerging pollutants', or 'emerging contaminants'. However, although some are novel, often it is the effects of these substances which are emerging. The website of the US Environment Protection Agency (USEPA) states: 'These are often generally referred to as "contaminants of emerging concern" (CECs) because the risk to human health and the environment associated with their presence, frequency of occurrence, or source may not be known' (USEPA, undated (a)). Sauvé and Desroisiers (2104) agree that 'contaminants of emerging concern' is a more appropriate description, and that is the terminology used herein, whilst recognizing that 'emerging pollutants', though less accurate, is widely used.

These substances can be defined and classified in different ways. Geissen *et al.* (2015, p. 58) include in their definition synthetic and naturally occurring chemicals, which may not be monitored but which may impact on human health or ecology. They note that:

> [i]n some cases, release of emerging pollutants to the environment has likely occurred for a long time, but may not have been recognized until new detection methods were developed. In other cases, synthesis of new chemicals or changes in use and disposal of existing chemicals can create new sources of emerging pollutants.

This chapter will begin with an explanation of some of the main categories of pollutants, the uses to which they are put and the potential consequences for the health of humans and aquatic ecosystems. It will then examine the approaches to their management, related to each of these consequences; including broad water quality approaches and specific rules relevant to both drinking water and wastewater treatment.

Definitions and types of emerging pollutants

CECs may be used in a wide variety of products. Balderacchi *et al.* (2014) include antiseptics, antioxidants, corrosion inhibitors, flame retardants, gas propellants, plasticizers, pharmaceuticals (prescribed, non-prescribed and drugs of abuse), solvents, stimulants (such as caffeine) and surfactants; and a range of 'personal care products' including fragrances and sunscreens. Pesticides and biocides are an important group with obvious potential toxicity. Some are combustion byproducts. Some are naturally occurring and others synthetic, and they include organic and inorganic compounds (Geissen *et al.* 2015; Deblonde *et al.* 2011). Individual products may include a number of constituent chemicals, and the effects of combinations of these chemicals, and their effects at low dosage and in combination, may not be well understood. Nanoparticles and microbeads are also defined as emerging pollutants, and many of these are a component (sometimes *the* component) of concern; for example, in personal care products. Further, they may behave differently, and have different environmental effects, than the same compounds or elements at larger scale (which may themselves be toxic; Mar *et al.* 2013).

There is an obvious relationship with the management of hazardous substances generally, not just in the water environment. Some hazardous substances are long-established, have well-recognized effects either individually or in combination and are managed accordingly in environmental and health and safety laws, but at lower doses or in combination, knowledge of effects may still be emerging. For example whilst the harmful effects of some metals, such as mercury or lead, have long been recognized, the impact of trace quantities of metals is now also being recognized. Heavy metals are often addressed in the same literature as CECs, and managed under the same monitoring and control regimes.

What is clear is the scale of the problem. Von der Ohe *et al.* (2011) state that there are 14 million chemicals in existence, and 100,000 are produced on an 'industrial scale'. Schriks *et al.* (2010) also note that more than 100,000 chemicals are registered in Europe; and that 300 million tonnes of synthetic compounds are discharged annually. The NORMAN network is a network of reference laboratories, research centres and related organizations for monitoring of 'emerging environmental substances' and was established by the European Union (EU) in 2005. In 2015, its list of individual substances of concern has 1,036 entries (NORMAN list 2015).

Impacts and consequences

Chemicals enter the environment in different ways. They may be emitted to the atmosphere – for example, during combustion, evaporation or in propellants. They may end up in landfill, which may be better or more poorly regulated. They may be discharged into wastewater systems, through industrial processes or domestic wastewater (including household cleaners, personal care products and unused medicines), or may be metabolized by humans or animals (pharmaceuticals). Human wastes may (or may not) enter a wastewater system and may (or may not) be subject to treatment; animal wastes may enter watercourses without treatment (or may be diverted to a wastewater treatment plant). Discharges to air and to land may end up in the water environment, where, depending on solubility, the compounds may be taken up by aquatic lifeforms or end up in sediments. If in sediments, they may remain for considerable periods of time, but may find new pathways to new receptors if the sediments are disturbed – for example, by dredging or otherwise affecting the structure of a river. The chapter will look only briefly at the management of air pollution and solid waste, but will focus on management of contaminants in the water environment.

In cases where severe negative consequences on human health (and ecosystems) have become manifestly obvious, such as the use of certain early generation pesticides, there have been outright bans on specific chemicals on human health grounds. Where these substances are persistent and bio-accumulative, they may continue to be detected in human and animal tissue over prolonged periods of time. Many other substances which are recognized as toxic, persistent and bio-accumulative continue to be manufactured and used for different purposes around the globe, with (variable) management frameworks intended to ensure the safety both of those using them directly, and those who will be exposed to them once released to the environment.

Schriks *et al.* (2010) investigated 50 'chemicals of emerging concern', of which ten had established values for drinking water; they derived provisional guideline values for the others and then assessed the prevalence of these in the waters of the Meusel and the Rhine, concluding that there was a 'significant safety margin' between the guideline values and the actual prevalence. However, the determination of safe doses tends to be assessed in relation to individual compounds rather than combinations, in relation to adults rather than children and to identify short-term acute effects rather than possible chronic effects of long-term low-dose exposure. Pal *et al.* (2010) identified dozens of studies of the occurrence of various pharmaceuticals in the US and the EU between 2006 and 2009, in the water environment, in wastewater and in terms of ecotoxic effect, and noted that the effects of many compounds are poorly understood, especially at low levels and in combination.

A wide-ranging review of 'hazardous chemicals of concern' in the freshwater environment was undertaken by the European Environment Agency (EEA) in 2011 (European Environment Agency 2011). This review noted that many chemicals can affect aquatic life and human health, and that evidence is emerging as to the additive and cumulative effects. In 2013 the EEA, with the EU's Joint Research Council (European Environment Agency/Joint Research Council 2013), assessed the relationship between environment and human health, including chapters on chemical pollution, nanotechnology and the water environment. One conclusion was that such complex challenges require systematic policy solutions. This chapter will attempt to assess whether such solutions are yet forthcoming.

Many CECs may also be endocrine disrupters. Endocrine disrupters interfere with the actions of hormones and, in the last decade, much more information has emerged as to their potential impacts on aquatic life and on human health. Most obviously, contraceptives are endocrine disrupters, but so too are many other compounds, including a wide range of pharmaceuticals as well as numerous other substances and products. In the US, the EPA's first list of substances for screening as potential endocrine disrupters in 2006 had 67 compounds, and the second list in 2013 contained 109. These include pesticides as well as pharmaceuticals (USEPA undated (b)). In 2012 the World Health Organization (WHO) and the United Nations Environment Programme (UNEP) issued their second report on endocrine disrupters (Bergman *et al.* 2012) and found that some 800 chemicals were 'known or suspected' to interfere with hormone receptors, synthesis or conversion. This report also notes the increase of endocrine disorders in both wildlife and humans. 'Intersex' fish are well documented (see, for example, Jobling and Owen 2013). The WHO/UNEP report identifies the following human health impacts as being on the rise and linked to endocrine disrupters: low sperm counts; genital malformations; adverse pregnancy outcomes; neuro-behavioural disorders; endocrine-related cancers; obesity; and type 2 diabetes.

International frameworks for chemical management

The sound management of chemicals is relevant, *inter alia,* to environmental management frameworks and to sustainable development, as well as human health. It was mandated in Agenda 21 (UN 1992, chapter 19) and the Johannesburg Plan of Implementation (UN 2002, para. 23). The sound management of chemicals is also noted in the new sustainable development goals (SDGs) (UN General Assembly 2015, para. 34). The SDGs make specific provision in Goal 3 ('healthy lives and wellbeing') but also in the 'water' goal (Goal 6) and Goal 12 ('sustainable consumption and production').

There are numerous initiatives, and relevant international law, at global and regional levels. UNEP is responsible for three related global 'Chemicals Conventions' – the Basel Convention on Transboundary Movements of Hazardous Wastes (UNEP 1989); the Rotterdam Convention on the Prior Informed Consent Procedure for Certain Hazardous Chemicals and Pesticides in International Trade (UNEP 1998); and the Stockholm Convention on Persistent Organic Pollutants (UNEP 2001). At regional level, for example, the UN Economic Commission for Europe (UN/ECE) manages the Geneva Convention on Long Range Transboundary Air Pollutants (UN/ECE 1979) and its protocols, including one on Persistent Organic Pollutants. Not every chemical managed under these Conventions will still be of 'emerging concern' today; the Geneva Convention was agreed as a response to better understanding of the impacts of oxides of sulphur in causing acid rain in the 1960s and 1970s. But many of substances of emerging concern will also be hazardous substances to which these broad frameworks are relevant. Further, both sets of conventions regularly review their protocols, and listed substances, to address new concerns identified by the science literature. Health and safety legislation is also relevant – for example, the International Labour Organization's Conventions on Safety in the Use of Chemicals at Work (ILO 1990) and on the Prevention of Major Industrial Accidents (ILO 1993).

Similarly, there are numerous policy initiatives at different levels. The Strategic Approach to International Chemicals Management (SAICM) was adopted by UNEP's Governing Council in 2006 in Dubai, with the agreement of 140 countries. SAICM has adopted resolutions on *inter alia*, endocrine disruptors, nanoparticles and chemicals in products, and is currently considering the impacts of pharmaceuticals (SAICM/UNEP/WHO, undated). The UN/ECE has developed a Globally Harmonized System of Classification and Labelling of Chemicals (UN/ECE 2003) which is open to all states who wish to participate; and states were encouraged to do so by the Johannesburg Declaration (UN 2002, para. 23). Regular revisions to this system again enable the management of new compounds, or newly recognized effects.

Clearly, it is not possible to address every aspect of the management of chemicals generally in this chapter. But it is worth noting that air pollutants fall to land, and contaminate water; and poor management of solid waste also causes water pollution in different ways, most obviously from leachate from landfills or from historically contaminated sites. The problem is multi-faceted, but the chapter will focus on the management of pollutants directly into the water environment; the management of drinking water quality; and the management of wastewater; and will make particular reference to the EU. This is partly because the EU is a major manufacturer of chemicals, and partly because of a highly developed suite of water and environmental laws. The chapter will also draw some comparisons with law in the (Federal) US.

The EU and chemical management

In 2009, the EU had one third of the global chemicals market, with 29,000 manufacturers and 1.2 million employees (European Commission 2009a). The REACH Regulation (European Commission 2006a) on the Registration, Evaluation, Authorisation and Restriction of Chemicals, established a European Chemical Agency and operates on the principle of 'no data no market'. Producers and importers of 1 tonne or more of chemicals must register them with the European Chemical Agency. There is an obligation to share data; industry should manage risk by providing information on the effects of constituent chemicals, and there is a recognition that there is insufficient information. Some chemicals are especially problematic ('substances of very high concern') and these should be progressively phased out as substitutes become available. In 2013, more than 9,000 registration dossiers had been submitted under REACH (European Chemicals Agency, undated). The EU also has a Regulation on the Classification, Packaging and Labelling of Substances and Mixtures (European Commission 2008), which in turn implements the UN/ECE's Classification System. As these are Regulations, they apply directly in Member States without further transposition into domestic law. The EU is also part of the SAICM.

The EU has extensive legislation relating to both industrial air pollution (for example, the Industrial Emissions Directive, European Parliament and Council 2010) and to waste management (for example, Framework Directive on Waste, European Parliament and Council 2008b; Landfill Directive, European Parliament and Council 1999). These may implement the international and regional instruments noted above; made in the form of Directives, they subsequently need to be transposed into domestic law by Member States. There is a European Pollutants Release and Transfer Register (European Commission 2006b) covering both industrial emissions and waste. The Industrial Emissions Directive applies to emissions to all environmental media and a wide range of industrial activities and processes.

Water quality management in the EU

The Water Framework Directive (WFD) (European Parliament and Council 2000) sets a framework for water policy in the EU's Member States. It requires a system of river basin planning, managing surface water and groundwater together and taking a 'combined approach' to point source and diffuse pollution, utilizing both environmental quality (ambient) standards and emission limit values. It has an overall objective of 'good' ecological status for surface waters, which entails the management of chemical quality, but also a biological assessment, and the management of the hydrology and morphology of the river. The ecological status of surface waters is assessed by a complex set of measures of different types, and, if deficient, should be actively improved by Member States to achieve 'good', unless a series of exemptions and extensions apply (Art. 4, and see also Chapter 4, this volume). States should establish a programme of measures in their river basin management plans to improve the status of waterbodies that are not yet 'good' (Art. 11). There is an explicit focus on supporting aquatic life under the WFD and the biological assessment looks at the diversity distribution and age of fish populations. The WFD, then, is the overarching mechanism for addressing the wider impacts of CECs on the aquatic environment.

The WFD works with other relevant legislation, mainly but not exclusively water legislation, some of which is especially important to CECs. Some of these were already in existence and some have been developed since the WFD. Prior legislation includes the Drinking Water Quality Directive (European Parliament and Council 1998) and the Urban Wastewater Treatment Directive (European Council 1991), each of which will be considered separately below. The WFD requires states to map 'protected areas', including drinking water abstraction points, with the specific goal of minimizing subsequent treatment (Art. 7). The EEA specifically notes drinking water protection zones as a way of minimizing the effects of harmful chemicals on human health (European Environment Agency 2011).

Pollutants in groundwater

The Groundwater Directive (GWD) (European Parliament and Council 2006) was made under the WFD (Art. 17) and sets EU-wide quality standards for nitrates and pesticides (Annex I). The former is derived from EU legislation on nitrates (European Commission 1991a) and the latter from pesticides legislation, addressed below. Member States are also required to identify substances in groundwater of especial concern in that state (or at river basin level), for which that state should set its own 'threshold values' (Art. 3). A list of pollutants for which Member States were to consider establishing threshold values is provided in the Directive (Annex II), including metals and two synthetic substances (Trichloroethylene, PCE, and Tetrachloroethylene, TCE, widely used as solvents). The Annexes are subject to revision – in 2013 and every six years (Art. 10) – and an EU FP7 project, GENESIS, looked *inter alia* at emerging pollutants to make recommendations over these revisions. Balderacchi *et al.* (2014) recommended that the two synthetics in Annex II should be moved to Annex I and have EU-wide quality standards established, as these two substances are 'excellent indicators of the groundwater pollution by multi-source diffuse-type urban pollution'. These pollutants especially derive from current and historic industrial sources and poorly regulated waste management facilities, and may find their way into ground (and surface) water by leaching, by deposition from air or through surface water and road drainage systems. Balderacchi *et al.* (2013) also noted that pharmaceuticals and personal care products (and caffeine, which they describe as 'ubiquitous') are indicators of contamination from household waste, either from wastewater, or via the use of sewage sludge on land; whilst benzene and MTBE are indicators of fuel and pollution from vehicles. They noted that, although two synthetics are identified in Annex II, Member States are setting threshold values for 62.

Pollutants in surface waters

The Priority Substances Directive (PSD) (European Parliament and Council 2008a, 2013) is also made under the WFD (Art. 16) and creates EU-wide quality standards for surface waters for a list of (now, in the 2013 revision) 45 substances and groups of substances. Their use should be progressively reduced (WFD Art. 17). These include 21 priority hazardous substances, the discharge of which should be ceased or phased out. The list of substances is also inserted into Annex X of the WFD. The PSD sets maximum allowable concentrations (or sometimes, annual averages) for background concentrations in water, or sometimes, in biota or sediment. The substances in the PSD include pesticides, solvents and metals. In general, these substances are likely to be toxic, persistent and bio-accumulative. Treated as priority substances, and included in the list, are a small subset of pollutants for which the EU had already set standards, including mercury and cadmium. In addition, Member States may designate pollutants of especial concern nationally – 'specific pollutants' – and manage these with the PSD list. As well as human health, the chemical standards in place in the PSD are intended to safeguard aquatic ecosystems.

The PSD is the obvious vehicle for managing the risks to the water environment of emerging pollutants. The Commission was required to review the list by 2013 and periodically thereafter (WFD Art. 16, PSD Art. 8) and these revisions were specifically intended to allow identification of chemicals that should be placed on that list. In 2011, von der Ohe *et al.* published an analysis of 500 chemicals that could potentially be added to the 33 in the original 2008 list or be designated as specific pollutants by Member States; and proposed new methods for risk assessment to prioritize actions.

Although the 2013 revisions added several substances to the list, several others that were proposed were instead placed on a 'watch list', to be further monitored (Art. 8, PSD 2013), with the aim of developing a 'strategic approach' by 2015, and measures by 2017. The substances on the 'watch list' include the main constituent of many contraceptives and a commonly used anti-inflammatory. Like the PSD List, these would be categorized as CECs. The proposal caused much opposition by Member States (Council of the European Union 2012) and the decision not to place at least some of these substances on the List was in part because of the cost of requiring their removal at wastewater treatment plants (EEA/JRC 2012, ch. 6).

Drinking water quality

The purpose of the Drinking Water Quality (DWQ) Directive, and similar instruments in other jurisdictions, is to safeguard human health. The Directive establishes a list of mandatory parameters for DWQ for all but the smallest supplies (less than 10m^3/day or serving less than 50 people). In turn, these reflect the guidelines published by the World Health Organization (WHO 2004). Provision of water of the acceptable quality normally requires treatment, though there is (not just in the EU) recognition that protection of drinking water catchments is a way of reducing, if not eliminating, treatment costs. The WHO suggests water safety planning as a holistic approach from source to tap, including source protection; the WFD requires source protection; in the US, the New York scheme for the protection of the Catskills is one of the oldest, and perhaps the most famous example (Appleton 2002). This approach will also contribute to the safeguarding of aquatic ecosystems, and other ecosystem services.

The WHO guidance is extensive; both it and the DWQ Directive already establish safe levels for some individual substances that may be of 'emerging concern'. However, the concerns raised at the beginning of this chapter, that the additive and cumulative effects of some substances might manifest at much lower levels, remains valid. In addition, many substances are not routinely monitored in drinking water, or the tests available will not detect the level of substance that is present. Although one would expect – and it is usually the case – that standards for treated drinking water would be tighter than standards for the water environment, Balderacchi *et al.* (2013) note that, in Canada, standards for some substances in the freshwater environment, designed to protect aquatic health, are lower than the corresponding drinking water standards in the EU. The science continues to develop, and to drive policy and law, but the timings are uncertain and the feedback and feedforward loops are imperfect.

Wastewater treatment

Insofar as the substances in question are found in, for example, pharmaceuticals, personal care products, or cleaning products, then in most of the developed world, their residues will find their way into a domestic wastewater system, which may be subject to various levels of treatment, in plant ranging from large municipal processes to septic tanks and other small onsite treatments. In much of the developing world, even if there is a sewerage system to remove wastewater, as much as 90 per cent of domestic wastewater may not be treated but simply discharged to a watercourse (UNEP 2010). Alternatively, domestic human waste may be managed by non-waterborne sanitation, which is effectively a solid waste problem that may still result in pollution of water, depending on the disposal arrangements. Industrial wastewater is likely to be subject to a trade effluent regime and, increasingly, businesses are looking to be more water (and wastewater) efficient, treating their own effluent either as part of their water supply demand management, or to reduce the costs of the trade effluent consent, or both. Nonetheless, in the developing world an estimated 70 per cent of industrial wastewater also goes untreated (UNEP 2010). Yet wastewater is a valuable resource that should be not just treated but also recaptured and reused, to close loops in the water cycle as well as protect human and ecosystem health (for some recent valuation work, see UNEP 2015).

Wastewater contains valuable nutrients and can be a source of energy, but also pathogens, metals and other contaminants, including substances of emerging concern as well as substances where the concerns are well established. The extent to which any group of contaminants is removed will vary, depending on the levels of treatment. It is expected that states will (at least) have a basic set of chemical standards to protect both human health and aquatic life (Helmer and

Hespanol 1997). These may be ambient standards for receiving waters, as in the EU's PSD, or emission standards for plant, or both, and would be applied through some sort of permit system. In many jurisdictions these apply to both municipal wastewater treatment plant and to any industry pre-treating its effluent and discharging directly under an individual permit.

In the EU, there is specific legislation, the Urban Waste Water Treatment Directive (UWWTD) (European Council 1991) requiring a norm of 'secondary treatment' as defined and working with the WFD. The UWWTD has entailed significant capital investment by Member States. Essentially, it requires collection systems and treatment at various levels, for biodegradeable wastewater (human waste and also, for example, food waste). Collection systems are not required where this would be excessively costly or would produce no environmental benefit (Art. 3), but the same level of environmental protection must be provided (for example, by individual systems). There was a staged system of implementation, beginning with the largest 'agglomerations' by population equivalent. Small communities (or industries) with population equivalent of less than 2,000 do not need to implement the UWWTD but instead must have 'appropriate treatment' sufficient to meet all other relevant EU quality standards (Art. 7). Larger communities discharging to coastal or estuarine waters may be designated as 'less sensitive' and be subject to only 'primary treatment' (Art. 6).

The usual level of treatment required is 'secondary treatment', as defined (biological treatment by digestion process). Annex I sets standards for three parameters: biochemical oxygen demand, chemical oxygen demand and total suspended solids. Where the discharge is made into 'sensitive' waters (subject to eutrophication), additional 'more stringent' (tertiary) treatment may be required, such as ultraviolet treatment, and standards are also set for total phosphorous and total nitrogen. Both sludge and treated wastewater should be reused 'wherever appropriate' (Arts. 14, 12). As evidence builds around the emerging effects of pollutants, whether in combination or at low levels, and for both human health and aquatic ecology, the implications for both reuse of water and use of sludge also grow.

If the PSD was to be amended to include more pharmaceuticals or personal care products, or to establish more stringent limits, then this would have implications for the nature and extent of wastewater treatment. During the discussions on the revision of the PSD, in a commentary piece in *Nature*, Owen and Jobling (2012) noted that the cost of introducing appropriate technology to strip out ethinyl estradiol (a commonly used constituent in the contraceptive pill) completely from wastewater might amount to £30bn for England and Wales, and called for a public debate. Their focus was on aquatic ecology; they noted that fish species had been seen to 'collapse' in a Canadian lake with the introduction of the relevant hormone at 'vanishingly small levels'. The debate did not happen – or at least, only amongst the few: the chemicals industry, water services sector, environmental regulators and policymakers. But Owen and Jobling have also suggested that, given the immense difficulties with establishing cause and effect, and the huge numbers of endocrine-disrupting synthetic chemicals in the environment, it might be appropriate to use the well-documented impacts on fish (and the precautionary principle regarding human health) as an indicator and a driver to take action.

The specific technology giving rise to the £30 billion figure was the use of granular activated carbon filters. These are sometimes used in drinking water treatment plant, and are one of the most expensive forms of drinking water treatment. They would be used, for example, where there is a risk of contamination by certain pesticides. In a wastewater context, their use would require both capital and maintenance expenditure well in excess of current costs. However, in a more detailed analysis for the EEA, Jobling and Owen (2013) also examined other novel or additional treatments for wastewater, and other relevant pollutants. They found that good results could also be obtained from other 'tertiary' treatments, including ozonation and also sand filters.

These might have additional capital costs, but the operational and maintenance costs are not as high as granular activated carbon filters. They found that the extent of dilution is important, and therefore relatively densely populated countries with small river systems, such as England, were particularly at risk. They also noted that in many other European countries, unlike the UK, tertiary treatment is much more common in municipal works. Their research was primarily focused on one oestrogen, which is particularly difficult to degrade and particularly problematic for fish; so they also considered the possibility of the development of a different product, with different active ingredients, and the possibility of product control as a solution will be returned to further below. However, more effective treatment of municipal wastewater would address a wide range of pharmaceuticals, personal care products, other household products and also some industrial and agricultural contaminants, insofar as these are going through such treatment plant. Some of the health effects of endocrine disruptors have recently been confirmed (EurActiv 2016a).

Other studies carrying out field work have also suggested that tertiary treatments (ozonation, but also advanced oxidation, filtration and activated carbon) produced far more effective removal of a range of problematic products and substances than secondary treatment alone. Schaar *et al.* (2010) investigated the effect on 29 substances (pharmaceuticals, personal care products and commonly occurring endocrine disruptors) of an additional ozonation process subsequent to biological treatment and to denitrification. They found that a retention time in excess of ten days (during the biological process) was helpful, but that subsequent ozonation produced much better results. Rosal *et al.* (2010) found similarly good results for ozonation in a study of 70 substances at a Spanish treatment works.

Much work is also underway on the management of nanoparticles in wastewater treatment – which may interfere with, or simply not be treated by, conventional treatment processes (Yang *et al.* 2013; Mar *et al.* 2013). As 'nanoparticles' describes size, rather than indicating function or effect, it could be noted that whilst many of these are a component (sometimes the component) of concern in, for example, a personal care product, equally nanotechnology is itself being used in developing treatment options.

Whichever technique is used, there will be additional – potentially substantial – costs for water services suppliers and, ultimately, the public served by these systems. In the EU, the costs of implementing the UWWTD as it currently stands were high for the then 15 EU Member States following its introduction in 1991, and are an ongoing problem for the 13 new entrants since (European Commission 2013a). Although the UWWTD as such is probably unique, the need to have appropriate treatment of wastewater, that protects both human and aquatic health, is common to all countries, whatever their stage of development. Meantime, in any jurisdiction, very small, onsite systems such as septic tanks, which are often privately owned and maintained, will offer only biological treatment; and wastewater that finds its way into watercourses through surface water drainage, or by direct runoff, will not be treated at all. These last two routes include urban diffuse pollution, containing household and industrial chemicals and combustion byproducts, especially from traffic, and agricultural runoff.

Agricultural pollution

Agricultural pollution is a large topic, including the management of fertilizers, pesticides and veterinary pharmaceuticals, as well as management of soil (to address erosion, salinity, carbon management). Forestry, and other rural land uses, may also deposit fertilizers or impact on soil quality or carbon deposits, but in terms of 'emerging' concerns, pesticides and pharmaceuticals are most relevant. Similarly, each of these may also be used in urban and domestic contexts, but at scale, agricultural use is more relevant here. It has been estimated that more than 5,000 tonnes of antibiotics and 5 tonnes of hormones are employed in the European meat production (Balderacchi *et al.* 2014); Schriks *et al.* (2010) estimate that 140 million tonnes of fertilizer and 'several million' tonnes of pesticides are applied to land each year. Large pig and poultry production in the EU is covered by the Industrial Emissions Directive which should entail the proper management of all wastes from these activities, but smaller-scale production or extensively reared livestock may impact directly on watercourses, potentially causing pollution by nutrients and pathogens as well as metabolized pharmaceuticals.

In terms of pesticides, which may include herbicides, biocides, fungicides and insecticides, some 400 active substances are in use in Europe (Balderacchi *et al.* 2013), with 39 compounds being monitored at national level under the GWD, including 16 which are no longer sold, but which are persistent, especially in sediment or groundwater, and bio-accumulative. In the EU the DWQ Directive sets a maximum limit of 0.1 ug/L for (the active ingredient of) any individual pesticide, and a maximum in total of 0.5 ug/L. The same limit is found in the GWD Annex I, as an EU-wide standard. There have been restrictions on the use of these products for some time; the current Regulation (European Commission 2009b), along with implementing Regulations, enables the harmonized assessment and where necessary restriction of these products, whilst a separate Directive addresses their safe use (European Parliament and Council 2009). Very recently, the European Parliament decided to reapprove glysophate, despite concerns over its carcinogenic properties, but for seven years instead of the customary 15 (EurActiv 2016b).

New generation pesticides such as neo-nicotinoids are less persistent but bring different and novel problems, especially concerns on effects on bees and other pollinators. These are being scrutinized in many countries; the EU has brought in some restrictions (European Commission 2013b), and their use has been challenged successfully in the US Supreme Court (*Pollinator Stewardship Council v USEPA* 2015). Court action was also raised in Ontario, Canada (CBC News 2014), in both cases by associations of bee-keepers. The EU (specifically, the European Food Safety Agency) has published new guidance (EFSA 2014) and an Opinion on the science (EFSA 2012).

'Sulfoxaflor' was the specific neonicotinoid subjected to the US challenge. The US Court held that the EPA's studies, on which consent for sulfoxaflor was based, were flawed; and that it needed to collect more data. It is troubling to note that the EU restrictions apply to three other specific neonicotinoids (clothianidin, thiamethoxam and imidacloprid), which are still permitted in the US; yet the EU has recently authorized sulfoxaflor, finding it less likely to cause harm.

These rules apply to 'plant protection products' – that is, for agriculture and similar uses; other biocides, for example insecticides, anti-fouling paint or disinfectants, are managed under the REACH system. In an agricultural context, especially, all states should have domestic legislation and policy, such as best practice guidance, to safeguard workers as well as consumers. In their work on prioritization and risk assessment, von der Ohe *et al.* (2011) note that 74 per cent of the substances they identified as 'high' or 'very high' risk were pesticides; that pesticides cause contamination despite the system for pre-market approval; and that new active ingredients for pesticides are introduced on the market faster than monitoring programmes can be developed.

Regulatory frameworks in the US

An analysis of the regulatory frameworks in the US specific to pharmaceuticals (Eckstein 2015) looked *inter alia* at the principal Federal legislation, as well as some state rules. As with this chapter, he also reviewed a wide range of scientific literature identifying the problem and the weaknesses of the current solutions. He found that there were a wide variety of approaches and standards, but no systematic management of most pharmaceuticals. He also examined initiatives to promote return of unused drugs to pharmacies and other suppliers, which can reduce the volume of discharge to sewers, but this is only a small part of the problem.

Unsurprisingly, US legislation creates control regimes performing similar functions to those applicable in the EU. The Resource Conservation and Recovery Act (US Congress 1976) manages hazardous substances and waste, with a 'cradle-to-grave' approach, but excludes wastewater which is covered by the Clean Water Act (US Congress 1972). Eckstein notes the difficulties with hazard assessment of the large numbers of active ingredients in pharmaceuticals (and other hazardous substances covered by the Act). Other Federal US rules also relevant to the broad topic would include the Clean Air Act (US Congress 1970), managing industrial air pollution, and the Insecticide, Fungicide, and Rodenticide Act (US Congress 1996), managing pesticides. Of course, all of the principal Acts from the 1970s have been much amended, and are supplemented and implemented by regulation, guidance and state legislation.

The Clean Water Act manages the use and quality of freshwaters, using similar tools and mechanisms to the EU WFD and national rules. States should designate quality standards, and allowable water uses, within their jurisdiction, and discharges are regulated by the EPA or by state authorities. The system establishes total maximum daily loads for specific pollutants and requires permits for discharges. However, the Clean Water Act does not set standards for most pharmaceuticals and the permitting system does not address these.

The Safe Drinking Water Act establishes limits for listed substances similar to the EU DWQ Directive. The subsidiary regulations establishing the list can also mandate specific treatment technologies. Eckstein notes that the EPA can develop a 'candidate list' of substances not yet subject to a standard, and that this was recently done, with 104 substances. However, although 287 pharmaceuticals were identified for possible inclusion, all but one were subsequently removed. This seems very similar to the discussions on the revision of the EU PSD list. The only pharmaceutical included was nitro-glycerine, placed on the list for its explosive, rather than medicinal, uses.

Conclusions

There is little evidence in the legal literature of any state or jurisdiction (in the EU, the US or elsewhere) making comprehensive provision for the management of pharmaceuticals or other CECs. As discussed above, some may be 'caught' by general controls on discharges to fresh or marine waters; or disposal of solid wastes (not just sewage sludge) to landfill; or management of industrial air pollution. They may be managed by rules on wastewater treatment, or identified in drinking water standards. But, despite the science evidence of impacts on both human health and aquatic ecosystems, there is little policy response as to how to manage cumulative effects or effects at very low doses. In many cases the availability of appropriate monitoring techniques is limited, and the assessment of risks and hazards struggles to keep pace with the rapid development of new compounds and products.

Usually, as with all waste management issues, tackling the problem at source is the best approach. Yet, although citizens may be concerned about substances and products in the environment or impacting on health, source management is particularly difficult for pharmaceuticals or personal care products. Any suggestion of limiting either is likely to be very challenging for policymakers. Attempts to challenge and restrict the use of pesticides have also proved very difficult. The scientific 'burden of proof' is one issue, and the precautionary principle seems sorely neglected.

From the research surveyed in this chapter, one approach would be to introduce tertiary treatment for wastewater. Yet this would be very costly, and would not address surface water runoff (urban or rural), or wastewater that is not treated, or emitted only to small onsite systems. Other possibilities are to work with manufacturers, and consumers, to identify new products which are less problematic – yet the constant introduction of new products into the market is part of the problem. In some ways, the problem seems insurmountable and there are no easy solutions. Meantime, the public debate still needs to happen. Perhaps a combination of high-profile environmental campaigns and increased concerns over human health will keep the issue high enough on the public agenda to maintain the interest of policymakers, as well as scientists.

References

Appleton, A. (2002) *How New York City Used an Ecosystem Services Strategy Carried out Through an Urban-Rural Partnership to Preserve the Pristine Quality of Its Drinking Water and Save Billions of Dollars* [online]. Available from: http://www.ourwatercommons.org/sites/default/files/New-York-preserving-the-pristine-quality-of-its-drinking-water.pdf [accessed 25 November 2015].

Balderacchi, M. *et al.* (2013) Groundwater pollution and quality monitoring approaches at the European level. *Critical Reviews in Environmental Science and Technology* 43(4), pp. 323–408 [online]. DOI: 10.1080/10643389.2011.604259 [accessed 25 November 2015].

Balderacchi, M. *et al.* (2014) Does groundwater protection in Europe require new EU-wide environmental quality standards? *Frontiers in Chemistry*, 2 June [online]. http://dx.doi.org/10.3389/fchem.2014.00032 [accessed 25 November 2015].

Bergman, A. *et al.* (eds) (2012) *State of the Science of Endocrine Disrupting Chemicals 2012* UNEP / WHO [online]. Available from: http://www.unep.org/pdf/WHO_HSE_PHE_IHE_2013.1_eng.pdf [accessed 18 August 2016].

CBC News (2014) *Canadian Beekeepers Sue Bayer and Syngenta Over Neonicotinoid Pesticides* [online]. Available from: http://www.cbc.ca/news/technology/canadian-beekeepers-sue-bayer-and-syngenta-over-neonicotinoid-pesticides-1.2754441 [accessed 27 November 2015].

Council of the European Union (2012) *Note from General Secretariat to Delegations. Subject: Current legislative proposals: Proposal for a Directive on environmental quality standards in the field of water policy (priority substances). Information from the Presidency*, Interinstitutional File: 2011/0429 (COD) Council of the European Union, Brussels.

Deblonde, T., Cossu-Leguille, C. and Hartemann, P. (2011) Emerging pollutants in wastewater: a review of the literature. *International Journal of Hygiene and Environmental Health* 214(6), pp. 442–8 [online]. DOI:10.1016/j.ijheh.2011.08.002 [accessed 24 November 2015].

Eckstein G. (2015) Drugs on tap: managing pharmaceuticals in our nation's waters. *New York University Environmental Law Journal* 23, pp. 37–90.

EFSA (2012) Scientific opinion on the science behind the development of a risk assessment of plant protection products on bees. *EFSA Journal* 10(5), p. 2668 [online]. Available from: http://www.efsa.europa.eu/sites/default/files/scientific_output/files/main_documents/2668.pdf [accessed 27 November 2015].

EFSA (2014) EFSA guidance document on the risk assessment of plant protection products on bees. *EFSA Journal* 11(7), p. 3295 (updated 2014) [online]. Available from: http://www.efsa.europa.eu/en/efsajournal/pub/3295 [accessed 27 November 2015].

EurActiv (2016a) *Endocrine Disruptors' Link to Infertility Confirmed* [online]. http://www.euractiv.com/section/health-consumers/news/endocrine-disruptors-link-to-infertility-confirmed/ [accessed 29 April 2016].

EurActiv (2016b) *Parliament Agrees to Re-authorise Glyphosate, Demands Restrictions* [online]. http://www.euractiv.com/section/science-policymaking/news/parliament-agrees-to-re-authorise-glyphosate-demands-restrictions/ [accessed 29 April 2016].

European Chemicals Agency (undated) *Chemicals Registered Under REACH* [online]. Available from: http://echa.europa.eu/view-article/-/journal_content/title/2-923-more-chemicals-registered-by-industry-under-reach [accessed 24 November 2015].

European Commission (2006a) EC/1907/2006 Regulation on the Registration, Evaluation, Authorisation and Restriction of Chemicals.

European Commission (2006b) Regulation on the European Pollutant Release and Transfer Register 166/2006/EC.

European Commission (2008) EC/1272/2008 Regulation on the Classification, Packaging and Labelling of Substances and Mixtures.

European Commission (2009a) *Towards 2020: Making Chemicals Safer: The EU's Contribution to the Strategic Approach to International Chemicals Management* [online]. Available from: http://ec.europa.eu/environment/chemicals/reach/pdf/publications/saicm_09.pdf [accessed 24 November 2015].

European Commission (2009b) Regulation (EC) No 1107/2009 Concerning the Placing of Plant Protection Products on the Market.

European Commission (2013a) Seventh Report on the Implementation of the Urban Waste Water Treatment Directive (91/271/EEC) COM (2013) 574 final; And Commission Staff Working Document, SWD (2013) 298 final [online]. Both available from: http://ec.europa.eu/environment/water/water-urbanwaste/implementation/implementationreports_en.htm [accessed 26 November 2015].

European Commission (2013b) Commission Implementing Regulation (EU) No 485/2013.

European Commission (2013c) Commission Regulation (EU) 283/2013.

European Commission (2013d) Commission Regulation (EU) 284/2013.

European Council (1991) Urban Wastewater Treatment Directive 1991/271/EEC.

European Parliament and Council (1998) Drinking Water Quality Directive 1998/83/EC.

European Parliament and Council (1999) Landfill Directive 1991/31/EC.

European Parliament and Council (2000) Framework Directive on Water Policy 2000/60/EC.

European Parliament and Council (2006) Groundwater Directive 2006/118/EC.

European Parliament and Council (2008a) Priority Substances Directive 2008/105/EC.

European Parliament and Council (2008b) Framework Directive on Waste 2008/98/EC.

European Parliament and Council (2009) Directive 2009/128/EC Sustainable Use of Pesticides.

European Parliament and Council (2010) Industrial Emissions Directive 2010/75/EU.

European Parliament and Council (2013) Priority Substances Directive 2013/39/EU.

European Environment Agency (2011) *Hazardous Substances in Europe's Fresh and Marine Waters* EEA Technical Report 8/2011.

European Environment Agency/Joint Research Council (2013) *Environment and Human Health* EEA Report 5/2013; Report EUR 25933 EN.

Geissen, V. *et al.* (2015) Emerging pollutants in the environment: a challenge for water resource management. *International Soil and Water Conservation Research* [online] 3(1), pp. 57–65 [accessed 24 November 2015].

Helmer, R. and Hespanol, I. (eds) (1997) *Water Pollution Control* (WHO publication) [online]. Available from: http://www.who.int/water_sanitation_health/resourcesquality/watpolcontrol.pdf [accessed 25 November 2015].

ILO (1990) Convention Concerning Safety in the Use of Chemicals at Work (No. 170) (Entry into force: 4 November 1993) Adoption: Geneva, 77th ILC session (25 June).

ILO (1993) Prevention of Major Industrial Accidents Convention, 1993 (No. 174) (Entry into force: 3 January 1997) Adoption: Geneva, 80th ILC session (22 June).

Jobling, S. and Owen, R. (2013) Ethinyl oestradiol in the aquatic environment. *Late Lessons from Early Warnings: Science, Precaution, Innovation*. EEA Report 1/2013 (Volume B Chapter 13).

Mar, R. *et al.* (2013) Fate of zinc oxide and silver nanoparticles in a pilot wastewater treatment plant and in processed biosolids. *Environmental Science and Technology* 2014(48), pp. 104–12. dx.doi.org/10.1021/es403646x

NORMAN List (2015) NORMAN list 2015 final [online]. Available from: http://www.norman-network.net/?q=node/19 [accessed 24 November 2015].

von der Ohe, P.C. *et al.* (2011) A new risk assessment approach for the prioritization of 500 classical and emerging organic microcontaminants as potential river basin specific pollutants under the European Water Framework Directive. *Science of the Total Environment* 409(11), pp. 2064–77 [online]. Available from: http://dx.doi.org/10.1016/j.scitotenv.2011.01.054 [accessed 26 November 2015].

Owen, R. and Jobling, S. (2012) Environmental science: the hidden costs of flexible fertility. *Nature* 485(7399), p. 441.

Pal, A., Yew-Hoong Gin, K., Yu-Chen Lin, A. and Reinhard, M. (2010) Impacts of emerging organic contaminants on freshwater resources: review of recent occurrences, sources, fate and effects. *Science of the Total Environment* 408, pp. 6062–9.

Pollinator Stewardship Council; American Honey Producers Association; National Honey Bee Advisory Board; American Beekeeping Federation; Thomas R. Smith; Bret L. Adee; Jeffery S. Anderson, Petitioners, v US Environmental Protection Agency; Bob Perciasepe, Respondents, Dow Agrosciences Llc, Respondent-Intervenor. United States Court of Appeals for the Ninth Circuit No. 13-72346 [online]. Available from: http://cdn. ca9.uscourts.gov/datastore/opinions/2015/09/10/13-72346.pdf [accessed 26 November 2015].

Rosal, R. *et al.* (2010) Occurrence of emerging pollutants in urban wastewater and their removal through biological treatment followed by ozonation. *Water Research* 44, pp. 578–88.

SAICM/UNEP/WHO (undated) *Strategic Approach to International Chemicals Management* [online]. Available from: http://www.saicm.org/images/saicm_documents/saicm%20texts/New%20SAICM%20Text%20 with%20ICCM%20resolutions_E.pdf [accessed 24 November 2015].

Sauvé, S. and Desrosiers, M. (2014) A review of what is an emerging contaminant. *Chemistry Central Journal* 8(15) [online]. http://journal.chemistrycentral.com/content/8/1/15 [accessed 24 November 2015].

Schaar, H. *et al.* (2010) Micropollutant removal during biological wastewater treatment and a subsequent ozonation step. *Environmental Pollution* 158, pp. 1399–404.

Schriks, M. *et al.* (2010) Toxicological relevance of emerging contaminants for drinking water quality. *Water Research* 44, pp. 461–76.

UN (1992) Agenda 21: An Agenda for the 21st Century A/Conf. 152/126.

UN (2002) Report of the World Summit on Sustainable Development Incorporating the Johannesburg Declaration and Plan of Implementation A/Conf.199/20s.

UN/ECE (1979) Geneva Convention on Long Range Transboundary Air Pollutants 18 IILM 1979.

UN/ECE (2003) Globally Harmonized System of Classification and Labelling of Chemicals [online]. Available from: http://www.unece.org/trans/danger/publi/ghs/ghs_welcome_e.html [accessed 24 November 2015].

UNEP (1989) Basel Convention on Transboundary Movements of Hazardous Wastes 28 ILM 1989.

UNEP (1998) Rotterdam Convention on the Prior Informed Consent Procedure for Certain Hazardous Chemicals and Pesticides in International Trade 38 ILM 1991.

UNEP (2001) Stockholm Convention on Persistent Organic Pollutants 40 ILM.

UNEP (2010) *The Greening of Water Law: Managing Freshwater Resources for People and the Environment.* Nairobi: UNEP.

UNEP (2015) *Economic Valuation of Wastewater: The Cost of Action and the Cost of No Action.* Nairobi: UNEP.

UN General Assembly (2015) *Transforming Our World: The 2030 Agenda for Sustainable Development* A/RES/70/1.

US Congress (1970) Clean Air Act 42 U.S.C. §7401 et seq.

US Congress (1972) Clean Water Act of 1972, 33 U.S.C. §1251 et seq.

US Congress (1974) Safe Drinking Water Act of 1972, 42 U.S. C. §300 et seq.

US Congress (1976) Resource Conservation and Recovery Act of 1976, 42 U.S.C. §6901 et seq.

US Congress (1996) Insecticide, Fungicide, and Rodenticide Act, 7 U.S.C. §136 et seq.

USEPA (undated (a)) *Contaminants of Emerging Concern* [online]. http://water.epa.gov/scitech/cec/ [accessed 24 November 2015].

USEPA (undated (b)) *Endocrine Disruption* [online] Available at http://www2.epa.gov/endocrine-disruption [accessed 24 November 2015].

WHO (2004) *Guidelines for Drinking Water Quality 3rd Edition* [online]. Available from: http://www.who. int/water_sanitation_health/dwq/gdwq3/en/ [accessed 25 November 2015].

Yang, Y., Zhang, C. and Hu, Z. (2013) Impact of metallic and metal oxide nanoparticles on wastewater treatment and anaerobic digestion. *Environmental Science: Processes Impacts* 15, p. 39. DOI: 10.1039/c2em30655g

8

THE HUMAN RIGHT TO WATER

Inga T. Winkler

The human right to water under international law

Process of recognition

The discussions around a right to water gained momentum in the late 1990s. This was supported by a general trend of increasing attention to social and economic rights after the end of the Cold War. At the same time, the rise in attention to the right to water was linked to tensions around private sector participation in the provision of water services in many countries that mobilized civil society and brought water issues to the fore. Even though the human right to water does not catagorically rule out private sector participation, and the two questions should be considered separately (Albuquerque and Winkler 2010), they have been linked in mobilization efforts.

As a result of this increasing attention, the Committee on Economic, Social and Cultural Rights (CESCR), the UN treaty body responsible for monitoring the implementation of the International Covenant on Economic, Social and Cultural Rights (ICESCR), issued its General Comment No. 15 on the right to water in late 2002 (CESCR 2003). The General Comment defined the right to water as the right 'to sufficient, safe, acceptable, physically accessible and affordable water for personal and domestic uses' (ibid., ¶ 2). It argued that the right to water is 'essential for securing an adequate standard of living' and that it is also 'inextricably related to the right to the highest attainable standard of health (Art. 12, ¶ 1) and the rights to adequate housing and adequate food (Art. 11, ¶ 1)' (ibid., ¶ 3).

The General Comment is a statement by experts providing an authoritative interpretation of the ICESCR. It also provided an impetus for further developments at the political level (not only on water, but also on sanitation). In 2007, the Human Rights Council requested the High Commissioner for Human Rights to prepare a report on 'the scope and content of the relevant human rights obligations related to equitable access to safe drinking water and sanitation under international human rights instruments' (HRC 2007), where she concluded that 'it is now time to consider access to safe drinking water and sanitation as a human right' (ibid., ¶ 66). Following this study, the Human Rights Council created a Special Procedures mandate on water and sanitation (HRC 2008), and Catarina de Albuquerque, as the first mandate-holder, took up her functions as 'Independent Expert on the issue of human rights obligations related to access to safe drinking water and sanitation' in November 2008. The mandate's title indicates that, at the time, there was not yet political agreement on the status of the human right to water. Spain and Germany, as the two main drivers behind the initiative, envisaged introducing a resolution in the Human Rights Council in 2011 that would explicitly recognize the human right to water and sanitation. However, even before that, in the summer of 2010, Bolivia introduced a resolution in the UN General Assembly that led to the political recognition of the 'right to safe and clean drinking water and sanitation as a human right that is essential for the full enjoyment of life and all human rights' (United Nations General Assembly 2010, ¶ 1) even though the resolution was not adopted by consensus. In September 2010, the Human Rights Council affirmed the human right to water and sanitation in a resolution adopted by consensus (HRC 2010). At the political level, these resolutions brought the breakthrough for the recognition of the human right to water.

Legal foundations

While of great political significance, by themselves these resolutions do not provide the legal foundation for the human right to water. However, the Human Rights Council resolution stresses that 'the human right to safe drinking water and sanitation is derived from the right to an adequate standard of living and inextricably related to the right to the highest attainable standard of physical and mental health, as well as the right to life and human dignity' (ibid., ¶ 3). While Art. 11 ICESCR does not mention water, it provides a broad guarantee of the human right to an adequate standard of living. It mentions several components of an adequate standard of living, including food and housing, and uses an open formulation that leaves room for other unnamed components. Indeed, an adequate standard of living cannot be realized without access to water for personal and domestic use. The human right to water is therefore to be understood as an implicit component of the human right to an adequate standard of living (Winkler 2012, pp. 41ff.). More recent human rights treaties, in particular the Convention on the Elimination of all Forms of Discrimination against Women (CEDAW), support this interpretation: Art. 14(2)(h) CEDAW explicitly lists water as one of the components necessary for adequate living conditions.

Apart from being an implicit component of the right to an adequate standard of living, the right to water is inextricably linked to a range of other human rights. Access to safe water is one of the main underlying determinants of the right to health (CESCR 2000, ¶ 11), and water is also linked to the right to housing, the right to life and many other human rights (Winkler 2012, pp. 45ff.).

Normative content

One of the latest resolution on the human right to water and sanitation by the Human Rights Council specifies that the right to water 'entitles everyone, without discrimination, to have access to sufficient, safe, acceptable, physically accessible and affordable water for personal and domestic use' (HRC 2013). Using this definition, which is based on the work of the CESCR and the Special Rapporteur, the content of the right to water can be determined by the criteria of availability (sufficient quantity), safety, acceptability, accessibility and affordability.

Water must be available to everyone in sufficient quantities to satisfy personal and domestic needs. This includes water for drinking as well as water for personal and household hygiene, including bathing, washing and cleaning. Specific requirements vary due to climatic and geographic conditions, individual health conditions, age and other factors, and the human rights framework stresses the need to accommodate and meet individual requirements. Still, international recommendations on water quantity can provide broad guidance: the World Health Organization estimates that under most circumstances domestic needs can be met without health risks with a daily amount of about 100 litres per person (Howard and Bartram 2003, p. 7). While lower quantities are sometimes suggested, these fail to guarantee personal and household hygiene, compromise human health and thus cannot be understood as corresponding to requirements under the human right to water.

Apart from water in sufficient quantities, water safety is just as essential for ensuring human health. Water must not pose a threat to human health (CESCR 2003, ¶ 12). The WHO *Guidelines for Drinking-Water Quality* specify that safe drinking water must not 'represent any significant risk to health over a lifetime of consumption, including different sensitivities that may occur between life stages' (WHO 2011, p. 1) and put forward limits for a large number of substances that may be harmful. In addition to safety as such, water must also be acceptable to users. If water is safe, but not acceptable, users may turn to alternative unsafe sources, demonstrating the importance of an acceptable odour, colour and taste (CESCR 2003, ¶ 12(b)).

In addition, water must be accessible to all users on a reliable and continuous basis. Again, this criterion stresses the need to meet individual requirements, including of persons with disabilities, older persons, children, people with health conditions, pregnant women, among others. General Comment No. 15 by the CESCR stipulates that 'water must be accessible within, or in the immediate vicinity, of each household, educational institution and workplace' (ibid., ¶ 12(c)(i)).

Finally, water services must be affordable, which is often overlooked in current policy-making and tariff-setting (SR WatSan 2015, ¶ 1). All too often, the assumption is that people will somehow pay for water. And given that water is essential for human survival and health, people will go to great lengths to pay for water services. The human rights framework does not generally require that water services are provided free of charge, yet it does require that services are affordable. People must not be put in a position where they compromise the realization of other human rights, including health, housing, education or food, in order to pay for water services. In this regard, the human rights framework stipulates important parameters for designing tariff schemes and ensuring both social and financial sustainability – for instance, by using cross-subsidies (ibid., ¶ 7).

Obligations to realize the human right to water

The human rights framework links rights with obligations. Individuals have inalienable human rights, including the human right to water, and states, as the primary duty-bearers, have the obligation to realize these human rights. State obligations are commonly described as obligations to respect, to protect and to fulfil the right to water.

The Special Rapporteur on the human rights to water and sanitation has explained that '[t]he obligation to respect the rights to water and sanitation requires states to refrain from action that will unjustifiably interfere with their enjoyment. This obligation is of immediate effect' (SR WatSan 2014a, ¶ 17). States must not directly interfere with access to water, for instance through unjustifiable disconnections (ibid., ¶ 18), and they must not pollute, divert or deplete water resources that households rely on for personal and domestic uses (ibid., ¶ 20).

Not only states, but also third parties may have a significant impact on the realization of the human right to water. In this regard,

> [t]he obligation to protect requires States to enact and enforce necessary protections of the rights to water and sanitation to protect individuals from human rights abuses by third parties. Such obligation is generally considered to be of immediate effect, although

in some cases it will take time and resources to develop the necessary institutional capacity and frameworks. Non-State actors, including private actors and international organizations, also contribute to the realization of human rights and, conversely, their action or inaction may also lead to human rights abuses. Where private actors are involved in the provision of water and sanitation services, their role comes with human rights responsibilities.

Ibid., ¶ 25

Among other measures, states must put in place adequate regulatory measures to ensure that third parties do not exploit water resources to the extent that it would cause deprivation or harm to individuals relying on these water resources for the realization of their human rights. Likewise, states must develop and enforce regulation to protect people from contamination of water resources.

Finally, '[t]he obligation to fulfil requires States parties to adopt the necessary measures directed towards the full realization of the right to water' (CESCR 2003, ¶ 26). The obligation to fulfil may be the most complex obligation, and the one surrounded by most misconceptions. The human right to water does not require states to provide services to all people directly, nor does it require states to achieve its full realization immediately.

Above all, the obligation to fulfil is an obligation to facilitate access and to enable people to provide for themselves. Individuals are expected to contribute to the enjoyment of their right to water with their own means (for instance, by paying water tariffs), but states have to create the enabling environment through the adoption of legislation, policies, regulations and programmes and creating appropriate institutions for their implementation (ibid., ¶ 26). However, where people do not have the capacity to provide for themselves (for instance, in conditions of detention) or do not have the means to do so for reasons beyond their control, the state's obligation turns into an obligation of direct provision (ibid., ¶ 25).

The human rights framework acknowledges that the right to water cannot be fully realized overnight. States' obligations are obligations of progressively realizing the human right to water. This means that states must move towards the goal of full realization as expeditiously and effectively as possible, taking deliberate, concrete and targeted steps, using the maximum of their available resources (CESCR 1990, ¶ 2). Hence, it does not leave the realization of human rights to the state's discretion, while it recognizes that the full realization of human rights is a long-term process that needs to take account of resource and other constraints.

The broader human rights framework relevant to water law and policy

Apart from the human right to water as such, the broader human rights framework also has implications for water law and policy. This section will discuss other relevant human rights, including the human right to sanitation, and human rights principles.

A note on the human right to sanitation

The previously-discussed General Assembly and Human Rights Council resolutions on water and sanitation address the human right to water and sanitation as one combined human right. Similarly, the mandate of the Special Rapporteur addresses both water and sanitation. Yet, in comparison with water, the human right to sanitation has received much less attention, and in practical terms progress on access to sanitation is lagging behind (JMP 2015, p. 5). Overall, there is a lack of policies, strategies, financing and other measures to improve access to sanitation. While there is the perception of water and sanitation being linked to each other, this does not necessarily have to be the case, and both the Special Rapporteur and the CESCR have argued that sanitation has distinct characteristics that warrant its understanding as a distinct human right (CESCR 2011, ¶ 7). If sanitation is not understood as a distinct human right, efforts are likely to fail to address the specific challenges in its implementation, both in technical terms and in terms of changing behaviour around sanitation. Moreover, without a specific focus on sanitation, sanitation is likely to continue to be treated as an add-on to water that does not get the attention that is urgently needed (Winkler 2016). Water and sanitation should, therefore, be understood as two distinct human rights as components of the right to an adequate standard of living. Recognizing the distinctiveness of sanitation, the UN General Assembly acknowledged water and sanitation as two separate human rights in its resolution from 2015 (United Nations General Assembly 2015, ¶ 1).

Recognizing sanitation as a distinct human right does not imply that sanitation should be considered without its linkages to water, health, housing and other fields. In fact, human rights law stresses that all human rights are indivisible, inter-dependent and inter-related. For the field of water law and policy, sanitation becomes relevant in several ways. Where water-borne sanitation is used this has an impact on the amount of water required by households as flushing toilets uses relatively large quantities of water, which must be considered in decisions on water allocation. Even more importantly, adequate sanitation and wastewater management is essential for ensuring water quality. Where sewage is not adequately treated, it may contaminate large quantities of fresh water. And where on-site sanitation facilities are not adequately managed, where pits overflow, where contents of septic tanks leak into groundwater, or where pits are emptied but their contents are dumped into the environment, this leads to pollution of water sources and may contaminate people's drinking water. In this regard, a holistic approach to water and wastewater governance is essential (Zimmer *et al.* 2014).

Other relevant human rights

Water is used not only for personal and domestic uses guaranteed by the human right to water, but also for many other uses, including food production through agriculture and food processing, industrial purposes, power generation and, at a smaller scale, cultural and religious activities. At a global scale around 70 per cent of global water use occurs in the agricultural sector, around 20 per cent in the industrial sector and less than 10 per cent is used in households (World Water Assessment Programme 2009, p. 99).

In turn, beyond the human right to water as such, the realization of many other human rights depends on water. The right to food cannot be realized without water in sufficient quantities. The right to work of many individuals is dependent on economic development, which in turn requires water, whether for small-scale livelihood activities or large-scale industrial development. Religious and cultural practices, including of indigenous peoples, which require access to water, are also protected by human rights. Finally, environmental processes require water and the right to a healthy environment is recognized in a range of regional and national human rights instruments and constitutions (Winkler 2012, pp. 155ff.).

Establishing these linkages, however, does not imply that all water used in agriculture is directed to the realization of the human right to food, or that all household water use is protected under the human right to water. Luxury uses such as watering lawns, filling up swimming pools or washing cars do not fall under human rights guarantees. Similarly, water-intensive production of some agricultural crops, such as high-value fruit as well as the production of bio-fuels, does not contribute to the realization of the right to food or goes beyond its requirements. Linkages between water used in industrial processes and the right to work may be even looser, and in many cases large-scale industrial or agricultural development may also harm human rights. The reliance on water for the realization of human rights, the possible violations of human rights and the inter-connectedness of different human rights need to be carefully considered on a case-by-case basis (Winkler 2012, pp. 207ff.).

Human rights principles

In addition to the standards determined through different human rights, cross-cutting human rights principles have an impact on water law and policy. Non-discrimination and equality is a fundamental human rights norm that relates to all other human rights. Specific instruments such as CEDAW, the Convention on the Rights of Persons with Disabilities (CRPD), the International Convention on the Elimination of all Forms of Racial Discrimination (ICERD) and the Convention on the Rights of the Child (CRC) include protection for specific individuals or population groups. At a more general level, Art. 2(2) ICESCR includes a broad guarantee of non-discrimination, obliging states parties 'to guarantee that the rights enunciated in the present Covenant will be exercised without discrimination of any kind as to race, colour, sex, language, religion, political or other opinion, national or social origin, property, birth or other status'. Art. 2(1) and Art. 26 of the International Covenant on Civil and Political Rights (ICCPR) include similar guarantees of non-discrimination.

> [D]iscrimination constitutes any distinction, exclusion, restriction or preference or other differential treatment that is directly or indirectly based on the prohibited grounds of discrimination and which has the intention or effect of nullifying or impairing the recognition, enjoyment or exercise, on an equal footing, of Covenant rights.
>
> *CESCR 2009, ¶ 7*

Discrimination does not have to be intentional. Human rights law also requires states to address discriminatory impacts of certain laws, policies or other measures with the aim of achieving substantive equality. States must take measures to reduce and ultimately eliminate conditions that cause and entrench inequalities (SR WatSan 2014c, Principles, p. 13). This does not mean that all people must be treated identically. Equality relates to the equal enjoyment of human rights, but to achieve such equal enjoyment states must accommodate and embrace differences between people (ibid., p. 12).

Participation, transparency and access to information are further key human rights principles. A number of human rights instruments cover the right to participation. The Universal Declaration of Human Rights stipulates in Article 21(a) that everyone has the right to take part in the government of their country. The Declaration on the Right to Development states in Article 2(3) that participation should be 'active, free and meaningful'. Article 25(a) of the ICCPR guarantees the right 'to take part in the conduct of public affairs, directly or through freely chosen representatives'. More recent human rights treaties, including CEDAW, CRC and CRPD, spell out more detailed guarantees of participation and address the challenges that particular individuals might face in taking part in participatory processes (SR WatSan 2014b, ¶ 11 ff.). Furthermore, Arts. 6–8 of the UNECE Convention on Access to Information, Public Participation in Decision-Making and Access to Justice in Environmental Matters (Aarhus Convention) guarantee the right to participate in decisions on the development of laws and the establishment of plans, programmes and policies.

Based on this understanding of participation, states are obliged to ensure that participation is active, free and meaningful. Opportunities must be provided for people to take part in decision-making on policies, programmes and activities on water at all levels that have an impact on their lives. Transparency and access to information are essential in their own right, but are also necessary for participation to be meaningful and informed (SR WatSan 2014c, Principles, p. 35).

Accountability is the central tenet of the human rights framework. States are accountable for meeting their obligations to realize human rights. Mechanisms need be put in place to ensure accountability, which must be accessible, affordable, timely and effective (ibid., Justice, p. 39). They include complaints mechanisms at different levels such as service providers, regulators or administrative bodies; national human rights institutions; parliamentary review committees; petition committees; and, ultimately, recourse to the courts and international human rights mechanisms (ibid., pp. 25ff.). Accountability is also essential in relation to third parties (private companies, informal sector, non-governmental organizations, among others), whether involved in the provision of services, using water resources, or discharging wastewater.

While not traditionally included in a human rights analysis, sustainability is gaining recognition as a human rights principle and is linked to the principle of non-retrogression in the realization of the human right to water (ibid., Principles, pp. 75ff.). Sustainability requires the uninterrupted and long-term enjoyment of human rights related to water. It is essential to ensure that not only present generations enjoy the benefits of water and sanitation, but also that future generations are catered for. Sustainability can only be ensured through the protection and conservation of ecosystems to ensure water quality and safeguard people's health (ibid.).

Implications for water law and policy

The human rights framework does not prescribe approaches to water law and policy. However, it does set important parameters both in terms of substantive and procedural requirements.

Prioritization of water uses

Prioritizing water uses from a human rights perspective requires ensuring that everyone's basic human requirements are met. Above all, this relates to water for personal and domestic use as guaranteed by the human right to water. In contrast to other water uses, personal and domestic use stands out because there are no alternatives for realizing the right to water; water cannot be replaced, and people require direct access to water (Winkler 2012, pp. 207ff.). Yet, such an approach does not mean prioritizing the domestic sector as such: not all household water uses are part of the human right to water. For this reason, certain forms of rationing or restrictions, especially if targeting certain water uses, can be necessary and desirable. The State Water Board in California, for example, has adopted explicit water conservation rules such as forbidding or otherwise restricting the watering of lawns or washing of cars (California State Water Resources Control Board 2016).

In addition, priorities in water allocation relate to the realization of other human rights such as the right to food and cultural rights, among others. Here again, the resources needed to realize the right to food cannot be equated with the entire water use in agriculture. Rather, water allocation for agriculture needs to prioritize basic human requirements to sustain livelihoods and cultural rights of subsistence and small-scale farmers and realize their right to food (Winkler 2012, pp. 164ff.). The design of irrigation projects is also crucial in order to work towards the realization of human rights of people living in poverty in rural areas. Prioritizing poor households' and farmers' water needs can contribute in substantial ways to better health and livelihoods of people living in these circumstances. And yet, questions are more complex in this context as there are alternatives for realizing the right to food. People can access food through markets, for example, in which case direct access to water is not required. Therefore, the extent to which such water use needs to be prioritized for the realization of human rights depends on the specific context.

Redressing disadvantages in access to water

Particular groups and individuals are often disadvantaged in decisions on water governance. These include women and girls who spend their time collecting water where supply is inadequate. People living in informal settlements, marginalized ethnic groups, or indigenous communities are especially disadvantaged in water governance arrangements that do not focus specifically on human rights. Subsistence farmers, including female farmers, might not be recognized as legitimate stakeholders in decisions on water management, especially where land tenure is precarious (Water Governance Facility 2012, p. 8). Also, specific geographic regions seem to be more disadvantaged in water management practices, as they tend to be out of the focus of planners and policy-makers. These include the large and growing peri-urban areas of major cities in the global South, housing mostly poor residents in substandard housing, but equally offering locations to polluting industries. Human rights require making water governance relevant for these individuals and groups and targeting interventions towards them to redress existing disadvantages.

Participation in decision-making

All stakeholders and, in particular, the people concerned by decisions, must be given opportunities to participate in decision-making about use of water and water management. Such participation must go beyond information-sharing and superficial consultation. It must be active, free and meaningful, providing opportunities to actually influence decisions. States must put into place mechanisms to ensure participation, in particular of those traditionally lacking voice, including people living in poverty, women, indigenous peoples and ethnic minorities, avoiding processes that are captured by a few well-established non-governmental organizations or local elites. Mechanisms for participation should reach out to people at all levels of society, taking into account constraints that might prevent them from attending. This would mean organizing meetings close to where people live, or work, in all relevant regions of the country, organizing meetings during hours when people are available, using local languages, organizing parent- and child-friendly meetings, using organizations of which people are already members as platforms for undertaking such meetings and other measures, among other channels (SR WatSan 2014c, Principles, p. 64). While certain aspects of water governance require technical expertise, such inputs must be balanced with the needs and preferences of people concerned by decisions, taking into account local solutions (ibid., p. 62).

Transparency and access to information are important prerequisites for enabling people to participate. Relevant information should be made publicly available. Information should be widely disseminated and made available in all relevant languages via multiple channels to ensure accessibility. This requires providing information in easily accessible formats so that people without the respective technical background can access and understand it. Transparency is also needed regarding existing policies and measures. Only when, for instance, current priorities in the allocation of resources are understood can these be scrutinized and assessed for eventual necessary changes.

Integrating human rights in water laws and policies

To integrate human rights, states must reform water law and other relevant legislation, policies and strategies and introduce mechanisms that protect human rights as necessary. Human rights are relevant at all levels, from local regulations and ordinances, national level policies and legislation, to international water law. As states revise constitutions and legislation, many are explicitly recognizing the human right to water. Such recognition needs to be translated into concrete reforms of existing legal frameworks that integrate human rights standard and principles (ibid., Frameworks). This requires a detailed assessment of existing legal and policy frameworks. For instance, in some cases customary water rights may protect traditional water use by indigenous groups, whereas in other cases they may disadvantage marginalized groups by reinforcing traditional power structures. The human rights framework does not prescribe which specific solutions states choose, but whatever solutions are adopted must reflect human rights standards and principles.

Human rights are also relevant in a transboundary context. While international human rights law in that context is still developing, the Maastricht Principles on Extraterritorial Obligations of States in the area of Economic, Social and Cultural Rights have systematically elaborated on human rights obligations in extraterritorial contexts (Maastricht Principles 2012). Extraterritorial human rights obligations and international water law apply in the same context, and human rights law helps to interpret international water law. International water law has been codified in the 1997 United Nations Convention on the Law of Non-Navigational Uses of International Watercourses. Art. 10(2) sets forth that a conflict between uses of an international watercourse shall be resolved 'with special regard being given to the requirements of vital human needs'. The human rights framework reinforces this prioritization, requiring not simply 'special regard', but, further, an absolute priority of basic human needs protected by human rights. More recently, the 'Berlin rules on water resources' have taken significant steps in integrating the international water law and human rights law (Winkler and Phan 2014, p. 15).

Conclusion

The human right to water is part of international human rights law. It has gained significant attention over the last decade and can be defined as entitling everyone 'without discrimination, to have access to sufficient, safe, acceptable, physically accessible and affordable water for personal and domestic use' (HRC 2013). Beyond the human right to water itself, other human rights that rely on water in their realization also have implications for water law and policy. This includes the human rights to sanitation, food, work and other human rights, as well as the human rights principles of non-discrimination and equality, participation, access to information, transparency and accountability. The human rights framework does not prescribe what approaches states adopt in decisions on water laws and policies, but it does set significant parameters that are relevant at all levels, from local to international law. It requires that states prioritize basic human needs, it requires redressing disadvantages that marginalized individuals and groups experience, and it requires that decisions on water law and policy are taken in a participatory manner. As such, integrating human rights in water law and policy will help achieve results that benefit all people.

References

Albuquerque, C. de and Winkler, I.T. (2010) Neither friend nor foe: why the commercialization of water and sanitation services is not the main issue in the realization of human rights. *Brown Journal of World Affairs* 17(1), pp. 167–79.

California State Water Resources Control Board (2016) Regulations for Drought Emergency Water Conservation. Available from: http://www.waterboards.ca.gov/water_issues/programs/conservation_portal/docs/emergency_reg/final_reg_enacted.pdf

CESCR (Committee on Economic, Social and Cultural Rights) (1990) General Comment No. 3: The Nature of States Parties' Obligations (Art. 2, ¶ 1 of the Covenant), UN Doc. E/1991/23.

CESCR (2000) General Comment No. 14: The right to the highest attainable standard of health (Article 12 of the International Covenant on Economic, Social and Cultural Rights), UN Doc. E/C.12/2000/4, 11 August.

CESCR (2003) General Comment No. 15: The right to water (arts. 11 and 12 of the International Covenant on Economic, Social and Cultural Rights), UN Doc. E/C.12/2002/11, 20 January.

CESCR (2009) General Comment No. 20, Non-discrimination in economic, social and cultural rights (Art. 2, ¶ 2, of the International Covenant on Economic, Social and Cultural Rights), UN Doc. E/C.12/GC/20, 2 July.

CESCR (2011) Statement on the Right to Sanitation, UN Doc. E/C.12/2010/1, 18 March.

Convention on Access to Information, Public Participation in Decision-Making and Access to Justice in Environmental Matters (Aarhus Convention), 25 June 1998, entered into force 30 October 2001, United Nations Treaty Series, vol. 2161, 447.

Convention on the Elimination of All Forms of Discrimination against Women (CEDAW), 18 December 1979, entered into force 3 September 1981, United Nations Treaty Series, vol. 1249, 13.

Convention on the Law of Non-Navigational Uses of International Watercourses (UNWC), 21 May 1997, entered into force 17 August 2014, Doc. A/51/869.

Convention on the Rights of Persons with Disabilities (CRPD), 13 December 2006, entered into force 3 May 2008, United Nations Treaty Series, vol. 2515, 3.

Convention on the Rights of the Child (CRC), 20 November 1989, entered into force 2 September 1990, United Nations Treaty Series, vol. 1577, 3.

Howard, G. and Bartram, J. (2003) Domestic Water Quantity, Service Level and Health, World Health Organization, Geneva, WHO/SDE/WSH/03.02.

HRC (Human Rights Council) (2007) Report of the UN High Commissioner for Human Rights on the scope and content of the relevant human rights obligations related to equitable access to safe drinking water and sanitation under international human rights instruments, UN Doc. A/HRC/6/3, 16 August.

HRC (2008) Resolution, Human rights and access to safe drinking water and sanitation, UN Doc. A/HRC/RES/7/22, 28 March.

HRC (2010) Human rights and access to safe drinking water and sanitation, Resolution, UN Doc. A/RES/HRC/15/9, 6 October.

HRC (2013) The human right to safe drinking water and sanitation, UN Doc. A/HRC/RES/24/18, 8 October.

International Convention on the Elimination of all Forms of Racial Discrimination (ICERD), 7 March 1966, entered into force 4 January 1969, United Nations Treaty Series, vol. 660, 195.

International Covenant on Civil and Political Rights (ICCPR), 16 December 1966, entered into force 23 March 1976, United Nations Treaty Series, vol. 999, 171 and vol. 1057, 407.

International Covenant on Economic, Social and Cultural Rights (ICESCR), 16 December 1966, entered into force 3 January 1976, United Nations Treaty Series, vol. 993, 3.

International Law Association (2004) Berlin rules on water resources law. *Report of the 71st Conference.* London: International Law Association, p. 334.

JMP (Joint Monitoring Programme, WHO and UNICEF) (2015) Progress on Sanitation and Drinking Water, 2015 Update and MDG Assessment.

Maastricht Principles on Extraterritorial Obligations of States in the area of Economic, Social and Cultural Rights (2012) 29 February.

SR WatSan (Special Rapporteur on the human right to safe drinking water and sanitation) (2014a) Report on common violations of the human rights to water and sanitation, UN Doc. A/HRC/27/55, 30 June.

SR WatSan (2014b) Report on participation in the realization of the human rights to water and sanitation, UN Doc. A/69/213, 31 July.

SR WatSan (2014c) Realizing the Human Rights to Water and Sanitation, A Handbook.

SR WatSan (2015) Report on affordability of water and sanitation services, UN Doc. A/HRC/30/39, 5 August.

United Nations General Assembly (1948) Universal Declaration of Human Rights, A/810 Res 217 (III), 10 December.

United Nations General Assembly (1986) Declaration on the Right to Development, A/Res/41/128, 4 December.

United Nations General Assembly (2010) The Right to Water and Sanitation, Resolution, UN Doc. A/RES/64/292, 3 August.

United Nations General Assembly (2015) The Human Rights to Safe Drinking Water and Sanitation, Resolution, UN Doc. A/Res/70/169, 17 December.

Water Governance Facility (2012) Human rights based approaches and managing water resources. Exploring the potential for enhancing development outcomes (No. 1). Stockholm: SIWI.

WHO (2011) *Guidelines for Drinking-Water Quality*, 4th edn. Geneva: WHO.

Winkler, I.T. (2012) *The Human Right to Water: Significance, Legal Status and Implications for Water Allocation*. Oxford: Hart.

Winkler, I.T. (2016) The human right to sanitation. *University of Pennsylvania Journal of International Law* 37(4), pp. 1331–1406.

Winkler, I.T. and Phan, H.L. (2014) Über Grenzen hinweg – Die Bedeutung des Menschenrechts auf Wasser für grenzüberschreitende Gewässer. *Journal of Law of Peace and Armed Conflict* 27(1), pp. 17–27.

World Water Assessment Programme (2009) *United Nations World Water Development Report 3: Water in a Changing World*. London and Paris: Earthscan and United Nations Educational, Scientific and Cultural Organization.

Zimmer, A., Winkler, I.T. and de Albuquerque, C. (2014) Governing wastewater, curbing pollution, and improving water quality for the realization of human rights. *Waterlines* 33(4), pp. 337–56.

9

GOVERNANCE AND REGULATION OF WATER AND SANITATION SERVICES PROVISION

Richard Franceys and Paul Hutchings

Governance and regulation

The focus of this chapter is the governance and regulation of water supply and sanitation (WATSAN), a subset of overall water governance which includes water for agriculture, water for energy and water in the environment. At its broadest level, water governance has been defined as 'the range of political, social, economic and administrative systems that are in place to develop and manage water resources, and the delivery of water services, at different levels of society' (Rogers and Hall 2003). Here we deal with the particular challenge of developing and regulating appropriate legislative and institutional frameworks within which WATSAN services can be produced, delivered, monitored and assured with respect to quality and price, the ultimate goal being that every citizen-consumer can access an affordable yet sustainable WATSAN service, respecting the globally agreed human right to these services.

Placing the governance and regulation challenges in context, it is helpful to reflect on the scale of capital and recurrent investment required to meet the needs of the unserved and the underserved around the world. The numbers can be staggering and, as one observer suggests, focusing on the middle and high-income market: 'You could sink a couple of trillion dollars into [the water sector] and still not quench the need to invest' (Gasson 2016). The latest World Bank study on the costs of meeting the minimum needs of the 2030 sustainable development goal (SDG) targets on drinking water, sanitation and hygiene states that the 'total capital cost of meeting SDG targets 6.1 and 6.2 is USD$114 billion per year' which is $1.7 trillion over the 15-year period for capital expenditure alone (Hutton and Varughese 2016). It is in this context that the need for good governance, and effective regulation of service providers who have to absorb this amount of capital efficiently, has to be addressed. The chapter aims to consider the reality of this challenge by drawing on examples from around the world. These examples help illustrate the tension at the heart of most efforts to promote global development: that those countries which need good water governance the most, to ensure the delivery of the necessary capital investment in order to deliver quality water and sanitation services for all, are the countries which are poorest and have the lowest institutional capital.

Reflecting for a moment on a basic definition, we understand that 'governance is ultimately concerned with creating the conditions for ordered rule and collective action' both within and beyond the government system (Stoker 1998). The OECD has a recommended programme of research into achieving this within the water sector.[1] They propose a 'multi-level approach' that helps set the scene for subsequent discussions (OECD 2012). It makes an important distinction between the vertical dimensions of governance, reflecting the linkages between lower and higher levels of government, and the horizontal dimensions of governance, related to co-operation between regions and municipalities. This provides a framework for analysing a mega-trend in water governance (and public service delivery more generally) which is the supposed shift from more centralised or vertical systems to more decentralised or horizontal governance approaches (noting exceptions which will be discussed). At the core of the OECD framework are prescriptions about the need for governance to be *effective, efficient* and to necessarily involve *trust* and *engagement*. They argue that effectiveness requires clear roles and responsibilities, appropriate scales within basin systems, policy coherence and capacity; whilst efficiency is believed to need data and information, financing, regulatory frameworks and innovative governance. Trust and engagement needs integrity and transparency, stakeholder engagement, trade-offs across users and monitoring and regulation. The challenge that remains is how to 'operationalise' these numerous principles? What sort of regulation and service provider models are most likely to deliver good water and sanitation according to these principles?

There are a wide variety of options as to which entity delivers which part of the overall WATSAN services package. Similarly, there are different financing models and oversight models. However, the key constraining factor is the nature of networked services, which are the water pipes and sewer networks which supply and collect water and wastewater. These networked services are necessarily a monopoly activity, to be undertaken by a single entity, whether public, community or private. The reason for this monopoly is that pipe and sewer networks are expensive to construct ('capital intensive'), but, because of very long asset lives, are then able to deliver relatively low-cost water. The governance of water supply and sewerage (and its legislative framework) therefore requires not only a focus on the sustainability of water resources, the starting point for water delivery to households and businesses, the treatment and resulting quality of that water supplied, as well as the cleanliness of the resulting collected wastewater as it is returned to the aquatic environment, but also of the effective economic management of that service. Economic regulation, the overseer of service providers' financial management, attempts to ensure that citizen-consumers, usually also customers, are not overcharged by the monopoly service provider, whether through unrecognised or tolerated inefficiency (also 'producer capture'), through professional-led 'gold-plating' (over-investment in fixed assets) or through 'profiteering', whether public or private.

This balancing of interests in governance and regulation is best shown in Figure 9.1, which is a reworking of the World Bank's 2004 'accountability triangle' (WDR 2004). Although first prepared in the context of a more general World Development Report, we find that it is helpful in high-income countries as well as lower-income settings. In this version we have placed the citizen-consumer at the apex, with government as policy-maker and utility service providers forming the 'foundation' to support services to consumers. World Bank describes the 'short accountability route' being directly between consumer/customer and the service provider, where it is presumed (hoped?) that the provider is incentivised to deliver good services at the right price. Because of the danger of monopoly providers being less focused on the needs of customers, the concept of the 'long route' to accountability sees citizen-consumers being able to influence their services (and the environmental context in which they deliver) through their influence

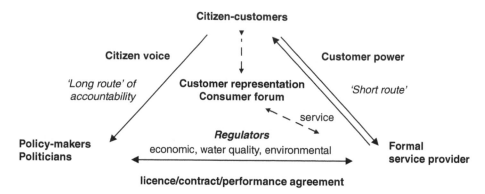

Figure 9.1   Accountability triangle
Source: After WDR 2004 and predecessors

on elected representatives and government policy-makers, who then establish the legal frame-work within which the service providers operate.

To facilitate responsiveness to customer's needs, when the 'long route' indeed takes too long, some models see the benefit in having a form of official customer representation through a 'quango' (quasi autonomous non-governmental organisation) such as the Consumer Council for Water in England and Wales or the WaterWatch groups in Zambia. Such bodies, established by government or regulator, have duties to focus on the services obtained by consumers, helping and perhaps adjudicating with customers in their complaints, and promoting the customer inter-est in any price-setting process, all recognising the absence of a more normal, market-oriented, protection of consumers where it is possible to change suppliers.

This approach, both regulation and consumer protection, can also be seen as a function of centralisation of water services provision. England and Wales have one utility per 2.8 million people, whereas the figure for France is approximately one per 5,000 and community providers in India function at one per 1,000–2,000. The efficiencies presumed to be achieved by the cen-tralisation of monopoly service providers require the additional vertical governance structures of regulation and customer watchdogs which decentralised systems do not require. In the decen-tralised approach, the Mayor or Sarpanch has everyday experience of the quality of water ser-vices being received and problems particular consumers may be experiencing, whether through leaks or poor quality water or inability to pay. But the asymmetries in information persist. How does the outsider know just what the underlying costs of providing that service are? This classic example of 'principal-agent' theory suggests a need for some form of independent monitoring of costs and performance even in the most decentralised, close to the consumer, horizontal gov-ernance set-ups.

In an ideal world, the generally preferred public sector service provider (preferred ideologi-cally in many places) would simply be an extension of the government and its policy-making, and there would only need to be a single entity supporting services to consumers. In reality it has been found to be advantageous to have a clear organisational separation between govern-ment, as policy-maker and representative of the citizens, and its service delivery agents, to achieve the required governance principles of enhanced transparency and accountability. Ensuring that in countries where there are many service providers they do indeed deliver according to the agreed policy and standards leads to the need for licensing and monitoring entities to oversee this organisational separation. Hence, in our accountability diagram, the role of regulator/monitors to oversee the relationship between government and service provider – in the first instance to ensure environmental quality as well as water potability. Such entities can be, and usually are, departments of government. However, there is also now recognised to be the need for some form of 'economic regulation' – that is, an overseer of the commercial and consumer-oriented delivery progress of the service provider.

Noting, as above, the capital-intensive nature of networked utilities with respect to timely investment in fixed asset development, extension and enhancement, there is the concomitant need for timely investment in renewing such fixed assets. Then there is a need for government, representing the citizen-consumer in this situation, to ensure that not only are these fixed assets delivered and maintained in a cost-effective manner, but also that they are operated effectively, equitably and efficiently for the benefit of all consumers. This requires oversight of both performance and costs, with a similar oversight of the means of financing these capital and operating expenditures. The most common route to financing has been for government to fund capital expenditure (via taxation and/or borrowing supported by taxes) with tariffs paid direct by consumers to finance operating expenditure and capital maintenance.

In practice, this balance has not always worked (has usually not worked!) in that governments are unable to manage investment in the timely expansion of assets as populations and cities grow. This is largely due to the myriad competing societal needs for tax-based financing and the envisioned political risks of governments setting an appropriate level of tariffs to deliver the necessary level of operating expenditure, not to mention capital maintenance expenditure, for the system. Note, again, that the public sector, like any organisation, does not necessarily function in the best interests of the public, but, over time, can tend towards favouring the interests of its own staff – what is known as 'producer capture'. It is for these reasons that 'economic regulators', even in public systems, deliver a critical function when given sufficient autonomy to act between government and the service provider entities. A regulator needs a level of independence from government in order to give an objective view of service quality, costs and benefits so as not to be overly captured by political interests, which usually relate to keeping prices of services too low, to the detriment of long-term service provision. Being realistic, every 'independent' regulator has to operate within a political environment if both their organisation and their personal roles are to survive, therefore independence is never an absolute. But we believe that one of the key roles of a regulator is to 'protect' the service provider from government demanding levels of investment and service whilst denying that provider the funding and tariff levels which are needed to enable such provision.

Interestingly, in the recent OECD survey of economic regulators (OECD 2015), such a function was not recognised. The main reasons given to justify the establishment of a water regulator (based on over 30 responses from regulators worldwide) were, in order of perceived importance: to protect the public interest, to make service providers more accountable, as part of a broader process of regulatory reform, to accompany a privatisation process, in response to an international commitment and, finally, to curb corruption. We suggest that service providers need the protection of a regulator so as to ensure the revenue levels necessary to finance the capital-intensive water sector. Indeed, when England and Wales economic regulator Ofwat was established in 1989, one of its two main statutory duties was to ensure that water providers could finance service provision. Protection of customer interests was only added in subsequent legislation.

Lower-income societies may well not want, or may well not benefit from, a proliferation of different entities as described here. However, the functions described need to be incorporated in any governance and regulatory approach. There is no single model of regulatory office, every country has to adapt the principles to suit its own structure of governance. There can be regulation by: government department; performance agreement; contract; competition or fair trading authorities; or by temporary advisors or 'expert panels' comprising legal, financial and technical expertise, regulating primarily during price-setting periods. Similarly, these 'semi-autonomous' regulators can function at city level, at state or province level, nationally with the added variation of either functioning only as a water and sanitation regulator, such as NWASCO in Zambia, or as a multi-utility regulator concerned with power as well as water such as PURC in Ghana or RURA in Rwanda (which also has responsibilities in overseeing communications and media, transport and energy as well as water and sanitation). Again, there are models where the regulator is empowered to act as a single person, rather like a judge in a court of law, and others where some form of regulatory board is seen as more effective in 'de-personalising' any decision.

Ensuring service to the poor is a critical element of governance and regulation, something we will come back to later in this chapter. But we suggest that the purpose of having a public water provider (or a private provider operating under a public licence) is to serve all the public, and particularly the poorest of the public who are most at risk from public health challenges, recognising that the rich can, and often do, afford to pay for their own services, irrespective of government-mandated support. Governance and regulation of water and sanitation, therefore, should be seen as inherently pro-poor in its construction, noting that the exact form of organisations, and their separate and overlapping activities and responsibilities, are much less important than the societal wealth available to finance and deliver ongoing networked services.

Ultimately, the work of the regulator is to come to a conclusion on the numbers in the service provider's business plan. That is, the level of investment to be allowed in any period and the reasonable operating and capital maintenance costs, with the expectation for ever-increasing efficiency in both capital and operating delivery. If the resulting tariff is 'too high', then the regulator, through consultations with policy-makers, service providers and, ideally, customers, reduces the investment plans or seeks for greater efficiency savings etc. Whereas if the resulting tariff is 'too low', then the regulator must work with those stakeholders to get the tariff to a viable level that allows sufficient investment in the sector. For more detailed information please see 'The Lisbon Charter: Guiding the Public Policy and Regulation of Drinking Water Supply, Sanitation and Wastewater Management Services' which seeks to 'lay out the basic principles for good public policy and effective regulation of drinking water supply, sanitation and wastewater management services, declaring the respective rights, duties and responsibilities of the governments and public administration, regulatory authorities, service providers and users' (IWA 2015).

Regulations and standards

If regulating is the art of balancing between competing interests, just as a judge in a court of law applies agreed laws or standards, a regulatory system needs accepted standards of water quality, wastewater effluents, efficiency indicators and customer service performance in order to come to objective decisions. Regulations refer specifically to 'the rules that emanate from governments and public administration and are enforceable by regulatory authorities or regulators' (ibid.). A key aspect of governance delivering for people is, therefore, the setting of standards which can be both the targets to aspire to, in the case of the SDGs and the progressive realisation of human rights in lower-income countries, and the performance floor below which services should not fall in higher-income states.

The global targets of the SDGs, in this context SDGs 6.1 and 6.2, have the aim of delivering 'By 2030, universal and equitable access to safe and affordable drinking water for all' and 'access to adequate and equitable sanitation and hygiene for all and end open defecation, paying special attention to the needs of women and girls and those in vulnerable situations'. Monitoring of these targets refers to 'Percentage of population using safely managed drinking water services' and 'safely managed sanitation services'. These objective 'standards' deliberately give a high-level focus on 'safely managed' service delivery, a concept which incorporates many sub-sets of standards for management but does not specify too much too soon with regard to water quality, for example. With no city in south Asia, for example, achieving 24x7 continuous water supply, there can be no immediate regulatory expectation for potable water for all.

In higher-income countries, which have already achieved universal coverage, the standards for water quality, wastewater recycling etc. become much more relevant, and ever more detailed, representing the next level of achievement to be ensured. Typically, it is the World Health Organization standards, developed over years through professional inputs from many countries, which form the basis for water quality and wastewater recycling targets. The WHO water quality standards (WHO 2011) give 'international norms on water quality and human health in the form of guidelines that are used as the basis for regulation and standard setting, in developing and developed countries world-wide'. Countries then either adopt the WHO standards totally or adapt them to suit their own situation, as in England and Wales, where the Drinking Water Inspectorate reports that 'Many of the standards come from the 1998 European Drinking Water Directive and are derived mainly from the recommendations of the World Health Organization' (DWI 2009). WHO also has standards for wastewater recycling – that is, wastewater treatment and discharge to the environment (WHO 2006). Again, countries, through their environment agencies, normally have their own variations on these standards to be achieved, as in the USA's Environmental Protection Agency (EPA): 'Effluent Guidelines are national regulatory standards for wastewater discharged to surface waters and municipal sewage treatment plants. EPA issues these regulations for industrial categories, based on the performance of treatment and control technologies' (EPA undated).

In some countries the setting and enforcing of specific water and wastewater quality standards are functions of a single regulatory body, but the most common approach in higher-income countries is to separate the highly specialised and scientific focus on water quality and wastewater discharges from the economic aspects. This is slightly more costly in terms of government agencies, but the additional transparency delivered through entities having to formally respond to each other's interests in the public domain is beneficial.

Beyond drinking water and wastewater quality standards, regulators need 'key performance indicators' which act as standards or benchmarks against which they can measure the performance of the service providers. This should include assessing customer satisfaction levels, but can also involve more technical benchmark measures such as the following ones proposed by the East and Southern Africa Water and Sanitation Regulators Association (ESAWAS 2015): water (80–90 per cent) and sewerage (80–90 per cent) coverage; hours of supply (16–20); non-revenue water (physical and commercial losses, 20–25 per cent); O&M cost coverage (100–149 per cent); bill collection efficiency (90–95 per cent); metering ratio (95–99 per cent); and staffing efficiency (5–8/1,000 connections). There are many more possible indicators with the International Benchmarking Network for Water and Sanitation Utilities (IBNET)[2] information recommended as the most useful global overview.

Service providers

The challenge of service provision is to ensure that expensive societal investments in fixed assets for water and sanitation are managed, within the context of governance and regulatory oversight, in such a way that all can benefit. The physical structure of the systems necessarily leads to provision by a monopoly. In describing the range of public, community, private service provider models it is first necessary to make the distinction between bulk provision, similar in manufacturing terms to producer and wholesaler responsibilities, with distribution – that is, retailing the service. Different approaches or models could well serve for each of wholesale and retail water supply as well as for 'retail' – that is, local – wastewater service collection and 'wholesale' wastewater treatment and recycling. We now consider the different models of service provision, looked at in more detail in Franceys 2001. Every one of these models has interesting and unique features, some of which models being directly beneficial to consumers and society, others simply developing in response to a particular time and context.

State providers

There are a range of public or state provider models, across the spectrum of centralisation and decentralisation and from government-managed to government-owned and autonomously managed. *Government ministry* or *department* provision, which delivers centralised government control of investment and service delivery, is commonly seen as not requiring additional regulatory input, but when it comes to *decentralised government department* or a *municipal department* (*regie directe*), to encourage autonomy and responsibility at district or municipal level, central government then begins to want some form of oversight, to at least have the information as to what is being spent for what purpose and what service are consumers getting as a result. One of the challenges of municipal department service provision has been the tendency for municipalities to see water supply as a source of cash revenue which then can be used to subsidise other municipal services. The resulting lack of finance for capital renewal and expansion usually leads to a decline in service standards. Hence the need for *devolved municipal control,* whereby the municipal department has some level of autonomy in managing its assets whilst still under the overall control of local elected politicians. Where local government is seen as too weak, *semi-autonomous utility departments or authorities* (*regie autonome*) are developed to ensure managerial autonomy on a wider scale. This leads on to *autonomous, corporatised utility boards* (*regie personnalisee*), which act like a private business with full managerial, financial and legal autonomy, but where the Board and shareholders are the government (Blokland *et al.* 2000). It is not uncommon to find a transition between these models over time as governments search for more effective and efficient service providers, without apparently losing control through any of the forms of private sector participation described below. Abiwu (2016) describe the ongoing changes in the Ghanaian public sector as they seek better models over the last century. This story indicates that there is no quick-fix solution, apart perhaps from the unaffordable 'heart transplant' of significant privatisation, and that all developments in governance, institutional development, regulation and change management are typically a long drawn-out process with no end in sight.

Non-state providers and contract models

There are a range of community management approaches, both formal and informal. This generally starts with the volunteerism typical for community management of handpumps that are managed by a village water and sanitation committee, often referred to as a community-based organisation (CBO). Yet as systems become more sophisticated more professional forms of 'community-based management' become more effective and efficient and this is where registered trusts or societies or water users associations, acting on a 'not for profit' basis take on responsibility to manage the (usually government- or donor-funded) assets. This then moves up in sophistication towards different forms of professional contracting arrangements.

Privatisation

In the most recent version of the *Water Yearbook*, David Lloyd Owen's (2015) overview of the level of private sector involvement around the world, there are an estimated 1,140 million people 'currently served to some extent by [formal, large] private sector contracts. This compares with, for example 563 million people as being identified as served by the private sector in 2005 and 335 million in 2000.' This section describes the most common models; for a fuller investigation of the development of the variously labelled privatistation, private sector participation, or even public private partnerships, please see Marin 2009.

Service agreements or contracts

The simplest form of non-state or private involvement is through small-scale agreements or contracts to undertake specific tasks within a relatively short period. Many of these tasks could be, and often have been, undertaken by the primary utility provider. However, management finds that it is more effective use of their time to focus on their core responsibilities of water and sanitation provision rather than, for example, managing security staff on all their sites, organising in-house vehicle maintenance or delivering new water connections (Sansom *et al.* 2003). The first examples have the additional advantage to some managers of both reducing the number of staff with public sector pay scales and benefits whilst delivering an improved 'staffing ratio' (staff per 1,000 connections), which has become such a focus of the funding agencies. The latter example can also be illustrative of the governance drive in some countries to facilitate competition in service provision wherever possible.

It is not only conventional private sector companies contracted to the public utility who undertake service contracts. For example, the sector recognises communities coming together as 'community contractors' (also NGO contractors) to undertake specific tasks, perhaps excavating trenches for new water distribution pipelines, or constructing street drains or taking contractual responsibility for bill collection. There are also what are known as micro enterprises: the small-scale independent providers (SSIPs), who operate in local water abstraction and distribution, selling through vending and kiosks, as well as pit latrine emptying, solid waste collection). Such SSIPs can be informal, self-selecting or franchised – that is, given formal rights, often exclusive, to serve customers in particular area under 'delegated management contracts' but then with duties then to ensure water quality and limit *prices.*

Management contracts

Moving up the spectrum of complexity of tasks and value of contract, we find 'services contractors' (*prestation partielle*) who receive fees per task, perhaps for installing new connections, leak detection and leak repair, sewer pipe-jetting. A management contractor (*gerance*) takes responsibility for the overall service delivery, usually over five to seven years, receiving fees proportional to the physical output of fixed assets, collecting tariffs according to agreed levels. A shared-profit management contract, a results-based performance contract (*regie interesse*), incentivises the contractor through fees proportional to output, plus bonus or shared profits due to out-performance of agreed targets. The agreed targets are a form of *ex ante* regulation, clear performance indicators having been set for a relatively short period with an accepted level of profit-sharing. A good example of a recent management contract can be seen in Algiers, where, at a significant price, the contractor Suez has transformed services in the city over a four-year period, managing the government's new asset investments as well as the operation and maintenance of water and sewerage services to the city whilst the government continues to subsidise the tariffs paid.

Investment contracts

The most common form of an investment contract is a build, operate, transfer (BOT) agreement whereby a private contractor takes responsibility to build a new asset for the water service provider, usually a water treatment or wastewater treatment facility on a 'greenfield' site. The contractor finances the building of the asset through its own financing arrangements (bonds and bank loans) and then runs the plant for a specified period before transferring it to the service provider in good condition. The service provider has paid for that service for the duration of the contract under a specified 'take or pay' agreement whereby payment is made irrespective of the actual demand for water from the plant. This is believed to deliver the best balance between risks that are controllable by the contractor and those under the control of the contracting service provider.

The initial idea of a BOT has been extended (at least in proliferation of initials as opposed to any significant development of the model) into a BOOT, which included ownership through the length of the contract; or with an additional T to represent training, or a D for design, as in design build operate; or with an R to represent rehabilitate operate transfer; or with an F to represent finance (though that was always assumed anyway); or even, to conclude this alphabet soup, reverse BOT, where there is public financing but the contractor takes the risk of making it all work.

Lease contracts and joint ventures

These include 'lease contracts' (*affermage*), whereby the private contractor leases the fixed assets from the public asset holder, operates the service and bills customers, paying an agreed amount of the tariff to the asset holder but, different from a management contract, taking the commercial risk of supplying sufficient water and efficiently collecting sufficient tariffs to pay their own costs as well as the asset holders. Such contracts are very common in France, though now under the challenge of 're-municipalisation'. It is this increasing risk in operation which defines the different types of private sector involvement as we move across the public–private partnership (PPP) spectrum. In an 'enhanced lease' the lessor also takes responsibility for investing some of their own capital, in distribution extensions – for example, solving a municipality problem when they have no access to additional investment capital.

In a 'joint venture' the public sector jointly owns the private service provider along with a parent private company, to ensure control and sharing of any profits, with the resulting contract designed to operate as if it were a private-only entity.

Concessions

Under a 'concession', a once popular approach to meet the investment and service delivery reform in large cities (Buenos Aires, Manila and Jakarta being classic examples), the concessionaire takes on all responsibilities for the likely 30-year length of the concession, investing in and operating and extending services to the whole of the specified service area, accepting the risk that the needed high investments at the start of the concession will be financed by the 'profits' (returns to capital employed) during the remainder of the concession. The example of Buenos Aires, where government cancelled the concession approximately ten years into the 30-year period, caused very significant financial losses to the concessionaire, despite an award of hundreds of millions of dollars by the International Centre for Settlement of Investment Disputes, sums yet to be paid (ICSID 2015). This and other examples have led the major international water companies to be very wary of new concessions.

Divestiture

In a 'divestiture', popular only in Chile and England, the public company is transferred to private sector ownership, either through sale to a private body or sale of shares. Any or all profits from operations are returned to shareholders who own/purchased/constructed the fixed assets under regulated licences (25-year 'rolling' licences in England and Wales) or in perpetuity. This has been the real driver for economic regulation.

Regulating for services for all

The move towards commercialisation of water services has, by default rather than design, nudged too many service providers to focus on serving the middle- and higher-income consumers, to deliver the necessary revenues. This has led the poorest in too many countries to lose the opportunity to access clean water through nearby public standposts, paying instead to intermediary vendors at a volumetric rate often ten times the official price. Similarly, in high-income countries this tendency has led to 'cost-reflective' tariffs necessary to deliver environmental as well as public health benefits that are simply unaffordable to the poorest 20 per cent of the population. The task of a water provider, whether publicly or privately managed, as we described earlier, is to ensure the public health benefits of potable water and effective sanitation to all the public, with an in-built bias towards the poor.

With the post-divestiture 40 per cent real increase (above the rate of inflation) in water tariffs in England and Wales, the government responded in its policy-making role by banning disconnections for non-payment of water tariffs. The water companies had to respond in various ways to ensure adequate revenue flow continued to fund the service to all, but the result has been a further indirect or 'hypothetical', according to Ofwat, increase on the tariffs of those who do pay of $27 per household per year to cover those bad debts (Ofwat 2015), approximately 5 percent of the annual average water bill. More positively, the regulator, following the policy of government, has required service providers to develop social tariffs to support those on social welfare benefits, in addition to initiatives, such as WaterSure, which all providers have developed to give financial support to metered household customers who use large amounts of water for essential purposes. Following a recent regulatory requirement of enhancement of customer involvement in decision-making, those customers surveyed with respect to social tariffs agreed to only a very small increase in their own bills in order to cross-subsidise the poorest, often suggesting that it was the role of government to undertake that task, not the water companies through tariff increases.

This is, indeed, the model developed by the government of Chile, with support from regulator Superintendencia de Servicios Sanitarios (SISS), which:

> has achieved extraordinary regulatory achievements since 1990, with the country moving from a situation where 93% of people were supplied by public companies to one where none are, and where over the same period of time potable water coverage has gone from 98% to 100%, sewerage coverage from 85% to 96% and wastewater treatment from 18% to 100%.
>
> *Stedman 2014, p. 30*

In this model, the government takes responsibility for ensuring that the poorest are partially supported in the payment of their water bills (increasing due to the significant improvements to the service) leaving the now private service providers to continue doing what they are presumed to be best at – that is, supplying water.

In low-income countries, the limited access to urban water services leads the poorest to necessarily overpay through vendor suppliers, both formal and informal. The increased cost of water due to the poorest having to pay for the cost of the vendor directly, whilst the supplier avoids their duty to fund a distribution network in the informal settlements, is inequitable. There have been a number of initiatives by private service providers (Jacobs and Franceys 2008) to deliver innovative solutions to the slums, many of which have now been taken up by public providers, often supported by donors. Indeed, some governance systems have responded through the technological solution of electronic pre-paid meters, which have the benefit of allowing consumers to access water at the lifeline tariff through a 'robot' which is programmed to sell at any time and at a reduced price (Heymans *et al.* 2014). The high cost to the service provider of the imported robot could perhaps be better used, in development terms, funding people to undertake the service with a much cheaper level of support from the provider. Regulators have a duty (we argue they should have a more explicitly stated duty) to ensure that service providers deliver water to all; in the informal areas this requires innovative, 'organic' approaches which complement the 'mechanical' standardised approaches necessary to ensure potable water through a conventional pipe network. Service providers have responded by developing 'pro-poor units' (WSP 2009) whereby a special team of sociologists and engineers come together to extend services to the slums – sometimes through pipe connections direct to single-room households (Franceys and Jalakam 2010). This then requires regulatory approval to ensure easy payment terms to obtain the convenience and health benefits of a household connection.

In all of these approaches, the role of the regulator can be considered as analogous to the role of a referee in a football match, the job of the referee being to ensure that the rules of the game are followed but not being so rule-bound that the game grinds to a halt too often. The teams in our analogy are the government, as policy-maker, and the service provider, the regulator above all having the task of keeping that interaction effective. The crowd represent the consumers who have to pay to be entertained by the ongoing match, with any consumer representative forum being the 'biased linesmen', always signalling offside on behalf of the consumers. We hope for a 'win–win, score-draw' in the match, but our analogy also recognises that in many parts of the world low-income citizens might not be able to afford the entry ticket to the ground, playing their own unrefereed game on wasteland outside. When the match is taking place on a non-level playing field between a premier league club (government) and a hobbled and blindfolded third league team (a typical low-income country service provider?) then it is not possible for the referee to make much of a difference.

But conventional regulation is not required in every setting in order to ensure good governance of improved water and sanitation for all. The classic, and complex, 'incentive-based economic regulation' of privatised service providers through the Ofwat approach in England and Wales, supported by the quality regulators of the Drinking Water Inspectorate and Environment Agency, has worked extremely well in that, by giving a five-year 'contract' (price agreement), the suppliers are able to keep a significant proportion of any additional efficiency savings that they deliver during the five-year review period. This benefit to shareholders drives further efficiency into service delivery with the additional profits shared with consumers (through reduced tariffs) at the next price setting.

However, such incentive-based regulation is extremely time-consuming, data demanding and can lead to apparently overly high profits for quite some time, a real challenge to governance. Alternative approaches of 'rate of return regulation', whereby the private supplier is allowed a fixed profit margin each year, may lead to similarly high governance and transaction costs of auditing accounts, and give no incentive to the supplier to out-perform. If these are the challenges of regulating private suppliers, the public providers are disadvantaged in having no real

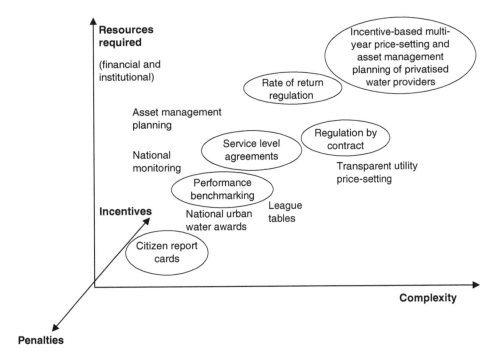

Figure 9.2 The regulatory spectrum

incentives or penalties for good and poor performance. This is an indication that over-investment in economic regulation of public providers might not be cost effective, but regulators such as WASREB in Kenya manage to deliver for consumers through using annually reported key performance indicators compiled into an 'impact' report (WASREB 2015). Key to the report's success are the league tables of best and worst performers, along with prizes awarded in public each year. Even though there is no financial incentive to out-perform, there is an impressive professional driver to be seen to be doing better in front of professional peers.

Where there is only a single national urban supplier, such as National Water and Sewerage Corporation in Uganda, there are moves now to introduce a regulator. The corporation's managing directors have long felt that their performance contract with the host Ministry is a more cost-effective method of regulation, when coupled with their own incentive-based staff contracts. These 'internal delegated management contracts', signed with their own regional managers, supported by significant financial incentives to staff (if the team out-performs, all receive additional payments; if there is under-performance only the manager suffers financially) have the effect of making senior management the economic regulator of their own internal service providers.

Manuel Alvarinho, President of the Water and Sanitation Regulatory Council (CRA), Mozambique, explains regulation in his context (Stedman 2014, p. 31).

> The job of being a regulator in Africa is worst because the concept of independent regulation depends very much on the context of where the regulator operates. There is very often a feeling of isolation. In the Africa region, regulation is a strange animal – we need to put together the strange animals. Communication is very important for us; isolation means lack of access to knowledge and experience. We are a small family but we are growing in practical terms, doing peer exercises, benchmarking. We cannot

benchmark internally; the only solution is benchmarking Maputo against Nairobi, Dar-es-Salaam and so on.

He also warned: 'it is very important that we keep defending the independence of the regulator. It is difficult at country level. Regulators say they are independent, then you find a minister shares the board.'

For further information on economic regulation please see Groom *et al.* (2006), Katko *et al.* (2012) and Rouse (2013).

Conclusion

The world has come up with a remarkable variety of solutions to delivering improved water and sanitation to all – with a similar, if smaller, variety of means of regulating that delivery. Overall, it is the principles of governance of the water which need to be recognised in order to ensure the progressive realisation of services to all.

High-income countries, with their very different models, have the wealth and institutional capital to make any of these models deliver appropriate services at a reasonable cost. However, in our earlier study *Regulating Water and Sanitation for the Poor* (Franceys and Gerlach 2008), we concluded that, in the 'typical' low and middle income country (LMIC), regulatory outcomes are very obviously shaped by the political context. Public and social service objectives, widespread poverty and intensifying scarcity of resources accentuate the interplay of technical and social – and therefore political – dimensions of water services. The transformation of the role of the state, implicitly required for effective and efficient regulation, has rarely been achieved. As a result, regulatory agencies in LMICs, born into delicate economic and institutional situations, operate in politically charged environments. The process of regulation, which involves many stakeholders, could be more adequately described as a subtle, perhaps impossible, balancing act on the part of a regulatory agency to achieve the outputs desired by customers and society, as against the inputs that customers and governments are willing to contribute. This implies the need for political leadership to recognise that the regulatory role requires a greater degree of personal leadership and discretion than is generally perceived reasonable in what is often presumed to be only an administrative, accounting-based tariff-setting process.

Nevertheless the research findings suggest that economic regulation of water services can be an effective mechanism for LMIC governments to institutionalise their commitment to universal service and consumer protection whilst also promoting incentives for efficiency and effectiveness. A clearer definition of a relevant and flexible universal service obligation (USO), and the mandate to require service providers to achieve that USO, is a crucial regulatory tool and therefore a requirement for government policy. However, the extent to which benefits for vulnerable members of society can be realised depends on regulators themselves recognising and addressing the realities faced by the poor. This has often been a step beyond their initial remit but, to their credit, our ongoing studies suggest that many have indeed taken that step. The economic regulator in Kenya now has a 'pro-poor' indicator against which public service providers have to report each year, and through which all their peers, as well as civil society, consumers, politicians and policy-makers, now have the necessary information to hold them to account. Governance and regulation works.

Notes

1 OECD Programme on Water Governance. http://www.oecd.org/env/watergovernanceprogramme.htm
2 International Benchmarking Network for Water and Sanitation Utilities. www.ib-net.org

References

Abiwu, N. (2016) Private sector involvement in urban water supply management, Ghana, unpublished PhD thesis, Cranfield University.

Blokland, M., Braadbaart, O. and Schwartz, K. (2000) *Private Business, Public Owners: Government Shareholdings in Water Enterprises.* The Hague, The Netherlands: Ministry of Housing, Spatial Planning, and the Environment.

Davison, A., Howard, G., Stevens, M., Callan, P., Fewtrell, L., Deere, D. and Bartram, J. (2005) *Water Safety Plans: Managing Drinking-Water Quality from Catchment to Consumer.* Geneva: WHO.

DWI (2009) *Drinking Water Safety Guidance to Health and Water Professionals.* London: Drinking Water Inspectorate.

EPA (undated) *Effluent Guidelines.* https://www.epa.gov/eg

ESAWAS (2015) Regional benchmarking of large water supply and sanitation utilities 2013/14 Report, October. Eastern and Southern Africa Water and Sanitation Regulators Association, Lusaka, Zambia.

Franceys, R. (2001) *Public Private Community Partnerships in Urban Services for the Poor Theme Paper: Water, Sanitation and Solid Waste, RETA 5926.* Asian Development Bank. Available from: http://citeseerx.ist.psu.edu/viewdoc/summary?doi=10.1.1.197.3982

Franceys, R. and Gerlach, G. (2008) *Regulating Water and Sanitation for the Poor: Economic Regulation for Public and Private Partnerships.* London: Earthscan.

Franceys, R. and Gerlach, G. (2010) Regulating water services for all in developing economies. *World Development* 38(9), pp. 1229–40.

Franceys, R. and Jalakam, A. (2010) 24x7 water supply is achievable. *Water and Sanitation Program Field Note.* Washington, DC: World Bank.

Gasson, C. (2016) Where's water in a non-Piketty economic environment? *Global Water Intelligence.* https://www.globalwaterintel.com/insight/where-s-water-in-a-non-piketty-economic-environment (accessed 4 February 2016).

Groom, E., Halpern, J. and Ehrhardt, D. (2006) Explanatory notes on key topics in the regulation of water and sanitation services. Water supply and sanitation sector board discussion paper series, Paper No. 6, June, World Bank.

Heymans, C., Eales, K. and Franceys, R. (2014) The limits and possibilities of prepaid water in urban Africa: lessons from the field, World Bank Group WSP Report, Washington, DC.

Hutton, H. and Varughese, M. (2016) *The Costs of Meeting the 2030 Sustainable Development Goal Targets on Drinking Water, Sanitation, and Hygiene, Water and Sanitation Program.* Washington, DC: World Bank.

ICSID (2016) Suez, Sociedad General de Aguas de Barcelona S.A. and Vivendi Universal S.A. v. Argentine Republic (ICSID Case No. ARB/03/19), decision taken 9 April 2015, International Centre for Investment Disputes, Washington, DC, reported Suez Environnement Press Release, 9 April 2015, Paris.

IWA (2015) The Lisbon Charter: Guiding the Public Policy and Regulation of Drinking Water Supply, Sanitation and Wastewater Management Services. IWA.

Jacobs, J. and Franceys, R. (2008) Better practice in supplying water to the poor in global PPPs. *Municipal Engineer*, Institution of Civil Engineers 161(ME4), pp. 247–54.

Katko, T., Juuti, P. and Schwartz, K. (2012) *Water Services Management and Governance: Lessons for a Sustainable Future.* London: Portland.

Kishimoto, S., Lobina, E. and Petitjean, O. (2015) *Our Public Water Future: The Global Experience with Remunicipalisation.* London: PSIRU.

Lloyd Owen, D. (2015) *Indepth Water Yearbook Your Guide to Global Water Industry Data: 2014–15.* London: Arup.

Macheve, B, Danilenko, A., Abdullah, R., Bove, A. and Moffitt, L. (2015) *State Water Agencies in Nigeria: A Performance Assessment.* Directions in Development. Washington, DC: World Bank.

Marin, P. (2009) *Public Private Partnerships for Urban Water Utilities, A Review of Experiences in Developing Countries.* Washington, DC: World Bank.

Nickson, A. and Franceys, R. (2003) *Tapping the Market.* London: Palgrave Macmillan.

OECD (2012) 'Seven Gaps': Water Governance in OECD Countries: A Multi-Level Approach. http://www.oecd.org/gov/regional-policy/OECD-Principles-on-Water-Governance-brochure.pdf

OECD (2015) *Managing Water for Cities.* Paris: OECD.

Ofwat (2015) Affordability and debt 2014–15. December, Birmingham.

Rogers and Hall (2003) *Effective Water Governance.* www.un.org/waterforlifedecade/waterandsustainabledevelopment2015/governance_frameworks.shtml

Rouse, M. (2013) *Institutional Governance and Regulation of Water Services: The Essential Elements*, 2nd edn. London and New York: IWA.

Sansom, K., Franceys, R., Njiru, C. and Morales-Reyes, J. (2003) *Contracting Out Water and Sanitation Services*, Vols I and 2. Loughborough: WEDC, Loughborough University, 82 pp.

Stedman, L. (2014) The crucial role of regulators in water and sanitation provision. *Water21*, December. IWA.

Stoker, G. (1998) Governance as theory: five propositions. *International Social Science Journal* 50(155), pp. 17–28, March.

WASREB (2015) A performance review of Kenya's water services sector 2013–2014, 8/2015. Water Services Regulatory Board, Nairobi.

WDR (2004) *Making Services Work for Poor People*. World Development Report. Washington, DC: World Bank.

WHO (2006) *Guidelines for the Safe Use of Wastewater, Excreta and Greywater, Volume 1. Policy and Regulatory Aspects*. Geneva: WHO.

WHO (2011) *WHO Guidelines for Drinking-water Quality*, 4th edn. Geneva: WHO.

WPP (2010) *Water Governance Principles and Practice Water Sector Governance in Africa: Volume I Theory and Practice*. Tunis: Water Partnership Program (WPP) of the African Development Bank.

WSP (2009) Setting up pro-poor units to improve service delivery: lessons from water utilities in Kenya, Tanzania, Uganda and Zambia. *Field Note September, Water and Sanitation Program*. Washington, DC: World Bank.

10

LEGAL ASPECTS OF FLOOD MANAGEMENT

Andrew Allan

Introduction

The purpose of this chapter is to examine some of the ways in which law is relevant to flood management, and to set out a number of the most recent key developments in the ways that national governments have attempted to manage floods.

The law relating to floods cuts across many different areas: disaster management and emergency response; tort; civil defence; water; urban planning; coastal zone management; and land use, to name just a few. There has historically been an assumption among some that the study of the legal or governance aspects of flood management should be seen as a 'soft' approach to flood management, compared with the 'hard' infrastructural focus favoured by engineers, but this is rather more binary than the reality. In fact, as will be seen below, law is as directly relevant to 'soft' approaches such as natural flood management as it is to 'hard' solutions such as infrastructure development, the latter requiring authorisations provided under legislation, for example; neglecting the legal aspects of the 'hard' solutions may, in fact, be a contributory *cause* of flooding rather than helping prevent it.

Much of the legislation expressly dedicated to flood management is concerned with the allocation of responsibilities across institutions, delimiting institutional functions and financial issues, and clarifying questions of liability. It is also greatly concerned with the definition of triggers that necessitate particular responses, obligations and rights. Historically, the prevailing view of those promulgating flood legislation was that flooding was always bad, insofar as it could damage property and cause loss of life. Flood regulation is generally regarded as an ecosystem service (see, for example, Haines-Young and Potschin 2010), and therefore a benefit for humanity. In fact, the impacts of flooding are both negative and positive: flooding may be necessary for providing nutrient-rich sediment needed for agriculture (Egyptian flood irrigation techniques outlasted ditch-based irrigation practised by other ancient societies because of sediment provision and the avoidance of salinisation) and can be a critical factor in the recharging of aquifers. Floods also provide ecological triggers for some migratory fish and have important ecosystem rejuvenation characteristics (APFM 2013). Legal frameworks must allow, facilitate and accommodate these positive elements while minimising the impacts of the more negative aspects.

Although floods take place across multiple scales, this chapter will focus on legal frameworks that apply at the national level and below, but, other than with respect to the European Union, will not address the issues that arise in the context of international water and transboundary floods (see, instead, Rieu-Clarke 2008).

The ways in which the various facets of law interact with flooding changes over time, and may be disaggregated in many different ways. For simplicity, this chapter will adopt a structure based on hazard, risk and vulnerability as this corresponds best with much of the scientific analysis that has been done on floods, and this provides a useful framework in which to examine the law. The following analysis does not pretend to be exhaustive, but seeks to highlight some of the key elements of each, and those areas where legal aspects may be of most note for the future.

Hazard

There are a number of different types of flooding, each with unique combinations of causal elements. Although there is no universally accepted typology of floods, Barredo (2007) suggests that floods are generally grouped into three categories: river floods, flash floods and storm surge, to which can be added the further groupings of groundwater floods (Younger 2007, pp. 180–1), ice-jam floods, dam and levee failure floods, debris, landslide and mudflow floods. A further distinction can be drawn between extensive long-lasting floods and those that are local and sudden. Combined events are also possible: for example, flash floods causing river floods downstream (Barredo 2007) and, in Bangladesh, saltwater flooding incidents occur largely as a result of cyclone activity, but the fact that annual river flooding can inundate up to 60 per cent of the whole country (Salehin *et al.* 2007) makes flooding there highly complex. Floods may therefore take a multitude of forms, and this creates pressure on legal frameworks to respond appropriately.

Defining floods is difficult (Jones 1997) partly because scientific understanding of floods sees increased flows as simply part of a hydrological continuum. From a societal perspective, a flood is defined by its impact on property or human life, and from the standpoint of the law, a flood is only a flood if it has triggered a legal response by an affected party. Howarth's distinction between 'natural inundation' and 'flood' is instructive (Howarth 2002). Early flood legislation tended to avoid defining them – for example, there is no definition in the 1928 Act in the USA (United States Congress 1928) or in the Flood Prevention (Scotland) Act of 1961. This is no longer the case, with primary legislation now incorporating definitions that differ across jurisdictions. It may be that this has been driven, at least in part, by the growing demand for insurance: the National Flood Insurance Program (NFIP) in the USA contains a definition that is drawn directly from the insurance context, being the: 'general and temporary condition where two or more acres of normally dry land or two or more properties are inundated by water or mudflow' (see NFIP website at www.floodsmart.gov).

In the European Union, Art. 2 of the Floods Directive defines floods thus: 'the temporary covering by water of land not normally covered by water. This shall include floods from rivers, mountain torrents, Mediterranean ephemeral water courses, and floods from the sea in coastal areas, and may exclude floods from sewerage systems' (European Parliament and Council 2007).

The mapping of flood hazard is essential for the understanding of which areas are prone to flooding, and under what circumstances. The Floods Directive requires that Member States prepare flood hazard maps based principally on river basin districts or relevant coastal areas, identifying those areas where flooding is most likely. Areas where the likely return period of flooding is more than or equal to 100 years are deemed medium risk in the directive's categorisation (European Parliament and Council 2007, Art. 6). Hazard maps must not only include scenarios for flooding caused by precipitation events, but should also take account of sudden events such as the failure of flood defences (as happened in New Orleans under the influence of Hurricane

Katrina) and of dams, glacial lakes and storage facilities. Maintenance of dykes and dams (Vorogushyn *et al.* 2010) may fall below optimal levels in many countries, and problems with reporting and monitoring levels may undermine what appear to be strong legal measures. Dams and storage reservoirs are important elements of a flood management strategy (see Tarlock 2012 for the US example), but the consequences of their breach or overtopping must be incorporated in hazard maps. In mountainous regions such as the Himalayas, understanding the extent of the hazard posed by glacial lake outburst floods (GLOFs) is also critical, but these phenomena unfortunately coincide strongly with relatively weak state capacity to produce hazard maps and make them publicly available. Iceland is the exception to this in terms of both its capacity to deal with such floods and their relative frequency, with hazard mapping of its equivalent of GLOFs, *jökul-hlaups*, a priority. Interpretation of flood hazard maps connects directly with the legal and planning contexts because progressively more restrictive limitations on land use can be applicable as event frequency and magnitude increase (and vice versa), and this can be linked directly to, for example, colour-coded hazard probability zones on the map, with colour codes based on the expected return period of a particular magnitude.

The question of data availability will be addressed below, but requiring inundation maps for dam failure should be mandatory for both private and public operators (for an example of how this can work in practice, see Victoria State Government 2013). Hazard maps do not commonly include inundation areas in the event of dam failure, but this is because such events are viewed as too rare. Emergency plans will normally be required at the environmental impact assessment (EIA) stage for dam construction, indicating planning and implementation procedures, in the event of a breach, monitoring processes and the potential area of inundation clearly set out (see, for example, Mouvet *et al.* 2001). This may limit the proportion of the public who have access to such information, but ordinarily emergency planning procedures implemented by the dam's operators should include appropriate coordination and information availability for those living in the inundation area. There may be more general concerns with making this sort of information more widely available – for example, specifically those linked to the potential for informing terrorism.

The duration, velocity, extent and depth of flood events can be influenced by a number of anthropogenic factors. These include land use within the flood zone, such as urbanisation that reduces infiltration capacity, and upstream of the flood zone, including, for example, de- and afforestation, soil compaction and agricultural intensification (Forbes *et al.* 2015; Wheater and Evans 2009). Flood plain zoning is intended, in part, to limit construction in flood zones, as this will affect the extent of flooding and drainage capacity. Interference with drainage channels will also have an effect on the speed with which flood waters are able to disperse – that this remains problematic is highlighted by the fact that much of the case law on flooding, in the UK at least, has concerned culverts (Howarth 2002). Waterlogging can be a significant problem with respect to dykes and polders, where drainage channels may be inadequate or their maintenance may be neglected. Flood events take place but water is unable to drain properly (Kobayashi and Porter 2012; Nicholls *et al.* 2016), and this can be especially problematic when saltwater inundation has taken place, as is the case in Bangladesh, for instance (Nicholls *et al.* 2016), with longer-term impacts being exacerbated by the extended harm to poldered agricultural land and freshwater supplies.

While flood hazard may be affected by direct anthropogenic activities, the indirect influence of humanity is also manifested through climate change. Flood risk is increasing in some areas, due to the effects of climate change on, for example, sea level rise and on weather patterns that are resulting in more intense rainstorms and more intense storms (see, for example, IPCC 2014, p. 8). The IPCC *Fifth Assessment Report* identifies a number of approaches that might be used for managing individual risks of climate change, and although it does not mention the quality of legal, institutional and policy frameworks explicitly, it is clear that these are fundamental for the achievement of the approaches listed (e.g. early warning systems, hazard mapping etc., ibid., p. 15). This is all especially true with respect to deltaic areas that are vulnerable not only to sea level rise, but also to natural subsidence and to the influence of upstream uses and impoundments of rivers that affect sediment supply, erosion patterns and sediment trapping. Recent research has also indicated that the effects of socio-economic developments may be proportionately much greater than those resulting from climate change itself (Winsemius *et al.* 2016), and although the study is limited to river flooding alone, this gives impetus to the idea that governments potentially have a great deal of scope to alleviate the consequences of future flooding through the choice of appropriate adaptation responses and policy direction.

The expected increase in the number of intense storms and precipitation events in certain parts of the world highlights the difficulty in quantifying the hazard from flash floods. These are the result of intense rainfall over a small area over a short period of time (less than six hours according to the US Geological Survey) (Barredo 2007, p. 131). They are normally more common in hilly and mountainous areas, but flat land can be vulnerable too, if the conditions are right. Spain has been particularly badly affected in Europe since 1950 (ibid., p. 141). Flash floods are of interest mainly because of their disproportionate representation in flood casualty figures: Barredo's study of European flood events between 1950 and 2005 indicates that 40 per cent of the casualties of flooding have been as a result of flash floods (Barredo 2007; and Marchi *et al.* 2010), and the Asian Development Bank indicates that the figure is higher in China, with 70 per cent of the casualties of flooding coming from flash floods (Kobayashi and Porter 2012, p. 6). The difficulty in forecasting flash floods, and the urgency with which this capacity is needed, is underlined by the extensive research that has been funded to that end (see, for example, Quevauviller 2011).

Problematic modelling is not limited to flash flooding. The extent to which conventional water storage dams are incorporated into flood hazard mapping has been addressed above. Dams may, however, be designed for a number of purposes (one of which is, of course, flood amelioration) and recent events have drawn attention to dams that are designed to store mining waste, so-called tailings dams. The most notorious incidents in recent memory occurred at Mount Polly in Canada and Baia Mare in Hungary, with the latest in November 2015 with the Mariana tailings dam failing in Brazil. Although it is unclear at this stage why it happened and what the long-term consequences might be, the village of Bento Rodrigues was overwhelmed by the flood released after the Fundão dam ruptured, killing at least 19 people (Kiernan 2016). Prosecutors are now seeking damages to cover the costs of remediation, but with media sources suggesting that enforcement and lax monitoring contributed to the collapse, a quick solution seems unlikely.

The problems associated with tailings dams have been understood for many years (see, for example, ICOLD 2001), but efforts to improve emergency management and early warning procedures have not stemmed the increasing number of dam collapses. The World Information Service on Energy maintains a non-exhaustive list of tailings dams collapses, and their impacts, as far back as 1960 (at http://www.wise-uranium.org/mdaf.html; see also Kossoff *et al.* 2014), cataloguing around 100 incidents, not all of which entailed flooding *per se*, but all involving inundation of some sort. Tailings dams are particularly problematic with respect to flooding for a number of reasons: modelling the impact of their failure is difficult, partly because the nature of the debris held behind the dams makes the direct comparison with conventional dams inappropriate. A reliable methodology is required in order to inform understanding of the potential impact of such a failure (Rico *et al.* 2007). Tailings dams vary enormously – for example, in terms of height, design, material, storage volume, and the nature of the material stored behind them – and their scale increases with the life of the mine they serve. Making predictions about the nature of the impact of a failure in terms of the area affected is therefore difficult, but what is known is that they are more likely to fail than conventional water dams because of their characteristics (including the use of local materials for fill, for example, and a lack of clear regulations as to the design of tailings dams) (Rico *et al.* 2008). Application of the rarity standard used for conventional dams is therefore neither practical nor realistic.

Aside from these issues and the potential for the long-term damage to the environment caused by the toxicity of tailings, the interface with flood management more generally may be uncertain at best. In re-examining the two definitions of 'flood' given above, it will be noted that the EU limits its regulation to situations where there is inundation by water only: it does not include mudflow as the USA does. The impact of development restrictions is clearly highly relevant here, but not all jurisdictions have the capacity to enforce limitations on land use. Even in richer countries that nominally seek to restrict land use on areas prone to flooding, economic and political considerations can easily over-ride flood hazard mitigation priorities. The opposite is true in poorer countries, where the poor will develop and live on land that is potentially most vulnerable to flood hazard, land slips and mudslides. This will be discussed further below, with respect to vulnerability.

With respect to coastal flooding, the Dutch response to centuries of flooding, and more especially the catastrophic flood of 1953, was the construction of delta works that protect inland areas from coastal inundation. With natural subsidence and an increase in the population and level of economic activity in the area prone to flooding, the risk of flooding has gone up substantially since the 1953 flood. The approach now in the Netherlands is to protect against floods through the annually revised Delta Programme (mandated under the Delta Act on Flood Risk Management and the Freshwater Supply (States General of the Netherlands 2011)) and to flood-proof urban development (Van Alphen 2015). Similar approaches are also now at various stages of development and planning in both Vietnam and Bangladesh, though their efficacy has not yet been tested in these different contexts. More broadly, the need for levees and floodwalls to be kept properly maintained has been strongly underlined by the Hurricane Katrina experience in New Orleans, with robust monitoring regimes needed to ensure that the state of maintenance is understood. See, for example, Verchick (2015) on the uncertain condition of storm surge levees in the United States. In many countries, floodplains are the most fertile land, and therefore most suited for agricultural cultivation. Efforts to restrict construction in these areas are therefore likely to fail – Assam in India is a case in point – and the impact of flooding correspondingly inflated.

Risk

While flooding is a hazard, flood *risk* can be defined as a combination of the severity of a particular event with the probability of its occurrence, as mediated by the social vulnerability of the human system affected (Brooks 2003). Art. 2 of the Floods Directive uses a slightly different interpretation, with less overt focus on the vulnerability of the affected population: '[a] combination of the probability of a flood event and of the potential adverse consequences for human health, the environment, cultural heritage and economic activity associated with a flood event'.

The basic assumption underpinning many of these analyses is that flood risk is a function of the severity of a particular event and the chances of it actually happening, combined with a quantification of its impact on the human and physical systems affected. This latter element must necessarily include an understanding of what the UNISDR calls the 'characteristics and circumstances of a community, system or asset that make it susceptible to the damaging effect of a hazard' (quoted in Smith 2013, p. 53): vulnerability.

Leaving aside the question of vulnerability to the following section, the impacts of flood include direct losses such as immediate economic losses, loss of life and treatment costs. Indirect losses might also include a measure of economic and social disruption, and potentially also premature death and longer-term health problems (ibid., p. 25). Direct impacts can also include the mortality rate that follows flood events due to the resulting spread of disease (Smith 2013). This latter effect can be, in some cases, more significant than the immediate results of a flood: in the case of the Banqiao dam collapse in China, when 175,000 people died following what has been described as a one in 2,000-year event, the flood itself caused only one sixth of the death toll, with the vast majority dying as a result of famine and disease afterwards (Fish 2013). Floods can potentially cause an increase in transmission of water-borne disease (e.g. diarrhoeal disease, leptospirosis) and vector-borne disease (e.g. malaria), among others (WHO 2006). Establishing appropriately robust legal frameworks is a major problem as they need to be capable of mitigating impacts from floods, minimising the vulnerability of affected populations and putting in place emergency response frameworks that have the momentum to provide support for disease control and medical relief for some time following the event itself.

Between 2000 and 2015, the total damage from floods globally was just under US$430 billion, giving an annual average of around $27 billion (Guha-Sapir *et al.* 2016). Hallegatte *et al.* (2013) suggest that this could rise to $63 billion per year by 2050. During the same period, almost 90,000 people lost their lives as a result of flood events (Guha-Sapir *et al.* 2016). Urban development is a significant element in this rise, and this is being exacerbated by the general global trend towards urbanisation. Projections suggest that 66 per cent of the world's population will live in urban areas by 2050 (compared with 54 per cent in 2014 (UN DESA PD 2014)). In addition, most major urban centres lie on, or in close proximity to, bodies of water (Jha *et al.* 2011). This will have a significant impact on flood risk as the potential impact of flood events will be massively increased.

Reducing the impacts of flooding may be done a number of different ways. The first is through the development of infrastructure designed to contain flood flows and storm surges. This has been fundamental to the US approach, securing protection through the use of levees, and allowing construction on protected floodplains (Tarlock 2012). Engineered solutions have also been key elements of flood management strategies historically in Japan (Takahashi 2011), China (Kobayashi and Porter 2012) and the Netherlands (to name a few only).

Flood protection measures will not prevent damage in all eventualities, however, as there will always be events that exceed infrastructural capacity. The costs involved in constructing and maintaining this infrastructure will continue to increase as urbanisation progresses and sea levels continue to rise (see, e.g., Jonkman *et al.* (2013) specifically on coastal flood infrastructure and sea level rise). Such infrastructure may also have significant impacts on local ecosystems (Nicholls *et al.* 2016). Furthermore, there are fundamental problems with the idea of 'protecting' an area through the use of levees, as the risk to 'protected' areas is actually magnified because more people build on it, and when levees break (as they do), the impact is much larger (Tarlock 2012). Progressive urban development over time may affect run-off patterns and undermine the effectiveness of water control infrastructure (Takahashi 2011). The realisation that engineered solutions can never provide unlimited levels of protection has led to greater focus on other approaches, and a reassessment of the need to contextualise flood management within water resources management on a basin scale more broadly. The Chinese response to disastrous floods in 1998, in its '32-word' policy, was to directly acknowledge the role of land use management in exacerbating the impact of floods, and consequently focus heavily on afforestation and natural flood retention areas (Kobayashi and Porter 2012).

One of the alternative management tools that has been receiving much greater attention is natural flood management. This focuses on slowing or storing flood waters using natural features rather than 'hard' infrastructural interventions. It seeks to balance natural capacity with existing land uses such that, rather than replacing floodworks, it enhances these existing defences and management in a cost-effective way – there is no binary choice between hard and 'soft' protection/prevention methods. It also allows provision of some degree of flood protection in areas where risk may be low or where there is frequent small-scale flooding (Forbes *et al.* 2015). In Scotland, new legislation has mandated the use of natural flood management in implementing the Floods Directive (Flood Risk Management (Scotland) Act 2009 s.20 (Scottish Parliament 2009)).

In order to put the natural flood management systems in place, arrangements may need to be agreed with landowners that effectively limit the uses to which their land may be put. This may involve afforestation to slow run-off, or simply ensuring that land is left unused and undeveloped so that flood waters can accumulate there. Ideally (to avoid problems associated with repeated negotiations), these land use restrictions must remain in force for successive owners although the duration of the applicable restriction may be dependent on the relevant land use tenure (Law Commission 2014, p. 83). Conservation easements (also described as covenants or burdens depending on jurisdiction) restrict the use to which private landowners can put their own land (or parcels thereof), in order to protect the interests of neighbouring landowners, or more broadly, the public interest (see, e.g., Reid 2013). In return for financial compensation payments, landowners have been encouraged to modify the use of their land so that it is managed in a way that mitigates flood risk, or to agree to allow particular areas of land to be subject to flooding (see, e.g., Law Commission 2014). It may be that public interest considerations might justify repeated temporary flooding of particular parcels of land, but this will be jurisdiction-specific, and can be messy (see Tarlock 2012 for the US situation on takings).

The question of impact was historically addressed with respect to damage to property and loss of life, but, as noted above, this has been adapted over time to incorporate the environment itself. In the EU, the Floods Directive is tied directly to water resource management and its associated legislation, the 2000 Water Framework Directive (WFD) (European Parliament and Council 2000). The latter is primarily concerned with water quality, conjunctive management at basin level of ground and surface waters, and the achievement of environmental objectives. Echoing the disaster risk management framework that will be examined further below, the Floods Directive emphasises in its preamble that flood risk management plans should focus on prevention, protection and preparedness. Art. 7(2) requires Member States to prepare flood risk management plans at the WFD-linked river basin district level in the first instance. Like the WFD and its demand that environmental objectives are met, flood risk management objectives should be established that address 'the reduction of potential adverse consequences of flooding for human health, the environment, cultural heritage and economic activity, and, if considered appropriate, on non-structural initiatives and/or on the reduction of the likelihood of flooding'. When this was transposed in Scotland, a duty was imposed on Ministers to, among other things, 'promote sustainable flood management' (Water Environment and Water Services (Scotland) Act, 2003, s.2(4)(b)(i) (Scottish Parliament 2003)). Interpretation of the term 'sustainable flood management' was elaborated by technical groups (the National Technical Advisory Group, and latterly the Flood Issues Advisory Committee), and this focused primarily on enhancing resilience through four inter-connected elements: awareness; avoidance; alleviation; and assistance (see Spray *et al.* 2009).

The difficulties inherent in floodplain zoning have been alluded to above. Ensuring that the building of new properties within floodplains is restricted can have clear beneficial impacts on flood risk. The impacts of flooding may change over time, and this is driven, in part, by physical and climatic factors, such as changes in precipitation patterns or the geomorphology of a particular water source. Risk may be transferred across areas prone to flooding as a result of new construction and this may not only exacerbate the risk to properties already suffering from flooding, but also create risk for properties hitherto unaffected by flooding. As a consequence, curtailing building in flood risk areas is a popular approach, although the rigorousness with which it is applied may be affected by other drivers, such as the need to construct strategic infrastructure. Political drivers such as population pressures, urbanisation and the attractiveness of floodplain land are also important, and these may over-ride efforts to minimise floodplain construction: the *Financial Times* estimated, at the end of 2015, that 7 per cent of new houses were being built annually on floodplains, in defiance of the Environment Agency (Allen and Bounds 2015; see also Harvey 2016), which does not have a veto. The fact that risk changes over time highlights the need to regularly review and potentially revise flood risk management plans.

Vulnerability

As with 'hazard' and 'risk' above, there is no single accepted definition of 'vulnerability' when it comes to flooding. It is a key element of risk (Kobayashi and Porter 2012), but analysis of vulnerability can be made temporally, taking account of the period before, during and after a flood event (Balica *et al.* 2012), with the latter referring specifically to resilience (see also Smith 2013). In the analysis of Balica *et al.* (2012), resilience is a mitigating element in the overall calculation of vulnerability, offsetting the problems caused by the hazard itself and a community's susceptibility to it. Legal frameworks are widely accepted as being influential with respect to vulnerability (Handmer and Monson 2004).

The vulnerability of those living in areas that are liable to flooding varies dramatically, but the poor are often disproportionately affected. This is evident in Bangladesh (Nicholls *et al.* 2016) and in the consequences of Hurricane Katrina on the poorest areas of New Orleans (Gabe *et al.* 2005). A comprehensive examination of the role of law in each of the many elements of vulnerability cannot be undertaken here, but a few key aspects can be addressed, especially with respect to susceptibility of human populations and their resilience post-flood.

The first relates to the availability of information. In situations where people have the luxury of choosing where they construct their homes and locate their businesses based on factors beyond immediate necessity, the availability of accurate flood hazard maps will have a major influence on their choice. In richer countries, this is facilitated through dynamic online hazard maps, like those mandated under the Flood Directive – examples can be found at http://map. sepa.org.uk/floodmap/map.htm for Scotland, http://www.risicokaart.nl/en/ for the Netherlands and, in the non-EU context, http://dnrm-floodcheck.esriaustraliaonline.com.au/floodcheck/ shows hazard maps for Queensland in Australia. These are the same maps that are used by lending institutions and planning authorities, and may be used to limit financing for construction or property purchase, or to prohibit construction altogether. Developments in early warning systems also serve to minimise harm to human health, with mobile communications technology in both rich and relatively poor countries potentially using text messaging services to raise alarms.

Over the past 15 years or so, a more institutionalised and holistic approach to reducing vulnerability has been taking root, with the advance of disaster risk management frameworks, where flooding is incorporated into broader legislation that deals with disaster management for all types of event (whether earthquake, tsunami, flood or, in some cases, civil emergency situations). There was a flurry of disaster management legislation around the turn of the twenty-first century, driven in part by the Hyogo Framework for Action of 2005 (United Nations 2005) and its successor, the Sendai Framework (United Nations 2015). These include: South Africa in 2002; Queensland in 2003; Bangladesh, India and Sri Lanka in 2005; Canada in 2007; Pakistan and the Philippines in 2010. These are all fundamentally based on a number of key principles:

- prevention
- mitigation
- preparedness
- response and
- rehabilitation/recovery.

The approach recognises that complex vertical and horizontal integration across scales and sectors, and across time, is required if the impacts of flood events are to be minimised, and that 'a continuous and integrated multi-sectoral multi-disciplinary process of planning and implementation of measures' is needed (Disaster Management Act, 2002, South Africa, s.1 (Parliament of South Africa 2002)).

From the perspective of 'response', there is normally a hierarchy of institutional responses based on the scale of the event, so that local institutions, which are normally at the implementation end of the flood management process, are not swamped by events that overwhelm their capacity. It is effectively the direct application of the principle of dynamic subsidiarity, decision making at the lowest appropriate level in the circumstances: if circumstances do change, institutional responses will upscale with the event.

The language used in the Floods Directive does not tally directly with the terminology used above. The directive focuses primarily on reducing the impact of floods on the environment and on society, but the measures to be incorporated into the flood risk management plans do not expressly concentrate on the susceptibility or resilience of communities. That does not mean, however, that Member States will not adopt measures that reduce susceptibility or increase resilience – the Scottish transposition referred to above highlights the perceived need to frame flood risk management measures within resilience (Spray *et al.* 2009).

These two approaches – incorporating flood risk management within water resources management; and addressing flood risk management in a broader disaster risk management framework – are not mutually exclusive, but evidence does not suggest that efforts in the disaster risk management (DRM) context are necessarily being coordinated in line with the needs of water resources management at the basin level. Of the countries noted above that have disaster risk management legislation, South Africa might be considered one of those most likely to have connected the two, but in reality this has not been the case (Humby 2012).

One other method of reducing the impact of floods on society that is increasingly being examined is the provision of insurance, whether by private or public institutions. Traditionally, insurance has been used most extensively in wealthier countries, but the industry is expanding from what is admittedly a very low base in developing nations – for example, recent floods in Kashmir created losses of almost $16 billion, but insured losses were only around $236 million (Parvaiz 2015). Initiatives that introduce elements of flood insurance in very poor countries are being piloted currently (for example by Oxfam) in line with cultural norms and affordability concerns.

Policy decisions must be taken by national governments regarding the apportionment of the costs of insuring against risk of flooding. Should individual homeowners shoulder the burden alone, or should this risk be subsidised in some way? In the UK, the FloodRe scheme (under the

Flood Reinsurance (Scheme Funding and Administration) Regulations 2015), which came into force in April 2016, effectively facilitates affordability for those who would otherwise pay very high premiums, by spreading the cost of insurance across all householders. A levy is taken from all household insurance, which is then consolidated to create a separate resource pool. Insurers that become involved in the scheme will then provide flood insurance as normal under the bundled property insurance process, but the flood risk element would be passed on to a specialist flood reinsurer, FloodRe, which would then cover flood losses from the pool created by the general levy. This would keep flood insurance affordable for those most at risk, through a cross-subsidy from all property owners. The UK is one of the few countries to include flood insurance as an integral part of buildings insurance. Such 'bundled' approaches are comparatively unusual (Lamond and Penning-Rowsell 2014). The National Flood Insurance Program in the USA has been fraught with problems for many years, but the cumulative effects of a general lack of interest or incentive for property owners to insure, for developers to avoid risky development and for banks to enforce mandatory insurance requirements seems to have pushed the programme over the point of no return, and it awaits a final reckoning by government (Tarlock 2012).

There are other specifically legal aspects of vulnerability and resilience, although these are not often incorporated in their analysis. The first relates to questions of liability generally for damage caused by flooding, and the question of how feasible it may be to hold public authorities and private institutions or individuals accountable for failures in fulfilling their obligations. These are directly relevant to questions of resilience because the burdens of flooding cannot be appropriately distributed unless they are spread equitably. While there has been case law at the extreme end of credibility (causation in the claim that flash floods in Rapid City were the result of cloud seeding using table salt was sadly never tested, but legal action was dismissed – see Dennis 2010), Takahashi (2011) describes a spate of actions taken by property owners against river managers in the 1970s for negligent water resource management. Most notoriously, the federal Flood Control Act of 1928 expressly relieves the United States government of any liability for damage or harm caused by flooding. Thus, while the US Corp of Engineers was judged to be grossly incompetent following levee breaches resulting from Hurricane Katrina, they could not be found to be financially liable (Nossiter 2008).

The second legal aspect relates to the increasing importance of human rights legislation with respect to seeking redress for the damage caused by flooding. In *Marcic v Thames Water* (2003), for reasons unrelated to human rights, the appellant did not ultimately succeed in their case against Thames Water for frequent flooding caused by the latter's sewerage system. The House of Lords did, however, agree that there had been a breach of Art. 8 of the European Convention on Human Rights on the 'right to respect for private and family life, his home and correspondence', and Art. 1 of the First Protocol on the peaceful enjoyment of possessions.

In a 2012 case in the European Court of Human Rights, Mrs Kolyadenko and five others succeeded in their claim against the Russian Federation, following the release of water from the Pionerskoye dam in response to unusually heavy rainfall in Vladivostok in 2001 (*Kolyadenko and others v Russia* 2012). The claimants' properties were flooded as a consequence of the poor maintenance of the channel that was supposed to act as a conduit for flood waters. Claims were made under a number of headings, but for the purposes of this analysis, those made under Art. 2 (right to life), Art. 8 and Art. 1 of the First Protocol are the most relevant. These claims largely succeeded, with a causal relationship being drawn between the neglect of the channel and the flooding. Arguments from the Russian government that the flooding was simply the result of a natural event for which they could not be blamed were rejected by the court. The question of liability for what can be perceived as a natural event is one that has underpinned much of the case law on flooding.

In South Africa, human rights have been connected directly to the question of response, with the Constitutional Court stating that democratically elected governments are obliged to provide relief to the victims of disasters (*Minister of Public Works & Ors v Kyalami Ridge Environmental Association & Anor* 2001).

Conclusions

The overwhelming conclusion that can be taken from an analysis of flood management globally is that the days of reliance on infrastructural solutions alone are over. Global efforts to reduce the terrible effects of flooding clearly indicate that flood protection can never be absolute, but that effective planning, risk management, emergency response and rehabilitation can mitigate the impacts.

Flood management demands an institutionally and sectorally integrated response that can work effectively across multiple scales over time. This is an immensely challenging problem that will only get more difficult as global change continues and flood risk increases. Implementation capacity varies tremendously across jurisdictions, and the effectiveness of legal frameworks to accommodate changing circumstances and the multitude of factors that need to be considered is always going to be variable.

There appear to be two tracks being followed with respect to trying to integrate some of these considerations with flood management: the disaster risk management approach; and the consolidation of flood risk management with water resources management. These are both welcome, and it is especially gratifying to see that, with respect to the former especially, it does seem possible to implement in poorer countries. The experience of Bangladesh in its disaster risk management policy, strategy and standing orders demonstrates that the level of integration required can be achieved in ways that have seen the impacts on human lives of flooding reduced drastically over the past ten years or more. This has not been true with respect to water resources management more generally, and it does not yet appear that what might be called the DRM and WRM approaches are being coordinated sufficiently, even though they are mutually complementary. A further delineation can be seen between the impacts of floods on humans directly (in terms of loss of life and of property) and the environmental impacts. These appear consolidated in the Floods Directive, but it may be that definitions of flooding need to be expanded in some cases to ensure that the full consequences of inundation from sources other than water are managed appropriately. With tailings dams collapsing so frequently, robust approaches are needed to ensure the hazards they create for the environment and for communities are considered fully.

Recent experience of the use of human rights legislation suggests this might be a promising route for those seeking redress for harm caused by flooding. These cases suggest that the use of human rights legislation may become more important in the future. It is easy to conceive of cases brought by those adversely affected by flooding in instances where there has been, for example, inappropriate urban development, poor implementation of flood zoning or illegal deforestation.

Despite the magnitude of problems associated with flooding, it is clear that the need for integrated and holistic responses is being taken on board progressively by national governments. This gives hope that in the longer term legal frameworks that affect and are affected by flood events will improve in effectiveness and reduce flood impacts on the environment and society.

References

Allen, K. and Bounds, A. (2015) UK building 10,000 homes a year on flood plains. *Financial Times*, 28 December. Available from: www.ft.com

APFM (Associated Programme on Flood Management) (2013) www.apfm.info/

Association of British Insurers/Scottish Government (2008) Joint Statement on the Provision of Flood Insurance. Available from: https://www.abi.org.uk/~/media/Files/Documents/Publications/Public/Migrated/Flooding/Statement%20of%20principles%20Scotland.pdf

Balica, S., Wright, N. and van der Meulen, F. (2012) A flood vulnerability index for coastal cities and its use in assessing climate change impacts. *Natural Hazards* 64, pp. 73–105.

Barredo, J. (2007) Major flood disasters in Europe: 1950–2005. *Natural Hazards* 42, pp. 125–48.

Brooks, N. (2003) Vulnerability, risk and adaptation: a conceptual framework. Tyndall Centre Working Paper no. 38. Available from: www.tyndall.ac.uk

Dennis, A. (2010) Cloud seeding and the Rapid City flood of 1972. *Journal of Weather Modification* 42, pp. 124–6.

European Parliament and Council (2000) Directive 2000/60/EC of the European Parliament and of the Council of 23 October 2000 establishing a framework for Community action in the field of water policy. *Official Journal of the European Union* 327, 22 December, pp. 1–73.

European Parliament and Council (2007) Directive 2007/60/EC of the European Parliament and of the Council of 23 October 2007 on the assessment and management of flood risks. *Official Journal of the European Union* 288, 6 November, pp. 27–34.

Fish, E. (2013) The forgotten legacy of the Banqiao Dam collapse. *The Economic Observer*, 8 February. http://www.eeo.com.cn/ens/2013/0208/240078.shtml

Forbes, H., Ball, K. and McLay, F. (2015) *Natural Flood Management Handbook*. Stirling: SEPA. Available from: http://www.sepa.org.uk/media/163560/sepa-natural-flood-management-handbook1.pdf

Gabe, T., Falk, G., McCarty, M. and Mason, V. (2005) Hurricane Katrina: social-demographic characteristics of impact areas. *CRS Report for Congress*, 4 November. Available from: http://www.tidegloballearning.net/sites/default/files/uploads/crsrept.pdf

Guha-Sapir, D., Below, R. and Hoyois, Ph. (2016) *EM_DAT: International Disaster Database*. Brussels: Université catholique de Louvain. http://www.emdat.be/database

Haines-Young, R. and Potschin, M. (2010) The links between biodiversity, ecosystem services and human well-being. In D. Raffaelli and C. Frid (eds), *Ecosystem Ecology: A New Synthesis*. BES Ecological Reviews Series. Cambridge: Cambridge University Press.

Hallegatte, S., Green, C., Nicholls, R. and Corfee-Morlot, J. (2013) Future flood losses in major coastal cities. *Nature Climate Change* 3, pp. 802–6.

Handmer, J. and Monson, R. (2004) Does a rights based approach make a difference? The role of public law in vulnerability reduction. *International Journal of Mass Emergencies and Disasters* 22, pp. 43–59.

Harvey, F. (2016) Build on flood plains despite the risks, say UK government advisers. *Guardian*, 27 January. www.theguardian.com

Howarth, W. (2002) *Flood Defence Law*. Crayford, Kent: Shaw and Sons.

Humby, T. (2012) *Analysis of Legislation Related to Disaster Risk Reduction in South Africa*. Geneva: IFRC. Available from: http://www.ifrc.org/PageFiles/41164/1213900-IDRL_Analysis_South%20Africa-EN-LR.pdf

ICOLD (International Committee on Large Dams) (2001) Tailings dams: risks of dangerous occurrences; lessons learnt from practical experiences, ICOLD Committee on tailings dams and waste lagoons. *Bulletin 121*. Available from: http://www.unep.fr/shared/publications/pdf/2891-TailingsDams.pdf

IPCC (2014) *Climate Change 2014: Synthesis Report. Contribution of Working Groups I, II and III to the Fifth Assessment Report of the Intergovernmental Panel on Climate Change*. Core writing team, R.K. Pachauri and L.A. Meyer (eds). Geneva: IPCC.

Jha, A., Lamond, J., Bloch, R., Bhattacharya, N., Lopez, A., Papachristodolou, N., Bird, A., Proverbs, D., Davies, J. and Barker, R. (2011) Five feet high and rising: cities and flooding in the 21st century. Policy Research Working Paper 5648. Washington, DC: World Bank.

Jones, J.A.A. (1997) *Global Hydrology*. London: Longmans.

Jonkman, S., Hillen, M., Nicholls, R.J., Kanning, W. and van Ledden, M. (2013) Costs of adapting coastal defences to sea-level rise: new estimates and their implications. *Journal of Coastal Research* 29(5), pp. 1212–26.

Kiernan, P. (2016) Mining dams grow to colossal heights, and so do the risks. *Wall Street Journal*, 5 April.

Kobayashi, Y. and Porter, J. (2012) Flood risk management in the People's Republic of China: learning to live with flood risk. Asian Development Bank, Manila. Available from: http://www.adb.org/sites/default/files/publication/29717/flood-risk-management-prc.pdf

Kolyadenko and others v Russia (2012) 17423/05, ECHR 338.

Kossoff, D., Dubbin, W.E., Alfredsson, M., Edwards, S.J., Macklin, M.G. and Hudson-Edwards, K.A. (2014) Mine tailings dams: characteristics, failure, environmental impacts and remediation. *Applied Geochemistry* 51, pp. 229–45.

Lamond, J. and Penning-Rowsell, E. (2014) The robustness of flood insurance regimes given changing risk resulting from climate change. *Climate Risk Management* 2, pp. 1–10.

Law Commission (2014) Conservation Covenants, LAW COM No. 349. Available from: www.gov.uk

Marchi, L., Borga, M., Preciso, E. and Gaume, E. (2010) Characterisation of selected extreme flash floods in Europe and implications for flood risk management. *Journal of Hydrology* 394, pp. 188–33.

Marcic v Thames Water (2003) 37 EHRR 28.

Minister of Public Works & Ors v Kyalami Ridge Environmental Association & Anor (2001) ICHRL 33, 29 May.

Mouvet L., Mueller R.W. and Pougatsch, H. (2001) Structural safety of dams according to the new Swiss legislation. Proceedings of the ICOLD European Symposium, Geiranger, Norway.

Nicholls, R., Hutton, C., Lazar, A., Allan, A., Adger, N., Adams, H., Wolf, J., Rahman, M., Salehin, M. and Lawn, J. (2016) Integrated assessment of social and environmental sustainability dynamics in the Ganges–Brahmaputra–Meghna delta, Bangladesh. *Estuarine, Coastal and Shelf Science* forthcoming.

Nossiter, A. (2008) In court ruling on floods, more pain for New Orleans. *New York Times*, 1 February.

Parliament of South Africa (2002) Disaster Management Act, no. 57 of 2002.

Parliament of the United Kingdom of Great Britain and Northern Ireland (2015) Flood Reinsurance (Scheme Funding and Administration) Regulations 2015, no. 1902.

Parvaiz, A. (2015) Indian Kashmir flood sparks wave of property insurance sales. Reuters 13 July. http://www.reuters.com/article/us-india-flood-insurance-idUSKCN0PN1AK20150713

Quevauviller, P. (2011) Adapting to climate change: reducing water-related risks in Europe – EU policy and research considerations. *Environmental Science and Policy* 14, pp. 722–9.

Reid, C.T. (2013) Conservation covenants. *Conveyancer and Property Lawyer* 77(3), pp. 176–85.

Rico, M., Benito, G. and Diez-Herrero, A. (2007) Floods from tailings dam failures. *Journal of Hazardous Materials* 154, pp. 846–52.

Rico, M., Benito, G., Salgueiro, A.R., Diez-Herreiro, A. and Perreira, H.G. (2008) Reported tailings dam failures: a review of the European incidents in the worldwide context. *Journal of Hazardous Materials* 152, pp. 846–52.

Rieu-Clarke, A. (2008) A survey of international law relating to flood management: existing practices and future prospects. *Natural Resources Journal* 48, p. 649.

Salehin, M., Haque, A., Rahman, M.R., Khan, M.S.A. and Bala, S.K. (2007) Hydrological aspects of 2004 floods in Bangladesh. *Journal of Hydrology and Meteorology* 4(1), pp. 33–907.

Scottish Parliament (2003) Water Environment and Water Services (Scotland) Act, 2003 asp 3.

Scottish Parliament (2009) Flood Risk Management (Scotland) Act 2009.

Smith, K. (2013) *Environmental Hazards*, 6th edn. London: Routledge.

Spray, C.S., Ball, T. and Rouillard, J. (2009) Bridging the water law, policy, science interface: flood risk management in Scotland. *Water Law* 20, pp. 165–74.

States General of the Netherlands (2011) Delta Act on Flood Risk Management and the Freshwater Supply 2011 (Deltawet waterveiligheid en zoetwatervoorziening)

Takahashi, Y. (2011) Flood management in Japan in the last half-century. *Institute for Water Policy Serial* No. IWP/WP/No.1/2011.

Tarlock, A.D. (2012) United States flood control policy: the incomplete transition from the illusion of total protection to risk management. *Duke Environmental Law and Policy Forum* 23, p. 151.

UN DESA PD (United Nations, Department of Economic and Social Affairs, Population Division) (2014) World Urbanization Prospects: The 2014 Revision, Highlights (ST/ESA/SER.A/352).

United Nations (2005) Hyogo Framework for Action 2005–2015: Building the Resilience of Nations and Communities to Disasters, 22 January. A/CONF.206/6.

United Nations (2015) Sendai Framework for Disaster Risk Reduction 2015–2030, 18 March. A/CONF.224/CRP.1.

United States Congress (1928) Flood Control Act of 1928, ch.569, 45 Stat. 534.

Van Alphen, J. (2015) The Delta Programme and updated flood risk management policies in the Netherlands. *Journal of Flood Risk Management*. DOI: 10.1111/jfr3.12183

Verchick, R. (2015) Katrina's lessons: learned and unlearned. *Houston Chronicle*, 28 August.

Victoria State Government (2013) Management of Flooding Downstream of Dams, Version 1.0, 6 February. Available from: http://www.ses.vic.gov.au/em-sector/em-planning/em-partners-resources/management-of-flooding-downstream-of-dams.pdf

Vorogushyn, S., Merz, B., Lindenschmidt, K.-E. and Apel, H. (2010) A new methodology for flood hazard assessment considering dike breaches. *Water Resources Research* 46, W08541.

Wheater, H. and Evans, E. (2009) Land use, water management and future flood risk. *Land Use Policy* 26S, S252–S264.

Winsemius, H., Aerts, J., van Beek, L., Bierkens, M., Bouwman, A., Jongman, B., Kwadijk, J., Ligtoet, W., Lucas, P., van Vuuren, D. and Ward, P. (2016) Global drivers of future river flood risk. *Nature Climate Change* 6, pp. 381–5.

World Health Organization (2006) *Communicable Diseases Following Natural Disasters: Risk Assessment and Priority Interventions*. Geneva: WHO.

Younger, P. (2007) *Groundwater in the Environment: An Introduction*. Oxford: Blackwell.

11

WATER ALLOCATION AND MANAGEMENT DURING DROUGHT

A. Dan Tarlock

Drought risks and allocation models

As the Israelites were about to enter the Promised Land, God cursed them with drought: 'I will make the sky above you as hard as iron, and your soil as hard as bronze, so that your strength shall be spent in vain and your land will bear no crops and its trees no fruit' (Leviticus 26: 19). We now understand that disasters such as droughts are not acts of divine retribution but a combination of natural events and human choice (Wisner *et al.* 2003). Still, countries with 'bad hydrology' (Briscoe 2009) face the risk of short- and long-term droughts, and these risks will be exacerbated by global climate change. The Intergovernmental Panel on Climate Change (IPCC) has consistently warned that many arid and semi-arid countries face the risk of decreased water supplies (IPCC 2014).

Law cannot make rain, but it can help mitigate the adverse impacts of drought in two fundamental ways. First, water can be allocated among competing uses to promote the fair, efficient and secure enjoyment of water. If this is done, users will have incentives to plan for drought by investing in technology that uses water efficiently as well as being prepared to adapt to droughts, through, *inter alia*, the use of markets or the withdrawal of land from agricultural production to reallocate water. Second, the law can manage water use during periods of drought by deciding how the pain of shortage should be shared among all users. Water law has always been a risk allocation system because entitlements are of necessity incomplete for both physical reasons and the social need to share the use of water among a wide range of public and private uses (Schorr 2012). Thus, risk is an inherent element of all water entitlements. The primary risks include 1) droughts which necessitate the curtailment of licence allocations, 2) curtailment due to wasteful use and 3) curtailment due to an over-riding public interest.

There are a variety of ways to allocate water. In many parts of the world, water, especially groundwater, is allocated by capture, leading to the tragedy of the commons (Kulkarni *et al.* 2015). In totalitarian or highly central centralized regimes water was allocated by administrative fiat. The former Soviet Union allocated water among the Central Asian republics to complete the tsarist project of intensive agriculture in the three downstream republics on the Amu Darya. Large reservoirs were built in Kyrgyzstan and Tajikistan, which were operated primarily for downstream irrigation supply. In the official water governance structure of the Soviet Union, all water resources were controlled by the Union-wide Ministry of Melioration and Water

Management (MinVodKhoz). In Central Asia, initially a regional agency (SredAzVodKhoz) was responsible for the whole Aral Sea basin and also received orders from Moscow. Later, corresponding ministries in the Soviet republics were established, but their responsibilities were mainly restricted to the implementation of the decisions of the central MinVodKhoz in Moscow. The agency pursued an integrated basin–wide water and energy management approach in which each republic fulfilled a particular function. Water allocation was standardized, with fixed schedules for republics, provinces and districts (Water Unites n.d.). However, in most countries vulnerable to drought, water is allocated by 1) customary practice, 2) judicial decisions or 3) the issuance of water use licences by a state administrative agency. In each case, the allocation system creates private entitlements and protects them from interference by other users. This chapter describes these three forms of allocation, with the primary emphasis on administrative allocation through water licensing.

Customary allocation

Water users in small areas may agree among themselves on how available water will be shared among users during a drought. This can be done outside the supervision of the state or via the state's creation of entitlements that recognize customary practices. Customary practices can evolve into a *de facto* water code, as happened in places such as Peru. Customary water allocation generally has the two elements of state-created water entitlements that confer only the right to use, as opposed to the ownership of water (Ramazzotti 2008): 1) quantified use rights and 2) an enforcement mechanism. Customary rights still play an important part in water allocation in the Middle East (Caton 2013) and in Africa and Asia (Schlager 2005).

Customary rights have been recognized for indigenous people, and often form a parallel system of water rights in countries which recognize them. The most studied example is the United States' Supreme Court decision in *Winters v US* (1908). In an unprecedented decision, the Court held that the reservation had an implied irrigation water right with a priority as of the date of the 1888 Treaty modifying the reservation (Schurts 2008), and thus the right was superior to all state appropriative rights. The right is a quasi-appropriative and riparian right. It has a priority date, but it does not depend on exercise. It can lay dormant until reservation decisions to assert it. In other parts of the world, customary rights recognized by the state are often limited to indigenous peoples. In Canada, Australia (*Spinifex v Western Australia* 2000; Poirier and Schartmueller 2012), Chile (Barrera-Hernández 2005), South Africa and the United States customary rights for indigenous peoples have been recognized by judicial decision or legislation.

Judicial allocation

Both the Continental European civil law and the English common law recognize the Roman law right to use water as incident to the ownership of land adjacent to a stream rather than a state licence (*Colquhoun's Trustees v Orr Ewing* 1877) and initially relied on courts to create and enforce water entitlements. The doctrine of riparian rights posits that all owners of land along a stream have equal and correlative rights to use water in the stream. The common law has influenced the water law of Australia, Canada, New Zealand and the United States, as well as Roman-Dutch law in South Africa (Kidd 2009; Tewari 2009). French civil law recognized the absolute right of a property owner to use water arising on his land, but the courts limited this right in the interest of other users, basically aligning civil with common law (Caponera and Nanni 1992). Judicial allocation has long been criticized as inefficient because it is too uncertain to encourage investment and conservation. The incoherence of the common law doctrine makes it almost impossible to determine whether the use will withstand judicial scrutiny if challenged by another user.

Two reforms have adapted the common law to an industrial and agricultural society, but they have not solved the incoherence problem. The common law developed to resolve conflicts among mill owners, and thus propounded that theory that a riparian could only use the current of the river but had to leave it unimpaired for downstream riparians. This blocked irrigation withdrawals and carry-over storage reservoirs (*Herminghaus v Southern California Edison Co.* 1926). The United States and other arid countries abandoned the natural flow theory and substituted a reasonable use rule (Constitution of the State of California Article X, Section 2 1928), although England adhered to the theory until a licensing system was introduced. The second reform is to try and introduce the protection of prior uses into the common law. Under the common law, if another use interferes with a prior use, the victim must bring a lawsuit and face the risk that its use will be curtailed because no element of priority is civil or common law riparianism. The United States Restatement of Torts (Second) makes the protection of prior uses a relevant factor, but judicial protection of prior uses is erratic (*Edmonson v Edwards* 2003; *Harris v Brooks* 1955). In places such as the United States, the common law is primarily important to resolve non-consumptive water disputes such as the use of small lakes.

Administrative allocation

Due to its incoherence, most Anglo-American legal systems have replaced the common law with an administrative licensing system, although many withdrawals are exempted. For example, England introduced the country's first licensing system in 1963 and has progressively expanded it (Evans and Howsam 2004). The first generation of water allocation legislation had three primary objectives: 1) the assembly of accurate hydrological data to facilitate project planning and allocation, 2) the creation of a relatively secure system of public and private entitlements and 3) the construction of multiple-purpose projects with carry-over storage to permit the enjoyment of those entitlements in times of scarcity and to generate low-cost electricity. These remain important functions, but in the past two decades the allocation of water during drought periods has become more complex due to three changed conditions. The first changed condition is climate change. The second is the protection and restoration of aquatic ecosystems. This reflects the environmentalist criticism of large-scale multiple-purpose projects (World Commission on Dams 2000) and the growing appreciation of the functions and services that free-flowing rivers provide (Tarlock 2014a). The third changed condition is the growing pressure to subject water allocation and use to greater market discipline.

Modern water allocation systems generally have some combination of the following characteristics, although they are seldom all found in any one national law:

1   The assertion of the state's power to control of the allocation and reallocation of water used by public and private entities.
2   A permit or licensing system that creates private relatively secure, quantified entitlements and special procedures to declare rivers and groundwater basins closed to new uses.
3   The establishment of rules for the use of water and the distribution of the pain of shortages (World Wildlife Federation 2007).
4   Special protections, either water reserves or public rights, for *in situ* environmental uses.
5   Procedures to allow market reallocation of water.
6   The limitation of groundwater mining.

7   Standards for water planning with an increasing recognition that water plans need to factor in possible adaptations to global climate change (GCC) which threatens to alter rainfall patterns and create more extreme cycles of flood and drought, especially in arid countries.
8   Procedures for the enforcement of entitlements.
9   The coordination of water quantity and quality regulation.
10 The development of more inclusive decision-making processes.

Four foundational issues in licensing systems

Any permit system must deal with four foundational issues: 1) the basis of state power to regulate private use, 2) the status of pre-existing rights, 3) exempt uses and 4) the water available for licensing.

The basis of state power to manage water use

Almost all countries, regardless of the form of the economy or form of government, recognize that individual entitlements to the use of water must be created because the alternative is an open access regime which will lead to the tragedy of the commons through waste and over use (Shah 2014). Most countries also recognize that the state has a strong interest in regulating the use of water.

Modern state regulation is a product of the late nineteenth and early twentieth centuries (Caponera 2003). In common and civil law countries modern water rights are a product of the nineteenth century, when state power to define and regulate property rights was much contested as either a violation of natural rights or classic liberalism (Hodgson 2006). Water entitlements were initially tied to a specific parcel of land (Getzler 2004), and land owners claimed that they owned both surface and underground waters. Today, almost all countries have adopted the common law view that water entitlements are incomplete property rights limited to the use of the water rather than ownership of the water itself (Caponera 2003). Water rights are usufructuary and correlative (Caponera 2003; Ware 1905). Each user's right is subject to the rights of other similarly situated users on a stream or over an aquifer and is limited to the privilege to use, rather than ownership, of the body of water.

The incomplete, usufructuary nature of water rights has allowed states to declare themselves the 'owner' of waters. Common law states borrowed from Roman law and declared water *res nullius*, asserting a trust over the 'unowned' resource. Civil law countries built on another aspect of Roman law, the distinction between public and private waters, to assert 'ownership' over large amounts of public water. Other countries, such as Israel, Iran and Colombia, simply declared the state the owner of all water and thus all uses required a government licence (Caponera 2003). Modern state ownership asserts that 1) water rights are limited to the privilege to use water as opposed to individual ownership of streams and aquifers and 2) that access to water requires state permission in the form of a permit or licence. State ownership can also be a basis for the recognition of trust duties of state allocation (Blumm and Guthrie 2012).

Countries with declarations of public or state ownership include Armenia,[1] Brazil,[2] China,[3] Italy,[4] Kazakhstan,[5] Morocco,[6] Mozambique,[7] Senegal,[8] South Africa[9] and Yemen.[10] Chile is the most prominent example of a country that has gone in the opposite direction. The 1857 Civil Code establishes that waters are 'national goods for public use'.[11] However, in 1981 Chile adopted a new water code that created perpetual property rights in water.[12]

The status of pre-existing rights

The treatment of pre-existing water rights is a crucial problem when a regulatory scheme is introduced. If pre-existing rights are grandfathered, the state's discretion to allocate water for new uses or manage drought shortages may be constrained. In states which constitutionally protect property rights, the state cannot simply extinguish pre-existing rights (Pienaar and van der Schyvif 2007; Kluig 1997). Pre-existing rights are generally protected when a regulatory scheme is introduced, but the level varies. The most common form is to convert pre-existing private or customary rights to state licences.

The Australian state of Victoria converted riparian rights to statutory rights in 1866 (Caponera and Nanni 1992). And this technique has been followed in England (Hodgson 2006), in the Australian states (Water Management Act 2000) and in the Canadian provinces. The same process was followed in Spain's 1985 Water Act, which protected pre-existing rights to use public waters for 75 years and protected prior riparian rights in other waters. But, the Act provided that any increase in historic use would require a new licence. The limitation of prior licences to 75 years was upheld against the argument that it was an expropriation (STC 1988). South Africa's post-apartheid 1998 Water Act (National Water Act 1998) gives prior rights more limited protection; 'existing lawful water uses'[13] actually used prior to the Act to continue, which suggests that the state may limit pre-existing uses (Stein 2002).

The same technique can be applied to prior water licences when the state reforms its law to adapt to changed conditions. In New South Wales, Australia, the Water Management Act 2000 converted existing licences issued under the Water Act 1912 into new licences. In Alberta, Canada, the 1996 Water Act provided that existing licences, which are known as 'deemed' licences, continued to be valid (Water Act 2000).

Exemptions

Most licence systems have exemptions. The most common exemption is for small abstractions for basic human needs and subsistence agriculture. Exemptions are generally based on volumetric criteria or on the purpose, usually domestic use and stock raising (Hendry 2015). However, exemptions can limit the state's discretion to manage droughts. Exemptions may also mean that licence holders will have to bear a disproportionate share of the curtailment burden during droughts (Van Koppen and Schreiner 2014).

Water available for licensing

The amount of water for licensing can be limited for two reasons: 1) the exhaustion of the average annual supply and 2) the establishment of minimum environmental flows. Many water licensing or permit scheme can be used to prevent overuse by closing an over-allocated basin to new abstractions. Alberta, Canada, closed the Bow, Oldman and South Saskatchewan sub-basin systems. New water allocations have to be obtained through water allocation transfers (Hipel *et al.* 2013). Closing groundwater basins to new uses is a way to curb mining, even though safe yield of a basin is not a scientific standard. Groundwater is more difficult to regulate.

Historically, an *in situ* use was not considered a 'use' of water, and no entitlements could be obtained for this purpose. The environmental movement changed this thinking (Tarlock 2014b) and has led to the recognition of the need for *in situ* uses such as minimum flows to protect and restore aquatic ecosystems. There are four basic techniques: 1) a water reserve can be created which sets a floor on the water available for a licence, 2) *in situ* use entitlements can be created for either public agencies or private individuals, 3) the licensing process can be used to deny new licences because the use would compromise public environmental and other values, 4) existing entitlements can be curtailed to protect environmental values. For example, South Africa's Water Act[14] requires the creation of environmental reserves to 'protect aquatic ecosystems in order to secure ecologically sustainable development and use of the relevant water resource'.[15] Under a reserve system, during a drought consumptive abstractions may have to be curtailed to maintain the reserve.

The practical and constitutional implications of curtailment to protect environmental flows have yet to be explored outside Australia. The country has implemented an ambitious aquatic ecosystem protection scheme in its major basin, the Murray–Darling, because the basin faces downstream saline intrusion from upstream agricultural withdrawals. Initially, a cap on water withdrawals was imposed, but after the federal states failed to implement it fully, in 2007 innovative Commonwealth water legislation was enacted (Water Act 2007) which required the Murray–Darling Basin Authority to prepare a strategic plan for the integrated and sustainable management of water resources in the basin to manage the basin's water resources. This means that private entitlements will have to be curtailed, and to achieve this objective, water trades, the voluntary retirement and sale of irrigation rights are being used. But, in the interest of fairness, Australia has combined subordination with compensation.

Allocation criteria and shortage allocation

Entitlement characteristics and limitations

Water should be allocated to serve two basic functions. First, a balance should be struck between public and private use. Second, entitlements should incorporate two seemingly inconsistent criteria. They should provide sufficiently certain allocations to induce investment in water use, but be capable of variation in the event that availability falls short of entitlements. Water entitlements generally have the following characteristics: 1) a fixed quantity, 2) the calendar periods in which the water may be used, e.g. the irrigation season, 3) the duration of the right, 4) the location of the abstraction point, 5) inherent limitations of the exercise of the right and 6) transfer conditions (Abernethy 2005).

The distribution of the risks of shortages

Water allocation laws have long had to provide criteria for sharing shortages when it is not possible to satisfy all rights. GCC makes the incorporation of shortage risk sharing an essential component of any allocation law. Shortage risk sharing can be incorporated into water licences in four basic ways: 1) priority, 2) preferences, 3) *pro rata* cut backs and 4) *ad hoc* administrative or judicial mandated reduction. Sharing can be administered in two basic ways. The rules can either be enforced by courts or administrative agencies.

Prior appropriation versus pro rata sharing

All licence systems have to protect to some degree prior rights, but priority allocation is associated with the United States doctrine of prior appropriation. 'Hard' prior appropriation allocates a stream's water budget in the order that licences are issued. In times of shortage, the most junior right is curtailed first, the second next and so on. The risk of shortage curtailment is assigned completely to most recent right holders who can be required to bear the full costs of senior calls

(Tarlock 2014b); there is no pro rata sharing. The justification for the system is that strict enforcement of priority schedules provides fair notice to junior users that they are required to bear the full costs of senior calls. Thus, junior water right holders have strong incentives to use the market to reallocate water or to take other adaptive measures such as investment in more efficient water use technologies. Water law has generally not adopted the technology-forcing principle that underlies modern pollution control law. The law has also not encouraged conservation because many systems embody the principle of use it or lose it. Thus, there are disincentives to conserve water, but these disincentives are being off-set by GCC, which is forcing many users to anticipate more limited available supplies, thus creating an incentive to use what water will be available more efficiently or temporary fallowing. The sting in prior appropriation is lessened by the doctrines that uses must be beneficial, non-wasteful and can be lost if the use is abandoned.

'Hard' or United States prior appropriation has not been widely adopted. Only the province of Alberta has expressly adopted it.[16] 'Softer' versions exist. When the Commonwealth of Australia was formed and water laws adopted, prior appropriation was expressly rejected. However, Australia's subsequent licence system developed into a *de facto* priority system and, beginning in the 1980s, the federal states began to reform their water laws to define rights in terms of volumetric allocations (Haisman 2005). This system defines more clearly the risks of shortages both before and during a drought (Young 2014). In New South Wales, the initial licence, *inter alia*:

a.   must recognize and be consistent with any limits to the availability of water that are set (whether by the relevant management plan or otherwise) in relation to the water sources to which the regime relates, and

b.   must establish rules according to which access licences are to be granted and managed and available water determinations to be made, and

c.   must recognize the effect of climatic variability on the availability of water, and

d.   may establish rules with respect to the priorities according to which water allocations are to be adjusted as a consequence of any reduction in the availability of water.[17]

Strict priority enforcement can be both harsh and inefficient. Junior water users often push back and, not surprisingly, states are experienced in ways to avoid priority 'calls' (Tarlock 2014b). For example, there is precedent for the rule that diversions must be reasonable, and it is possible to argue that amount of water extracted is disproportionate to the user's needs. Junior right holders can pay the senior right holder to fallow its land (MacDougal 2015), either to use the 'saved' water or to store it to use in future droughts. Much water is often distributed by user associations, such as irrigation districts. These districts often hold the actual water right and distribute water through contractual entitlements. This gives the district the flexibility to impose uniform *pro rata* cuts on its members.

Soft: priority recognition

In practice, most licensing schemes operate by *de facto* priority principles; a new use will only be approved if existing uses are not harmed (Johnson 2014). For example, the New South Wales Water Management Act[18] provides that:

> [a] water use approval is not to be granted unless the Minister is satisfied that adequate arrangements are in force to ensure that no more than minimal harm will be done to any water source, or its dependent ecosystems, as a consequence of the proposed use of water on the land in respect of which the approval is to be granted.

However, most other national licensing laws lack a clear specification of what priorities will be recognized during a drought (Salman and Bradlow 2006). For example, Mexico's national Water Act creates concessions for water users associations, but shortage problems are often worked out informally among smaller users on the ground (Palerm-Viqueira 2015).

China has relatively effective ways to set use reduction targets at the national, provincial or basin level, but the lack of a clear risk allocation permit system makes it difficult to curtail uses effectively at the user level. China's 2002 water law reflects the country's historic anti-litigation tradition and expressly provides that both individual water user units and disputes between regions will be first resolved through consultation and mediation by higher-level government, with litigation as a last resort (Global Water Partnership 2015). For example, China limited withdrawals from the Yellow river in 1998, but a 2011 study found that:

> In implementing the Yellow river water quantity regulations, there exist some localities which do not put into practice the water quantity allocation and dispatch plan, and exceed the allocation limits in using water resulting from inter-provincial flows not according with control limits.
>
> *YRCC 2011*

China has implemented a red-line policy which determines the total amount of surface and groundwater available in each basin and thus caps water use within the basin (Global Water Partnership 2015).

Preferences

Preferences designate certain which must be satisfied before all other uses. The most common preference is for the use of small quantities of subsistence use from the necessity to obtain a licence. Article 13 of the Albanian Water Law, for example, provides that 'Everyone has the right to use surface water resources freely for drinking and other domestic necessities and for livestock watering without exceeding its use beyond individual and household needs' (available from: faolex.fao.org/docs/texts/alb6343E.doc). The New South Wales, Australia Water Management Act has a preference priority system:

a) local water utility access licences, major utility access licences and domestic and stock access licences have priority over all other access licences,
b) regulated river (high security) access licences have priority over all other access licences (other than those referred to in paragraph (a)),
c) access licences (other than those referred to in paragraphs (a), (b) and (d)) have priority between themselves as prescribed by the regulations,
d) supplementary water access licences have priority below all other licences.[19]

However, in general, preferences play a limited role in water allocation, but the growing recognition of a human right to water could change their role (Winkler 2012).

Market reallocation

Water has long been considered a free good but it has often been tied to a specific location. In the 1960s, economists argued that many early allocations of water, especially for low-value crop irrigation, are inefficient and that the market or subsidy reduction should be used to reallocate water (Hartman and Seastone 1970). More recently, water entitlement transfers have been proposed as promising drought and GCC adaptation strategy because the market can relatively move lower- to higher-valued uses in times of shortages.

Israel is an example of reallocation through subsidy reduction. Over the past three decades, the state water authority shifted much of the scarce water supplies from agriculture to urban use, in part by reducing the subsidies for the former's rates (Kislev 2013; Hadas and Gal 2014). The net result of the efficiency critique of water allocation is the concept of 'water marketing.' Australia, Chile, Spain (National Water Commission 2012) and the United States are the primary countries that have embraced water markets. Australia authorized transfers when state water laws were reformed, starting with Victoria's 1989 Water Act.[20] New South Wales followed in 2000.[21] Chile allows the transfer of registered water rights (Vergara Blanco 2014).

Water transfers are more complicated than land transfers because the transaction costs are higher (MacDonnell *et al.* 1990). Land can be transferred without regard to the effect on neighbouring property holders, but for water rights, especially under prior appropriation, an appropriative water right cannot be transferred unless there is no injury to junior water right holders. Thus, only the portion of the water consumed as opposed to the portion returned to the river can be transferred (*Green v Chaffee Ditch Company* 1962). Experts must determine the range of affected water right holders, the amount of water actually beneficially used by the sellers and the amount of return flow to which junior water right holders are legally entitled. Riparian rights can be transferred, but the transferee cannot bind other riparians (*State v Abfelbacher* 1918), and in many jurisdictions the process for moving water from one watershed to another remains unclear (Abrams 1983). Transfers can be permanent or short term. A water entitlement can either be leased or sold outright (Brewer *et al.* 2007).

State 'adjustment' of pre-existing rights

Serious, prolonged droughts put pressure on the state to redistribute water. Water right holders can resist redistribution in countries that afford constitutional protection to water rights. Constitutions do not, however, protect entitlement holders against limitations inherent in their title. One possible limitation is the public trust. The public trust is an American doctrine which expanded the Roman and English common law doctrine that navigable or public rivers were subject to a public servitude of navigation into a doctrine that limits the power of the state to grant private rights that threaten to destroy the resource and gives the state the power to curtail previously granted rights (*Ill. Cent. R.R. v Illinois* 1892). In addition to the United States, the Supreme Court of India has recognized the doctrine (*M.C. Mehta v Kamal Nath* 1997). Other countries which have recognized the doctrine include the Philippines (*Metro. Manila Dev. Auth. v Concerned Residents of Manila Bay* 2011) and Kenya (*Waweru v The Republic* 2006). Australia is implementing a strategy to do this in the Murray–Darling basin,[22] although there is no explicit mention of the public trust. South Africa incorporated the public trust into its new water law by creating an environmental reserve as a cap on diversions. A leading United States water and public trust scholar has suggested that 'courts in states with ecological and/or evolutionary public trust doctrines could engage in a form of judicial adaptive management by adjusting private and other rights in water resources in response to climate change impacts' (Kundis Craig 2009).

Conclusion

Countries with 'bad hydrology' will face the increased risk of more prolonged droughts due to GCC. A water allocation regime that encourages the efficient use of water to mitigate the adverse consequences of a drought, distributes the burden of shortage fairly and strikes a balance between consumptive and non-consumptive uses is an essential element of any national drought adaptation strategy. No national water allocation regime is currently completely designed to meet these challenges. However, many existing national water allocation regimes offer important precedents for countries seeking to reform their water laws to meet the challenges of an uncertain water future.

Notes

1 Water Code of Armenia, Art. 4.
2 National Water Management Policy Act, Law no. 9.433, 8 January 1997.
3 Xianfa Art. 9.
4 Law 5 January 1994, no. 36, Art.1. See Nicola Lugaresi (2000), *Rethinking Water Law: The Italian Case for a Water Code.* Trento: Department of Legal Sciences, pp. 104–7.
5 Constitution of the Republic of Kazakhstan, Art. 6.
6 Law no. 10-95 on Water, Art. 6 (1995).
7 Law no. 16/91, Art. 1 (1991).
8 Law no. 81-13 (1981).
9 National Water Act 36 of 1998 § 3.
10 The Constitution of the Republic of Yemen, Art. 8.
11 Cód. Civ., Art. 595 (1857).
12 Código de Aguas [Cód. Aguas] [Water Code], 29 October 1981, Diario Oficial de Chile [DO]. Constitución Política de la República de Chile [CP], Art. 19, no. 24 ¶ 2. See Carl Bauer (2004), *Chilean Water Law as a Model for International Water Law Reform.* Abingdon: Routledge.
13 *Id.*, at §§ 21 and 32.
14 National Water Act 36 of 1998 (South Africa). For a brief history of South African water law, see Mike Muller (2010), *Lessons from South Africa on the Management and Development of Water Resources for Inclusive and Sustainable Growth.* https://ec.europa.eu/europeaid/sites/devco/files/erd-consca-dev-researchpapers-muller-20110101_en.pdf
15 National Water Act 26 of 1998 (South Africa) §§1xviii(b) and 12–13.
16 R.S.A. cW-3.5 (elec. 2007).
17 New South Wales Water Management Act 2000 § 20(2).
18 Water Management Act 2000 s 96 (Austl.).
19 Water Management Act 2000 s 58 (Austl.).
20 Water Act 1989 (Vict.) s 46G (Austl.).
21 Water Management Act 2000 (NSW) ss 71L-Z (Austl.).
22 The Water Act 2007 (Cth).

References

Abernethy, Charles L. (2005) Governing institutions for sharing water. In B.R. Bruns, C. Ringler and R. Meinzen-Dick (eds), *Water Rights Reform: Lessons for Institutional Design.* Washington, DC: FWPRI, pp. 55, 61–2.

Abrams, Robert H. (1983) Interbasin transfer in a riparian jurisdiction. *William and Mary Law Review* 24, p. 591.

Barrera-Hernandez, Lila. (2005) Indigenous peoples, human rights and natural resource development: Chile's mapuche peoples and the right to water. *Annual Survey of International and Comparative Law* 11(1).

Blumm, Michael C. and Guthrie, Rachel D. (2012) Internationalizing the public trust doctrine: natural law and constitutional and statutory approaches to fulfilling the Saxion vision. *UC Davis Law Review* 45, p. 741.

Brewer, Jedidiah, Glennon, Robert, Ker, Alan and Libecap, Gary (2007) Transferring Water in the American West: 1987–2005. *University of Michigan Journal of Law Reform* 40, p. 1021.

Briscoe, John Water (2009) Security: why it matters and what to do about it. *Innovations* 4, p. 3.

Caponera, Dante A. (2002) Existing systems of water law in the world. In Slavko Bogdanovič (ed.), *Legal Aspects of Sustainable Water Resources Management.* 3, 4.

Caponera, Dante A. (2003) *National and International Water Law and Administration*. The Hague, The Netherlands: Kluwer.

Caponera, Dante A. and Nanni, Marcella (1992) *Principles of Water Law: National and International*, 2nd edn. Boca Raton, FL: CRC Press.

Caton, Steven C. (2013) History of sustainable water use in Yemen. In S.C. Caton, *Middle East in Focus: Yemen*. Santa Barbara, CA: ABC–CLIO, p. 284.

Colquhoun's Trustees v Orr Ewing (1877) 4R HL 116, p. 127.

Dellapenna, Joseph W. and Gupta, Joyeeta (eds) (2009) *The Evolution of the Law and Politics of Water*. New York: Springer, p. 87.

Easter, William and Huang, Qiuqiong (2014) Water markets: how do we expand their use? In W. Easter and Q. Huang (eds), *Water Markets for the 21st Century. What Have We Learned?* Global Issues in Water Policy 11. Dordrecht: Springer.

Edmonson v Edwards (2003) 111 S.W.3d 906 (Mo.App.).

Evans, B. and Howsam, P. (2004) A critical analysis of the riparian rights of water abstractors within England and Wales. *Journal of Water Law* 16, p. 90.

Getzler, Joshua (2004) *A History of Water Rights at Common Law*. Oxford: Oxford University Press.

Global Water Partnership (2015) China's water resources management challenge: the 'three red lines'. Technical Focus Paper, pp. 14–15.

Green v Chaffee Ditch Company (1962) 371 P.2d 775 (Colorado).

Hadas, Efrat and Gal, Yoav (2014) Inter-sector water allocation in Israel, 2011–2050: urban consumption versus farm usage. *Water and Environment Journal* 28, p. 63.

Haisman, Brian (2005) Impacts of water rights reform in Australia. In B.R. Bruns, C. Ringler and R. Meinzen-Dick (eds), *Water Rights Reform: Lessons for Institutional Design*. Washington, DC: FWPRI, p. 113.

Harris v Brooks (1955) 283 S.W.2d 129 (Ark.) (prior rice irrigator's diversions curtailed to protect a subsequent marina from lower lake levels).

Hartman, Loyal, M. and Seastone, Don (1970) Water transfers. In National Water Commission, *Water Policies for the Future*. Washington, DC: National Water Commission, pp. 260–1.

Hearne, Robert and Donoso, Guillermo (2014) Water markets in Chile: are they meeting needs?. In R. Hearne and G. Donoso, *Water Markets for the 21st Century*. New York: Springer, p. 103.

Hendry, Sarah (2015) *Frameworks for Water Law Reform*. Cambridge: Cambridge University Press, p. 55.

Herminghaus v Southern California Edison Co. (1926) 252 p. 607 (Cal.).

Hipel, Keith W., Fang, Liping and Wang, Lizhong (2013) Fair water resources allocation with application to the South Saskatchewan river basin. *Canadian Water Resources Journal* 38(47), p. 53.

Hodgson, Stephen (2006) *Modern Water Rights: Theory and Practice*. 92. Rome: Food and Agriculture Organization.

Ill. Cent. R.R. v Illinois (1892) 146 US 387.

IPCC (2014) *Climate Change 2014 Synthesis Report Summary for Policy Makers*. Geneva: IPCC, pp. 7–8 ('very high confidence' that droughts will disrupt food production).

Johnson, Rosa Maria Formiga (2014) *Water Resources Management in Brazil: Challenges and New Perspectives* (April–June). http://www.worldbank.org/content/dam/Worldbank/Feature%20Story/SDN/Water/events/Rosa_Formiga_Johnson_Presentacion_Ingles-3.pdf

Kidd, Michael (2009) South Africa: the development of water law. In J.W. Dellapenna and J. Gupta (eds), *The Evolution of the Law and Politics of Water*. New York: Springer, p. 87.

Kislev, Yoav (2013) Water in agriculture. In Nir Becker (ed.), *Water Policy in Israel: Context, Issues and Options*. New York: Springer, p. 51.

Kluig, Heinz (1997) Water law reform under the new Constitution. *Human Rights and Constitutional Law Journal of South Africa* 5, p. 5.

Kulkami, H. *et al.* (2015) Shaping the contours of groundwater governance in India. *Journal of Hydrology: Regional Studies* 4, p. 172.

Kundis Craig, Robin (2009) Adapting to climate change: the potential role of state common law public trust doctrines. *Vermont Law Review* 43, p. 852.

Leviticus 26: 19.

M.C. Mehta v Kamal Nath (1997) 1 S.C.C. 388.

MacDonnell, Larry J. *et al.* (1990) *The Water Transfer Process as a Management Option for Meeting Changing Demands*. Boulder, CO: University of Colorado.

MacDougal, Douglas (2015) Irrigation and drought in the Northwest and the potential for market-based reallocation of water to protect high-value crops. *Marten Law* (14 July). http://www.martenlaw.com/newsletter/20150714-drought-reallocation-water-high-value-crops

Metro. Manila Dev. Auth. v Concerned Residents of Manila Bay (2011) G.R. No. 171947-48, 574.

National Drought Mitigation Center (2016) *What is Drought?*. http://drought.unl.edu/DroughtBasics/ WhatisDrought.aspx

Pienaar, Gerrit. and van der Schyff, Elmarie (2007) The reform of water rights in South Africa. *Law, Environment and Development Journal* 3(2), p. 179.

Poirier, Robert and Schartmueller, Doris (2012) Indigenous water rights in Australia. *Social Science Journal* 49(3), pp. 317–24.

National Water Act (1998) Act 36.

National Water Commission (2012) *Water Banks in Mexico*. http://www.conagua.gob.mx/bancosdelagua/ SGAA-4-12A-English.pdf

Palerm-Viqueira, Jacinta (2015) *Water Rights and Institutions in Mexico*. http://siteresources.worldbank.org/ EXTWAT/Resources/4602122-1213366294492/5106220-1213804320899/10.2_Water_Use_ Rights_Mexico.pdf

Ramazzotti, Marco (2008) Customary water rights and contemporary water legislation: mapping the interface, 5 July. FAO Legal Papers Online No. 78.

Salman, M.A. and Bradlow, Daniel D. (2006) *Regulatory Frameworks for Water Resources Management: A Comparative Study*. Washington, DC: World Bank.

Schlager, Edella (2005) Getting the relationship right in water property rights in water rights reform: lessons for institutional design. In B.R. Bruns, C. Ringler and R. Meinzen-Dick (eds), *Water Rights Reform: Lessons for Institutional Design*. Washington, DC: FWPRI, pp. 27, 37–43.

Schorr, David (2012) The Colorado doctrine: water rights, corporations, and distributive justice on the American frontier. http://papers.ssrn.com/sol3/papers.cfm?abstract_id=2172432

Schurts, J. (2000) *The Winters Doctrine in its Social and Legal Context, 1880s–1930s*. Norman: University of Oklahoma Press.

SCRA (2008) 661 (SC, 18 December) (Phil.).

Shah, Tushaar (2014) Groundwater governance and irrigated agriculture. Global Water Partnership Technical Committee Background Papers No. 19.

Spinifex v Western Australia (2000) FCR 1717 (Austl).

State v Apfelbacher (1918) 167 N.W. 244 (Wis.).

STC (1988) 29 November (BOE, No. 307) (Spain).

Stein, Robyn (2002) *Water Law in a Democratic South Africa: A Country Case Study Examining the Introduction of a Public Rights System*. http://scholar.law.colorado.edu/cgi/viewcontent.cgi?article=1065&context= allocating-and-managing-water-for-sustainable-future

Tarlock, A. Dan (2011) The Legacy of Schode v Twin Falls Land and Water Company: the evolving reasonable appropriation doctrine. *Environmental Law* 42, p. 37.

Tarlock, A. Dan (2014a) The transformation of water. In K.H. Hirokawa (ed.), *Environmental Law and Contrasting Ideas of Nature: A Constructivist Approach*. Cambridge, MA: Cambridge University Press, p. 248.

Tarlock, A. Dan (2014b) *Law of Water Rights and Resources*. Eagan, MN: Clark Boardman Callaghan, ch. 5.

Tewari, D.D. (2009) A detailed analysis of evolution of water rights in South Africa: an account of three and a half centuries from 1652AD to present. *Water SA* 35, p. 693.

van Koppen, Barbara and Schreiner, Barbara (2014) Priority general authorizations in rights-based water use authorization in South Africa. In M.J. Neal (Patrick), A. Lukasiewicz and G.J. Syme (eds), *Supplemental Issue Why Justice Matters in Water Governance. Water Policy* 16(S2), pp. 59, 63. London: IWA.

Vergara Blanco, Rafael (2014) *Derecho de Aguas*. Chile: Editorial Jurídica de Chile, p. 257.

Ware, Eugene. R. (1905) *Roman Water Law: Translated from the Pandects of Justinian*. St Paul, MN: West.

Water Act, Revised Statutes of Alberta 2000, Ch. W-3 § 18.

Water Law of the People's Republic of China (2002) Art. 56-57 (promulgated by the Standing Committee of the National People's Congress, 21 January 1988, revised 29 August 2002, effective 1 October 2006).

Water Management Act 2000 (NSW) s 396 abolishes all riparian rights to the flow of a stream.

Water Unites (n.d.) *Water Usage and Water Management in the Soviet Union*. http://www.waterunites-ca.org/ themes/8-water-usage-and-water-management-in-the-soviet-union.html

Winkler, Inga T. (2012) *The Human Right to Water: Significance, Legal Status and Implications for Water Allocation*. Oxford: Hart.

Waweru v The Republic (2006) 1 KLR 677, 677 (HCK) (Kenya).

Winters v US (1908) 207 US 564.

Wisner, Ben, Blaikie, Piers, Cannon, Terry and Davis, Ian (2003) *At Risk: Natural Hazards, People's Vulnerability and Disasters*, 2nd edn. Abingdon: Routledge. See also United Nations and World Bank (2010) *Natural*

Hazards, Unnatural Disasters: The Economics of Effective Prevention. https://www.gfdrr.org/sites/gfdrr.org/files/nhud/files/NHUD-Report_Full.pdf

World Commission on Dams (2000) *Dams and Development: A New Framework for Decision-Making.* Abingdon: Routledge.

World Wildlife Federation (2007) *Allocating Scare Water: A Primer on Water Allocation, Water Rights, and Water Markets.* http://assets.wwf.org.uk/downloads/scarce_water.pdf

Young, Michael D. (2014) Designing water abstraction regimes for an ever-changing and ever-varying future. *Agricultural Water Management* 145, p. 32.

YRCC (Huangshuihui) (2011) 'Huanghe shuiliang tiaodu tiaoli' zhiding yu shijian [Formulation and implementation of the 'Yellow River water quantity dispatching regulations'], 14 August. Available from: http://www.yellowriver.gov.cn/zlcp/kjcg/kjcg07/201108/t20110814_103291.html (accessed 18 November 2015).

12

STAKEHOLDER ENGAGEMENT FOR INCLUSIVE WATER GOVERNANCE

Aziza Akhmouch and Delphine Clavreul

Introduction: recent trends

Water demand is expected to more than double by 2015 as the world's population heads towards 9 billion people, 4 billion of which will live in severely water-stressed basins (OECD 2012a). Despite pressures on this crucial resource, people often have little or no opportunities to participate in water-related decisions that affect them. Water crises were ranked as the first global risk in terms of impact by the 2015 World Economic Forum.[1] The OECD argues that, in many circumstances, such crises are primarily governance crises, whereby the problem goes beyond hydrology, infrastructure and financing; it is about who does what, at which scale, how and why (2015b). Therefore, future water challenges will generate fierce competition between users and trade-offs will need to be managed, at the least cost for society. Engaging the broad range of stakeholders holds the promise of improving acceptance and trust in water governance, and reducing the potential for conflicts over water issues.

The last two decades have seen a change of paradigm whereby public policy making has trended away from the old 'top-down hierarchical model' to a more transparent and holistic model that involves public, non-state actors, including the private sector and not-for-profit organisations. The water sector has also undergone this change: the traditional role of 'governments' as the single decision making authority was shaken by repeated calls for multi-level, polycentric governance. This transition acknowledges the important role that stakeholders from different institutional settings can contribute to water management.

This shift is demonstrated by the rising profile of stakeholder engagement in the global arena. All World Water Fora highlighted the critical role of multi-actor partnerships (Marrakech 1997); participatory approaches (The Hague 2000); alliances, networks and dialogues (Kyoto 2003); co-ordination across levels of government (Mexico 2006); the critical role of vulnerable and marginalised groups (Istanbul 2009); the need for multi-stakeholder platforms (Marseille 2012);[2] and the necessity to put stakeholder engagement into practice and fit to local realities (Daegu 2015).[3]

In that context, international instruments, both hard and soft, have proliferated. The 1992 Rio Declaration on Environment and Development introduced the emerging public involvement norms to the Agenda 21 in developing, implementing and enforcing environmental laws and policies, including the management of freshwaters. Further legislation, at the regional level, elaborated on these principles. This is the case of the 1998 United Nations Economic Commission for Europe (UNECE) Convention on Access to Information, Public Participation in Decision making and Access to Justice in Environmental Matters (known as the 'Aarhus Convention')[4] and subsequent regulations and participatory principles, notably at the European Union level (e.g. 2003/4/EC, 2003/35/EC and Article 14 of 2000/60/EC).[5] Inclusive approaches to water governance also gained traction in the 1990s, when the political agenda moved from 'technical' supply-driven and infrastructure-led solutions towards greater demand management. The generalisation of the concept of integrated water resources management (IWRM) featured issues of co-ordination across different interests, sectors and levels. The newly adopted sustainable development goals include multiple references to water, governance and stakeholder engagement through not only a dedicated goal on water (no. 6) and a target on local participation, but also several governance-related goals referring to inclusiveness, gender equality, capacity building, policy coherence, multi-stakeholder partnerships, data, monitoring and accountability.

The last decades have also seen an evolution of semantics from 'participation' to 'engagement' and from 'public' to 'stakeholder', thus showing a scope wider than 'civil society' and project-based approaches. Overall, more progress has been made in engaging stakeholders in water *resource* management than in water *service* delivery. This is probably due to the fact that the integrated water resource management paradigm spurred the creation of river basin organisations where stakeholders could take part in planning and related discussions, while for water supply and sanitation engagement has often been restricted to handling customers' complaints.

A framework to engage stakeholders for inclusive water governance

The OECD defines stakeholder engagement as the process by which any person or group directly or indirectly affected by water policy and/or having the ability to influence the outcome positively or negatively is involved in related activities, decision making and implementation processes (OECD 2015a). Water governance refers to the range of political, institutional and administrative rules, practices and processes (formal and informal) through which decisions are taken and implemented, stakeholders can articulate their interests and have their concerns considered, and decision makers are held accountable for water management (OECD 2015b).

Despite quite an extensive literature on stakeholder engagement in the water sector, there has been a lack of evidence-based analysis and policy tools to guide decision makers in setting up result-oriented engagement processes that, ultimately, contribute to tangible water outcomes. As threshold questions, decision makers may enquire who should be engaged, for what purpose, under which form and how engagement processes can be sustained. For engagement processes to be relevant, a careful balance between what they try to achieve, the resources they require and whether they succeed in reaching the intended objectives is required.

The OECD attempted to provide responses by carrying out a survey on stakeholder engagement in water governance across 215 stakeholders,[6] and collecting 69 case studies (OECD 2015a). Policy recommendations were provided, building on an analytical framework to assess external and internal *factors* that trigger engagement processes; the *categories* of stakeholders that ought to be considered; *obstacles* and risks to mitigate; *mechanisms* available to manage multiple players; and type of *costs* and *benefits* (monetary or not) incurred by engagement. This chapter presents the policy guidance derived from this empirical work through the lens of six basic principles (Table 12.1).

Table 12.1 OECD principles on stakeholder engagement in water governance

Inclusiveness and equity	Principle 1: Map all stakeholders who have a stake in the outcome or that are likely to be affected, as well as their responsibility, core motivations and interactions
Clarity of goals, transparency and accountability	Principle 2: Define the ultimate line of decision making, the objectives of stakeholder engagement and the expected use of inputs
Capacity and information	Principle 3: Allocate proper financial and human resources and share needed information for result-oriented stakeholder engagement
Efficiency and effectiveness	Principle 4: Regularly assess the process and outcomes of stakeholder engagement to learn, adjust and improve accordingly
Institutionalisation, structuring and integration	Principle 5: Embed engagement processes in clear legal and policy frameworks, organisational structures/principles and responsible authorities
Adaptiveness	Principle 6: Customise the type and level of engagement as needed and keep the process flexible to changing circumstances

Inclusiveness and equity

Stakeholder mapping is a stepping stone to understand how the water sector is organised in terms of functions and responsibilities, and appropriately determine who should be engaged in decision making and implementation. Identifying stakeholders, their role and influence is important to help decision makers understand the balance of power, hotspots, champions and requisite trade-offs.

OECD (2015a) distinguishes three main categories of stakeholders: 1) traditional actors, with recurrent roles in water policy design and implementation; 2) emerging stakeholders, which have become dynamically active and influential in water governance; and 3) under-represented stakeholders, who are often omitted or remain unheard (Figure 12.1).

New players have gained interest and influence in water governance. While the role of the private sector has been inclined to focus on companies delivering water supply and sanitation, business has been paying increasing attention to water governance, especially to cope with regulatory risks and to secure water allocation. In parallel, citizens and water user associations have greater scrutiny over political decisions on water. As risks of floods intensify, property developers should also be considered, given that spatial development generates long-term liabilities and financial implications on water, such as compensation for the loss of green areas and water amenities. They can play an important role in harnessing new sources of finance and contributing to the development of solutions to manage floods. Institutional investors (e.g. pension funds, insurance companies, mutual funds) have also begun to factor environmental, social and governance issues into their decision making process and are investing more and more in water infrastructure and utilities.

Some stakeholder categories tend to be excluded. These comprise women (as the primary users of water in many parts of the world, for domestic consumption, subsistence agriculture and health), youth (as the future generation that will need to solve issues related to water), the rural and urban poor (as the main consumers in informal urban and rural settlements) and indigenous and aboriginal communities. Nature and other non-consumptive users are also often absent from engagement processes. Greater efforts to encourage innovation are called for to connect and

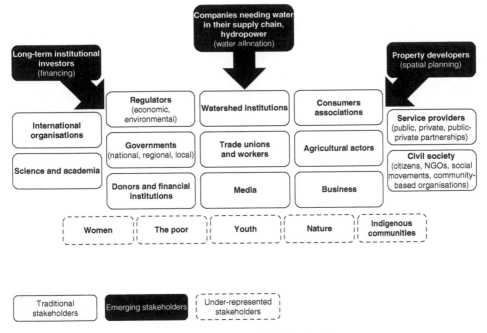

Figure 12.1 Traditional, new and under-represented stakeholders in the water sector
Source: OECD (2015a)

engage with these groups or individuals, who do not always come forward on their own. In order to gain a more balanced picture, it is very important to include these minority or 'less-vocal' stakeholders, beyond formal engagement channels.

A deeper understanding of stakeholders' aspirations in water governance can help manage their expectations. Identifying how different categories of stakeholders are contributing to improving water governance can shed light on why they should or wish to be involved in decision making and implementation. In turn, it helps decision makers understand the extent of influence stakeholders have over the decision making process and limit disappointment. As an example, the Hydro Sustainability Assessment Forum was created by the International Hydropower Association to evaluate the sustainability of hydropower projects through audits, interviews and in-depth analysis. It gathered 14 stakeholders that all had clear and specific reasons to be part of the process: developing countries were involved because hydropower plays an important role in meeting national electricity requirements; developed countries were interested in supporting development assistance programmes related to hydropower; banks were motivated by the need to manage reputational risk in relation to financing hydroelectric projects; environmental non-government organisations were motivated by the desire to avoid poor practices in hydroelectric projects of addressing environmental issues; and social non-government organisations wanted to ensure that future hydroelectric projects would properly address issues of resettlements or indigenous rights.

Engagement processes should also take into account the question of 'scale'. For stakeholder engagement to truly reflect the inherent multi-level nature of water, it is critical for decision makers to assess at which scale stakeholders are carrying out their roles and responsibilities and to diagnose governance gaps hindering effective co-ordination. For example, the Ontario's Great Lake Strategy was established to tackle deteriorating ecosystems and new invasive species. The Strategy implied engagement at different levels from municipalities to non-governmental organisations, industries and businesses, as well as First Nations communities, and resulted in strong relationships and partnerships for protecting the watersheds over the long term.

Analysing who interacts with whom can also help to draw a clear picture of how different stakeholders are connected. OECD's analysis shows important variations in how frequently stakeholders interact with one another in the water sector. Most notably, it reveals that interactions tend to take place in silos, especially in the case of governments, watershed institutions, civil society, business and academics. Decision makers should analyse the networks within which actors are embedded to understand their interrelations, assess the degree of cohesion among them and determine who are the 'integrators' and the 'dividers', and who remains secluded. In turn, it provides some indication as to their degree of influence on one another.

Finding the right balance between inclusiveness and empowerment of stakeholders is also important. Engagement processes (and related mechanisms) need to accommodate the needs of stakeholders with varying levels of interests and resources to ensure inclusivity and accessibility. Risks related to consultation capture from better-organised groups of actors and lobbies, to the detriment of unheard voices, deserve careful consideration. Equity between present and future generations in a perspective of sustainability should also be promoted.

Stakeholder mapping is a strategic tool for decision makers. However, mappings are specific to given places and times, across which stakeholders change and adapt. Such exercises should be iterative, transparent, regularly assessed and adjusted. Therefore, formulating rules on how to ascertain that engagement processes remain 'fit-for-target' is important to guide decision making and implementation.

Clarity of goals, transparency and accountability

Successful stakeholder engagement comes from a real understanding of the rationale that underlies it. Evidence from the OECD survey identifies two types of drivers. On the one hand, *structural* drivers can be clustered into four broad categories: 1) climate change will affect water availability and resilience of water infrastructures, with different levels of impacts across the world; 2) economic and demographic trends will drive water demand, in particular in cities, and affect the capacity of governments to respond (e.g. their ability to mobilise public funds); 3) socio-political trends, such as recent developments in EU and UN standards towards more attention to adaptive governance; and 4) innovation and technologies will stimulate greater connectivity and new relationships, in particular related to web-based communication avenues. On the other hand, *conjunctural* drivers also influence stakeholder engagement, primarily debates around water-related policy reforms and projects; crises, change or emergency-driven situations such as floods and droughts; legal requirements such as the Law that created the National Water Agency in Japan in 1962, which requires the national water agency to carry out stakeholder engagement processes in all projects; and competition over water resources.

Clarifying the objective and level of engagement is key to building mutual understanding and trust and for informed stakeholders to provide quality contributions in line with expectations. There are several levels of engagement depending on intentions and expected outcomes. Different typologies of engagement and participation have been discussed in the literature. A well-known categorisation is the 'ladder of citizen participation' developed by Arnstein (1969), which identifies eight levels or 'rungs', ranging from manipulation (the lowest in the group of non-participation steps) to citizen control (the highest step and highest degree of citizen power).

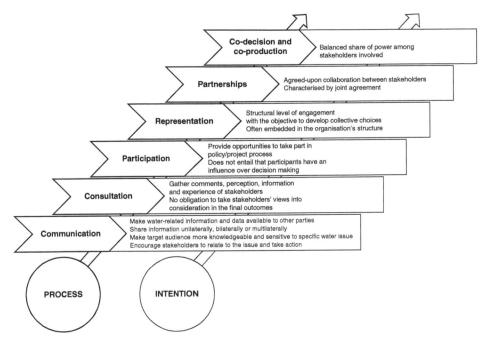

Figure 12.2 Levels of stakeholder engagement
Source: OECD (2015a)

This range shows that there is a significant gradation of citizens' participation. Further studies and more recent theories have overcome Arnstein's paradigm that considered participation as an end in itself rather than as a means (Wehn *et al.* 2014). Recent typologies include Pretty (1995) – 'typologies of participation'; Fung (2006) – 'democracy cube'; and UNDP Water Governance Facility, Stockholm International Water Institute, Water Integrity Network (2013) – 'levels of engagement'. Building on this stock of practices and concepts, the OECD distinguishes six levels of stakeholder engagement (Figure 12.2).

A lack of clarity on the use of stakeholders' inputs can result in mistrust on the part of involved stakeholders. Stakeholders may feel misled or manipulated, tend to lose interest and their motivations deflate, resulting in consultation 'fatigue' or 'frustration'. Therefore, the authority responsible for taking decisions, and its willingness to take stakeholders' ideas on board in doing so, should be clearly identified to enhance confidence in the value of the process. Transparency and accountability in how the engagement process is designed and implemented (e.g. through stakeholder mapping methods) is crucial to improve credibility and legitimacy, and to build trust among the stakeholders involved. Diligent work is necessary to ensure that the engagement process is fair and equitable.

Capacity and information

Engagement comes at a cost, be it monetary or not. Though leadership, clear goals and stakeholder salience are important for the development and promotion of stakeholder engagement processes, their effectiveness and sustainability will depend on available funds to take decisions and carry out the implementation. This is particularly true in times of tight budgets. In other words, the costs of producing or implementing an engagement process are likely to be a critical factor in determining whether and, more importantly, how stakeholders are engaged in water-related decision making. Funding is essential, regardless of whether the goal of the engagement process is to inform, consult, engage, collaborate, or empower.

Logistical expenses related to meeting venues or support material (e.g. publication of brochures, launch of online platforms etc.), staff time and the development of information platforms such as website acquisition and maintenance can raise high costs and may deter decision makers from engaging stakeholders. Therefore, the need to secure funding to sustain engagement is urgent, particularly in the context of financial stress experienced by many governments recently.

Supporting a two-way information sharing through consistent and appropriate communication channels is also key to stakeholder engagement. Access to timely and understandable information and technical expertise in the right format and sufficiently on time is important to realistically and effectively participate. In Portugal, for example, the Water and Waste Services Regulation Authority (ERSAR) developed a very innovative mobile app to provide customers with information about service utilities' performance in 278 municipalities, including data, indicators for drinking water quality and tariffs. Information asymmetries when one or several categories of stakeholders have more or better information than the others can damage the process. Government authorities and service providers, as well as researchers, all dedicate significant resources to collect and produce data on water resources (quality, quantity etc.), water services (service performance, state of infrastructures etc.) and water-related disasters (meteorological projections etc.).

Compiling and sharing meaningful and consistent information at various levels can be problematic for several reasons. For instance, different and sometimes incompatible methods can prevent certain stakeholders from using the data produced by others. Also, the quality of the data is sometimes questionable if the necessary control mechanisms are not integrated into monitoring programmes, or if the information is not well documented. Overcoming such information asymmetries can help to bridge the gap between science, policy and action. Overall, policy makers rarely use the research results; and the science–policy interface remains rather weak in terms of guiding decision making and implementation. Focusing efforts towards more adequately generating and sharing information between decision makers and scientists can contribute to better alignment of science contributions and policy aspirations in decision making arenas, providing a robust foundation for evidence-based decision making.

The interpretation and application of these resources and information also require competences and capability development at all levels to enable sustainable stakeholder engagement (e.g. skills, social learning). Appropriate skills are also necessary to set up and facilitate engagement processes and ensure expected outcomes. This requires dedicated human resources trained in mediation, communication, use of technologies etc.

Efficiency and effectiveness

Measuring whether public and institutional resources, including stakeholders' time and efforts, are properly used can help determine whether the engagement process was successful and strengthen the accountability of decision makers. Assessing the process and outcomes of stakeholder engagement can provide insight on the effectiveness of an engagement process in the early stages of design and preparation (*ex ante* evaluation), or during the engagement process itself, or after (*ex post* evaluation). Evaluation can also anticipate and manage some risks that the process may face (e.g. divergent perspectives regarding flood defence measures between land planners, property owners and government authorities; or regarding water resource allocation between farmers, industries and environmentalists).

Such evaluations can be carried out by the actors targeted in the process, by the decision makers who set up the process, and/or by third parties, and can resort to a large diversity of tools. Multi-stakeholder meetings help to collect feedback on the level of performance of engagement processes; evaluation reports record the process (successes, failures, lessons learnt) and allow for analysis to improve future engagement processes; polls and surveys can provide insights on the levels of satisfaction. Fact-based and perception-based indicators are increasingly advocated as a tool for measuring the impact of stakeholder engagement. Caution is needed as they can be contentious both in theory and in practice. It can be argued that complex processes of social change should not be reduced to metrics, and that the process of defining indicators and analysing the implications of the results can be highly complex and political. Some outcomes of engagement processes can be intangible (such as improved relationships or a sense of empowerment) and both quantitative and qualitative indicators should be employed to review the engagement process even though indicators can indeed be informative measurements of complex systems.

Involving stakeholders in water projects and policy processes can raise costs and generate benefits. There are overall positive considerations for stakeholder engagement to improve outcomes and build consensus. Nevertheless, decision makers and project leaders have begun to criticise engagement processes as too costly, expensive in time and money, and as subject to capture by specific lobby groups in their own interests.

The OECD clusters long-term benefits of stakeholder engagement into four types (Figure 12.3): 1) *acceptability and sustainability*, in terms of effective implementation of water policy and projects, proper enforcement of regulation, political acceptability and ownership of decisions and outcomes; 2) *social equity and cohesion*, which is related to trust, confidence and customer satisfaction, as well as corporate social responsibility; 3) *capacity and knowledge*

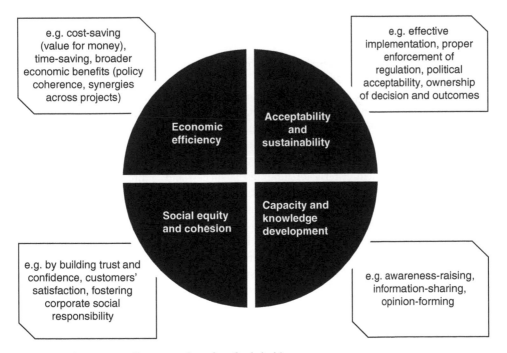

Figure 12.3 Four types of long-term benefits of stakeholder engagement
Source: OECD (2015a)

development, which emanates from raising greater awareness, sharing information and forming opinions; and 4) *economic efficiency*, as it can assist in optimising cost-saving, value for money and time-saving, as well as broader economic benefits from policy coherence and synergies across sectors and projects.

Different costs and benefits accrue to different stakeholder groups at different times and require managing trade-offs to ensure successful engagement processes and outcomes. The sustainability of stakeholder engagement will not only depend on the net difference between aggregate costs and benefits, but also on how they are distributed between stakeholders, and on stakeholders' willingness to bear them. Also, water policy reforms and large projects can induce important adjustment costs, especially in the short term, while the benefits of such initiatives may only become visible in the long term. It is crucial to reflect upon this and determine the appropriate trade-offs related to this dual temporality.

Results from evaluations should be disclosed to increase accountability, provide insight on the success of the engagement process in reaching its intended objectives and learn from experience to improve practice in the future. Transparent reporting to stakeholders on the results and outcomes of the engagement process is essential to build support and trust. Giving feedback to the participants on how their inputs have influenced the final decisions can help keep them on board. This is needed not only at the end of an engagement process, but throughout to ensure inclusiveness in terms of: 1) how information is made accessible; 2) how stakeholders are targeted and involved; and 3) resources allocated to the process. Failure to do so may lead to cynicism among stakeholders and decreased willingness to repeat such experience. At the same time, demonstrating to stakeholders that engagement is inclusive and worthwhile helps prevent backlash and consultation fatigue. In France, for example, the evaluation of stakeholder engagement activities of the inter-municipal public water service provider of Alsace-Moselle showed a steady increase in confidence in the service providers from elected officials (95 per cent of satisfaction rate) and users (80 per cent).

But evaluating stakeholder engagement is a difficult task for a number of reasons:

- First, there is a lack of comprehensive frameworks with agreed-upon evaluation methods and reliable measurement tools.
- Second, there is a wide variety of situations in the design and goals of engagement processes. Therefore, evaluation frameworks should be general enough to apply across different types of processes yet specific enough to have value for learning and practice.
- Third, stakeholder engagement is an inherently complex and value-loaded concept.

There are no widely held criteria for judging the success and failure of engagement efforts. Some evaluations focus on the intrinsic benefits of stakeholder engagement but disregard its instrumental outcomes. Others focus on the instrumental outcomes for stakeholders, communities, policy and governance. As a result, systematic comparison of engagement processes and methods in water governance is challenging. Decision makers need to move towards more comprehensive and methodical evaluations of stakeholder engagement to improve understanding of where, when, why and how these processes work and do not work. Evaluation can help decision makers understand what type of engagement works, under which conditions, for which results.

Overall, evaluating the efficiency and effectiveness of stakeholder engagement contributes to good governance. It provides information that can improve decision making and implementation on a number of fronts: first, it can reduce information gaps on engagement processes and encourage greater effectiveness. Second, it can provide the evidence base to select policy strategies, resource allocation and actors. Finally, it accounts for results, and it allows for learning, adjusting and improving (OECD 2015d).

Institutionalisation, structuring and integration

There is no water governance without governance at large. Similarly, there can be no effective stakeholder engagement without proper incentives for bottom-up and inclusive policy making. A clear set of rules, platforms and vehicles for doing so is key to move from reactive to proactive action.

The flexibility associated with project- or issue-based stakeholder engagement has made it a preferred option for many decision makers rather than engaging in more systematic inclusive approaches, apart from noticeable exceptions (e.g. the 'Polder approach' in the Netherlands for consensus-building). Incidental stakeholder engagement consists of setting up *ad hoc* mechanisms such as workshops, hearings, panels or campaigns to gather stakeholders around a specific issue. These engagement processes are often time-bound, limited in scope and related to the modalities of a specific water project or policy process. They require just enough human and financial resources to sustain the process, but often end with the implementation or evaluation of the project or policy.

However, more structural forms of stakeholder engagement are developing in the water sector. New legislation, guidelines and standards on public participation at various levels related have spurred the emergence of more formalised forms of stakeholder engagement. The Chesapeake Bay Program, for example, was created in the United States in 1983, based on an agreement that formally involves federal and state agencies, local governments, non-profit organisations and academic institutions. They operate through committees, goals implementation teams, workgroups and action teams to achieve restoration and protection of the bay. Also, the EU Flood Directive requiring public participation mechanisms to ensure citizens' involvement in the flood management cycle is another example. Organisations are referring to such participation in their overarching principles and policy. Increasingly, either because of legal requirements or on a voluntary basis, public authorities, service providers, regulators, basin organisations and donors have included requirements for co-operation, consultation or awareness-raising in their operational rules and procedures. In the private sector, in particular, more and more companies have embraced corporate social responsibility, facilitated through standards and common codes of conduct, and have adopted formal ways of engaging stakeholders (OECD 2001).

To avoid a 'tick-the-box' approach that promotes the minimum level of engagement required (e.g. information sharing) without making the maximum benefit from it, legal requirements and frameworks need to set out clear mandates on the exact form of stakeholder engagement required and the matter concerned. Institutionalisation should also be flexible to take into consideration place-based needs and changing circumstances while fostering a change in daily practices, professional skills and culture of decision making. In turn, it can help mitigate the risks of engagement 'fatigue' and/or 'capture'. Provisions for stakeholder engagement should be aligned coherently and holistically across the water chain and policy domains related to water.

Adaptiveness

Stakeholder engagement tools and mechanisms work differently across places, times, objectives and stages of the policy/project cycle. The nature of engagement varies at each stage of a given water policy or project, and so does the likelihood of impact on the decision making process. Similarly, selecting engagement mechanisms should take into account the categories of stakeholders targeted and the context in which they are implemented. Any stakeholder engagement effort will need to be flexible and resilient enough to adapt to these circumstances.

First, stakeholder engagement should be tailored to the development stage of water policy and projects. At the *design* stage, initiatives such as consultations and fora are often used to identify expectations or needs. At the *implementation* stage, partnerships in the form of water stewardships, for example, can bring together the private sector and governments to work jointly on water resources conservation. At the *evaluation* stage, surveys are common practice to assess outcomes and levels of satisfaction.

Second, stakeholder engagement has to be designed and managed in line with intended objectives. Stakeholder engagement can be a goal in itself (normative-democratic approach), a means to more efficient and legitimate water-related decision making; or an instrument to fulfil objectives that go beyond the water sector (e.g. empowerment of marginalised groups). These objectives rely on different types of mechanisms and players. For instance, to harness water-related knowledge, policy makers can set up water information systems whereas solving conflicts over water resources allocation requires the active involvement of those affected to identify the appropriate trade-offs and build consensus.

Third, decision makers should tailor existing mechanisms to specific categories of stakeholders. Careful attention to cultural habits, levels of education and material means is needed to select the appropriate tool. Capacity development is often used by international organisations, financial actors and donors who focus a large part of their activities on training, education and empowerment efforts to strengthen capacities and skills. Also, consultation as part of the regulatory process, is a tool favoured not only by regulators but also by national governments when looking to engage with stakeholders.

Finally, stakeholder engagement should be fitted to places. Some tools should be tailored to geographical specificities. For instance, the National Water Agency in Brazil developed an innovative National Management Pact as a contract for co-operation across levels of government, in which objectives and implementation means are tailored to the specificities of each of the 27 states. A series of multi-stakeholder workshops at state level helped identify a variety of challenges and led to specific water management targets, as well as dedicated funds, for each state based on their own consensus and vision of the future. Other tools are easier to use in urban areas than rural ones. Stakeholders living in urban areas tend to have more access to media and ICT-based tools like websites and social platforms, for instance. Urban settings can offer more opportunities to bring together groups and people from different backgrounds. Tools such as town hall meetings, citizen juries and public hearings concerning specific water issues are particularly appropriate for urban areas, although they have been used effectively in rural settings as well. Rural stakeholders are often disadvantaged due to the lack of communication infrastructure and the general tendency to focus more on the interests and concerns of urban actors. Tools that can be particularly useful in bringing the voice of rural citizens to the table include traditional media (e.g. radio channels) and community meetings.

It is crucial for decision makers to carefully align mechanisms with the level of engagement targeted and the context in which the engagement takes place. The effectiveness of the engagement process also relies on the capacities and resources needed, including knowledge, know-how and funding (travel expenses to attend a meeting, necessary technological settings).

Water governance systems are complex and in flux, where change is dynamic and often unpredictable. New methods are also being continually developed and require new skills and capacities. Engagement processes therefore need to enable multiple stakeholders to respond and adapt to uncertainty, and should remain flexible to manage risks and resilient to adapt to changing environments. There is not a single optimal mechanism for stakeholder engagement but a menu of options, for which the pros and cons need to be weighed up very carefully. Lessons can be learnt from failure in engagement approaches in terms of management of complexity and how to trigger fundamental change, but also from success stories.

Conclusion: ways forward

In a rapidly changing and connected world where climate change, population growth, urban development, rising water need for energy and food, natural disasters and water shortage are likely to damage societies and the environment, stakeholders must be empowered to act together to shape water governance. Setting up the enabling environment for collective management of water can allow actors to face these challenges together, and to successfully meet the needs of current and future generations.

Inclusive and consultative processes are no panacea for addressing all of the challenges facing the water sector, but they can contribute to more effective decision making and implementation. Stakeholders play a crucial role in determining the outcome of a given policy or project, and can be part of the solution. They can initiate and support it, but they can also oppose efforts, attempt to block them or divert them to serve their own aims. Stakeholder engagement provides opportunities to share objectives, experiences and responsibilities, and to be more supportive of solutions that will be reached while voicing and addressing concerns and interests.

Critical aspects of good governance should guide stakeholder engagement frameworks. First, *equity*: fair and equitable access to engagement opportunities is key to ensure a balanced and representative process that takes into account diverse ideas and opinions. Second, *transparency*: being transparent and open about the ways to identify stakeholders, choose engagement mechanisms and define the objectives pursued can help to raise interest among stakeholders and to develop an understanding of and support for the final decisions. Third, *accountability*: it is not sufficient to provide platforms for stakeholders to share their ideas; decision makers must also clearly demonstrate how these ideas are taken into account. Procedural transparency and timely disclosure of information, including alternative solutions, are therefore critical to ensure the legitimacy of decision making processes and their outcomes. Fourth, *trust*: engagement processes may bring together groups with opposing views who fear that their views will not be taken into account. Showing participants what the intention of the process is and how their input will be considered is important to ensure productive discussions and exchange of opinions. It is also important that decision makers are able to trust the quality and value of input from non-technical experts.

To support the implementation of the six principles presented in this chapter, the OECD has developed a tentative Checklist for Public Action to help decision makers fostering inclusive decision making and implementation through a self-assessment questions about framework conditions, a list of tools and international best practices, and tentative indicators to identify areas for improvement. Next steps include mainstreaming such indicators into a broader framework of OECD water governance indicators that will support the implementation of the OECD *Principles on Water Governance* (OECD 2015b). A best practice database on water governance, including a dimension on stakeholder engagement, is also under creation to foster experience-sharing based on success and pitfalls to avoid.

Notes

1 See World Economic Forum (2015).
2 See OECD (2012b).

3 See OECD (2015c).
4 The Aarhus Convention emphasises three areas – transparency, participation and accountability – and requires its 39 signatory countries to incorporate minimum requirements for each area into their laws and institutions. The convention relies on enforceable rights of citizens, including procedural rights and the human right to a healthy environment.
5 For instance, EU Directive 2003/4/EC calls for public access to environmental information, while Directive 2003/35/EC mandates public participation for drawing up certain plans and programmes related to the environment and for amending public participation and access to justice. Similarly, Article 14 of the Water Framework Directive requires member countries to encourage the active involvement of interested parties in the implementation of the directive.
6 Details of the survey's sample of respondents and methodology are provided in OECD (2015a).

References

Abelson, J. and Gauvin, F.P. (2006) Assessing the impacts of public participation: concepts, evidence and policy implications. Research Report P/06, Canadian Policy Research Networks, Ottawa, Canada.

Arnstein, S.R. (1969) A ladder of citizen participation. *Journal of the American Institute of Planners* 35(4), pp. 216–24.

Brown, M. and Wyckoff-Baird, B. (1992) Designing integrated conservation and development projects. Biodiversity Support Program, Washington, DC.

Carr, G., Bloschl, G. and Loucks, D.P. (2012) Evaluating participation in water resource management: a review. *Water Resources Research* 48(11), November. http://dx.doi.org/10.1029/2011WR011662

Dahl, R.A. (1994) A democratic dilemma: system effectiveness versus citizen participation. *Political Science Quarterly* 109(1), pp. 23–34. www.jstor.org/stable/2151659

Fung, A. (2006) Varieties of participation in complex governance. *Public Administration Review* 66(Supplement S1), pp. 66–75. http://dx.doi.org/10.1111/j.1540-6210.2006.00667.x

Moss, T. and Newig, J. (2010) Multi-level water governance and problems of scale: setting the stage for a broader debate. *Environmental Management* 46(1), pp. 1–6. http://dx.doi.org/10.1007/s00267-010-9531-1

OECD (2001) *Corporate Social Responsibility: Partners for Progress*. Paris: OECD. http://dx.doi.org/10.1787/9789264194854-en

OECD (2005) *Evaluating Public Participation in Policy Making*. Paris: OECD. http://dx.doi.org/10.1787/9789264008960-en

OECD (2011) *Water Governance in OECD Countries: A Multi-level Approach*. OECD Studies on Water. Paris: OECD. http://dx.doi.org/10.1787/9789264119284-en

OECD (2012a) *OECD Environmental Outlook to 2050: The Consequences of Inaction*. Paris: OECD. http://dx.doi.org/10.1787/9789264122246-en

OECD (2012b) Condition for success 1 'Good governance': Synthesis report of Target 1 Stakeholders' engagement for effective water policy and management as prepared by the Good Governance core group for the 6th World Water Forum, 12–17 March, Marseille. Available from: www.worldwaterforum6.org/uploads/tx_amswwf/CS1.1__Stakeholder__s_engagement_for_effective_water_policy_and_management_Report.pdf

OECD (2014) *Water Governance in the Netherlands: Fit for the Future? OECD Studies on Water*. Paris: OECD. http://dx.doi.org/10.1787/9789264102637-en

OECD (2015a) *Stakeholder Engagement for Inclusive Water Governance*. Paris: OECD. http://dx.doi.org/10.1787/9789264231122-en

OECD (2015b) *OECD Principles on Water Governance*. Paris: OECD. http://www.oecd.org/governance/oecd-principles-on-water-governance.htm

OECD (2015c) Outcomes of the governance sessions at the 7th World Water Forum, 12–17 April, Daegu-Gyeongbuk, Korea. Available from: http://www.oecd.org/gov/regional-policy/Outcomes-governance-7thForum.pdf

OECD (2015d) OECD Water Governance Indicators. Scoping note presented at the 6th meeting of the OECD Water Governance Initiative, 2–3 November, Paris.

Pretty, J. (1995) Participatory learning for sustainable agriculture. *World Development* 23(8), pp. 1247–63.

Rowe, G. and Frewer, L.J. (2005a) A typology of public engagement mechanisms. *Science, Technology, and Human Values* 30(2), pp. 251–90. http://dx.doi.org/10.1177/0162243904271724

Rowe, G. and Frewer, L.J. (2005b) Evaluating public participation exercises: A research agenda. *Science, Technology, and Human Values* 29( )4, pp. 512–56. http://dx.doi.org/10.1177/0162243903259197

Smith, D.H. (1983) Synanthrometrics: on progress in the development of a general theory of voluntary action and citizen participation. In D.H. Smith and J.Van Til (eds), *International Perspectives on Voluntary Action Research*. Washington, DC: University Press of America.

UNDP (United Nations Development Programme) Water Governance Facility, Stockholm International Water Institute, Water Integrity Network (2013) *User's Guide on Assessing Water Governance*. Denmark: United Nations Development Programme.

When, U., Rusca, M. and Evers, J. (2014) Participation in flood risk management and the potential of citizen observatories: a governance analysis. *Abstracts of the International Conference on Flood Management*, 16–18 September, Sao Paulo, Brazil.

World Economic Forum (2015) *The Global Competitiveness Report 2015–2016*. Geneva: World Economic Forum.

Yee, S. (2010) Stakeholder engagement and public participation in environmental flows and river health assessment. Australia–China Environment Development Partnership, Project code P0018, May.

13

MONITORING AND ENFORCEMENT

The United States Clean Water Act model

LeRoy C. Paddock and Laura C. Mulherin

Introduction

Regulation of water pollution in the United States is a complex enterprise under the United States Constitution (USC). States retained all powers not delegated to the Federal government at the time the country's Constitution was adopted. This meant that states, rather than the Federal government, are the regulator of first instance for water pollution. However, as early as 1899 Congress enacted the Rivers and Harbors Act under authority of Constitution's Commerce Clause to prevent disruption of navigation on 'navigable waters' of the United States. Section 13 of the Rivers and Harbors Act of 1899 (30 Stat. 1121) provides that the Secretary of the Army may permit the discharge of refuse into navigable waters, if its Chief Engineer determines that anchorage and navigation will not be injured thereby. In the absence of a permit, discharge of refuse is prohibited. Although used occasionally throughout the first half of the twentieth century, most efforts to deal with water pollution remained at the state level.

As water pollution became a more prominent concern, in part as a result of dramatic events such as the Cuyahoga River, near Cleveland, Ohio catching fire, the Federal government became more involved with water pollution issues beginning with the enactment of the Clean Water Act in 1948. However, the 1948 Act was quite limited, focusing on planning and technical assistance. Because water pollution concerns were increasing and because the 1948 Act and subsequent amendments did not provide clear regulatory authority for the Federal government, the Federal government looked back to the 1899 Rivers and Harbors Act to bring enforcement actions in the 1960s and early 1970s (Eames 1970, p. 1446). This allowed the Federal government to address some of the growing number of water pollution problems in the country but did not provide a comprehensive scheme for regulating water pollution and assuring a minimum level of water quality protection across the country.

The enactment of the Federal Water Pollution Control Act Amendments of 1972 (also referred to as the Clean Water Act or CWA) dramatically changed the landscape for water pollution permitting, monitoring and enforcement providing the framework for the standards now enforced in the United States. The Amendments moved the authority for permitting discharges of refuse (now 'pollutants') into navigable waters from the Army Corps of Engineers to the US Environmental Protection Agency (EPA). Amendments to the CWA in 1977 and 1987 completed the transition to the current national permitting, monitoring and enforcement requirements that establish a floor for water pollution regulation in the United States. These requirements continue to be based on the Federal government's authority to regulate commerce among the states (Polk 2013, p. 719). EPA's authority under the previous versions of the statute prevented EPA from proactively regulating water pollution (ibid., pp. 733–4). For example, the pre-1977 version of the CWA required the agency to wait until a discharge occurred and then react (ibid., p. 738; Copeland 2014). This situation changed with the enactment the 1977 and 1987 amendments.

The CWA regulatory scheme

The CWA prohibits the *discharge* of a *pollutant* from a *point source* into the *navigable waters of the United States* without a *permit* (see Polk 2013, p. 749; Coggins and Glicksman 2015, § 19.11). While disputes may arise as to what constitutes a 'discharge' – the '*addition* of any pollutant to navigable waters' (33 USC § 1362(12)) – this is normally not a significant issue. Pollutant is broadly defined under the Act to include: 'Dredged spoil, solid waste, incinerator residue, sewage, garbage, sewage sludge, munitions, chemical wastes, biological materials, radioactive materials, heat, wrecked or discarded equipment, rock, sand, cellar dirt and industrial, municipal, and agricultural waste discharged into water' (33 USC § 1362(6)).

Point sources are also expansively defined in the CWA, utilizing an illustrative list that includes 'any pipe, ditch, channel, tunnel, conduit, well, discrete fissure, container, rolling stock, concentrated animal feeding operation, or vessel or other floating craft, from which pollutants are or may be discharged' (33 USC § 1362(14)). Point sources are differentiated from non-point sources; non-point sources are discharges originating from 'diffuse sources' such as 'rainfall or snowmelt moving over and through the ground' (US Environmental Protection Agency, undated, *What is Nonpoint Source Pollution?*).[1] The distinction between point source and non-point sources is important since the permitting requirements, most monitoring requirements and most of the enforcement provisions of the CWA only apply to point sources (see 40 CFR §§ 122.21 (b) (scope of National Pollutant Discharge Elimination System (NPDES) program) and 122.3 (a)-(i) (excluding dredge and fill activities, non-point source discharges).[2]

The final key definition is that of 'navigable waters of the United States'. Because enactment of the CWA was based on the Federal government's authority under the Commerce Clause, the scope of the Act's authority is limited to 'navigable waters' that are 'channels of commerce'. However, navigable waters need not be navigable in fact. Instead, jurisdiction extends to many tributaries and 'adjacent' wetlands that can have an impact on waters that are navigable in fact. The nexus required between a navigable in fact water (which includes lakes and rivers) and nearby wetlands, tributaries, or other land types that can be categorized as a navigable water remains quite controversial. The US Supreme Court in *Solid Waste Agency of Northern Cook Cty. v Army Corps of Engineers* (531 US 159 (2001)) held in referring to one of its earlier decision that 'under the circumstances presented there, that to constitute "navigable waters" under the Act, a water or wetland must possess a "significant nexus" to waters that are or were navigable in fact or that could reasonably be so made' (*Rapanos v United States,* 547 US 715,759 (2006)). Recently, the EPA and the Army Corps of Engineers finalized new rules defining what waters have a sufficient nexus to waters that are or were navigable in fact such that the waters can be considered 'waters of the United States' triggering jurisdiction under the CWA (see 80 Fed. Reg. 37054, 29 June 2015). These rules were immediately challenged by a large number of organizations representing agriculture, developers and industry.

The CWA addresses non-point sources in a very different way. The Act requires states to establish 'water quality standards' for rivers and lakes within the state. These water quality standards are to be set at a level that preserves the designated use of the receiving water such as for recreation. Water quality standards address both effluent from point sources discharging pursuant to their permits and pollution from non-point sources. States are required by the CWA to periodically assess water bodies to determine if they meet water quality standards. If the combined point and non-point sources of pollution exceeds the water quality standard, the water body is considered 'impaired'. States are expected to establish 'total maximum pollution loadings' (TMDLs) for impaired waters that will restore them to the desired level of water quality. TMDLs can be achieved by a combination of further limiting discharges from permitted point sources, by limiting non-point source pollutants from entering the water body, or by creating trading systems that allow point sources to purchase pollutant reductions from other point sources or from non-point sources (water trading regimes) (see 33 USC § 1313). Because a significant portion of non-point source pollution in many locations results from agriculture or development activities, states have historically been reticent to impose regulatory limits on non-point sources, relying instead on voluntary reductions or water trading systems whenever possible.

Monitoring of non-point sources has been limited in the United States but new, low cost and web-linked in-stream sensors and new satellite sensing tools are beginning to emerge that make in-stream monitoring more feasible (see US Environmental Protection Agency, undated, *Advanced Nutrient Monitoring*). These developments may facilitate closer regulation of non-point sources and make water quality trading regimes more viable. Still, most monitoring and enforcement activity in the United States is focused on point source discharges. While non-point sources constitute a high percentage of the remaining water pollution issues in the United States, the remainder of this chapter will focus on point sources because that they are the principal focus of the CWA.

Methods of regulation

Permitting under the NPDES is the primary method of regulating discharges under the CWA (Copeland 2014, p. 6). Permit holders are allowed to discharge pollutants into navigable waters so long as the discharges comply with discharge (usually referred to as 'effluent') limitations set out in the permit (see 40 CFR § 122.5 (a)(1)).[3] Permits are mandatory; without a permit any discharge of a pollutant from a point source into a navigable water is illegal (see Copeland 2014, pp. 5–6; Vinal 1996, §3). The NPDES permit programme only applies to direct discharges to navigable waters. Discharges into wastewater treatment system are referred to as 'indirect' discharges. Indirect discharges are subject to a separate 'pretreatment' permitting programme that has its own set of discharge standards.

Setting discharge standards

The NPDES permit discharge standards correspond to specific discharge types regulated under the NPDES programme; conventional, toxic, or non-conventional discharges (see US Environmental Protection Agency, Office of Wastewater Management Water Permitting). Section 304(a)(4) of the CWA designates the following as 'conventional' pollutants: biochemical oxygen demand (BOD5), total suspended solids (TSS), faecal coliform, pH and any additional pollutants EPA defines as conventional. The Agency designated oil and grease as an additional conventional pollutant on 30 July 1979 (see 44 Fed. Reg. 44501). EPA has identified 65 pollutants and classes of pollutants as 'toxic pollutants', of which 126 specific substances have been designated 'priority' toxic pollutants. All other pollutants are considered to be 'non-conventional'.

EPA sets discharge limits based upon the technology available to control pollutant discharges. These standards vary depending upon the category of industry involved, the type of discharge and whether the facility is an existing or new facility. So called 'categorical' standards are developed for a range of industries such as publicly owned wastewater treatment facilities, pulp and paper mills, electroplating, steam electric power plants and many others. For conventional pollutants the standards are based on 'best practicable control technology currently available' for the category or class of industry. For toxic pollutants, 'best available technology economically achievable' for the category or class of industry. For new sources, 'best available demonstrated control technology' is required. EPA may consider cost and any non-water quality environmental impacts and energy requirements in setting new source standards (see 33 USC §§ 1311 and 1316; US Environmental Protection Agency, undated, *Effluent Guidelines Plan*, pp. 2–3).

To determine what constitutes the appropriate technology standard for an industry or a facility, EPA first gathers information on industry practices, technologies that are available to prevent or treat discharges, and the cost of utilizing the technologies. With this information EPA proposes a standard by publishing the proposal in the Federal Register, a daily publically available document that contains proposals for Federal actions. The agency then receives comments during the public review period and, after reviewing the comments, may publish a final rule in the Federal Register. Final rules can be challenged in Federal court and many of EPA's rules are challenged. The rule-making process is often both costly and time consuming, making it difficult to regularly revise rules, including the technology standards that are the basis for NPDES permits and their monitoring requirements.

The CWA's regulatory scheme utilized to set NPDES discharge limitations was designed to be 'technology forcing' in that the effluent standards are to be based on best available technology and periodically updated in order to meet the 'demands placed on those who are regulated by [the CWA] to achieve higher and higher levels of pollution abatement' (see Copeland 2014, p. 2; Grad 1997, p. 20; Albright 2012, p. 4). Although data on pollutant loading reductions from point sources is rather scarce, US EPA conducted the most comprehensive study of water quality improvements related to point source discharges in the late 1990s, about 20 years after the passage of the Clean Water Act. The study demonstrated that from 1973–95 the amount of biological oxygen demand (BOD) discharges from industrial point sources fell 40 per cent (Andreen 2013). The study also showed 'Direct industrial discharges of approximately 300 toxic water pollutants fell 72% between 1988 and 1996 from 164 million pounds per year to 45 million pounds per year' despite significant increases in GDP (Andreen 2004, p. 573, n. 268; see also Schroeder and Steinzor 2005, p. 63). Similarly, a study of water pollution along the Southern California Bight, which includes the city of Los Angeles, showed very significant reductions between 1971 and 2000 in several constituents despite a 25 per cent increase in wastewater volume. These decreases included a nearly 60 per cent reduction in suspended solids and BOD, more than 75 per cent reduction in total phosphorus, nearly 90 per cent reduction in cyanide, more than 80 per cent reduction in combined metals and more than 90 per cent reduction in total DDT and PCBs (Lyons and Stein 2008, p. 5). However, if technology standards are not periodically updated as anticipated by the law, the impact can be 'freezing' rather than 'forcing' technology. One study pointed out:

> During the past fifteen years, EPA has updated only one effluent limitation out of the top thirteen for industrial discharges of toxic water pollution. Old effluent limitations mean old technology, and opportunities to take advantage of technological advances are lost.
>
> *Andreen and Jones 2008, p. 15*

Delegation to states and tribes

Congress intended the federal effluent standards set by EPA under the CWA as a national minimum standard (or floor) for water pollution to prevent both industry from 'shopping for locations in the state with the most lenient pollution control approach' and states from trying to attract business by touting 'business friendly' environmental standards (a situation sometimes referred to as a 'race to the bottom'). Therefore, the states' role in implementing CWA regulations primarily focuses on enforcement and implementation of the federally established effluent standards (Grad 1997, p. 18). However, because states had historically regulated water pollution and because there are tens of thousands of facilities that require NPDES permits, the vast majority of the NPDES permits and enforcement of the permits occur at the state level (33 USC § 1342 (b) and (c); see also Copeland 2014, p. 7; Vinal 1996, § 3). This arrangement is referred to as 'cooperative federalism'. The CWA requires state governors to submit an application for delegation to EPA. The delegation request must include a statement from the state's Attorney General that the laws of the state provide 'adequate authority' to carry out the programme, including issuing permits, inspecting and monitoring facilities and enforcing the permit requirements (see 33 USC § 1342(b)). EPA has delegated CWA permitting and enforcement to 45 of the 50 states.

In most states, delegated programmes function quite well. However, in some states environmental organizations have raised concerns that states are not dedicating sufficient resources to permit and enforcement work under the CWA. EPA does have the authority to withdraw delegations if the state is not administering the programme in accordance with the requirements of the law (33 USC § 1342(c)). However, EPA has thus far not withdrawn any state programmes despite the petitions that it has received (US Environmental Protection Agency, undated, *State NPDES Program Withdrawal Petitions*). EPA also has authority to bring an enforcement action in cases where the Agency determines that a state enforcement action is inadequate; a process known as 'overfiling' (33 USC § 1319(a), 33 USC § 1342 (i)).[4] EPA has occasionally used its overfiling authority, although it does so sparingly since the process can cause significant strain on the state–federal relationship EPA relies on to implement the CWA.

Monitoring

The CWA includes specific language establishing point source monitoring requirements:

(a) the Administrator shall require the owner or operator of any point source to (i) establish and maintain such records, (ii) make such reports, (iii) install, use, and maintain such monitoring equipment or methods (including where appropriate, biological monitoring methods), (iv) sample such effluents (in accordance with such methods, at such locations, at such intervals, and in such manner as the Administrator shall prescribe), and (v) provide such other information as he may reasonably require; and

(b) the Administrator or his authorized representative (including an authorized contractor acting as a representative of the Administrator), upon presentation of his credentials –

　(i) shall have a right of entry to, upon, or through any premises in which an effluent source is located or in which any records required to be maintained under clause (A) of this subsection are located, and

　(ii) may at reasonable times have access to and copy any records, inspect any monitoring equipment or method required under clause (A), and sample any effluents which the owner or operator of such source is required to sample under such clause.

33 USC § 1318

In addition to the self-monitoring requirements set out in statute and in the NPDES permit, EPA and the states also monitor compliance through facility inspections. These inspections allow government agencies to take their own samples, to verify that monitoring equipment is functioning properly and to assure that appropriate records are being maintained. However, given the large number of permitted facilities, inspections are in most cases infrequent. EPA's Compliance Monitoring Strategy requires that country's nearly 7,000 'major facilities' (the largest municipal wastewater treatment facilities and the largest industrial discharges) receive a comprehensive inspection once every two years (see US Environmental Protection Agency 2014, p. 6). The 87,000 non-major facilities (that include municipal wastewater treatment facilities that discharge up to 1 million gallons per day) are required to be inspected once every five years (US Environmental Protection Agency, undated, *Effluent Guidelines Plan*, pp. 7–8). While facilities that are subject to EPA targeted enforcement initiatives may be inspected more frequently, the typical span between inspections indicates why self-monitoring, public reporting and citizen enforcement are such an important part of the CWA compliance system.

Discharge Monitoring Reports

Holders of 'NPDES permits [are required] to report their compliance to the agency' through Discharge Monitoring Reports (DMRs) (Polk 2013, p. 748). The EPA defines a DMR as 'the EPA uniform national form, including any subsequent additions, revisions, or modifications for the reporting of self-monitoring results by permittees' (40 CFR § 122.2). EPA also requires that 'DMRs must be used by "approved States" as well as by EPA [and that] EPA will supply DMRs to any approved State upon request' (40 CFR § 122.2). The specific content each permit holder is required to submit through a DMR varies depending on the conditions prescribed their permit. However, every permit must include:

(a) Requirements concerning the proper use, maintenance, and installation, when appropriate, of monitoring equipment or methods (including biological monitoring methods when appropriate), and

(b) Required monitoring including type, intervals, and frequency sufficient to yield data which are representative of the monitored activity including, when appropriate, continuous monitoring.

40 CFR § 122.48 (a)-(c)

Facilities holding NPDES permits are typically required to submit DMRs each month. DMRs traditionally had been submitted on paper forms to the states and then reported by the states to EPA to be made part of two information systems, the Permit Compliance System (PCS) and the Integrated Compliance Information System (ICIS) (US Environmental Protection Agency, undated, *Envirofacts*; US Environmental Protection Agency, undated, *Enforcement and Compliance History Online*).

Because data is maintained on separate systems in each state or tribe with a delegated CWA programme and then transferred to EPA, there had been a history of problems with data accuracy in the PCS system and in derivative systems such as ECHO. These data errors could show that a facility was not in compliance when in fact it had been in compliance or show the facility was in compliance when the correct data would have demonstrated non-compliance. As a result, EPA put a 'data correction' system in place to help alleviate the problem. A joint effort between EPA and the states to upgrade their information system was completed in 2012, allowing the older, less reliable, system to be retired. EPA, however, continues to maintain a data correction process (US Environmental Protection Agency, undated, *How to Report an Error*).

In September 2015, EPA adopted a new rule that will require all NPDES-regulated entities to electronically submit DMRs, as well as inspection information, violation determinations and enforcement actions within one year (US Environmental Protection Agency 2015a). The rule is designed to 'make reporting easier for NPDES-regulated entities, streamline permit renewals, ensure full exchange of basic NPDES permit data between states and EPA, improve decision-making, and better protect human health and the environment' (ibid.). The new reporting system is expected to save EPA, regulated entities, and states over $24 million per year (ibid.).

DMRs are publically available documents and, in fact, form the basis for many of the citizen suits that are discussed later in this chapter. In addition, EPA provides members of the public including facility operators and citizens with a DMR 'pollutant loading tool' that uses the terms of permits to help them understand 'who is discharging, what pollutants they are discharging and how much, and where they are discharging' (see US Environmental Protection Agency, undated, *Discharge Monitoring Report (DMR) Pollutant Loading Tool*).

Advanced monitoring and remote sensing

Water quality monitoring has historically been a problem in the United States, with most water bodies having only a few monitors to help government officials identify sources of pollutants, especially those sources not associated with permitted point sources. Advancements in low-cost pollutant sensor technology, remote sensing techniques such as the use of satellite-based monitoring devices, and information technology that allows data to be quickly communicated to regulatory agencies has begun to change this situation. The 2015 J.B. and Maurice C. Shapiro Environmental Law Symposium on Advanced Monitoring, Remote Sensing, and Data Gathering, Analysis and Disclosure in Compliance and Enforcement explored both some of the new technologies and the legal implications of these new technologies. The proceedings of the Symposium are available online[5] and articles from the symposium will be published in the 2015–16 volumes of the *George Washington University Journal of Energy and Environmental Law*.

Enforcement

The CWA includes a very robust set of enforcement tools, allowing both government and citizens to contribute to implementation of the law. These tools include administrative compliance orders, administrative penalty orders, civil judicial penalties, criminal sanctions and citizen suits.

Administrative orders

Typically, compliance orders are issued by states with delegated programmes. These orders require facilities to take actions to comply with their permits. If the EPA Administrator notifies a state of a violation and the state does not commence enforcement actions within 30 days, the Administrator is authorized to issue an administrative order requiring compliance with the permit (33 USC § 1319). One commentator notes that these orders can 'provide a quick, responsive, and flexible enforcement tool, particularly well-suited to remedying less egregious conditions. Indeed, the lack of such authority… may make violations of this nature virtually uncontrollable' (Schang and Abramson 2015, § 9.9). The EPA Administrator is also authorized, after consultation with the state in which a violation occurs, to issue an administrative order assessing penalties (33 USC § 1319(g)). Less serious violations (designated Class I violations by the CWA) may not exceed $10,000 *per violation*, with a maximum penalty for all violations subject to a single order of $25,000. The Administrator is required to notify the alleged violator of the proposed penalty and afford the person the opportunity for an informal hearing (33 USC § 1319(g)(2)(A)). More serious (Class II) violations have a cap of $10,000 *per day* of violation with a $125,000 ceiling for an order. The Administrator must afford the alleged violator with the opportunity to request a formal evidentiary hearing (33 USC § 1319(g)(2)(B)). For both Class I and Class II violations, the penalty amount must take into account:

> [the] nature, circumstances, extent, and gravity of the violation, or violations, and, with respect to the violator, the ability to pay, any prior history of such violations, the degree of culpability, economic benefit or savings (if any) resulting from the violation, and any such other matters as justice may require.
>
> *33 USC § 1319(g)(2)(C)*

Civil liability

The CWA authorizes the EPA to enforce the Act through the judicial system. The Administrator can commence a judicial action seeking 'appropriate relief' including a temporary or permanent injunction for any violation for which the Administrator has the authority to issue a corrective order. Jurisdiction lies in the Federal District Court where the defendant is located or doing business (33 USC § 1319(c)). Further, any person who violates the act is subject to civil penalties of up to $25,000 *per day for each violation* of any permit condition or limitation (33 USC § 1319(d)). These penalties can quickly add up to millions of dollars if violations occurred over a period of several days or if multiple violations are identified. Factors that a court must consider in determining civil penalties include:

> the seriousness of the violation or violations, the economic benefit (if any) resulting from the violation, any history of such violations, any good-faith efforts to comply with the applicable requirements, the economic impact of the penalty on the violator, and such other matters as justice may require.
>
> *33 USC § 1319(d)*

These factors are similar to, but not the same as, those that the Administrator is required to consider in assessing administrative penalties. The CWA operates as a 'strict liability statute' for purposes of civil violations meaning that government agencies or citizens seeking to have penalties judicially imposed need not demonstrate the violator intended to violate the Act to establish liability (Polk 2013, p. 755).

Even though the penalties are to be assessed by courts, the EPA typically settles civil penalty cases for final penalty amounts that are significantly less that the original penalty the agency proposed based on the number of days of violation and the number of violations. Settlements are negotiated based on a long-established settlement policy (US Environmental Protection Agency undated). In addition, EPA (and many of the states) allow some portion of a penalty (typically excluding the economic benefit associated with a violation) to be offset through commitments to what are called 'supplemental environmental projects' (SEPs). A SEP is an:

environmentally beneficial project or activity that is not required by law, but that a defendant agrees to undertake as part of a settlement of an enforcement action. SEPs are projects or activities that go beyond what could legally be required in order for the defendant to return to compliance, and secure environmental and/or public health benefits in addition to those achieved by compliance with applicable laws.

US Environmental Protection Agency 2015b

SEPs can often facilitate penalty settlements by allowing the defendant to see some benefits to the local community or to community organizations flowing from the settlement. Detailed criteria are set out in US EPA's Supplemental Environmental Projects Policy for the eligibility of projects for SEP funding and for the amount and type of penalties that can be offset by SEPs.

Criminal liability

Consistent with most major environmental law statutes, the CWA created a series of criminal enforcement penalties for violations of the Act. Violations that are the result of negligent conduct – that is, conduct involving failure to exercise due care – are subject to penalties of not less than $2,500 and not more than $25,000 per day of violation and one year in prison (33 USC § 1319(c)(1)). Crimes that are subject to one year or less in jail in the United States are typically referred to as 'misdemeanour' violations. Although misdemeanour violations may be charged in some cases, misdemeanour crimes are not used as a routine enforcement mechanism for the CWA, having been largely displaced by the administrative and civil penalties available under the Act (Schang and Abramson 2015, § 9:30). Administrative and civil penalties have a lower standard of proof (more probable than not, instead of beyond a reasonable doubt), are subject to strict liability (thus no failure of due care must be proven) and have higher available penalties. In addition, research indicates that use of minor criminal penalties as a principal enforcement approach for environmental violations may not be effective (Watson 2005, pp. 3–6).

In contrast to the misdemeanour penalty provisions of the CWA that are not a major factor in enforcement, the 'felony' crimes (one year or more in prison) created by the CWA are used to prosecute more serious violations of the Act, although criminal prosecutions account for less than 10 per cent of all Federal enforcement actions (see US Environmental Protection Agency, undated, *Enforcement Annual Results for Fiscal Year (FY) 2015*). 'Knowing' violations of several provisions of the CWA can be punished by a fine of not less than $5,000 per day of violation and not more than $50,000 per day of violation, and imprisonment for up to three years (33 USC. § 1319(c)(2)). Knowing conduct 'denotes crimes that require awareness of the nature of the act or omission, the results that will follow from an act or omission, or the circumstances indicating an act or omission' (Hansen 1990, p. 989). Although a defendant must be aware of the situation, the act does not need to be purposeful with intent to violate the law (ibid., p. 990).[6]

The CWA establishes more serious criminal penalties for circumstances where a person violates the Act and 'knows at that time that he thereby places another person in imminent danger of death or serious bodily injury' – so called 'knowing endangerment' crimes. These crimes are punishable by fines of up to $250,000 and imprisonment of up to 15 years (33 USC § 1319(c)(3)).

Both corporations and corporate officers may be held criminally liable. Corporations are included in the definition of person under the Act. Although corporations cannot be imprisoned, criminal penalties can be imposed and the stigma of a criminal conviction can have an adverse impact on a company's reputation. Further criminal conviction can result in a company being disqualified from government contracts (known as debarment) (33 USC § 1368(a)). Corporate officers can be prosecuted if they were directly involved in a violation, or if it can be demonstrated that the person held a position of responsibility in a corporation, had the ability to prevent a violation and failed to do so (Schang and Abramson 2015, §9:32). This later set of circumstances is often referred to as the 'responsible corporate officer doctrine'.

EPA employs sworn law enforcement agents to investigate environmental crimes and the US Department of Justice maintains an Environmental Crimes Division to assist US Attorneys in prosecuting environmental crimes cases. While a few states have dedicated environmental crimes staff, most states rely on local police and prosecutors to investigate and prosecute environmental crimes cases.

Enforcement of monitoring requirements

The integrity of monitoring results is critical to a system that relies heavily on self-monitoring and reporting. As a result, the CWA and other federal laws provide both civil and criminal penalties for tampering with monitoring equipment and for false reporting. If a person 'falsifies, tampers with, or knowingly renders inaccurate' any of the monitoring data or systems, EPA is authorized to enforce the violation by imposing fines of up to $20,000 (40 CFR § 122.41(j)(5)). In addition to the requirement that monitoring equipment and data remain uncorrupted, penalties can be imposed for falsely certifying the results of monitoring that are reported to the states or EPA. 'Knowingly' providing false information on 'any statement, representation, or certification in any record or other document' in connection with the permit issued under the CWA can constitute a criminal violation (40 CFR § 122.41(k)(2)). In addition, communicating false information to a Federal agency may constitute mail fraud or wire fraud under the United States criminal code. Mail and wire fraud includes intentionally depriving another of property or honest services via mail or wire communication (18 USC §§ 1341 and 1343). Penalties can include imprisonment for up to 20 years. To help assure the accuracy of self-reporting data, US EPA has periodically targeted CWA laboratories for enforcement scrutiny.

Citizen enforcement

Federal environmental laws adopted in the United States in the 1970s pioneered the idea of citizen enforcement. Under the CWA, 'citizens' may sue 'any person' including the 'United States, and any other governmental instrumentality or agency' (33 USC § 1365 (a)(1)–(2)) alleged to be in violation of '(A) an effluent standard or limitation under this chapter or (B) an order issued by the Administrator or a State with respect to a standard or limitation' (33 USC § 1365(h)). In order to bring a citizen suit, as a 'citizen' one must be 'a person or persons having an interest which is or may be adversely affected' (33 USC § 1365 (a) (1)). In addition to permit violators, citizen suits are also authorized to act against the EPA Administrator for failing to perform non-discretionary acts or duties (33 USC § 1365 (a)(2)).

Because the information that can establish violations of the law is available to the public in the form of DMRs, CWA citizen suits have become a prominent part of the CWA enforcement landscape. Several courts have held that information that demonstrates failure to comply with permit limitations contained in a DMR constitutes an 'admission', requiring no further proof to establish a violation has occurred. Citizen suit plaintiffs must provide notice to EPA, the state, and the alleged violator 60 days before filing a lawsuit. If EPA or the state in which the violation occurred file an enforcement action prior to the expiration of the 60-day period and are 'diligently prosecuting' the case, a citizen suite is precluded (33 USC § 1365(b)). Courts may award attorneys' fees to prevailing citizens, but any penalties that are assessed for the violations are payable to the United States.

In *Gwaltney of Smithfield, Ltd. v Chesapeake Bay Foundation*, 484 US 49, 52 (1987), the Supreme Court was asked to determine whether '§505(a) of the Clean Water Act [33 USC § 1365(a)]... confers federal jurisdiction over citizen suits for wholly past violations' of the NPDES permit programme. The court held that, in order to have standing to bring a citizen suit under the CWA, the violation could not be a 'wholly past' violation (*Gwaltney of Smithfield, Ltd. v Chesapeake Bay Foundation*, pp. 59–60). However, the court distinguished between an allegation that is 'wholly past' and allegations of 'on-going violations'. Thus, if a plaintiff makes a 'good faith allegation of [a] continuous or intermittent violation' then the plaintiff has standing under the CWA's citizen suit provision (*Gwaltney of Smithfield, Ltd. v Chesapeake Bay Foundation*, p. 64; see, generally, Buckman 1999).

Despite the ability of state or Federal enforcement actions to preclude citizen enforcement and the limitation on citizen enforcement to ongoing violations, citizen suites continue to play a very important role in enforcing the CWA. Between 20 and 40 citizen suites were filed each year between 2000 and 2011 (Gardner 2015, p. 22).

Conclusion

The US CWA model provides a number of lessons concerning the design of effective monitoring and enforcement programmes. First, it is important to have clear discharge standards that can be incorporated in permits. The categorical standards for both conventional and toxic pollutants in the CWA can easily be translated into specific discharge limits that are incorporated into permits. The one significant concern about the categorical standards system is the difficulty, at least under the US administrative law system, of updating the standards on a sufficiently regular basis to assure that old technology is not being locked in place instead of driving the adoption of new technology by regularly updating the categorical standards. The fact that non-point sources of water pollution that are not subject to a permit regime now make up about two-thirds of all remaining pollution in the United States indicates the importance of clear permit-based standards wherever it is possible.

Second, the incorporation of monitoring and reporting requirements in permits provides both government agencies and the public with important information about compliance with the discharge standards incorporated into the permit. Of particular importance are the monthly DMRs that must include test results that demonstrate whether or not the permit holder is complying with permit limits. The significant penalties, including serious criminal penalties, help assure the accuracy of the DMRs. The DMRs also provide agencies with information they need to enforce the permit limits and, importantly, provide the public with this information which is often considered by courts as an 'admission' for purposes of establishing responsibility for a violation. The very fact that the public has access to DMRs is likely to encourage regulated entities to take care that they are not in violation of their permit, especially for those regulated entities that are concerned about their environmental reputation. Especially for countries with limited inspection resources, self-monitoring requirements subject to clear criminal penalties for false reporting could play an important role in compliance programmes. If these requirements are made a condition of a permit and the reports made available to the public, they could stimulate compliance either through a desire of the reporting entity to avoid having to disclose information indicating a violations has occurred or, in countries that allow some type of citizen enforcement, by providing citizens with the information needed to prove a violation.

Third, the Act demonstrates the value of having a wide range of enforcement tools that can be adapted to the particular circumstances of each case. Administrative penalty authority has proven particular important and effective in dealing with the majority of routine violations. And, while misdemeanour criminal violations have not been effective in dealing with routine violations, the availability of felony criminal penalties to deal with knowing violations of the law, including false reporting of self-monitoring results, has proven important and effective in dealing with serious violations and in deterring violations by other organizations.

Finally, the citizen suite provisions of the Act provide a unique backstop to government enforcement by allowing citizens to bring their own enforcement actions if a government agency fails to act.

Notes

1 Coggins and Glicksman (2015), § 19.11. 'Nonpoint sources under the CWA are defined by default; everything that is not a point source is a nonpoint source.'

2 See Copeland (2014), describing regulation of non-point sources as done by states. Point and non-point sources are not the only categories of discharges regulated through the CWA; § 404 covers 'dredged or fill material' discharged into both wetlands and flowing waters. US Environmental Protection Agency, *Clean Water Act (CWA) Compliance Monitoring* (regulated by EPA and the US Army Corps of Engineers). Additionally, the CWA regulates oil spills and the discharge of hazardous substances into 'the waters of the US or adjoining shorelines'. US Environmental Protection Agency (2014). These regulations pertain to 'owners and operators of non-transportation-related oil facilities.' US Environmental Protection Agency (2014).

3 Section 122.5 (a)(1) describes how, except for limited circumstances, compliance with a permit 'constitutes compliance with sections 301, 302, 306, 307, 318, 403, and 405 (a)–(b) of CWA'.

4 33 USC § 1319(a), 33 USC §1342 (i) provides that 'Nothing in this section shall be construed to limit the authority of the Administrator to take action pursuant to section 1319 of this title.' See also Coggins and Glicksman (2015), § 19.2, describing EPA's 'concurrent' authority to 'enforce state issued permits'.

5 See 2015 J.B. and Maurice C. Shapiro Environmental Law Symposium videotape URLs:

Panel One: The State of Art of Technology 3/26 [9.00–12.30pm] https://video.law.gwu.edu:8443/ess/echo/presentation/f001d3f7-9175-4ac9-9af8-fc7dad3edd73

Panel Two: Advance Monitoring 3/26 [1.45–4:.00pm] https://video.law.gwu.edu:8443/ess/echo/presentation/db69ad01-6c73-480e-bd51-905d4348e327

Panel Three: Citizen Monitoring 3/26 [4.15–5.30pm] https://video.law.gwu.edu:8443/ess/echo/presentation/c0681895-f204-40be-ba34-fa68799bb83d

Panel Four: Information Gathering, Analysis, and Disclosure 3/27 [8.30–10.30am] https://video.law.gwu.edu:8443/ess/echo/presentation/d40ffc9c-0743-4078-bb3e-f8636304b2d2

Panel Five: State Perspectives 3/27 [10.45–12.15pm] https://video.law.gwu.edu:8443/ess/echo/presentation/0564ef89-a3d3-41b8-8620-5551e88f394f

Panel Six: Policy Perspectives 3/27 [1.30–3.30pm] https://video.law.gwu.edu:8443/ess/echo/presentation/86f73775-c1a9-4d2e-af0b-f027e640db5f

6 To 'knowingly' violate the CWA one is not required to have actual knowledge that their actions constitute a violation of the statutory provisions because, as held in *United States v Hopkins*, 53 F.3d 533, 537, 539-40 (2d. Cir. 1995), a person can violate § 1319(c)(2)(A) 'if the defendant's acts were proscribed, even if the defendant was not aware of the proscription'. The *Hopkins* court also held the same standard for 'knowingly' applied to § 1319(c)(4), which governs tampering with monitoring reports. *Hopkins*, 53 F.3d, at 541.

References

Albright, S.P. (2012) Emerging trends in the regulations of stormwater. *Texas Environmental Law Journal* 43, p. 1.

Andreen, W.L. (2004) Water quality today: has the Clean Water Act been a success?. *Alabama Law Review* 55, p. 537.

Andreen, W.L. (2009) Delegated federalism versus devolution: some insights from the history of water pollution control. In W.W. Buzbee (ed.), *Preemption Choice: The Theory, Law, and Reality of Federalism's Core Question*. Cambridge: Cambridge University Press.

Andreen, W.L. (2013) Success and backlash: the remarkable (and continuing) story of the Clean Water Act. 4 *George Washington Journal of Energy and Environmental Law* 4, p. 25.

Andreen, W.L. and Jones, S.C. (2008) *The Clean Water Act: A Blueprint for Reform.* Washington, DC: Center for Progressive Reform.

Buckman, D.F. (1999) *Requirement That There be Continuing Violation to Maintain Citizen Suit Under Federal Environmental Protection Statutes: Post-Gwaltney cases*, 158 A.L.R. Fed. 51.

Coggins, G.C. and Glicksman, R.L. (2015) Water pollution. *Public Natural Resources Law* 2, 2nd edn. New York: Clark Boardman Callaghan, ch. 19.

Copeland, C. (2014) *Clean Water Act: A Summary of Law*, Cong. Res. Serv., RL30030.

Eames, D.A. (1970) *The Refuse Act of 1899: Its Scope and Role in Control of Water Pollution, California Law Review* 58, p. 1444.

Gardner, P.J. (2015) *Citizen Suit Enforcement in a Mixed System: Evidence from the Clean Water Act*, 30 November.

Grad, F.P. (1997) *Treatise on Environmental Law.* Conklin, NY: Matthew Bender.

Hansen, K.M. (1990) Knowing environmental crimes. *William Mitchell Law Review* 16, p. 987.

Hirsch, R.M., Hamilton, P.A. and Miller, T.L. (2006) US Geological Survey perspective on water-quality monitoring and assessment. *Journal of Environmental Monitoring* 8, p. 512. http://water.usgs.gov/nawqa/jem.monitoring.displayarticle.pdf

Lyons, G.S. and Stein, E.D. (2008) How effective has the Clean Water Act been at reducing pollutant mass emissions to the Southern California Bight over the past 35 years?. *Environmental Monitoring and Assessment*, July. http://www.ncbi.nlm.nih.gov/pubmed/18568406

Polk, A.A. (2013) The Clean Water Act and evolving due process: the emergence of contemporary enforcement procedures. *Oklahoma Law Review* 65, p. 717.

Ryan, M. (ed.) (2002–3) *The Clean Water Act Handbook.* Chicago: ABA.

Schang, S.E., Stever, D.W. and Abramson, S.P. (eds) (2015) *Law of Environmental Protection.* New York: Thompson Reuters.

Schroeder, C.H. and Steinzor, R. (eds) (2005) *A New Progressive Agenda for Public Health and the Environment.* Durham, NC: Carolina Academic Press.

Thomas Jr, F.M. (1987) Citizen suits and the NPDES program: a review of Clean Water Act decisions. *Environmental Law Reporter* 17, p. 10050.

US Environmental Protection Agency (2009) *Interim Clean Water Act Settlement Penalty Policy.* http://www2.epa.gov/sites/production/files/documents/cwapol.pdf

US Environmental Protection Agency (2014) *Clean Water Act National Pollutant Discharge Elimination System Compliance Monitoring Strategy.* http://www2.epa.gov/compliance/clean-water-act-national-pollutant-discharge-elimination-system-compliance-monitoring

US Environmental Protection Agency (2015a) *Final NPDES Electronic Reporting Rule.* http://www2.epa.gov/sites/production/files/2015-09/documents/finalnpdeselectronicreportingrulefactsheet.pdf

US Environmental Protection Agency (2015b) *Supplement Environmental Projects Policy.* https://www.epa.gov/sites/production/files/2015-04/documents/sepupdatedpolicy15.pdf

US Environmental Protection Agency (undated) *Advanced Nutrient Monitoring.* https://www.epa.gov/water-research/advanced-nutrient-monitoring

US Environmental Protection Agency (undated) *Discharge Monitoring Report (DMR) Pollutant Loading Tool.* http://cfpub.epa.gov/dmr/

US Environmental Protection Agency (undated) *Enforcement and Compliance History Online.* https://echo.epa.gov

US Environmental Protection Agency (undated) *Enforcement Annual Results Numbers at a Glance for Fiscal Year (FY) 2015.* https://www.epa.gov/enforcement/enforcement-annual-results-fiscal-year-fy-2015

US Environmental Protection Agency (undated) *Envirofacts.* https://www3.epa.gov/enviro/

US Environmental Protection Agency (undated) *Effluent Guidelines Plan.* http://water.epa.gov/scitech/wastetech/guide/304m/upload/Final-2012-and-Preliminary-2014-Effluent-Guidelines-Program-Plans.pdf

US Environmental Protection Agency (undated) *How to Report an Error.* https://echo.epa.gov/help/how-to-report-error

US Environmental Protection Agency (undated) *State NPDES Program Withdrawal Petitions.* https://www.epa.gov/npdes/npdes-state-program-information

US Environmental Protection Agency (undated) *What is Nonpoint Source?.* http://water.epa.gov/polwaste/nps/whatis.cfm

US Goverment Accountability Office (2002) *Water Quality: Inconsistent State Approaches Complicate Nation's Efforts to Identify Its Most Polluted Waters.* http://www.gao.gov/products/GAO-02-186

Vinal, R.W. (1996) Proof of wrongful discharge of pollutant into waterway under Federal Clean Water Act. *American Jurisprudence Proof of Facts 3d* 36, p. 533.

Watson, M. (2005) The enforcement of environmental law: civil or criminal penalties?. *Environmental Law and Management* 17, p. 3.

PART 2

Transboundary water law and policy

14

THE TREATY ARCHITECTURE FOR THE GOVERNANCE OF TRANSBOUNDARY AQUIFERS, LAKES AND RIVERS

Alistair Rieu-Clarke

Introduction

Transboundary waters are certainly significant. There are an estimated 263 international rivers and lakes, and over 275 transboundary aquifers (Wolf *et al.* 1999; Puri and Aureli 2009). Almost half of the world's land mass is situated within transboundary river basins, 60 per cent of global freshwater flows are from transboundary waters and around 40 per cent of the word's population are dependent on waters shared across sovereign borders (Global International Water Assessment 2006). However, perhaps the most significant figure in terms of treaty practice is that 145 of the world's 196 countries share transboundary waters (Wolf *et al.* 1999). This means that, given the indivisibility of water resources, most states in the world must cooperate with their neighbours to ensure that waters are governed in an equitable and sustainable manner.

General statistics are encouraging. For instance, it is estimated that 3,600 international freshwater agreements have been adopted (Giordano and Wolf 2002, p. 6). The vast majority of these agreements focus on navigational issues, which is a reflection of the major use of international freshwaters during the nineteenth and early twentieth centuries (Dinar *et al.* 2007, p. 58). There are then believed to be over 400 treaties that relate to non-navigational uses of international watercourses (UNEP *et al.* 2002; Burchi and Mechlem 2005). However, these figures tell us little without knowing the geographical scope of each treaty and the types of activities that they cover. The purpose of this chapter is therefore to offer an overview of treaty coverage related to transboundary waters, both in terms of which rivers, lakes and aquifers are covered; and the functional scope of such instruments.

At the outset it should therefore be noted that the chapter will not seek to examine customary international law (general or regional-specific), or other sources of international law. Rather the primary focus will be on treaties as the most appropriate means by which states enter into cooperation arrangements concerning their transboundary waters.

Global and regional treaty practice related to transboundary waters

Significant efforts have been made at the global level to develop, or rather articulate, commonly agreed rights and obligations relating to transboundary waters.

The UN Watercourses Convention

In 1970, the UN General Assembly recommended that the International Law Commission should, 'take up the study of the law of the non-navigational uses of international watercourses with a view to its progressive development and codification' (see Chapter 15, this volume). At the time, the General Assembly justified the need for a global effort on *inter alia* an observation that, 'despite the great number of bilateral treaties and other regional regulations..., the use of international rivers and lakes is still based in part on general principles and rules of customary law' (UN General Assembly 1970). This call led to the adoption of the Convention on the Law of the Non-navigational Uses of International Watercourses (Watercourses Convention) (1997) on the 21 May 1997, which entered into force on 17 August 2014. The purpose of the UN Watercourses Convention, according to its preamble, is to, 'ensure the utilisation, development, conservation, management and protection of international watercourses and the promotion of optimal and sustainable utilisation thereof for present and future generations'.

As a framework instrument, the Convention seeks to supplement situations where there is no treaty arrangement in place for specific watercourses; where not all watercourse states are party to a specific watercourse treaty; or where a specific watercourse treaty does not fully cover the provisions contained within the Watercourses Convention. The Watercourses Convention is therefore designed in a way that seeks to supplement, rather than compete or replace, treaty arrangements adopted for specific watercourses. This is clearly stated in Article 3(1), which stipulates that, 'nothing in the present Convention shall affect the rights or obligations of a watercourse State arising from agreements in force for it on the date on which it became a party to the present Convention'. The UN Watercourses Convention (1997), under Articles 3, 4, 8 and 24, goes on to encourage states to consider harmonising their existing agreements with its provisions, and also encourages states to enter treaty and institutional arrangements in order to support the implementation of the Convention. Considerable discretion is therefore afforded to states in terms of whether and how they enter into such arrangements.

The content of the UN Watercourses Convention (1997) largely reflects customary international law (McCaffrey 2013). The primary substantive norms of equitable and reasonable utilisation and the duty to take all appropriate measures to prevent significant harm are mainly set out in Articles 5–7 and 10 of the Convention. The need to protect ecosystems of international watercourses, while arguably also part and parcel of the principle of equitable and reasonable utilisation, is afforded specific attention in Article 20 of the Watercourses Convention (1997). The instrument also puts particular emphasis on procedural aspects of cooperation. Key provisions in this regard include the duty to cooperate in the protection and development of an international watercourse (Articles 5(2) and (8)), the duty to regularly exchange data and information (Article 9), the duty to notify and consult over planned measures (Articles 11–19) and the duty to settle disputes in a peaceful manner (Article 33).

States have been slow to formally join the UN Watercourses Convention (1997) through accession, approval, acceptance or ratification. As of December 2015, there are 36 parties to the Convention (ibid.). The Convention required 35 parties for it to enter into force, which took place in 2014 following Vietnam's accession. Several reasons explain why this process has been slow (Dellapenna *et al.* 2013). However, despite this, the UN Watercourses Convention (1997) has been widely endorsed by states, international organisations and the International Court of Justice (Salman 2007), and interest among states in becoming party to the Convention is gaining momentum. Parties to the Convention represent states in Africa (Benin, Burkina Faso, Chad, Côte d'Ivoire, Guinea-Bissau, Morocco, Namibia, Niger, Nigeria, South Africa and Tunisia), Asia (Uzbekistan and Vietnam), Europe (Denmark, Finland, France, Germany, Greece, Hungary,

Ireland, Italy, Luxembourg, Montenegro, Netherlands, Norway, Portugal, Spain, Sweden and the United Kingdom) and the Middle East (Iraq, Jordan, Lebanon, Libya, Qatar, State of Palestine, Syria and Yemen).

The UNECE Water Convention

Another framework instrument increasingly taking on a global perspective is the UN Economic Commission for Europe Convention on the Protection and Use of Transboundary Watercourses and International Lakes (UNECE Water Convention) (1992), which was adopted on 17 March 1992 and entered into force on 6 October 1996.

Both the function and scope of the UNECE Water Convention (1992) is consistent with the UN Watercourses Convention (1997) (Tanzi 2015; Rieu-Clarke and Kinna 2014). Article 2(1) of the UNECE Water Convention (1992) obliges states to 'take all appropriate measures to prevent, control and reduce any transboundary impact'. Article 2(2) goes on to stipulate that states must ensure their transboundary waters are used 'in a reasonable and equitable way'. These two central provisions of the Convention, and their associated provisions, mirror the two substantive norms of equitable and reasonable utilisation and no significant harm, which are found in Articles 5–7 and 10 of the UN Watercourses Convention (1997). The importance of ecosystem protection is also recognised in various provisions of the UNECE Water Convention (1992), such as Article 2(2)(b) and (d). As with the UN Watercourses Convention (1997), the UNECE Water Convention (1992) puts a strong emphasis of the procedural aspects of cooperation, including joint monitoring and assessment (Articles 4 and 11), research and development (Articles 5 and 12), exchange of information (Articles 6 and 13), consultations (Article 10), warning and alarm systems (Article 14), mutual assistance (Article 15), the provision of public information (Article 16) and the settlement of disputes (Article 22).

In relation to the establishment of legal and institutional arrangements, and the relationships with older agreements, the UNECE Water Convention (1992) goes further than the UN Watercourses Convention (1997) by stipulating that states must enter into 'bilateral or multilateral agreements or other arrangements', where these do not yet exist, or adapt existing ones, where necessary to eliminate the contradictions with the 'basic principles' of the UNECE Water Convention. Along similar lines, the UNECE stipulates that joint institutional arrangement 'or "joint bodies" must be established to support the implementation of these arrangements' (Article 9(2)). However, while this is stronger wording than the UN Watercourses Convention (1997), states are still afforded considerable discretion in determining the form that any cooperative arrangements take (UNECE 2009, pp. 80–93).

Perhaps the most significant difference between the UN Watercourses Convention (1997) and the UNECE Water Convention (1992) is the institutional framework that is provided for in Part III of the latter instrument, which is lacking in the former. Under Part III the Parties have established a Meeting of the Parties, which meets every three years, and is supported by a Secretariat based in Geneva, and numerous working groups. Since the Convention's entry into force in 2006, this institutional framework has provided a highly effective means by which to support the implementation and development of the Convention. This has been done *inter alia* through the development of subsequent protocols on water and health, and civil liability, numerous guidance notes and recommendations, and various projects and pilot studies (Tanzi *et al.* 2015).

One of the most significant decisions taken by the Meeting of the Parties was to amend the UNECE Water Convention (1992) to make it open to non-UNECE Member States (Trombitcaia and Koeppel 2015). The amendment became operational in 2015, which means that the UNECE Water Convention (1992) now operates at a global level and, essentially, alongside the UN Watercourses Convention (1997). Given that both instruments are complementary, there has been a concerted attempt to promote both instruments as a package (Rieu-Clarke and Kinna 2014).

International Law Commission Draft Articles on the law of transboundary aquifers

At the global level an effort has also been undertaken to strengthen laws relating to transboundary aquifers. This work was initiated by the International Law Commission in 2003, and resulted in a set of Draft Articles on Transboundary Aquifers that were presented to the UN General Assembly in 2008.

The Draft Articles follow a similar approach to the UN Watercourses Convention (1997). The principles of equitable and reasonable utilisation, and the duty to take all appropriate measures not to cause significant harm, are reflected in Articles 4–6. The need to protect and preserve ecosystems is also included in Article 10. Procedural rules are a key feature of the Draft Articles, including provisions on the general obligation to cooperate (Article 7), the regular exchange of data and information (Article 8), monitoring (Article 13), management (Article 14), planned activities (Article 15) and emergency situations (Article 17). A key departure from the UN Watercourses Convention is the stipulation under Article 3 of the Draft Articles that, 'each aquifer State has sovereignty over the portion of a transboundary aquifer or aquifer system located within its territory'. This raises the question whether states can claim sovereignty over a resource that flows from one country to another (McCaffrey 2009).

A question remains over the status that the Draft Articles might take. In 2009, it was suggested that the General Assembly 'recommend to States concerned to make appropriate bilateral or regional arrangements for the proper management of their transboundary aquifers on the basis of the principles enunciated in the articles', and 'consider, at a later stage, and in view of the importance of the topic, the elaboration of a convention on the basis of the draft articles' (Eckstein and Sindico 2014). The decision on form has been deferred to 2016 (ibid.).

Water-related global conventions

As well as global instruments dedicated to transboundary waters, several other global instruments have some relevance to transboundary waters (Brels *et al.* 2008). Of note here are global environmental agreements, such as the Convention on Biological Diversity (1992), the UN Framework Convention on Climate Change (1992) and the Convention on Wetlands of International Importance (1971). Transboundary waters may also have an influence on the implementation of treaties relating to international investments regimes and human rights (Rieu-Clarke 2015).

Regional instruments

In addition to the UNECE Water Convention (1992), a notable regional instrument is the Revised Southern African Development Community Protocol on Shared Watercourses, which was adopted on 7 August 2000, and entered into force on 22 September 2003 (Revised SADC Protocol on Shared Watercourses 2000). This instrument is significant given that 70 per cent of water resources within the Sothern African Development Community (SADC) are shared between two or more states. The Revised SADC Protocol on Shared Watercourses (ibid.) was revised in order to ensure that it was in line with the UN Watercourses Convention (1997). The content therefore follows the global convention very closely, and word-for-word in places (Malzbender and Earle 2013).

However, a key distinction between both instruments is that the Revised SADC Protocol on Shared Watercourses (2000) establishes and is supported by an institutional framework that is tasked with its implementation and development. This institutional framework, under the auspices of SADC, has subsequently developed several soft law instruments in support of the 2000 SADC Protocol, including a Regional Water Policy (SADC 2005), a Regional Water Strategy (SADC 2006) and five-yearly regional strategy action plans, the first of which was approved by SADC in August 1998 (see, for example, SADC 2011).

While the UNECE Water Convention (1992) and the Revised SADC Protocol on Shared Watercourses (2000) represent the only two legal binding regional instruments on transboundary waters, various soft law instruments have been developed under other regional institutions, including the Regional Water Policy of the Economic Community of West African States (ECOWAS) (Garane and Abdul-Kareem 2013) and the plans initiated by the Inter-Governmental Authority on Development to develop a regional water policy and regional water protocol (IGAD 2015).

Treaty practice and Africa's transboundary waters

Africa faces several water-related challenges that constitute a threat to economic growth and sustainable livelihoods. Only 5 per cent of the region's land is irrigated, with the rest being rain-fed and therefore highly vulnerable to climate variability. Hydropower constitutes a major potential source of electricity, but only 10 per cent of the potential supply has been realised. Safe drinking water is also a major challenge that, in turn, affects the health and wellbeing of populations. Only 58 per cent of the population is believed to have access to safe drinking water. A major challenge in addressing these issues is to ensure that water is managed effectively at a transboundary level. The African continent is home over 64 transboundary rivers and lakes and 40 transboundary aquifers within Africa (Wolf *et al.* 1999; Puri and Aureli 2009). These transboundary waters represent over 80 per cent of the entire continent's water resources, and suggest that most states are reliant on transboundary waters.

Of the 64 transboundary river and lake basins, only 19 are covered by any basin-specific agreements (UNEP *et al.* 2002). Many of the agreements that are in place are also limited in either functional scope or the number of basin states party to them. For instance, only 14 out of a total of 26 transboundary rivers and lake basins have what could be described as a basin-wide agreement in place. An exception includes the Zambezi River, which is shared among nine states and, in 2004, adopted the Agreement on the Establishment of the Zambezi Water Commission. As well as establishing the Zambezi Water Commission, this agreement sets out both substantive and procedural rights and obligations that the states sharing the Zambezi are subject to. Similarly, the Convention on the Sustainable Management of Lake Tanganyika (2003) sets out detailed provisions on the management of the lake, including provisions not often found in transboundary agreements, such as detailed requirements relating to environmental impact assessment (Article 15), education and publication awareness (Article 16) and public participation in decision-making (Articles 17 and 19). The Protocol on the Sustainable Development of Lake Victoria (2003) adopts a similar approach, and is one of the few transboundary agreements to recognise the importance of gender mainstreaming with regard to decision-making, policy formulation and implementation of projects and programmes that fall within the purview of the Convention (Article 23).

Other basins have highly fragmented treaty arrangements. In the Nile, for example, 16 treaties have been adopted but they are mostly bilateral, despite the fact that the basin is shared by 11 riparian countries. Efforts to adopt a basin-wide agreement for the Nile resulted in the Agreement on the Nile River Basin Cooperative Framework (2009), which was negotiated by states of the Nile under the auspices of the Nile Basin Initiative. The Agreement on the Nile River Basin Cooperative Framework envisaged the establishment of a Nile Basin Commission, and sets out substantive rights and obligations for transboundary water sharing (Abseno 2013; Mekonnen 2010). However, only Ethiopia, Kenya, Uganda, Rwanda, Burundi, Congo and Tanzania have signed the agreement; and so far the Convention has only been ratified by Ethiopia, Rwanda and Tanzania.

A fragmented treaty landscape also applies to transboundary aquifers in Africa. Out of the 40 transboundary aquifers, only two are covered by specific treaty arrangements. In 2000 an agreement was adopted between Chad, Egypt, Libya and Sudan for the Development of a Regional Strategy for the Utilisation of the Nubian Sandstone Aquifer System (Burchi and Mechlem 2005, p. 4). The Agreement primarily sets out terms of reference for monitoring and exchange of groundwater data and information. The second agreement relates to the Northwest Sahara Aquifer system shared between Algeria, Libya and Tunisia. In 2002 these states adopted an agreement on the Establishment of a Consultation Mechanism for the Northwestern Sahara System (ibid., pp. 6–8). Besides these two aquifer-specific agreements, there have been recent efforts to better recognise the significance of groundwater in existing arrangements and ongoing initiatives.

Treaty practice and Asia's transboundary waters

Asia is home to 57 transboundary river and lake basins, which constitute 39 per cent of the continent's land surface (Wolf *et al.* 1999, pp. 404–8); and over 36 transboundary aquifers. There is considerable fragmentation in terms of the types of agreements adopted for transboundary waters in Asia.

For transboundary rivers and lakes, few agreements can be categorised as either basin-wide or comprehensive. Out of 57 transboundary river and lake basins, 24 are covered by treaty arrangements. Arguably the most comprehensive in scope is the Agreement on the Cooperation for the Sustainable Development of the Mekong River Basin (1995), which obliges its parties to, 'cooperate in all fields of sustainable development, utilisation, management and conservation of the water and related resources of the Mekong River Basin' (Article 1). However, a major constraint to achieving this aim is that the upstream states of the Mekong River (China and Myanmar) are not party to the Agreement. It might also be argued whether the design of the Mekong Agreement is sufficient to address some of the challenges faced within the basin (Rieu-Clarke and Gooch 2010).

Within the context of Central Asia, a major challenge is not so much the lack of agreements but rather the normative quality of the numerous instruments. In surveying the 14 agreements applicable to the Aral Sea Basin, Ziganshina (2013), for instance, concludes that, 'the problem lies in… the normative quality of these treaties, most of which have been adopted with no links to each other', and that the agreements, 'fall short in incorporating the contemporary principles of international water law and best water management practice, and neglect the significance of establishing a sound procedural system of transboundary water cooperation' (p. 167). A similar experience can be seen in the South Asian context. Sarfraz, for instance, points to the shortcomings of the Indus Water Treaty (1960), particularly in its failure to account for environmental norms (Sarfraz 2013).

Other Treaty practice in Asia has followed a similar pattern where, if agreements have been entered into, they tend to be quite narrow and limited in their scope. Examples include the 1996 Treaty Between India and Bangladesh Sharing the Ganges Waters at Farakka (Sands 1997); and the various bilateral instruments that China has entered into with its neighbouring states (Chen *et al.* 2013).

The only aquifer agreement that is in place in Asia is an Agreement between the Government of the Heshemite Kingdom of Jordan and the Government of the Kingdom of Saudi Arabia for the Management and Utilization of the Ground Waters in the Al-Sag/Al-Disi Layer (2015). The agreement sets out the main rights and obligations pertaining between the two states sharing the aquifer, and also establishes a Joint Technical Committee to oversee the implementation of the Agreement.

Treaty practice and Europe's transboundary waters

There are 69 international river and lake basins in Europe, together with an estimated 155 transboundary aquifers (Wolf *et al.* 1999; Puri and Aureli 2009). Of the 69 international river and lake basins, 28 lie solely within the borders of the European Union (EU) and a further 29 are shared between EU and non-EU states.

As well as the UNECE Water Convention (1992) being applicable at the regional level, EU Member States are also committed to the implementation of EU legislation related to water resources management (Rieu-Clarke 2008). The most significant piece of legislation in this regard, which covers both national and transboundary waters, is the EU Directive Establishing a Framework for Community Action in the Field of Water Policy (2000/60/EC).

Most transboundary rivers and lakes in Europe are covered by transboundary agreements, which in part can be attributed to the influence of the UNECE Water Convention (1992) and the EU Water Framework Directive (2000/60/EC). In order to comply with the terms of these regional instruments, basin agreements tend to be quite comprehensive both in terms of the watercourse states represented and their functional scope. An example can be seen in the case of the Convention on Cooperation for the Protection and Sustainable Use of the River Danube, which was adopted in 29 June 1994 and was ratified by 14 Danube states as well as the European Union (1994). The Convention offers a comprehensive set of substantive and procedural rules and principles related to the management of the Danube Convention, together with the establishment of an international commission that oversees the agreement's implementation.

Numerous other examples illustrate the significant track record of treaty adoption that European states have entered into. In some instances such agreements, like the case of the Danube (1994), have sought to adopt a basin-wide approach. Additional examples include the Treaty between Moldova and Ukraine on Cooperation in the Field of Protection and Sustainable Development of the Dniester River Basin (2012); the Convention on the Protection of the Rhine (1998); and the Agreement on the Protection of the River Scheldt (1994). An alternative approach has been for European states to enter into arrangements with their neighbours that cover several transboundary waters. Examples of such agreements include the Convention between Spain and Portugal for the Protection and Sustainable Use of the Spanish–Portuguese Hydrographic Basins (1998); and the Agreement Between Finland and Sweden Concerning Transboundary Rivers (2010).

Compared to the other regions of the world, Europe stands out in terms of both the level and scope of treaty practice pertaining to transboundary waters; although more could be done to embed groundwater issues within existing or additional treaty arrangements.

Treaty practice and North America's transboundary waters

There are a total of 19 river and lake basins, and 17 aquifers shared between Canada, the US and Mexico (Wolf *et al.* 1999; Puri and Aureli 2009). Of these transboundary resources 15 rivers and lakes are shared between Canada and the US, as well as seven aquifers; whereas four river and lake basins and ten aquifers are shared between the US and Mexico.

Treaty practice within North America is dominated by two key bilateral arrangements. The US and Canada have adopted the Boundary Waters Treaty (1909), Article VII of which established the International Joint Commission (IJC). The IJC has responsibility to investigate and make recommendations on questions or disputes referred to it by the states. The US and Canada have also adopted a series of other bilateral agreements that respond to key challenges and issues. A concerted effort to develop a treaty regime related to the Great Lakes has been undertaken. The states have adopted agreements Great Lakes Water Quality, which were updated in 2012 (Great Lakes Water Quality Agreement 2012). Other agreements adopted by the US and Canada on their transboundary rivers include the Columbia Treaty (1958), the Niagara River Treaty (1950), the Skagit River Treaty (1984), the St Lawrence Seaway Agreement (2005), the Agreement for Water Supply and Flood Control in the Souris River Basin (1989), the Lake of Woods Convention and Protocol (1925) and the Rainy Lake Convention (1938). This matrix of cooperative legal and institutional arrangements relating to US–Canada transboundary waters offers important mechanisms by which to ensure that such waters are utilised in an equitable and sustainable manner (Neir *et al.* 2009).

The first treaty relating to water allocation issues between US and Mexico was adopted on 21 May 1906 (Convention between US and Mexico 1906). The treaty stipulated that the US must deliver 74 million cubic metres per year to Mexico via the Rio Grande. Mexico, in turn, waived its rights to the waters of the Rio Grande between Juarez and Fort Quitman (Umoff 2008). This arrangement was revisited when a more comprehensive Treaty on the Utilisation of Waters of the Colorado and Tijuana Rivers and of the Rio Grande was adopted (1944). As well as setting out water allocation arrangements, the 1944 Treaty also revised the International Boundary Commission that was originally established in 1889, to the International Boundary Water Commission. Through 'minutes', the 1944 Treaty has evolved to cover issues including water quality. Strengthening cooperative arrangements pertaining to groundwater remains a major challenge for the US and Mexico, where border aquifers have experienced significant declines both in terms of quality and quantity (Carter 2015).

Treaty practice and South America's transboundary waters

South America is home to 38 transboundary rivers and lake basins, which make up 60 per cent of the continent's land surface (Wolf *et al.* 1999). In addition there are estimated to be 29 transboundary aquifers (Puri and Aureli 2009). Few agreements have been entered into for these transboundary waters. Out of the 38 transboundary rivers and lakes, only six are covered by treaty arrangements and many of those agreements lack sufficient detail to address cotemporary transboundary challenges.

The La Plata basin, shared between Brazil, Argentina, Paraguay, Bolivia and Uruguay, has the highest number of treaty arrangements (18) relating to it. Some of these agreement relate to specific projects, such as the Agreement between Argentina and Paraguay concerning the study of the Utilisation of the Water Power of the Apipé Falls (1961); or cover specific sub-basins, such as the Statute of the River Uruguay (1975). The only notable river basin-wide treaty is the Amazon Treaty (1978), which has been adopted by Bolivia, Brazil, Columbia, Ecuador, Guyana, Peru, Surinam and Venezuela. The Amazon Treaty led, in 2003, to the establishment of the

Amazon Cooperation Treaty Organisation to support implementation. However, a major constraint in the implementation of the Amazon Treaty is that it lacks key rules and principles that are commonly found in contemporary transboundary water treaty arrangements, including dispute resolution mechanisms, equitable and reasonable use and harm prevention provisions, and information exchange (Newton 2013).

An area that has witnessed notable cooperation between South American basin states is the adoption of the Guarani Aquifer Agreement (2010), between Argentina, Brazil, Paraguay and Uruguay. The agreement recognises the sovereign rights of the aquifer states to promote the management, monitoring and sustainable utilisation of the system, while not causing significant harm (Article 3). States are also obliged to promote the conservation and environmental protection of the system (Article 4). The Agreement also lays down certain procedural rules relating to the exchange of information, notification, consultation, negotiation and dispute settlement.

Conclusion: addressing fragmentation

This overview of treaty practice relating to transboundary waters demonstrates that there is considerable fragmentation within the system. At the global level, significant milestones have been the entry into force of the UN Watercourses Convention (1997) and the opening up of the UNECE Water Convention (1992) to non-UNECE Water Convention. While these events strengthen the global framework for transboundary water cooperation, a minority of states sharing transboundary waters has so far ratified one or both instruments. Much more therefore needs to be done to raise awareness of the benefits of both instruments and encourage more states to join them.

The chapter has demonstrated that, at the regional level, significant efforts have been undertaken to negotiate and adopt treaties relating to transboundary rivers, lakes and aquifers. However, the record is far from complete. Where multiple treaty arrangements are in place within a particular basin, that basin may still lack a comprehensive basin-wide arrangement that is able to foster a coherent approach. More alarming are the numerous transboundary waters that lack any cooperative arrangement. Such arrangements are particularly lacking within the case of transboundary aquifers, but also regional variations in terms of transboundary rivers and lakes are evident. More therefore needs to be done to strengthen the treaty architecture at the basin level, whether through basin-wide agreements or, where appropriate, bilateral arrangements covering several transboundary waters. The interface between surface water and groundwater also needs to be carefully considered. In some instances it might make most sense to establish a specific treaty arrangement for transboundary aquifers, such as in the case of the Guarani Aquifer, whereas in other contexts it might suffice to ensure that groundwaters are fully taken into account within existing river or lake basin arrangements.

References

Abseno, M. (2013) Nile River Basin. In F.R. Loures and A. Rieu-Clarke (eds), *The UN Watercourses Convention: Strengthening International Law for Transboundary Water Management*. London: Routledge.

Agreement between Finland and Sweden Concerning Transboundary Rivers (2009) 11 November (entered into force 1 October 2010) [online]. Available from: http://www.fsgk.se/2013/KORJATTU-VERSIO-24.6.2013_Finnish-Swedish-Transboundary-Rivers-Agreement-2009.doc-Finnish-Swedish-Transboundary-Rivers-Agreement-2009.pdf

Agreement between the Government of the Heshemite Kingdom of Jordan and the Government of the Kingdom of Saudi Arabia for the Management and Utilization of the Ground Waters in the Al-Sag/Al-Disi Layer (2015) 30 April [online]. Available from: http://www.internationalwaterlaw.org/documents/regionaldocs/Disi_Aquifer_Agreement-English2015.pdf [unofficial translation]

Agreement between the Government of Canada and the Government of the United States of America for Water Supply and Flood Control in the Souris River Basin (1989) 26 October [online]. Available from: http://faolex.fao.org/cgi-bin/faolex.exe?rec_id=012998&database=faolex&search_type=link&table=result&lang=eng&format_name=@ERALL

Agreement Concerning a Study of the Utilisation of the Water Power of the Apipé Falls (1958) 23 January [online]. Available from: https://treaties.un.org/doc/Publication/UNTS/Volume%20649/volume-649-I-9294-English.pdf

Agreement on Cooperation for the Protection and Sustainable Use of the Waters of the Spanish–Portuguese Hydrographic Basins (1998) 30 November. *Official State Bulletin [Portugal]* 37 (2000), p. 6703.

Agreement on the Cooperation for the Sustainable Development of the Mekong River Basin (1995) 5 April. *International Legal Materials* 35 (1995), p. 864.

Agreement on the Establishment of the Zambezi Watercourse Commission (2004) 13 July [online]. Available from: http://zambezicommission.org/newsite/wp-content/uploads/ZAMCOM%20agreement.pdf

Agreement on the Nile River Basin Cooperative Framework (2009) [online]. Available from: http://www.nilebasin.org/images/docs/CFA%20-%20English%20%20FrenchVersion.pdf

Agreement on the Protection of the River Scheldt (1994) 26 April [online]. Available from: http://iea.uoregon.edu/pages/view_treaty.php?t=1994-ProtectionScheldt.EN.txt&par=view_treaty_html

Brels, S., Coates, D. and Loures, F. (2008) *Transboundary Water Resources Management: The Role of International Watercourse Agreements in Implementation of the CBD* [online]. Available from: http://www.unwater.org/downloads/cbd-ts-40-en.pdf

Burchi, S. and Mechlem, K. (2005) *Groundwater in International Law.* Rome: FAO.

Carter N.T. (2015) *US–Mexico Water Sharing: Background and Recent Developments* [online]. Available from: https://www.fas.org/sgp/crs/row/R43312.pdf

Chen, H., Rieu-Clarke, A. and Wouters, P. (2013) Exploring China's transboundary water treaty practice through the prism of the UN Watercourses Convention. *Water International* 38(2), pp. 217–30.

Convention between Canada and the United States of America Providing for Emergency Regulation of the Level of Rainy Lake and the Level of Other Boundary Waters in the Rainy Lake Watershed (1938) 15 September (entered into force 3 October 1940) [online]. Available from: http://www.ijc.org/files/tinymce/uploaded/RainyLakeConvention1938_e.pdf

Convention and Protocol between His Britannic Majesty in Respect of the Domination of Canada and the United States for Regulating the Level of the Lake of the Woods (1925) 24 February [online]. Available from: http://www.ijc.org/files/dockets/Docket%203/Docket%203%20Convention%20and%20Protocol.pdf

Convention and Protocol Regarding Lake of the Woods (1925) 24 February [online]. Available from: http://www.lwcb.ca/BoardDesc/ConventionAndProtocolCanada1925.pdf

Convention between the United States and Mexico on Equitable Distribution of the Waters of the Rio Grande (1906) 21 May [Online]. Available from: http://www.ibwc.gov/Files/1906Conv.pdf

Convention on Biological Diversity (1992) 5 June (entered into force29 December 1993). *International Legal Materials* 31 (1992), p. 818.

Convention on Cooperation for the Protection and Sustainable Use of the River Danube (1994) 29 June (entered into force October 1998) [online]. Available from: https://www.icpdr.org/main/icpdr/danube-river-protection-convention

Convention on the Law of the Non-navigational Uses of International Watercourses (1997) 21 May (entered into force 17 August 2014). *International Legal Materials* 36 (1997), p. 700.

Convention on the Protection and Use of Transboundary Watercourses and International Lakes (1992) 17 March (entered into force on 6 October 1996). *International Legal Materials* 31 (1996), p. 1312.

Convention on the Protection of the Rhine (1999) 12 April [online] Available from: http://www.iksr.org/fileadmin/user_upload/Dokumente_en/convention_on_tthe_protection_of__the_rhine.pdf

Convention on the Sustainable Management of Lake Tanganyika (2003) 12 June [online]. Available from: http://www.ecolex.org/server2.php/libcat/docs/TRE/Full/En/TRE-001482.pdf

Convention on Wetlands of International Importance (1971) 2 February (entered into force 21 December 1975). *International Legal Materials* 11 (1972), p. 963.

Dellapenna, J.W., Rieu-Clarke, A. and Loures, F.R. (2013) Possible reasons slowing down the ratification process. In F.R. Loures and A. Rieu-Clarke (eds), *The UN Watercourses Convention: Strengthening International Law for Transboundary Water Management.* London: Routledge.

Dinar, A., Dinar, S., McCaffrey, S. and McKinney, D. (2007) *Bridges over Water: Understanding Transboundary Water Conflict, Negotiations and Cooperation.* New Jersey: World Scientific.

Draft Articles on Transboundary Aquifers (2008) Official Records of the General Assembly, 63rd Session, Supp. No. 10, UN Doc. A/63/10.

Eckstein, G. and Sindico, F. (2014) The law of transboundary aquifers: many ways of going forward, but only one way of standing still. *Review of European, Comparative and International Environmental Law* 23(1), pp. 32–42.

European Commission Directive Establishing a Framework for Community Action in the Field of Water Policy. 2000/60/EC.

Garane, A. and Abdul-Kareem, T. (2013) West Africa. In F.R. Loures and A. Rieu-Clarke (eds), *The UN Watercourses Convention: Strengthening International Law for Transboundary Water Management*. London: Routledge.

Giordano, M.A. and Wolf, A.T. (2002) The world's international freshwater agreements: historical developments and future opportunities. In UNEP, Oregon State University and FAO, *Atlas of International Freshwater Agreements*. Nairobi: UNEP.

Global International Water Assessment (2006) *Challenges to International Waters: Regional Assessments in a Global Perspective*. Nairobi: UNEP.

Great Lakes Water Quality Agreement (2012) 7 September (entered into force 12 February 2013) [online]. Available from: http://binational.net/wp-content/uploads/2014/05/1094_Canada-USA-GLWQA-_e.pdf

Great Lakes–St Lawrence River Basin Sustainable Water Resources Agreement (2005) 13 December [online]. Available from: http://www.glslregionalbody.org/Docs/Agreements/Great_Lakes-St_Lawrence_River_Basin_Sustainable_Water_Resources_Agreement.pdf

Guarani Aquifer Agreement (2010) 2 August. Available from: http://www.internationalwaterlaw.org/documents/regionaldocs/Guarani_Aquifer_Agreement-English.pdf [unofficial translation]

IGAD (2015) Regional Water Resources Policy Endorsed by Sectoral Ministries [online]. Available from: http://igad.int/index.php?option=com_content&view=article&id=1036:igad-regional-water-resources-policy-endorsed-by-sectorial-ministries&catid=43:agriculture-and-environment&Itemid=126

Indus Water Treaty (1960) 10 September [online]. Available from: http://wrmin.nic.in/writereaddata/InternationalCooperation/IndusWatersTreaty196054268637.pdf

Malzbender, D. and Earle, A. (2013) Southern Africa. In F.R. Loures and A. Rieu-Clarke (eds), *The UN Watercourses Convention: Strengthening International Law for Transboundary Water Management*. London: Routledge.

McCaffrey, S.C. (2009) The International Law Commission adopts draft articles on transboundary aquifers. *American Journal of International Law* 103(2), pp. 272–93.

McCaffrey, S.C. (2013) The progressive development of international water law. In F.R. Loures and A. Rieu-Clarke (eds), *The UN Watercourses Convention: Strengthening International Law for Transboundary Water Management*. London: Routledge.

Mekonnen, D.Z. (2010) The Nile Basin Cooperative Framework Agreement negotiations and the adoption of a 'water security' paradigm: flight into obscurity or a logical cul-de-sac?. *The European Journal of International Law* 21(2), pp. 421–40.

Neir, A.M., Klise, G.T. and Campana, M. (2009) The concept of vulnerability as applied to North America. In UNEP, *Hydropolitical Vulnerability and Resilience along International Waters: North America*. Nairobi: UNEP.

Newton, J. (2013) Amazon Basin. In F.R. Loures and A. Rieu-Clarke (eds), *The UN Watercourses Convention: Strengthening International Law for Transboundary Water Management*. London: Routledge.

Protocol for Sustainable Development of Lake Victoria (2003) 29 November [online]. Available from: http://www.internationalwaterlaw.org/documents/regionaldocs/Lake_Victoria_Basin_2003.pdf

Puri, S. and Aureli, A. (eds) (2009) *Atlas of Transboundary Aquifers*. Paris: UNESCO.

Revised SADC (Southern African Development Community) Protocol on Shared Watercourses (2000) 7 August (entered into force on 22 September 2003). *International Legal Materials* 40 (2001), p. 321.

Rieu-Clarke, A. (2008) The role and relevance of the UN Convention on the Law of the Non-navigational Uses of International Watercourses to the EU and its Member States. *British Yearbook of International Law*. London: Oxford University Press.

Rieu-Clarke, A. (2015) Transboundary hydropower projects seen through the lens of three international legal regimes: foreign investment, environmental protection and human rights. *International Journal of Water Governance* 3(1), pp. 27–48.

Rieu-Clarke, A. and Gooch, G. (2010) Governing the tributaries of the Mekong: the contribution of international law and institutions to enhancing equitable cooperation over the Sesan. *Pacific McGeorge Business and Development Law Journal* 22(2), pp. 193–224.

Rieu-Clarke, A. and Kinna, R. (2014) Can two global UN water conventions effectively co-exist? Making the case for a 'package approach' to support institutional coordination. *Review of European, Comparative and International Environmental Law* 23(1), pp. 15–31.

SADC (2005) Regional Water Policy, August [online]. Available from: http://www.sadc.int/files/1913/5292/8376/Regional_Water_Policy.pdf

SADC (2006) Regional Water Strategy, June [online]. Available from: https://www.sadc.int/files/2513/5293/3539/Regional_Water_Strategy.pdf

SADC (2011) Regional Strategic Action Plan on Integrated Water Resources Development and Management [online]. Available from: http://www.sadc.int/files/6613/5293/3526/Regional_Strategic_Action_Plan_-IWRM_III.pdf

Salman, M.A. (2007) The United Nations Watercourses Convention ten years later: why has its entry into force proven difficult? In F.R. Loures and A. Rieu-Clarke (eds), *The UN Watercourses Convention: Strengthening International Law for Transboundary Water Management.* London: Routledge.

Sands, P. (1997) Bangladesh–India: treaty on sharing of the Ganges waters at Farakka. *International Legal Materials* 36(3), pp. 519–28.

Sarfraz, H. (2013) Revisiting the 1960 Indus Water Treaty. *Water International* 38(2), pp. 217–30.

Statute of the River Uruguay (1975) 26 February [online]. Available from: http://www.internationalwater-law.org/documents/regionaldocs/Uruguay_River_Statute_1975.pdf

Tanzi, A. (2015) The Economic Commission for Europe Water Convention and the United Nations Water Convention: an analysis of their harmonised contribution to international water law [online]. Available from: http://www.unece.org/fileadmin/DAM/env/water/publications/WAT_Comparing_two_UN_Conventions/ece_mp.wat_42_eng_web.pdf

Tanzi, A, McIntyre, O., Kolliopoulos, A., Rieu-Clarke, A. and Kinna, R. (eds) (2015), *The UNECE Convention on the Protection and Use of Transboundary Watercourses and International Lakes: Its Contribution to International Water Cooperation.* The Hague: Brill.

Treaty Between Canada and the United States of America Concerning the Diversion of the Niagara River (1950) 10 October [online]. Available from: http://www.treaty-accord.gc.ca/text-texte.aspx?id=100418

Treaty Between Canada and the United States of America relating to the Skagit River and Ross Lake, and the Seven Mile Reservoir on the Pend d'Oreille River (1984) 2 April (entered into force 14 December 1984) [online]. Available from: http://ec.gc.ca/international/default.asp?lang=En&n=B2E4A354-1

Treaty Between the Government of the Republic of Moldova and the Cabinet of Ministers of Ukraine on Cooperation in the Field of Protection and Sustainable Development of the Dniester River Basin (2012) 29 November [online]. Available from: https://www.unece.org/fileadmin/DAM/env/water/activities/Dniester/Dniester-treaty-final-EN-29Nov2012_web.pdf

Treaty Between the United States and Great Britain Relating to Boundary Waters, and Questions Arising Between the United States and Canada (1910) 5 May [online]. Available from: http://www.ijc.org/en_/BWT

Treaty Between the United States of America and Mexico on the Utilisation of Waters of the Colorado and Tijuana Rivers and of the Rio Grande (1944) 3 February (entered into force 8 November 1945) [online]. Available from: http://www.ibwc.state.gov/files/1944treaty.pdf

Treaty on Amazonian Cooperation (1978) 3 July [online]. Available from: https://www.oas.org/dsd/Events/english/PastEvents/Salvador_Bahia/Documents/Amazonannexes.pdf

Treaty Relating to Cooperative Development of the Water Resources of the Columbia River Basin (1961) 17 January [online]. Available from: http://www.ccrh.org/comm/river/docs/cotreaty.htm

Trombitcaia, I. and Koeppel, S. (2015) From a regional towards a global instrument: the 2003 Amendment to the UNECE Water Convention. In A. Tanzi, O. McIntyre, A. Kolliopoulos, A. Rieu-Clarke and R. Kinna (eds), *The UNECE Convention on the Protection and Use of Transboundary Watercourses and International Lakes: Its Contribution to International Water Cooperation.* The Hague: Brill.

Umoff, A.A. (2008) An analysis of the 1944 US–Mexico Water Treaty: its past, present, and future. *Environs: Environmental Law and Policy Journal* 31(1), pp. 69–98.

UN Framework Convention on Climate Change (1992) 9 May (entered into force 21 March 1994). *International Legal Materials* 31 (1992), p. 849.

UN General Assembly (1970) Resolution 2669 (XXV), 8 December [online]. Available from: http://www.un.org/ga/search/view_doc.asp?symbol=A/Res/2669(XXV) [accessed 14 December 2015].

UNEP, Oregon State University and FAO (2002) *Atlas of International Freshwater Agreements.* Nairobi: UNEP.

Wolf, A.T, Natharius, J.A., Danielson, J.J., Ward, B.S. and Pender, K.K. (1999) International river basins of the world. *International Journal of Water Resources Development* 15(4), pp. 387–427.

Ziganshina, D. (2013) Aral Sea Basin. In F.R. Loures and A. Rieu-Clarke (eds), *The UN Watercourses Convention: Strengthening International Law for Transboundary Water Management.* London: Routledge.

15

THE EVOLUTION OF INTERNATIONAL LAW RELATING TO TRANSBOUNDARY WATERS

Stephen C. McCaffrey

Introduction

The first known treaty on any subject – in fact, the oldest known historical document – was concluded in *c.* 2450BC between the ancient Sumerian city-states of Umma (modern Umm al-Aqarib) and Lagash, both located in what is now southern Iraq. It is inscribed on what has become known as the Stele of Vultures, which is housed in the Louvre. The treaty brought to an end one of a number of wars between the two city-states over water diverted from the Euphrates and the border between them (Nussbaum 1954, p. 2; Jacobsen and Adams 1958, p. 1251).[1]

Water is essential to life. It also nourished the great ancient civilizations – known as the 'fluvial' civilizations – and drove their economies (Teclaff 1985, p. 15). These societies flourished not only in the river basins of the Nile, Tigris-Euphrates, Indus, Yellow and Yangtze, but also in Mexico and coastal Peru (Teclaff 1967, p. 15). Karl August Wittfogel argued in his book, *Oriental Despotism* (Wittfogel 1957), that the need to organize people to harness water for irrigation in arid areas led to the development of bureaucratic governmental structures. And Friedrich Berber observed, in his well-known work, *Rivers in International Law*, that 'the organization of the state as known to us over the last six thousand years had its origins in water rights' (Berber 1959, p. 1).

But after this rather auspicious beginning thousands of years ago, the law of the non-navigational uses of shared freshwater resources seems to have developed quite slowly until around the middle of the twentieth century (Weiss 2009; McCaffrey 1993, 2013). This impression may be due, at least in part, to the paucity of evidence of practice from parts of the world other than what is now Europe. But in the absence of other evidence the review of the development of international water law in this chapter will be based on the evidence that is available.

The chapter begins with a discussion of the priority accorded navigational uses over non-navigational ones in early agreements and state practice. The chapter then considers the current position on priorities among different kinds of uses. Finally, the chapter looks at indications of the evolution of international water law through other lenses.

The early dominance of navigation

What is clear from the available evidence of state practice, particularly from Europe, is that, apart from scattered diplomatic claims and the occasional bilateral treaty, non-navigational uses lagged far behind navigation in terms of the attention and importance given to each. This is certainly understandable in the context of Europe, which is generally well-watered and would thus tend to generate fewer concerns relating to water quantity than more arid regions. The importance of navigation in Western Europe is reflected in such foundational instruments as the Peace of Westphalia (1648), which opened the lower Rhine to free navigation (*Treaty of Peace between France and the Empire* 1648), and the Final Act of the Congress of Vienna (1815), which established a model for the regulation of navigation on international waterways that was followed in numerous instruments for over a century (Congress of Vienna 1815). The earliest entry in a Food and Agriculture Organization of the United Nations (FAO) compilation of some 2,000 legal instruments concerning water resources is a grant of freedom of navigation on the Rhine made in the year 805 by Charlemagne to a monastery (Food and Agriculture Organization of the United Nations 1978). In a study of over 2,000 international water agreements, Edith Brown Weiss notes that the percentage of such agreements dealing principally with navigation 'peaked in the period 1700–1930', while treaties focused on 'allocation and use issues were most significant as a percentage of total [international water] agreements negotiated during the period 1931–2000' (Weiss 2009, p. 163).[2]

While it came millennia later than the wars between Umma and Lagash over waters of the Tigris–Euphrates basin, what Herbert Smith found to be 'the first diplomatic assertion of any rule of international law upon the question' of non-navigational uses of international watercourses was a protest by Holland in 1856 concerning a Belgian diversion of water from the Meuse to serve the Campine Canal (Smith 1931, p. 137). As time passed, states began to show greater recognition of the increasing importance of non-navigational uses. A prominent illustration is the 1919 Treaty of Versailles which, in addition to declaring certain important rivers of Western and Eastern Europe to be international and open to merchant shipping, contained provisions on non-navigational uses, including hydropower, irrigation, fishing and water supply (*Treaty of Peace between the Allied and Associated Powers and Germany* 1919). Another multilateral treaty of similar vintage dealing with non-navigational issues is the 1923 Geneva Convention relating to the Development of Hydraulic Power Affecting More than One State (1923).[3] Indeed, the demands for water and power driven by the Industrial Revolution led to the development of the technology allowing the construction of large dams, as well as turbines to install in them for the production of hydroelectric power, by the late nineteenth century (United States Bureau of Reclamation 2016).[4] With large dams came the capacity to impound vast quantities of water and to manipulate the hydrology of entire rivers – potentially to the detriment of other riparian states. This was likely a driving force behind the 1923 Geneva Hydroelectric Power Convention (though this treaty never entered into force[5]).

Still, navigational uses remained the dominant concern of states in respect of international watercourses until later in the twentieth century. This set up an obvious potential conflict between navigational uses and the non-navigational ones made possible by large dams. But an expert group formed under the auspices of the League of Nations, the Commission of Enquiry, that produced the preparatory work for the 1921 Barcelona General Conference on Freedom of Communication and Transit, stated as follows in its report submitted in 1920: 'A hundred years ago waterways were principally used for purposes of navigation; today this is no longer invariably the case. Waterways nowadays frequently serve other purposes. . . . [F]rom this point of view the absolute priority of navigation is no longer invariably admissible' (League of Nations 1921).

This finding represents a tectonic shift from the historic priority given to navigation over other uses to a recognition that, at least under some circumstances, it should yield to non-navigational uses. Today it is very difficult to arrive at a general rule concerning priorities except, as will be seen in the following section, that no use enjoys inherent priority and such priorities as exist must be determined on a case-by-case basis.

The current position on priorities

Each international watercourse is unique, as are the states sharing it. Moreover, natural and human-related conditions do not remain static; they change over time. Therefore, whether a particular use of a given international watercourse enjoys priority over other uses must be determined by the states concerned in light of all relevant facts and circumstances. Such a determination would ordinarily be memorialized in an agreement or in the proceedings of a joint commission if one had been established. But, like the conditions on which it is based, the determination should not be fixed for all time but should rather be subject to modification at specified intervals to ensure that it remains consistent with current natural and human-related conditions.

The current legal position, that no one kind of use enjoys inherent priority, is reflected in Article 10 of the UN Watercourses Convention, which provides as follows:

Article 10
Relationship between different kinds of uses

1.   In the absence of agreement or custom to the contrary, no use of an international watercourse enjoys inherent priority over other uses.
2.   In the event of a conflict between uses of an international watercourse, it shall be resolved with reference to articles 5 to 7, with special regard being given to the requirements of vital human needs.

United Nations Convention on the Law of the Non-Navigational
Uses of International Watercourses 1997

If there is a conflict between different kinds of uses – such as navigation and hydropower production – it is to be resolved through the application of the principles of equitable and reasonable utilization (Articles 5 and 6) and prevention of significant harm (Article 7). The result of applying these provisions is likely to be a balanced outcome rather than one kind of use 'trumping' another kind entirely. And it will be tailored to the international watercourse concerned, and the states sharing it.

The one possible qualification to the 'no inherent priority' principle concerns 'vital human needs'. It makes sense that State A's polluting use, for example, should not make water undrinkable in State B. It could well be that, even apart from Article 10(2), such a situation would run afoul of the obligations of equitable and reasonable utilization and prevention of significant harm. Could pollution of an international watercourse that rendered the water undrinkable in a co-riparian state ever be equitable or, especially, reasonable, at least where the watercourse was relied upon for drinking water in the co-riparian state? But singling out vital human needs in a provision on priorities is justified by the fact that humans cannot live without water, while economic harm that may be caused by having to reduce another kind of use would generally not be life-threatening. A 'statement of understanding' adopted by the Working Group of the Whole in which the UN Convention was negotiated gives the following explanation of what is understood by the expression, 'vital human needs': 'In determining "vital human needs", special attention is to be paid to providing sufficient water to sustain human life, including both drinking water and water required for production of food in order to prevent starvation' (Report of the Sixth Committee 1997). This formulation is based on the International Law Commission's (ILC's) commentary on Article 10(2) (United Nations International Law Commission 1994, p. 109), which would in any event be relevant for the interpretation of that provision.

The ILC completed work on its draft articles in 1994 (ibid., p. 89, para. 22), well before a human right to water was widely recognized (McCaffrey 1992; General Comment No. 15 2002).

While human rights generally protect the individual from his or her government, Article 10(2) would have transboundary implications, protecting individuals in State B from acts by State A relating to an international watercourse the two states share. It thus provides protections that go beyond those inherent in a human right to water (but compare Leb 2012). Nevertheless, the concept of vital human needs should be informed by the human right to water, as recognized in General Comment No. 15 adopted by the United Nations Committee on Economic, Social and Cultural Rights in 2002 (General Comment No. 15 2002). Otherwise, protections available vis-à-vis one's government could exceed those available in a transboundary context. While this might be appropriate with respect to some human rights, it would not seem to be so for one so basic to life itself as the right to water.

Other indicia of evolution of the law

The gradual but complete displacement of navigation from the top of the list of priorities, in favour of an approach to determining which kinds of uses should receive priority in specific cases, is only one indicator of the manner in which international water law has evolved. There are a number of other respects in which regulation of the use of international watercourses has developed. This section will touch upon four of them, without intending to imply that those not covered are any less important. The four indicia of evolution to be discussed are the following:

- from the surface water channel to the system of waters;
- from piecemeal problem-solving to integrated management and development;
- from protection of fisheries to protection of fish; and
- from 'no harm' to equitable utilization.

These forms of evolution will be considered in turn.

From the surface water channel to the system of waters

Shared freshwater resources have historically been conceptualized as those that can be readily seen and used: rivers and lakes. And, with regard to rivers, most treaty and other practice has focused on the main stem of the watercourse rather than its entire basin, or watershed. This attitude is exemplified by Poland's position in the *River Oder* case (*Territorial Jurisdiction of the International Commission of the River Oder* 1929), that the two tributaries of the Oder at issue, which were located almost entirely within Polish territory, were not part of the international-ized Oder River under the Treaty of Versailles. (The Permanent Court of International Justice found that they were.) Groundwater, being out of sight, was largely out of mind. There was little understanding of it hydrologically, and until drilling technology developed significantly, accessing it was difficult and expensive if the aquifer containing it was located at any signifi-cant depth.

It must be recognized that the evolution in the way the regulated subject-matter is concep-tualized is ongoing. Many older agreements focusing only on the surface water in a particular river are still in force, and some states object on political grounds to a regulatory regime that would cover a drainage basin or watercourse system. But many modern agreements and other instruments adopt what may be called a 'system' approach, regulating not just the use of the water in the 'pipe', or channel, that runs along or across a border, but also that of associated water in tributaries and aquifers. Examples of this trend from different regions are the following:

Africa

- the Agreement concerning the Establishment of the Organization for the Management and Development of the Kagera River Basin (1977);
- the Agreement on the Action Plan for the Environmentally Sound Management of the Common Zambezi River System (1987);
- the Revised Protocol on Shared Watercourses in the Southern African Development Community (SADC) (2000);

The Americas

- the Treaty relating to Cooperative Development of the Water Resources of the Columbia River Basin (1961);
- the Treaty on the River Plate (La Plata River) Basin (1969);
- the Great Lakes Water Quality Agreement, as amended by Protocols of 1983 and 1987 (1978);

Asia

- the Agreement on the Cooperation for the Sustainable Development of the Mekong River Basin (1995);

Europe

- the EU Water Framework Directive (EU Water Framework Directive 2000).

All of these instruments take a holistic approach to the management of shared freshwater resources. It is now generally realized and accepted that attempting to manage only a part of a watercourse system is a futile endeavour and that, as will be seen below, to be truly effective, governing legal and management regimes must also cover land-based activities, resources and ecosystems associated with shared surface and groundwater.

From piecemeal problem-solving to integrated management and development

This trend parallels, to some extent, the one just discussed. But here we are speaking more of the increasing recourse to legal and institutional arrangements for the achievement of coordinated multipurpose river basin management, protection and development. The object of these regimes is to avoid conflicts between the multiple uses of shared river basins that are increasingly crowded by populations and industry. And, increasingly, such regimes will be necessary to adapt to and ameliorate the impacts of climate change. The tool for achieving these objectives is integrated river basin management and development (see, generally, United Nations, Department of Economic and Social Affairs 1970). An illustration of this approach is the 2000 EU Water Framework Directive, which calls on all EU Member States to adopt river basin management plans. The EU explains the rationale for this approach as follows:

> The best model for a single system of water management is management by river basin – the natural geographical and hydrological unit – instead of according to administrative or political boundaries. . . . For each river basin district – some of which will traverse national frontiers – a 'river basin management plan' will need to be established and

updated every six years, and this will provide the context for the co-ordination requirements [set forth in the Directive].

EU, Introduction to the new EU Water Framework Directive 2000

The integrated management approach contrasts with many early water treaties, which were often concluded for the purpose of dealing with a single issue, such as navigation (e.g. *Treaty of Paris* 1814, Art. 5; Final Act *Congress of Vienna* 1815, Arts. 108–16; *Navigation Act* 1857) or fishing (*Convention entre la Suisse, l'Allemagne et les Pays-Bas pour régulariser la pêche du saumon dans le bassin du Rhin* 1885; *Convention between Austria-Hungary, Baden, Bavaria, Liechtenstein, Switzerland and Württemberg Laying Down Uniform Provisions concerning Fishing in Lake Constance, with Protocol* 1893). It is all the more necessary where the upper and lower parts of a drainage basin are located in different states:

> The earlier and larger development [of a river basin]… particularly by agriculture, takes place in the hotter, more nearly level areas of the lower basin, and population charac-teristically develops more rapidly there than in the upper basin, because of the greater food supply and easier navigation of the river's lower reaches. But this lower area, in many instances, is dependent upon the construction of dams in the upper and more mountainous areas for its protection against floods, for the storage of water for use in dry seasons and in dry years, and, nowadays, for power generation.
>
> *Ely and Wolman 1967*

This situation, in which lower riparian states often develop their water resources before their upper riparian neighbours, and in which opportunities for such development differ and often complement each other, sets the stage perfectly for benefit-sharing among states sharing an international watercourse system, something that does much to negate the zero-sum game riparian states often find themselves having to cope with (e.g. Sadoff and Grey 2002, Phillips *et al.* 2006).

Integrated management of shared basins is best achieved through the establishment of joint commissions or other similar mechanisms, which riparian states have increasingly done. An early example of the use of a joint management mechanism is the 1909 Boundary Waters Treaty between the United States and Canada, which established the International Joint Commission (IJC) between the two countries (*Treaty between the United States and Great Britain relating to Boundary Waters, and questions arising between the United States and Canada* 1909, Art. III). The IJC is often pointed to as an example of a successful joint mechanism. Both of the global treaties on international watercourses, the UN (*United Nations Convention on the Law of the Non-Navigational Uses of International Watercourses* 1997) and ECE (*UNECE Convention on the Protection and Use of Transboundary Watercourses and International Lakes* 1992) Conventions, envision the establishment of joint commissions, demonstrating a recognition of their utility by the international commu-nity. The same is effectively true of the EU's Water Framework Directive, which provides for the assignment of river basins covering the territory of more than one Member State to an interna-tional river basin district (EU Water Framework Directive 2000, Art. 3(3)).

From protection of fisheries to protection of fish

The trend towards integrated river basin management parallels the growing realization that the protection of aquatic ecosystems is not only important for its own sake, but is also, in fact, benefi-cial to humans (McIntyre 2007; McCaffrey 2015, 2007 ch. 12). Scientific progress has demonstrated, perhaps now unsurprisingly, that a dead watercourse benefits no one. And better understanding of phenomena such as tipping points and feedback loops have underscored the critical role of prevention, in contrast to the more traditional practice of waiting for environmental problems to manifest themselves before taking corrective action.

While human capacity to pollute watercourses has mushroomed to proportions that could hardly have been imagined a century ago, even in the Middle Ages freshwater pollution had grown to the point that it has been held responsible for widespread epidemics such as the black plague (Sette-Camara 1984, p. 139). But treaty practice was, for the most part, directed at the protection of fisheries – that is, resources that were valuable to humans – rather than to protection of the fish themselves, their ecosystems, or even of the quality of water for human consumption (e.g. *Convention between France and Switzerland for the regulation of fishing in their frontier waters* 1904, Art. 17). An early exception is the 1909 Boundary Waters Treaty between Canada and the United States, which, in Art. IV(2), prohibits such water pollution as injured health or property on the other side of the boundary (*Treaty between the United States and Great Britain relating to Boundary Waters, and Questions Arising between the United States and Canada* 1909). Protection was attempted in these agreements through strict prohibitions – for example, of pollution harmful to fish, rather than through more flexible regulations or water quality standards – although the threshold of fish kills could, of course, be seen as a primitive standard.

It was only in the latter half of the twentieth century that agreements began to address protection of aquatic ecosystems. A leading example is the 1978 Agreement between Canada and the United States on Great Lakes Water Quality (*Great Lakes Water Quality Agreement* 1978) which states that its purpose is: 'to restore and maintain the chemical, physical, and biological integrity of the waters of the Great Lakes Basin Ecosystem'. Another example from the same period, also a bilateral treaty, is the 1975 *Statute of the River Uruguay*. Article 36 of that treaty provides: 'The Parties shall co-ordinate, through the Commission,[6] the necessary measures to avoid any change in the ecological balance and to control pests and other harmful factors in the river and the areas affected by it' (*Statute of the River Uruguay* 1975). And, according to Article 41(a) of the Statute, the parties undertake: 'To protect and preserve the aquatic environment and, in particular, to prevent its pollution, by prescribing appropriate rules and measures in accordance with applicable international agreements and in keeping, where relevant, with the guidelines and recommendations of international technical bodies' (*Statute of the River Uruguay* 1975). That these are obligations of due diligence was underlined by the International Court in the *Pulp Mills* case (*Pulp Mills on the River Uruguay* 2010).

The importance of protecting the fluvial environment more broadly is recognized in Article 20 of the UN Watercourses Convention in the following terms: 'Watercourse states shall, individually and, where appropriate, jointly, protect and preserve the ecosystems of international watercourses' (*United Nations Convention on the Law of the Non-Navigational Uses of International Watercourses* 1997, Art. 20.).

This is a broad provision that is not confined to protection of the ecology of the water in the watercourse only, but embraces the entire fluvial ecosystem. The commentary of the International Law Commission, which prepared the draft on the basis of which the UN Convention was negotiated, explains that the term 'ecosystem' refers to an 'ecological unit consisting of living and non-living components that are interdependent and function as a community' (United Nations International Law Commission 1994, pp. 280–1). This is of great importance in that poor land-use practices, such as grazing and deforestation in the vicinity of watercourses, can do great harm to life in the watercourse itself, and to life it sustains, as well as its banks and bed. This Article was the subject of 'complex negotiations' during the drawing up of the Convention, but the term 'ecosystem' was ultimately retained (Tanzi and Arcari 2001, p. 241).

From 'no harm' to equitable utilization

Holland's protest of Belgium's diversion of the Meuse in 1856, mentioned earlier, is typical of early diplomatic exchanges and treaties (e.g. *Treaty between the United States and Great Britain relating to Boundary Waters, and Questions Arising between the United States and Canada* 1909, Art. IV; *Exchange of Notes between the United Kingdom and Egypt* 1929, para. 4(b); and *Convention between Italy and Switzerland concerning the Regulation of Lake Lugano* 1955) in that it focuses on the avoidance of harm rather than allocation. The note contained the following statement:

> The Meuse being a river common both to Holland and to Belgium, it goes without saying that both parties are entitled to make the natural use of the stream, but at the same time, following general principles of law, each is bound to abstain from any action which might cause damage to the other. In other words, they cannot be allowed to make themselves masters of the water by diverting it to serve their own needs, whether for purposes of navigation or irrigation.
>
> *Smith 1931, p. 217*

While this note postdates the treaty between Umma and Lagash mentioned at the outset of this chapter by over 4,000 years, it is of interest because it has been characterized as the earliest known record of an assertion of such rights in the modern era (ibid., p. 137). Holland's protest is based on 'general principles of law', according to which 'each is bound to abstain from any action which might cause damage to the other'. Holland's claim is therefore in conformity with the maxim *sic utere tuo ut alienum non laedas* (so use your own as not to harm that of another). If the claim were given literal and strict effect, however, it would be highly problematic – especially in today's interconnected world, but even in the nineteenth century. For a categorical prohibition of the causing of harm would give absolute priority to the use claimed to be harmed, giving a state whose use was prior in time a veto over uses of co-riparian states that may harm the former state's uses. But state practice reveals a recognition that absolute entitlements are incompatible with the flexible cooperative relationships that are essential in respect of shared natural resources (see, in particular, *Convention concerning the Equitable Distribution of the Waters of the Rio Grande for Irrigation Purposes* 1906; see, generally, Bourne 1965).[7] Thus the so-called 'no-harm' principle has not generally been treated in practice as a prohibition of the causing of all factual harm (e.g. the *Trail Smelter Arbitration* 1941, 3 UNRIAA 1911 ('serious consequence', p. 1965); and *Lake Lanoux Arbitration* 1957 ('serious injury to the lower riparian State'[8]).[9] Instead, in keeping with the principle of equality of right, it has been seen as a prohibition of harm that is contrary to the rights of the harmed state. In the context of shared freshwater resources in particular, those rights would not extend to a categorical right to be free from all harm.

This proposition is supported by the '*Donauversinkung*' case between the German states of Württemberg and Prussia, on the one hand, and Baden, on the other, decided by the Staatsgerichtshof in 1927 (*Württemberg and Prussia v Baden* 1927).[10] In that case, the two applicant states sought relief from the phenomenon of the 'sinking of the Danube'. The court stated:

> The exercise of sovereign rights by every State in regard to international rivers traversing its territory is limited by the duty not to injure the interests of other members of the international community. . . The application of this principle is governed by the circumstances of each particular case. The interests of the States in question must be weighed in an equitable manner against one another. One must consider not only the

absolute injury caused to the neighboring state, but also the relation of the advantage gained by one to the injury caused to the other.

Württemberg and Prussia v Baden 1927, p. 131

Thus, the court recognized that the 'no-harm' principle of *sic utere tuo* does not constitute a categorical prohibition of harm but must be applied in a balanced way, to preserve what the United States Supreme Court has called the states' 'relative rights' (*Wyoming v Colorado* 1922, p. 484). The objective of this process is the achievement of an equitable apportionment of uses and benefits of the watercourse in a manner that is sustainable.

This regime is reflected in the UN Watercourses Convention, in particular in Articles 5–7 (*United Nations Convention on the Law of the Non-Navigational Uses of International Watercourses* 1997). Article 5(1) provides that states are to 'utilize an international watercourse in an equitable and reasonable manner', taking into account the interests of the co-riparian states concerned. Equitable and reasonable utilization requires taking into consideration all relevant factors, an indicative list of which is contained in Article 6. Article 7(1) provides that states are to 'take all appropriate measures to prevent the causing of significant harm to other watercourse States', but paragraph 2 of that Article recognizes that harm may nevertheless be caused and sets forth a process of dealing with the situation with 'due regard for the provisions of articles 5 and 6'. These rules are based on the practice of states, which recognizes that a categorical prohibition of harm is both untenable and unworkable, and that there must be flexibility in the legal regime of rights of co-riparian states.

The UN Watercourses Convention may be considered a codification of the basic principles of the law in the field,[11] which signals the emergence of equitable and reasonable utilization as the core principle. A state's perception that it has been harmed by uses in a co-riparian state will trigger the process of arriving at an equitable and reasonable resolution but will no longer be sufficient in and of itself to entitle the state to cessation of the allegedly harmful conduct.

Conclusion

In sum, international water law has evolved in many different ways over the course of the past century. This evolution has been driven by changing needs of society and by growing pressures from an increasing population and economic development. Navigation, while still important, has been supplanted in varying measures in different regions of the world by other forms of transport and its significance in relation to non-navigational uses has been diminished by the growing importance of the latter uses. There have been many developments within the field of non-navigational uses, as well. This chapter has considered only a selection of them, but it is hoped that those that have been discussed provide a window into the changing ways in which humanity is regulating the uses of international watercourses.

Notes

1 Girsu, the capital of the Lagash kingdom, and Umma 'had fought for generations over a fertile border district' in the lower Euphrates, Girsu being 'unable to prevent Umma, situated higher up the watercourse, from breaching and obstructing the branch canals that served the border fields'. Jacobsen and Adams 1958, p. 1251.

2 The year 2000 was the cut-off date for Brown Weiss's survey.

3 This treaty never entered into force.

4 'By the early 1900s, hydroelectric power accounted for more than 40 percent of the United States' supply of electricity.... The early hydroelectric plants were direct current stations built to power arc and

incandescent lighting during the period from about 1880 to 1895. When the electric motor came into being the demand for new electrical energy started its upward spiral. The years 1895 through 1915 saw rapid changes occur in hydroelectric design and a wide variety of plant styles built. Hydroelectric plant design became fairly well standardized after World War I' (United States Bureau of Reclamation).

5 The sensitivity of the issue is illustrated by the fact that the treaty was not ratified by any states sharing international watercourses, and that three of the states that did ratify were islands (Smith 1931, p. 199).

6 This refers to the Administrative Commission of the River Uruguay established under Art. 49 of the Statute (author's footnote).

7 This is the treaty that resolved the dispute between Mexico and the United States that gave rise to the infamous Harmon Doctrine of absolute territorial sovereignty.

8 Trans. from 1974 *Yearbook of the International Law Commission*, vol. 2, pt. 2, p. 196, para. 1064.

9 Both of these decisions allowed the project in question to proceed, though subject to adjustments that either had already been made (*Lake Lanoux*) or were to be made (*Trail Smelter*) by the proposing state.

10 The present discussion is based on the report in *Annual Digest*, years 1927 and 1928, p. 128, at pp. 131–2.

11 This is due both to the provenance of the Convention, which was based on 20 years' study by the International Law Commission, and to its treatment by the ICJ and in the literature. Four months after the Convention was concluded, it was relied upon by the ICJ in the *Gabčíkovo-Nagymaros Project* case 1997.

References

Agreement for the Establishment of the Organization for the Management and Development of the Kagera River Basin (1977) 1089 UNTS 165, entered into force 24 August 1977.

Agreement on the Action Plan for the Environmentally Sound Management of the Common Zambezi River System (1987) 26–28 May, 28 ILM 1109 (1988).

Agreement on the Cooperation for the Sustainable Development of the Mekong River Basin (1995) 5 April, 34 ILM 864 (1995).

Berber, F.J. (1959) *Rivers in International Law*. London: Stevens & Sons.

Bourne, C (1965) The right to utilize the waters of international rivers. *Canadian Yearbook of International Law* 3, p. 187.

Committee on Economic, Social and Cultural Rights (2002) *General Comment No. 15: The Right to Water*, 26 November. E/C.12/2002/11.

Congress of Vienna (1815) Arts. 108 and 109, Austria, France, Great Britain, Portugal, Prussia, Russia and Sweden, 1 MPT pp. 519, 567, 9 June.

Convention between Austria-Hungary, Baden, Bavaria, Liechtenstein, Switzerland and Württemberg Laying Down Uniform Provisions concerning Fishing in Lake Constance, with Protocol (1893) Legislative Texts and Treaty Provisions concerning the Utilization of International Rivers for Other Purposes than Navigation, UN Doc. ST/LEG/SER.B/12, Treaty No. 114, p. 403. 2 July.

Convention between France and Switzerland for the regulation of fishing in their frontier waters (1904) de Martens, Noveau Recueil, 2nd series, vol. 33, p. 501.

Convention between Italy and Switzerland concerning the Regulation of Lake Lugano (1955) 17 September, 291 UNTS 213.

Convention concerning the Equitable Distribution of the Waters of the Rio Grande for Irrigation Purposes (1906) 21 May, United States Treaty Series No. 455. Available from: http://www.ibwc.gov/Files/1906Conv.pdf

Convention entre la Suisse, l'Allemagne et les Pays-Bas pour régulariser la pêche du saumon dans le bassin du Rhin (1885) 30 June. Available from: https://www.admin.ch/opc/fr/classified-compilation/18850013/index.html

Council Directive 2000/60/EC of 22 December 2000 on Establishing a framework for the community action in the field of water policy.

Ely, N. and Wolman, A. (1967) Administration. In A. Garreston, R. Hayton and C. Olmstead (eds), *The Law of International Drainage Basins*. New York: New York University.

EU Water Framework Directive (2000) *Council Directive 2000/60/EC* of 23 October 2000 on the Water Framework. Available from: http://ec.europa.eu/environment/water/water-framework/index_en.html

EU (n.d.) Introduction to the new EU Water Framework Directive. Available from: http://ec.europa.eu/environment/water/water-framework/info/intro_en.htm

Exchange of Notes between the United Kingdom and Egypt (1929) 7 May, 93 LNTS 44.

Final Act of the Congress of Vienna (1815) Martens Recueil 2, p. 379, 9 June.

Food and Agriculture Organization of the United Nations (1978) Systemic Index of International Water Resources Treaties, Declarations, Acts and Cases by Basin. Rome: FAO.

Gabčíkovo-Nagymaros Project (Hungary/Slovakia) (1997) ICJ Rep. 7.

General Comment No. 15 (2002) The Right to Water (Articles 11 and 12 of the International Covenant on Economic, Social and Cultural Rights), UN Doc. E/C.12/2002/11, 26 November.

Geneva Convention relating to the Development of Hydraulic Power Affecting More than One State (1923) 36 LNTS 77.

Great Lakes Water Quality Agreement (1978) As amended by Protocols of 1983 and 1987, 22 November 1978, 30 UST 1383, TIAS 9257, as amended 16 October 1983, TIAS 10798 and 18 November 1987, TIAS 11551, 837 UNTS 213. Available from: https://treaties.un.org/doc/publication/unts/volume%201153/volume-1153-i-18177-english.pdf, consolidated in International Joint Commission, Revised Great Lakes Water Quality Agreement of 1978 (1988) Available from: http://www.ijc.org/files/tinymce/uploaded/GLWQA_e.pdf

Jacobsen, T. and Adams, T. (1958) Salt and silt in ancient Mesopotamia agriculture. *Science* 128(3334), p. 1251.

Lake Lanoux Arbitration (1957) (France v Spain), 12 UNRIAA 281.

League of Nations (1921) Barcelona Conference, Verbatim Records and Texts relating to the Convention on the Regime of Navigable Waterways of International Concern and to the Declaration Recognising the Right to a Flag of States Having No Sea-Coast, annex to Section IV (Draft Convention on the Regime of Navigable Waterways, text prepared by the Commission of Enquiry and submitted to the Conference). Report on the Draft Convention on the International Regime of Navigable Waterways. Presented to the General Communications and Transit Conference by the Commission of Enquiry, pp. 414, 415 (Geneva).

Leb, C. (2012) The right to water in a transboundary context: emergence of seminal trends. *Water International* 37, p. 640.

McCaffrey, S. (1992) A human right to water: domestic international implications. *Georgetown International Environmental Law Review* 5(1).

McCaffrey, S. (1993) The evolution of the law and international watercourses. *Austrian Journal of Public and International Law* 45(87).

McCaffrey, S. (2007) *The Law of International Watercourses*, 2nd edn. Oxford: Oxford University Press.

McCaffrey, S. (2013) The progressive development of international water law. In F. Loures and A. Rieu-Clark (eds), *The UN Watercourses Convention in Force: Strengthening International Law for Transboundary Water Management* London and New York: Routledge.

McCaffrey, S. (2015) Pollution of shared freshwater resources in international law. In S. Jayakumar *et al.* (eds), *Transboundary Pollution: Evolving Issues of International Law and Policy.* Cheltenham: Edward Elgar.

McIntyre, O. (2007) Environmental Protection of International Watercourses under International Law. London: Routledge.

Navigation Act (1857) (Danube) Martens Recueil, pt. 2, vol. 16, p. 75. 7 November.

Nussbaum, A. (1954) *A Concise History of the Law of Nations*, rev. edn. New York: Macmillan.

Phillips, D. *et al.* (2006) *Transboundary Water Cooperation as a Tool for Conflict Prevention and Broader Benefit Sharing.* Stockholm: Swedish Ministry of Foreign Affairs.

Pulp Mills on the River Uruguay (Argentina v Uruguay) (2010) ICJ Rep. 14.

Report of the Sixth Committee convening as the Working Group of the Whole (1997) *(11 April)*, A/51/869.

Revised Protocol Revised Protocol on Shared Watercourses in the Southern African Development Community (SADC) (2000), 7 Aug. 2000, available at http://www.internationalwaterlaw.org/documents/regionaldocs/Revised-SADC-SharedWatercourse-Protocol-2000.pdf.

Sadoff, C. and Grey, D. (2002) Beyond the river: the benefits of cooperation on international rivers. *Water Policy* 4(5), pp. 389–403.

Sette-Camara, J. (1984) Pollution of international rivers. *Recueil des Cours* (1984-III).

Smith, H. (1931) *The Economic Uses of International Rivers.* London: King & Son.

Statute of the River Uruguay (1975) 635 UNTS 91. 17 December.

Tanzi, A. and Arcari, M. (2001) *The United Nations Convention on the Law of International Watercourses.* Berlin: Kluwer.

Teclaff, L. (1967) *The River Basin in History and Law.* The Hague: Martinus Nijhoff.

Teclaff, L. (1985) *Water Law in Historical Perspective.* Buffalo, NY: William S. Hein.

Territorial Jurisdiction of the International Commission of the River Oder (1929) (Czechoslovakia, Denmark, France, Germany, Great Britain and Sweden/Poland) PCIJ, Ser. A, No. 23, 10 September.

Trail Smelter Arbitration (1941) (United States v Canada) 3 UNRIAA 1911.

Treaty between the United States and Great Britain relating to Boundary Waters, and Questions Arising between the United States and Canada (1909) 12 Bevans p. 319, 36 Stat. p. 2448, TS p. 548, 102 BFSP p. 137. 11 January. Available from: http://www.ijc.org/en_/BWT

Treaty of Paris (1814) Martens Recueil 2, p. 1 (1887). 30 May 1814.

Treaty of Peace between the Allied and Associated Powers and Germany (1919) Versailles, G. Martens, Recueil de traités 11, 3rd ser. 28 June.

Treaty of Peace between France and the Empire (1648) Art. XII, signed at Münster, 14(24) October, 1 CTS 271 (English trans. at p. 319).

Treaty on the River Plate Basin (1969) 875 UNTS 3. 14 August 1970.

Treaty relating to Cooperative Development of the Water Resources of the Columbia River Basin (1961) 17 January. Available from: http://www.ccrh.org/comm/river/docs/cotreaty.htm

United Nations Convention on the Law of the Non-Navigational Uses of International Watercourses (1997) 21 May, 36 ILM 700, A/RES/51/869.

United Nations, Department of Economic and Social Affairs (1970) E/3066/Rev.1: Integrated river basin development, report of a panel of experts.

UNECE Convention on the Protection and the Use of Transboundary Watercourses and International Lakes (1992) 31 ILM 1312, 17 March. Available from: http://www.unece.org/fileadmin/DAM/env/water/pdf/watercon.pdf

United Nations International Law Commission (1994) *Yearbook of the International Law Commission*, vol. 2, pt. 2.

United States Bureau of Reclamation (2016) *The History of Hydropower Development in the United States*. Available from%: http://www.usbr.gov/power/edu/history.html

Weiss, E. Brown (2009) The evolution of international water law. *Recueil des cours* 331(163), p. 235.

Wittfogel, K. (1957) *Oriental Despotism: A Comparative Study of Total Power*. New Haven, CT, and London: Yale University Press.

Württemberg and Prussia v Baden (1927) German Staatsgerichtshof 116 App., pp. 18–45. Annual Digest, years 1927 and 1928, p. 128.

Wyoming v Colorado (1922) 259 US 419.

16

INTERNATIONAL LAW AND TRANSBOUNDARY AQUIFERS

Gabriel E. Eckstein

Introduction

Growing demands for fresh water resources, coupled with declining supplies, have exacerbated water scarcity around the world. As a result, many nations are focusing their attention on ground water supplies in order to meet their societal, economic and environmental needs and objectives. Not surprisingly, they are finding that many of these subsurface water bodies are shared with their neighbours and pose unique transboundary political, social and legal issues. This phenomenon is particularly evident in the Middle East where the use and allocation of aquifers traversing political boundaries, such as the Mountain Aquifer shared by Israelis and Palestinians and the Nubian Sandstone Aquifer shared by Chad, Egypt, Libya and Sudan, continue to be a source of friction among overlying riparians.

At present, there are few concrete rules under international law that govern relations over cross-border ground water resources. The growing interest in transboundary aquifers, however, is spurring discussions and enquiry about the appropriate rules and mechanisms for exploiting and managing these buried treasures. Moreover, as nations around the world begin to extract (or intensify their withdrawals of) ground water from aquifers traversing national borders, they are beginning to explore and experiment with various approaches to address rights and obligations pertaining to these shared subsurface resources. As a result, while state practice in and scholarly analyses of the utilization of transboundary aquifers are still relatively sparse, trends and priorities are beginning to emerge.

In an effort to identify and characterize these trends and priorities, this study reviews the chief formal and informal mechanisms that have been proposed or implemented for the assessment, use, allocation and protection of transboundary ground water resources. The study begins by addressing the importance of transboundary aquifers as a source of fresh water for people and the environment. It then examines a number of formal and informal arrangements between nations for the assessment, use, allocation and protection of transboundary ground water resources. By studying these regimes, this chapter attempts to identify trends and priorities that may implicate the emergence of generally accepted international legal norms applicable to transboundary aquifers. This study concludes that while the law of transboundary aquifers is in an early stage of development, there is, nonetheless, a growing body of experience and practice suggesting the emergence of legal standards. Lastly, the chapter considers the gaps and shortcomings in the emerging international regulatory system and offers recommendations for the further development of the law.

Scope and significance of transboundary aquifers around the world

Transboundary ground water resources play a critical role in providing fresh water for people, industries, nations and the environment worldwide. This is especially true in the arid and semi-arid regions of the world where transboundary aquifers often serve as the primary or sole source of fresh water for human and environmental sustenance. Libya, for example, which has no meaningful surface water resources, obtains the majority of its fresh water – some 6.5 million cubic metres of water daily – from the Nubian Sandstone Aquifer, a vast underground reservoir that also underlies sections of Libya's neighbours: Chad, Egypt and Sudan (Watkins 2006). Similarly, Palestinians in the West Bank and Gaza obtain the great majority of their water from aquifers shared with Israel – the Mountain Aquifer underlying the West Bank and eastern Israel, and the Coastal Aquifer underlying Gaza's and Israel's Mediterranean coast (World Bank 2009). In addition, transboundary aquifers serve as the sole source of fresh water for many of the communities along the Mexico–United States border, including the Mexican cities of Puerto Palomas, Naco, Nogales, Sonoyta and Tecate, and their respective American sister cities of Columbus, Bisbee, Nogales, Lukeville and Tecate (Eckstein 2012).

Although the precise global significance and impact of these shared resources has escaped quantification, extrapolations from relevant studies and comparisons support the proposition that transboundary aquifers have become critical to human, economic and environmental sustainability worldwide. For example, ground water today is the most extracted natural resource on the globe and provides water for 20 per cent of irrigated agriculture, as well as more than half of humanity's fresh water for everyday uses such as drinking, cooking and hygiene (UN Educational, Scientific and Cultural Organization and World Water Assessment Programme 2003). Moreover, while 276 international watercourses traverse the world's international political boundaries (Wouters and Moynihan 2013), an ongoing study has identified more than 600 aquifers and aquifer bodies traversing the same frontiers (IGRAC 2015). Furthermore, while an estimated 40 per cent of the world's population reside in transboundary river basins around the world (UN Environmental Programme 2002), given that most domestic and internationally transboundary rivers have a hydraulically connected aquifer, and that there are scores of solitary fossil aquifers around the world that likewise traverse international political boundaries, it is logical to infer that a comparable if not larger number of people reside in the basins of transboundary aquifers globally.

Despite the relevance of transboundary aquifers to human existence, economic development and environmental sustainability, policy and legal attention to these subsurface resources is a relatively recent phenomenon. While over 3,600 treaties relating to the use of transboundary surface waters have been catalogued since 805CE, and over 400 since 1820 (Wolf 2002), the first agreement to focus exclusively on the management of a transboundary aquifer occurred in 1978 for the Genevese Aquifer along the French–Swiss border (Genevese Convention 2008).

Since then, greater attention has been focused on ground water resources traversing international boundaries and the legal, policy and political considerations that pertain to these international water bodies. In addition to the Genevese Convention, a small handful of formal and informal arrangements have been forged for a number of other transboundary aquifers, including: the Guarani Aquifer in South America, the Nubian Sandstone and North Western Sahara aquifer systems in Northern Africa, the Al-Sag/Al-Disi Aquifer shared between Jordan and Saudi Arabia, the Iullemeden and Taoudeni/Tanezrouft aquifer systems in West Africa, the Hueco Bolson Aquifer underlying the cities of Juárez and El Paso on the Mexico–US border and the Abbotsford–Sumas Aquifer between the US state of Washington and the Canadian province of British Columbia. In addition, transboundary ground water resources have featured prominently in the 1992 United Nations Economic Commission for Europe (UNECE) Convention on the Protection and Use of Transboundary Watercourses and International Lakes (1992 UNECE Water Convention) and the 1997 UN Convention on the Non-navigational Uses of International Watercourses (1997 UN Watercourses Convention).

Especially prominent in this evolutionary process are two documents proffering possible norms for administering transboundary ground water resources. The first are the 19 *Draft Articles on the Law of Transboundary Aquifers* (Draft Articles) prepared by the UN International Law Commission (UNILC) and submitted to the UN General Assembly (UNGA) in 2008. The second are the nine *Model Provisions on Transboundary Groundwaters* (Model Provisions) that were prepared under the auspices of the UNECE (Model Provisions 2014). While the outcome of the Draft Articles before the UNGA is still unresolved and the Model Provisions were intentionally proposed as exemplars, the two efforts represent significant milestones in the growing importance of transboundary aquifers on the international agenda. Concerns over aquifers shared by multiple nations are no longer secondary to those of surface water resources. Transboundary aquifers have come into their own and are now legitimate topics of international law, policy and relations.

Examples of cooperative mechanisms for transboundary aquifers

References to transboundary ground water resources have appeared in international instruments for more than 150 years. For example, an 1864 agreement between Portugal and Spain afforded both parties the common rights to springs located on the border (Treaty of Limits between Portugal and Spain 1864). Similarly, an 1888 agreement between the United Kingdom and France provided both parties the common rights to use the wells of Hadou, which lay on the newly created border of the Somali coast. All of these references, however, were secondary or even tertiary concerns under their respective agreement. It wasn't until the late twentieth century that transboundary aquifers began garnering an interest warranting individualized attention in both treaty-making and international law.

The following section summarizes the chief formal and informal mechanisms that have been proposed or implemented on a transboundary aquifer. All but three of these arrangements are exclusively focused on aquifers. While the other three arrangements address other water resources, they are included in this study because of their relevance and emphasis on a particular transboundary aquifer as a primary concern.

Formal agreements

The best known, and still the only treaty crafted to manage and specifically allocate the waters of a transboundary aquifer, is the *Convention on the Protection, Utilization, Recharge and Monitoring of the Franco-Swiss Genevese Aquifer* (Genevese Convention 2008). Originated in 1978 and revised in 2008, this singular arrangement addresses ground water quality, quantity, abstraction and recharge largely through the creation of a joint Genevese Aquifer Management Commission. While the Commission only has consultative status, its recommendations and technical opinions carry considerable weight in the management of the aquifer. In addition, the updated regime reasserts the Swiss artificial recharge obligations created by the original 1978 agreement, allocates expenses between the countries for the Swiss recharge efforts and places strict withdrawals limits on extraction in France (see Preamble, Articles 2.3, 8, 11–14 and *Annex to the Convention on the Inventory of the Recharge Equipment and Existing Extraction Works*). The Genevese Convention is particularly significant because it strikes a balance between state sovereignty and state responsibility in its management scheme, which is based almost exclusively on principles of transparency, good faith dealings and cooperation. Moreover, the agreement is unique in its structure as a treaty since it provides purely technical mechanisms for managing the shared aquifer and avoids any direct political, legal, or other reference to either country's sovereign rights to the aquifer or its waters (Genevese Convention 2008).

The newest arrangement for a transboundary aquifer is the 2015 *Agreement between the Government of the Hashemite Kingdom of Jordan and the Government of the Kingdom of Saudi Arabia for the Management and Utilization of the Ground Waters in the Al-Sag/Al-Disi Layer* (Al-Sag/Al-Disi Agreement 2015). In contrast to the Genevese Convention, the Al-Sag/Al-Disi Agreement was created for the limited purpose of restricting ground water extraction and protecting ground water quality. While it imposes no numerical limitations on abstraction, the agreement creates a 'Protected Area' or buffer zone in both countries from which ground water extraction is absolutely prohibited, as well as a broader 'Management Area' from which extractions are restricted exclusively for municipal purposes. In a similar restrictive vein, the agreement places a near absolute prohibition on ground water pollution within the Management Area. Also significant is the Al-Sag/Al-Disi Agreement's creation of a Joint Technical Committee (JTC), which like the Genevese Commission, does not have any decision-making authority (Al-Sag/Al-Disi Agreement 2015). While the JTC is responsible for monitoring both the quantity and quality of extractions, collecting and exchanging information, analysing collected data and submitting their findings to the competent authorities in both nations, it is yet unclear whether it will enjoy a strong consultative role as does the Genevese Commission.

Similar in concept to the Al-Sag/Al-Disi Agreement, the 1973 amendment to the 1944 *Mexico–US Treaty Relating to the Utilization of Waters of the Colorado and Tijuana Rivers and of the Rio Grande* known as *Minute 242* also focused on restrictions to ground water extractions on the border (Minute 242 1973). While the Minute was designed to address salinity levels in the Colorado River, paragraphs five and six deviated from that purpose and focused on ground water in the Arizona–Sonora border region near San Luis. Paragraph five limits ground water pumping in this region to specifically enumerated withdrawal targets, while paragraph six requires both countries to consult each other prior to pursuing new development of surface or ground water resources anywhere on the border that could have adverse transboundary impacts (ibid.). While paragraph five also referenced the future development of a border-wide ground water agreement, that arrangement has yet to be realized.

Taking a more generalist approach, the 2010 *Agreement on the Guarani Aquifer* – entered into by Argentina, Brazil, Paraguay and Uruguay – provides a basic framework for cooperating over the Guarani Aquifer (Guarani Agreement 2010). While the agreement references a number of broadly accepted substantive principles of international water law, including those of reasonable and equitable use (Articles 3 and 4) and of no significant harm (Articles 3, 6 and 7), it does not elaborate on their definition or implementation. For example, the agreement does not identify the factors relevant to assessing whether a particular use is reasonable and equitable. In the same vein, the Guarani Agreement references various procedural obligations using vague terminology and qualifications, including those for sharing information (Articles 8, 9 and 12), providing notification of planned measures that may result in a transboundary impact (Articles 9, 10 and 11), and the creation of a commission to oversee cooperation (Article 15). For example, in obligating the sharing of information, Article 8 qualifies the requirement by requiring the parties to 'proceed to adequately exchange technical information about studies, activities and works that contemplate the sustainable use of the Guarani Aquifer System water resources' (Guarani Agreement 2010). While these formulations might be regarded as creating ambiguous obligations, they may also be perceived as creating a necessarily flexible framework for cooperation (Sindico and Hawkins 2015).

The most provocative aspect of the Guarani Agreement is its endorsement of state sovereignty over portions of the aquifer that underlay each nation (Article 2). While some assert that this language harkens back to the long-discredited Harmon Doctrine and is scientifically and politically indefensible (McCaffrey 2011), others contend that the approach follows on the international notion of state sovereignty over natural resources and was necessary to achieve dialogue over this nascent topic (Villar and Ribeiro 2013). Notwithstanding its implications for evolving international law for transboundary aquifers, the Guarani Agreement has still not entered into force. While Uruguay and Argentina have ratified the instrument, Paraguay and Brazil have yet to do so.

Two additional arrangements must be considered when discussing formal mechanisms developed for addressing shared ground water resources. The first is actually a series of agreements entered into for the management of the Nubian Sandstone Aquifer in Northern Africa. The series begins with the 1992 *Constitution of the Joint Authority for the Study and Development of the Nubian Sandstone Aquifer Waters* (NSA Constitution 1992), which created a cooperative mechanism designed to collect and compile information on the aquifer, promote cooperation and develop common water management policies. That instrument was followed in 2000 by two agreements under the framework of a *Programme for the Development of a Regional Strategy for the Utilisation of the Nubian Sandstone Aquifer System* (NSAS Agreements 2000). Under the first agreement – *Agreement No. 1 – Terms of Reference for the Monitoring and Exchange of Groundwater Information of the Nubian Sandstone Aquifer System, done in Tripoli* – the four parties consent to share via an internet portal data that had been previously compiled in the Nubian Aquifer Regional Information System, as well as information on developmental aspects related to the aquifer, such as socio-economic data, environmental issues, drilling experiences, meteorological data and other data. Under the second agreement – *Agreement No. 2 – Terms of Reference for Monitoring and Data Sharing, done in Tripoli* – the parties agree to continuously update this information by monitoring the aquifer through specified studies, measurements and analyses (NSAS Agreements 2000).

The second noteworthy arrangement, which was implemented in 2002, is entitled *Establishment of a Consultation Mechanism for the Northwestern Sahara Aquifer System* (NWSAS Agreement 2002). This second North African arrangement creates a regime whose mandate is to 'coordinate, promote and facilitate the rational management of the NWSAS water resources'. As part of its duties, the Consultative Mechanism must: a) manage a hydrogeologic database and simulation model; b) develop and oversee a reference observation network; c) process, analyse and validate data relating to knowledge of the NWSAS; d) develop databases on socio-economic activities in the region related to water uses; e) develop and publish indicators on the NWSAS and its uses; f) promote and facilitate joint or coordinated studies and research; g) formulate and implement training programmes; h) regularly update the NWSAS model; and i) develop proposals for the continued evolution of the consultation mechanism.

Informal arrangements

Formal agreements are not the only evidence of trends and priorities in the development of customary international law.[1] State conduct in the form of informal arrangements can also serve as an indication of emerging state practice.

One of the more fascinating arrangements is the 2014 *Memorandum of Understanding for the Establishment of a Consultation Mechanism for the Integrated Management of the Water Resources of the Iullemeden, Taoudeni/Tanezrouft Aquifer System (ITAS)* entered into by Algeria, Benin, Burkina Faso, Mali, Mauritania, Niger and Nigeria (ITAS MoU 2014).[2] As noted by its title, the focus of this arrangement is on the creation of a Consultative Mechanism tasked with promoting cooperation over the management of both the Iullemeden Aquifer System and the Taoudeni/Tanezrouft Aquifer System. More specifically, the Mechanism is responsible *inter alia* for: conducting joint studies; formulating recommendations for harmonizing water-related legislative, institutional and management framework, as well as procedures and policies; and settling disputes between the parties (Article 5). In contrast to the purely consultative status of the Genevese Aquifer Management Commission and the JTC under the Al-Sag/Al-Disi Agreement, or the unclear status of the commission under the Guarani Agreement, the Consultative Mechanism under the ITAS MoU has legal personality and authority to contract, acquire and dispose of property, seek and obtain loans, gifts, and technical assistance and be a party in legal proceedings (Article 6). In addition, and in stark contrast to the Genevese Convention, the ITAS MoU explicitly relies on the well-known international water and environmental law principles of equitable and reasonable utilization, no harm, exchange of data and information, prior notification, protection of the environment, public participation, precautionary approach and polluter and user pays (see Articles 13–14, 18–20 and 22–4). Unlike the Guarani Agreement, which merely referenced similar international principles, the ITAS MoU offers considerable details on how these notions are to be construed and implemented. Finally, akin to the Genevese Convention and in contrast to the Guarani Agreement, the ITAS MoU emphasizes a balance between state sovereignty and state responsibility and avoids any explicit mention of sovereignty in its formulation (ITAS MoU 2014).

Another particularly unique informal arrangement is the 1999 *Memorandum of Understanding between the City of Juárez, Mexico Utilities and the El Paso Water Utilities Public Services Board (PSP) of the City of El Paso, Texas* (Juárez-El Paso MoU 1999). This mechanism is distinctive in that it was entered into by sub-national political entities without the oversight of the respective federal governments. While legally unofficial and unenforceable, the purpose of the Juárez-El Paso MoU is to encourage cooperation over the management and exploitation of the Hueco Bolson Aquifer and the Rio Grande River. It is also designed to facilitate the exchange of data and information, coordinate joint projects, develop compatible plans to secure water supplies and create an Executive Committee tasked with fulfilling the objectives of the MoU. Given its parochial origin and perspective, it is understandable that the Juárez-El Paso MoU makes no direct references to principles of international law or notions of sovereignty.

One other informal arrangement requires mentioning – the 1996 *Memorandum of Agreement Related to Referral of Water Right Applications* (BC-WA MoA) entered into by the Canadian province of British Columbia and the US state of Washington. Like the Juárez-El Paso MoU, the BC-WA MoA was adopted by sub-national political entities without the oversight of the respective federal governments. Moreover, like the Juárez-El Paso MoU, the BC-WA MoA was intended to encourage cooperation over a transboundary aquifer, the Abbotsford–Sumas Aquifer, along with other related fresh water resources. The arrangement, however, is distinctive in that it calls for cross-border prior consultation, comment periods, and exchange of information on water quantity allocations within each party's territory that 'could potentially significantly impact water quantity on the other side of the border' (BC-WA MoA 1996). To the extent that the BC-WA MoA applies to all surface water, ground water and reservoir waters, it facilitates a form of cross-border public participation over decision-making related to these shared water resources.

UNILC Draft Articles and UN ECE Model Provisions

The most profound milestones in the ongoing development of international law applicable to transboundary aquifers are the 19 Draft Articles prepared by the UNILC and the nine provisions under the UN ECE Model Provisions.

UNILC Draft Articles

In late 2008, following six years of intense work, the UNILC submitted this work-product to the UNGA. As a preliminary matter, the UNGA acknowledged the UNILC's efforts, issued a resolution commending the 19 Draft Articles to the attention of its Member States, and encouraged nations to take the Articles into account when entering into bilateral and regional arrangements pertaining to the management of transboundary aquifers (UNGA Resolution 2008).

Modelled largely on the 1997 UN Watercourses Convention, the chief substantive obligations include the well-respected international watercourse rules of equitable and reasonable utilization and no significant harm. In both instances, however, the principles are tailored to the unique qualities that differentiate surface waters from ground water resources. For example, the list of factors for assessing what constitutes equitable and reasonable utilization includes notions relevant to ground water resources, such as 'the natural characteristics of the aquifer or aquifer system', 'the contribution to the formation and recharge of the aquifer or aquifer system' and 'the role of the aquifer or aquifer system in the related ecosystem' (see Articles 5(1)(c), (d) and (i), UNGA Resolution 2008). Likewise, the no significant harm rule obligates aquifer states not to cause significant harm through 'activities other than utilization of a transboundary aquifer... that have, or are likely to have, an impact upon that transboundary aquifer' (see Article 6, UNGA Resolution 2008). This latter modification specifically relates to the distinct likelihood that non-aquifer utilization activities undertaken above or around aquifers and their recharge and discharge zones could have detrimental impacts on those subsurface water bodies. Activities contemplated by this provision include industrial and agricultural operations in the recharge zone, mining activities in the aquifer matrix, and construction, forestry and other activities that might affect the normal recharge or discharge processes (Eckstein 2007).

Other notions found in the Draft Articles include obligations to regularly exchange data and information, provide prior notification of planned activities, safeguard ecosystems, protect recharge and discharge zones, prevent pollution and monitor the aquifer (see Articles 8, 10, 11, 12, 13 and 15, UNGA Resolution 2008). Since their submission in 2008, the Draft Articles have appeared on the UNGA's agenda in 2011 and 2013. Each time, they were commended to the attention of UN Member States and tabled for a subsequent meeting. Interestingly, in the 2013, the UNGA commended the articles 'as *guidance* for bilateral or regional agreements and arrangements for the proper management of transboundary aquifers' (Report of the Sixth Committee 2013). While hardly a resounding endorsement, the 'guidance' characterization suggests an elevated status for the Draft Articles (Eckstein and Sindico 2014). The Draft Articles are slated to appear again on the UNGA's agenda in late 2016.

UN ECE Model Provisions

The Model Provisions on Transboundary Groundwaters (Model Provisions), which were finalized in 2012 under the auspices of the UNECE, were drafted to provide guidance to states party to the UNECE Water Convention on the Convention's relevance to ground water resources. While the Model Provisions were designed to be non-obligatory, the nine provisions were specifically aligned to the Water Convention and crafted to take advantage of the binding regime established under that instrument. As a result, the Model Provisions do not necessarily manifest new principles pertaining to transboundary ground waters. Rather, they provide interpretative guidelines for and facilitate the implementation of the UNECE Water Convention with regard to such water bodies (Tanzi *et al.* 2015).

In reviewing the Model Provisions, it is clear that the authors drew some of their inspiration from the Draft Articles. Like the Draft Articles, the Model Provisions give considerable credence to the substantive obligations of equitable and reasonable utilization and no significant harm (see Provision 1). They also endorse the procedural obligations to regularly exchange data and information, monitor, prevent pollution and provide prior notification of planned activities (see Provisions 3, 5, 6 and 8). The Model Provisions, however, also expand on the Draft Articles. For example, whereas the Draft Articles merely recommended the creation of joint institutional mechanisms to carry out the various objectives and obligations (see Articles 7 and 14), the Model Provisions mandate the creation of such bodies (see Provision 9). Moreover, the Model Provisions take the somewhat progressive steps of requiring that transboundary ground waters be used in a sustainable manner (see Provision 2) and mandating that transboundary ground and surface waters shall be managed in an integrated fashion (see Provision 4).

The status of international law for transboundary aquifers

The international law for managing and allocating transboundary ground water resources is still in a nascent state. There is yet no global instrument or series of customary norms that encapsulate the rules governing state conduct in this realm. Nevertheless, there is growing international interest in the subject matter, as well as an increasing number of formal and informal arrangements between nations over shared ground water resources. Taken as a whole, trends and priorities can be discerned that could yet result in customary norms of international law. While the extent of state practices relating to the management of transboundary aquifers is still rather limited, a review of the arrangements discussed above hints at the emergence of a number of norms.

Regular exchange of data and information

Possibly the most palpable and consistent conduct emerging from state practice is a procedural obligation to regularly exchange data and information over transboundary aquifers. Appearing in all but one of the arrangements discussed in this study, the duty is fundamental to the sound management and protection of transboundary aquifers. Absent such sharing of information, aquifer states are faced with the consequences of the 'blank map' syndrome whereby researchers on one or the other side of the border are able to characterize and describe only the portion of the aquifer located within their side (Sanchez *et al.* 2016). As a result, states are all too often unable to fully project and mitigate any deleterious cross-border consequences that might result from the utilization of a particular transboundary aquifer (Eckstein 2007).

In order to fulfil this duty, both logic and emerging state practice suggest that aquifer states should share on a continuing basis all available data and information on a transboundary aquifer. The precise type of material that must be shared, however, is not always spelled out in the various agreements. For example, the Al-Sag/Al-Disi Agreement simply refers to, 'The collection and exchange of information, statements and studies and their analysis' (Al-Sag/Al-Disi Agreement 2015), while the Model Provisions reference 'the exchange of information and available data on the condition of transboundary groundwaters' (Model Provisions 2014).

Nevertheless, it is obvious that the type of material that should be exchanged would have to pertain to the character, use and functioning of the aquifer. Building on this reasoning, Article 8 of the Draft Articles provides that such data and information should include material of a 'geological, hydrogeological, hydrological, meteorological and ecological nature and related to the hydrochemistry of the aquifers or aquifer systems, as well as related forecasts' (UNGA Resolution 2008).

Using more aquifer-specific and descriptive language, the BC-WA MoA provides that the parties shall 'cooperate in sharing relevant water quantity information necessary to provide management of those water resources' and, subject to any domestic legal restricting disclosure, 'commit to freely sharing and exchanging information on' water licences and permits, as well as applications for new and modification of existing licences/permits, and regional water availability and development studies (BC-WA MoA). Using equally focused language, the Juárez-El Paso MoU requires:

a) Sharing historical and current groundwater pumpage, sources of water, and water quality data.
b) Sharing technical support and information.
c) Sharing knowledge and experience regarding funding including, grants and/or loans, and determining the means by which to obtain such funds....
f) Sharing information and analysing issues related to population growth and the economy of the region to include in the regional planning processes, as well as focus on long-term needs (funding for new water resources) on both sides of the border.

Juárez-El Paso MoU 1999

Monitoring and generation of supplemental data and information

A corollary procedural obligation to the duty to regularly exchange data and information is the duty to generate supplemental data and information on an ongoing basis through monitoring and related activities. The obligation, which appears in a considerable majority of the arrangements considered in this study, acknowledges the need to maintain vigilance in managing a transboundary aquifer and, therefore, is also indispensable to fulfilling the duty to exchange data and information.

The Genevese Convention, for example, is conceptualized largely on the notion of monitoring and further developing information about the aquifer. For example, it explicitly references monitoring in its title. Moreover, chapter 4 of the agreement addresses 'Quantitative and Qualitative Monitoring of the Resource' and mandates periodic assessment of water quality and quantity as well as the exchange of that new information, while Article 17 requires the parties to 'maintain a monitoring network... intended for the issuance of warnings in the case of accidental pollution likely to affect the water quality of the aquifer'. Moreover, Article 10 mandates that 'data from the extractions shall be performed by each user and reported at the end of the year to all users', while Article 16 provides that water pollution analyses 'shall be made at regular intervals' (Genevese Convention 2008).

In a similar vein, Agreement No. 2 under the NSAS Agreements utilizes 'Monitoring and Data Sharing' in its title and explicitly focuses on developing and exchanging new data and information:

Hence, it is herewith agreed between the four countries... to monitor and share among them the following information:

- Yearly extraction in every extraction site, specifying geographical location and number of producing wells and springs in every site.
- Representative Electrical Conductivity measurements (EC), taken once a year in each extraction site, followed by a complete chemical analysis if drastic changes in salinity is [sic] observed.
- Water level measurements taken twice a year in the locations shown in the attached maps and tables. The proposed monitoring network is subject to changes upon the feedback of the National Coordinators of the concerned countries.

Using more general language, Provision 3 of the Model Provisions discusses monitoring in terms of 'quantity and quality of transboundary groundwaters'. The Provision, however, adds the critical requirements that the parties must harmonize their monitoring standards and methodologies, agree on assessment criteria and parameters to be regularly monitored and, where appropriate, link the monitoring of ground and surface waters (Model Provisions 2014).

The obligation to monitor and continuously generate additional data accords with the comparable duties imposed on riparians of transboundary surface waters. In his separate opinion in the *Case Concerning the Gabčíkovo-Nagymaros Project* before the International Court of Justice (ICJ), Judge Christopher Weeramantry argued for emergence of a principle of continuing environmental impact assessment. In that opinion, Judge Weeramantry opined that '[a]s long as a project of some magnitude is in operation, [an environmental impact assessment] must continue, for every such project can have unexpected consequences; and considerations of prudence would point to the need for continuous monitoring' (Gabčíkovo Case 1997, p. 111). More recently, in the *Case Concerning the Pulp Mills on the River Uruguay*, the ICJ asserted that 'once operations have started and, where necessary, throughout the life of the project, continuous monitoring of its effects on the environment shall be undertaken' (Pulp Mills Case 2010, p. 205). The Court again recognized that obligation in the combined decision on the cases concerning *Certain Activities Carried Out By Nicaragua in the Border Area* and *Construction of a Road in Costa Rica Along the San Juan River* (San Juan River Cases 2015, p. 60). While all three cases recognized this recurring obligation in the context of a transboundary watercourse, the logic utilized by the ICJ is equally and undeniably pertinent to all transboundary ground water resources.

Prior notification of planned activities

Another procedural obligation found in a majority of the above-noted instruments is the duty to provide prior notification of planned activities. Where a planned project has the potential to adversely affect either the territory of another aquifer state or the transboundary aquifer itself, the acting state is obligated to notify other aquifer states of its plans. The purpose of such obligations is to allow potentially affected states to evaluate the possible consequences and to seek an understanding or compromise with the acting state (Eckstein 2007).

While the precise procedures required under this concept vary among the instruments, the basic notions of prior notification are well accepted in international water law. Under Paragraph 6 of Minute 242, Mexico and the US agreed to 'consult with each other prior to undertaking any new development of either the surface or the groundwater resources, or undertaking substantial modifications of present developments, in its own territory in the border area that might adversely affect the other country' (Minute 242 1973). Similarly, Article 15 of the Draft Articles would require aquifer states to provide 'timely' notification 'accompanied by available technical data and information... to enable the notified State to evaluate the possible effects of the planned activities' (UNGA Resolution 2008). Indirectly emphasizing consultation in good faith, Article

11 of the Guarani Agreement imposes the additional obligation that the party proposing the actions that may have a transboundary impact must delay implementation of those measures for at least six months while negotiating with the potentially affected state (Guarani Agreement 2010). Moreover, Provision 8 of the Model Provisions mandates an environmental impact assessment for all planned activities that are likely to have a significant effect on transboundary ground water resources, and requires that the assessment be transmitted to all potentially impacted states upon request (Model Provisions 2014).

In contrast to the above instruments, the Iullemeden MoU proffers much more rigorous requirements and processes for notification. While Article 27 provides the basic prior notification obligation for 'activities, policies and strategies, plans, programs and projects proposed in the area, which may pose a risk to' the water resources of the transboundary aquifer or otherwise cause transboundary adverse impacts, Article 31 calls for 'technical data and information, including the results of any evaluation of the environmental and social impact' to accompany the notification and requires the notifying state to 'refrain from implementing or permitting the implementation of the planned measures' during a six-month review process. Article 32 authorizes the notifying state to proceed with the planned activity in the absence of a response to the notification within six months. Article 33 requires that states engaged in consultations and negotiations over planned measures must do so 'according to the principle of good faith, taking into account the legitimate interests of any other signatory State'. Article 34 permits potentially affected states to request a state engaging in planned measures to comply with the notification obligations and requires disagreements on such obligations to be pursued through consultation and negotiation. Finally, Article 34 allows planned measures to proceed without notifications in emergency situations (Iullemeden MoU 2014).

Creation of institutional mechanisms to facilitate or implement the arrangement

One of the most interesting trends perceived from the various arrangements is the creation of joint institutional mechanisms to carry out the objectives of the various regimes. This is particularly noteworthy because of the 276 rivers and lakes found on Earth, less than 40 per cent (105) employ some type of water management institution (Drieschova and Eckstein 2014). In contrast, of the handful of arrangements that have been implemented or proposed for a transboundary aquifer, all but two implement or propose some type of joint institutional mechanism. Moreover, the Draft Articles in Articles 7 (General obligation to cooperate) and Article 14 (Management), as well as the Model Provisions in Provision 9, clearly contemplate the creation of such mechanisms. While the structures and levels of authority granted these entities vary across the regimes, it remains clear that most aquifer nations that have entered into a cross-border arrangement recognize both the value of and the need for institutional and other cooperative mechanisms to facilitate and realize the sound and sustainable management of their shared ground water resources.

For example, the Model Provisions, Guarani Agreement, Al-Sag/Al-Disi Agreement and Juárez-El Paso MoU all call for the creation of an institutional mechanism to carry out the purposes of the respective agreements. The Model Provisions and Guarani Agreement provide the simplest iteration of this obligation and offer no additional instructions about the structure and operation of such an entity (Provision 9, Model Provisions 2014; Article 15, Guarani Agreement 2010). The Guarani Agreement, however, does provide in Article 15 that the mechanism would be established in accordance with Article VI of the 1969 Treaty of the Plata River Basin, and stipulates in Article 17 that it will be tasked with helping to resolve disputes by evaluation situations and formulating recommendations (Guarani Agreement 2010). Article 3 of the Al-Sag/Al-Disi Agreement offers slightly more detail and notes that the institutional mechanism is composed of representatives of the national water resources agencies in the two Member States, and that its mandate includes: 'The supervision and observation' of ground water levels, quality and extraction; 'The collection and exchange of information, statements and studies and their analysis' related to the aquifer; and the submission of such information and analyses to the two governments (Al-Sag/Al-Disi Agreement 2015). Likewise, the Executive Committee of the Juárez-El Paso MoU is tasked in Paragraph 2 with data sharing and project coordination obligations, and is also assigned to facilitate a number of locally specific activities, including completion of a feasibility study that was begun prior to implementation of the MoU (Juárez-El Paso MoU 1999).

In a similar vein, the Genevese Convention creates a commission whose purpose is to implement the agreement. The Genevese Aquifer Management Commission, however, has more extensive authorities than under the above-noted arrangements. Its mandate, for example, as described in Article 2, includes proposing an annual aquifer utilization programme, providing technical opinions on construction of new ground water extraction operations and modification of existing equipment, and performing audits of investment and operational costs related to the recharge installation. It is also responsible for overseeing waterworks and equipment construction (Article 5), recording water extractions (Article 6), collecting water level and quality data (Article 10) and establishing water quality analysis criteria (Article 16) (Genevese Convention 2008).

In contrast to the above five arrangements, where creation of a joint institution was an important albeit a secondary component to the agreement, the NWSAS Agreement, the Iullemeden MoU and the Constitution of the Joint Authority, by their very titles and purposes, were formulated and implemented specifically to create a joint cooperative mechanism. The NWSAS Agreement, for example, created a 'Consultative Mechanism' to 'coordinate, promote and facilitate the rational management of the NWSAS water resources' (Para. I, NWSAS Agreement 2002), while the Iullemeden MoU created an identically named mechanism 'to promote and foster cooperation between the Signatory States... based on solidarity and reciprocity for a sustainable, equitable, coordinated and collaborative use of the ITAS water resources' (Article 3, Iullemeden MoU 2014). While the NSA Constitution does not include a purpose statement, the 'tasks' outlined in Article 3 of the agreement are representative of the functions and responsibilities assigned to the mechanisms under each of these three agreements: collect and develop all data and information relevant to the shared aquifer; promote and facilitate additional studies; formulate proposals for the sustainable management of the aquifer; and undertake and facilitate appropriate training programmes and other mechanisms for the disseminating of information (NSA Constitution 1992).

Substantive obligations

While the above obligations may properly be described as procedural in nature, the various arrangements discussed here also endeavour to create a number of substantive responsibilities. The most prolific of these is the Iullemeden MoU, which commits Signatory States to such principles as: equitable and reasonable utilization, non-damaging use, sustainable development, ecosystem protection, precaution and polluter pays (Iullemeden MoU 2014).

A closer review of the arrangements, however, suggests no conclusive trends of cross-cutting substantive norms emerging from such practices of states. Varying references to adverse transboundary effects, impacts and harm are found in two formal instruments (Paragraph 6, Minute 242; Articles 6 and 7, Guarani Agreement), one informal mechanism (Article 20, Iullemeden

MoU) and both the Draft Articles (Article 6) and Model Provisions (Provision 1). While Minute 242 simply refers to possible transboundary 'adverse effect' in the context of prior notification, the others impose a due diligence obligation to prevent, control and reduce such impacts. Of these five arrangements, all except Minute 242 also refer to the cornerstone international water law principle of equitable and reasonable utilization (Article 13, Iullemeden MoU; Article 4, Guarani Agreement; Article 4, Draft Articles; and Provision 1, Model Provisions). However, only the Iullemeden MoU and the Draft Articles offer factors to be taken into account when assessing what uses may be deemed equitable and reasonable.

Aside from these few similarities, the lack of consistent appearance of additional principles in the various instruments and mechanisms reviewed in this study indicates that no other norms or obligations are trending towards customary status. Notwithstanding, as the practice of states pertaining to transboundary aquifers continues to evolve and new agreements are forged, this conclusion should be periodically re-evaluated.

Considerations for further development of the law

While surface and ground water resources are both integral components of the hydrologic cycle and share numerous similarities, ground water possesses a number of unique characteristics that must be considered carefully when contemplating regulatory tools for managing the resource. For example, the relatively slower flow rates of water through subsurface strata, as compared to water flow in rivers, can impair an aquifer's natural filtration abilities and, thereby, their capacities to reclaim and cleanse themselves of pollutants. As a result, ground water can be more vulnerable than surface water to agricultural, industrial and municipal pollution, as well as other sources of contamination (Eckstein 2007). Moreover, because of the geographic extent of most aquifers and the challenges associated with monitoring underground formations, the artificial reclamation of a polluted aquifer can be prohibitively complex and expensive. As a result, once an aquifer is contaminated, it may be rendered unusable for years, decades or longer (ibid.). Among other issues, this raises the question of whether the threshold for actionable harm should be different for transboundary ground water resources as compared to cross-border surface water bodies.

Furthermore, the 'functioning' of aquifers – which refers to how particular aquifers work or behave as aquifers – also must be taken into account when formulating appropriate regulatory mechanisms for the sound management of transboundary ground water resources. Aquifer functioning encompasses how subsurface strata can store and transport water, dilute wastes and other contaminants, provide a habitat for aquatic biota, serve as a source of fresh water and nutrients to aquifer-dependent ecosystems, and even provide geothermal heat. Each of these characteristics is dependent on the particular aquifer's structure, hydrostatic pressure, hydraulic conductivity, interaction with other geophysical phenomenon and mineralogical, biological and chemical attributes. Moreover, all of these traits may be interdependent to the extent that an aquifer's sustained operation as a dynamic hydrogeologic system depends on the continuation of a particular function or series of functions (Heath 2004). If any of these natural characteristics were to be impaired or destroyed, it could detrimentally affect the viability and integrity of the aquifer as a whole, as well as communities and ecosystems dependent on that aquifer. Accordingly, in order to manage a transboundary aquifer in ways that maximize both its utility and sustainability, regulatory mechanisms must take into account the functioning as well as the unique vulnerabilities and characteristics of each shared subsurface water body.

In addition, when contemplating appropriate regulatory mechanisms for the sound management of transboundary ground water resources, the recharge and discharge processes of each aquifer also require special attention. Recharge and discharge zones regulate the flow and quality of water moving into and out of aquifers. Hence, these processes, as well as the geographical area in which they operate, must be properly maintained and protected. In the case of recharge zones, this consists of ensuring both the quantity and quality of water flowing through the recharge zone and entering the aquifer. Thus, recharge zone protection might include restrictions on industrial and municipal developments in the recharge area, as well as constraints on agricultural activities that might contaminate the recharge area and, thereby, the aquifer. Similarly, discharge zones protection could include restrictions on construction and other activities that might inhibit the discharge process, water flow within the aquifer, the location of the water table, or the aquifer's natural cleansing abilities. Restrictions for both zones might also include limitations on mining activities that remove or modify the strata within the recharge or discharge area.

The above concerns do not reflect all of the characteristics, issues and gaps in knowledge that must be addressed. Other concerns and topics that should be considered include: the relevance of the principles of no significant harm and equitable and reasonable use to transboundary aquifers; if relevant, whether the no significant harm standard is subordinate or superior to that of equitable and reasonable use in the context of transboundary aquifers; mechanisms to harmonize metadata and methodologies produced by aquifer riparians pertaining to a shared aquifer; whether the exploitation of non-recharging aquifers, as compared to recharging aquifers, require a distinct legal and governance regime.

Conclusion

Transboundary ground water resources today play a critical role in providing fresh water for people, industries, nations and the environment worldwide. For billions of people, they serve as the bulwark against the challenges posed by expanding demands for fresh water and the declining supplies resulting from overexploitation and climatic changes. As a result, transboundary aquifers are now receiving greater international attention by overlying nations, non-governmental advocacy groups and UN entities. Moreover, many states around the world are beginning to pursue various strategies for their exploitation and management.

While the level of attention that these aquifers are receiving still pales in comparison with that paid to rivers and lakes, it is reasonable to expect that nations will continue to explore their transboundary aquifers. The value of these resources is undeniable, and growing water scarcity is driving many nations to investigate all new possibilities. As a result, it is also reasonable to expect that more states will engage their cross-border neighbours in an effort to collaborate and coordinate their activities. Moreover, as cooperation over transboundary aquifers expands and the number of formal and informal arrangements grows, as is certain to happen, trends and priorities will become more evident and will lead to the development of more definite customary norms for the management of transboundary ground water resources.

Notes

1 Customary international law refers to international law that is based on the accepted practices of nations rather than on codified rules. It emerges from the broad and consistent conduct of states that is undertaken by a belief that such behaviour is both legally appropriate and mandated (Brownlie 1998).
2 The ITAS MoU is considered here as an unofficial arrangement. Conceptualized as a Memorandum of Understanding, the ITAS MoU technically cannot be deemed a binding instrument. Yet, in its final provisions, it references the 'binding' nature of decision taken by the Consultative Mechanism (Article 47) as well as the need to ratify the MoU for it to enter into force (Article 53). Moreover, the details and

language used in the MoU suggest an intention by the parties to comply with the terms of the resulting agreement once it comes into force. Notwithstanding, as of this writing, only Benin, Mali, Niger, and Nigeria have signed the MoU and only Nigeria has processed the MoU's ratification within its domestic system (Nigeria Politics Online, 2014).

References

Articles, books and reports

Almássy, E. and Busás, Zs. (1999) *UN/ECE Task Force on Monitoring & Assessment, Guidelines on Transboundary Ground Water Monitoring*, Volume 1: *Inventory of Transboundary Ground Waters*. UN Economic Commission for Europe.

Brownlie, I. (1998) *Principles of Public International Law*, 5th edn. New York: Oxford University Press.

Burchi, S. (1999) National regulation for groundwater: options, issues and best practices. In S.M.A. Salman (ed.), *Groundwater: Legal and Policy Perspectives, Proceedings of a World Bank Seminar*. Washington, DC: World Bank, pp. 55–67.

Drieschova, A. and Eckstein, G. (2014) Cooperative transboundary mechanism. In J.C. Sanchez and J. Roberts, *Transboundary Water Governance: Adaptation to Climate Change*. Geneva: IUCN, pp. 51–79.

Eckstein, G. (2003) A hydrogeological approach to transboundary ground water resources and international law. *American University International Law Review* 19(2), pp. 201–58.

Eckstein, G. (2007) Commentary on the UN International Law Commission's Draft Articles on the Law of Transboundary Aquifers. *Colorado Journal of International Environmental Law and Policy* 18(3), pp. 537–610.

Eckstein, G. (2012) Rethinking transboundary ground water resources management: a local approach along the Mexico–US border. *Georgetown International Environmental Law Review* 25(1), pp. 95–128.

Eckstein, G. and Hardberger, A. (2008) State practice in the management and allocation of transboundary groundwater resources in North America. *Yearbook of International Environmental Law* 18, pp. 96–125.

Eckstein, G. and Sindico, F. (2014) The Law of Transboundary Aquifers: many ways of going forward, but only one way of standing still. *Review of European, Comparative and International Environmental Law* 23(1), pp. 32–42.

Hayton, R. and Utton, A. (1989) Transboundary groundwaters: the Bellagio Draft Treaty. *Natural Resources Journal* 29, pp. 663–722.

Heath, R.C. (2004) *Basic Ground-Water Hydrology, Water Supply Paper 2220*. US Geological Survey.

IGRAC (2015) *Transboundary Aquifers of the World: Special Edition for the 7th World Water Forum 2015*. Available from: http://www.un-igrac.org/download/file/fid/179 [accessed 15 December 2015].

International Court of Justice (1997) *Case Concerning the Gabčíkovo-Nagymaros Project*, Separate Opinion of Judge Weeramantry, Judgment of 25 September. Available from: http://www.icj-cij.org/docket/files/92/7383.pdf [accessed 22 December 2015].

International Court of Justice (2010) *Case Concerning the Pulp Mills on the River Uruguay*, Judgment of 20 April. Available from: http://www.icj-cij.org/docket/files/135/15877.pdf [accessed 15 December 2015].

International Court of Justice (2015) *Cases Concerning Certain Activities Carried Out By Nicaragua in the Border Area* and *Construction of a Road in Costa Rica Along the San Juan River, Judgment of 16 December 2015*. Available from: http://www.icj-cij.org/docket/files/152/18848.pdf [accessed 25 March 2016].

International Law Association (2004) *Berlin Conference on Water Resources Law*. Available from: http://internationalwaterlaw.org/documents/intldocs/ILA_Berlin_Rules-2004.pdf [accessed 15 December 2015].

McCaffrey, S. (2007) *The Law of International Watercourses*, 2nd edn. New York: Oxford University Press.

McCaffrey, S.C. (2011) The International Law Commission's flawed Draft Articles on the Law of Transboundary Aquifers: the way forward. *Water International* 36, pp. 566–72.

Nigeria Politics Online (2014) FG seals deal with 6 countries on water resources. *Nigeria Politics Online*, 29 October 29. Available from: http://nigeriapoliticsonline.com/fg-seals-deal-with-6-countries-on-water-resources/ [accessed 15 December 2015].

Puri, S. and Aureli, A. (2009) *Atlas of Transboundary Aquifers*. Paris: UN Educational, Scientific and Cultural Organization.

Regional Strategic Action Plan for the Nubian Aquifer System (2013) 18 September. Available from: http://internationalwaterlaw.org/documents/regionaldocs/Regional_Strategic_Action_Plan_for_the_Nubian_Aquifer.pdf [accessed 22 December 2015].

Report of the Sixth Committee (2013) *The Law of Transboundary Aquifers*. UN Doc. A/68/470, 19 November. Available from: http://www.un.org/ga/search/view_doc.asp?symbol=A/68/470 [accessed 15 December 2015].

Sanchez, R., Lopez, V. and Eckstein, G. (2016) Identifying and characterizing transboundary aquifers along the Mexico–US border: an initial assessment. *Journal of Hydrology* 535, pp. 101–19.

Sindico, F. and Hawkins, S. (2015) The Guarani Aquifer Agreement and Transboundary Aquifer Law in the SADC: comparing apples and oranges?. *Review of European Community and International Environmental Law* 24(3), pp. 318–29.

Stephan, R.M. (ed.) (2006) *Transboundary Aquifers: Managing a Vital Resource – The UNILC Draft Articles on the Law of Transboundary Aquifers*. Paris: UN Educational, Scientific and Cultural Organization.

Tanzi, A., *et al.* (eds) (2015) *The UNECE Convention on the Protection and Use of Transboundary Watercourses and International Lakes: Its Contribution to International Water Cooperation*. Leiden: Brill/Nijhoff.

UN Economic Commission for Europe (2014) Model Provisions on Transboundary Groundwaters, ECE/MP.WAT/40. Available from: https://www.unece.org/fileadmin/DAM/env/water/publications/WAT_model_provisions/ece_mp.wat_40_eng.pdf [accessed 22 January 2016].

UN Educational, Scientific and Cultural Organization and World Water Assessment Programme (2003) *Water for People, Water for Life, The United Nations World Water Development Report*. UN World Water Assessment Programme.

UN Environmental Programme (2002) *Atlas of International Freshwater Agreements*. Nairobi: UNEP.

UN General Assembly (2008) *Resolution on the Law of Transboundary Aquifers*, A/RES/63/124. Available from: http://internationalwaterlaw.org/documents/intldocs/UNGA_Resolution_on_Law_of_Transboundary_Aquifers.pdf [accessed 22 December 2015].

UNESCOPRESS (2008) UNESCO publishes first world map of underground transboundary aquifers. Press Release No. 2008-108, 22 October.

Villar, P.C. and Wagner, C.R. (2013) The Agreement on the Guarani Aquifer: cooperation without conflict. *Global Water Forum*. Available from: http://www.globalwaterforum.org/2013/09/02/the-agreement-on-the-guarani-aquifer-cooperation-without-conflict/ [accessed 22 December 2015].

Watkins, J. (2006) Libya's thirst for 'fossil water'. *BBC News*, 18 March. Available from: http://news.bbc.co.uk/2/hi/science/nature/4814988.stm [accessed 22 December 2015].

Wolf, Aaron T. (2002) *Atlas of International Freshwater Agreements*. Geneva: UN Environmental Programme.

World Bank (2009) *Assessment of Restrictions on Palestinian Water Sector Development*, Report No. 47657-GZ.

Wouters, P. and Moynihan, R. (2013) Benefit sharing in the UN Watercourses Convention and under international water law. In F.R. Loures and A. Rieu-Clarke (eds), *The UN Watercourses Convention: Strengthening International Law for Transboundary Water Management*. London: Routledge.

Yamada, C. (2011) Codification of the Law of Transboundary Aquifers (Groundwaters) by the United Nations. *Water International* 36(5), pp. 557–65.

Treaties, agreements and other international arrangements

Agreement between the Governments of Great Britain and France with regard to the Somali Coast. Done on February 1888. Reprinted in Oakes, A. H. and Maycock, W. (1897). *British and Foreign State Papers 1890–1891*. p. 672.

Agreement between the Government of the Hashemite Kingdom of Jordan and the Government of the Kingdom of Saudi Arabia for the Management and Utilization of the Ground Waters in the Al-Sag/Al-Disi Layer. Done in Riyadh, Saudi Arabia, 30 April 2015. Available from: http://internationalwater-law.org/documents/regionaldocs/Disi_Aquifer_Agreement-English2015.pdf [accessed 15 December 2015].

Constitution of the Joint Authority for the Study and Development of the Nubian Sandstone Aquifer Waters 1992. Available from: http://internationalwaterlaw.org/documents/regionaldocs/Constitution_of_the_Joint_Authority-Nubian_Sandstone_Aquifer.pdf [accessed 23 December 2015].

Convention on the Protection, Utilisation, Recharge and Monitoring of the Franco-Swiss Genevese Aquifer between the Community of the 'Annemassienne' region, the Community of the 'Genevese' Rural Districts, and the Rural District of Viry, on one part, The Republic and Canton of Geneva, on the other. Done in Geneva, 18 December 2007; in force on 1 January 2008. Available from: http://internationalwaterlaw.org/documents/regionaldocs/2008Franko-Swiss-Aquifer-English.pdf [accessed 22 December 2015].

Establishment of a Consultation Mechanism for the Northwestern Sahara Aquifer System (SASS). Done in Rome, 19-20 December; endorsed 6 January 2003 (Algeria), 15 February 2003 (Tunisia), 23 February 2003 (Libya). Available from: http://www.fao.org/docrep/008/y5739e/y5739e05.htm#bm05.2.1 [Accessed 15 December 2015].

Guarani Aquifer Agreement. Done in San Juan, Argentina, 2 August 2010. Available from: http://internationalwaterlaw.org/documents/regionaldocs/Guarani_Aquifer_Agreement-English.pdf [accessed 20 December 2015].

Memorandum of Agreement Related to Referral of Water Right Applications between the State of Washington as Represented by the Department of Ecology and the Province of British Columbia as Represented by the Minister of Environment, Lands and Parks. Done in 10 October 1996. Available from: http://internationalwaterlaw.org/documents/regionaldocs/Local-GW-Agreements/1996-BC-WA-Water-Right-Referral-Agreement.pdf [accessed 15 December 2015].

Memorandum of Understanding between City of Juárez, Mexico Utilities and the El Paso Water Utilities Public Services Board of the City of El Paso, Texas. Done on 6 December 1999. Available from: http://internationalwaterlaw.org/documents/regionaldocs/Local-GW-Agreements/El_Paso-Juarez_MoU.pdf [accessed 15 December 2015].

Memorandum of Understanding for the Establishment of a Consultation Mechanism for the Integrated Management of the Water Resources of the Iullemeden, Taoudeni/Tanezrouft Aquifer Systems (ITAS) (Algeria, Benin, Burkina Faso, Mali, Mauritania, Niger, Nigeria). Done in Abuja, Nigeria, 28 March 2014. Available from: http://internationalwaterlaw.org/documents/regionaldocs/Iullemeden_MOU-2014.pdf [accessed 22 December 2015].

Minute 242: Permanent and Definite Solution to the International Problem of the Salinity of the Colorado River. International Boundary and Water Commission. Done on 30 August 1974. Available from: http://www.ibwc.gov/Files/Minutes/Min242.pdf [accessed 15 December 2015].

Programme for the Development of a Regional Strategy for the Utilisation of the Nubian Sandstone Aquifer System (NSAS). Done in Tripoli, 5 October 2000. Available from: http://www.fao.org/docrep/008/y5739e/y5739e05.htm [accessed 22 December 2015].

Treaty of Limits between Portugal and Spain. Signed at Lisbon, 29 September 1864.

Treaty between the United States and Mexico on the Utilization of Waters of the Colorado and Tijuana Rivers and of the Rio Grande. Signed in Washington, DC, 14 November 1944.

UN Convention on the Law of Non-navigational Uses of International Watercourses, 1997. GA Res. 51/229, UN GAOR, 51st Sess., UN Doc. A/RES/51/229 (1997).

UNECE Convention on the Protection and Use of Transboundary Watercourses and International Lakes. Done at Helsinki, 17 March 1992. Available from: http://www.unece.org/env/water/pdf/watercon.pdf [accessed 22 December 2015].

17

SUBSTANTIVE RULES OF INTERNATIONAL WATER LAW

Owen McIntyre

Introduction

As the global freshwater crisis has come to be recognised as 'the new environmental crisis of the 21st century' (Brown Weiss 2013, p. 1), one might reasonably expect that competition between co-basin states for the right to use shared transboundary water resources will intensify significantly. Thus, the role of international water law in promoting effective hydro-diplomacy and equitable transboundary water resources management will become ever more critical. Though international water law is a relative newcomer as a discrete body of rules and principles, states have long engaged in formal cooperative arrangements over the use of shared international rivers and lakes. This is illustrated by the fact that an agreement on the apportionment of the waters of the Euphrates between the ancient Mesopotamian city-states of Umma and Lagash provides us with the oldest known example of an international treaty (Teclaff 1967, pp. 21–5), while the Rhine Commission provides the first example of a permanently consti-tuted international (inter-governmental) organisation (Teclaff 1991, pp. 48–51). However, in the modern practice of international law relating to shared water resources, states had until rela-tively recently been principally concerned with rights of navigation upon international rivers and lakes (Rieu-Clarke 2007, p. 392: PCIJ 1929, pp. 5–46), with the seminal Helsinki Rules, the first codification of international rules applying to the utilisation and protection of shared water resources, only being adopted by the International Law Association in 1966 (ILA 1966, p. 484). The first global conventional instrument on the use of shared international water resources, the UN Watercourses Convention (UNWC), was not adopted by the UN General Assembly until 1997 and only obtained the 35 ratifications required to enter into force in August 2014 (UN 1997). Similarly, measures to amend the 1992 United Nations Economic Commission for Europe (UNECE) Water Convention (UNECE 1992), so as to open it up to global member-ship, only entered into effect in 2013 (UNECE 2004; Fitzmaurice and Merkouris 2015, pp. 101–15). Despite the immense importance of groundwater resources for meeting human needs around the world, international law relating to the utilisation and protection of trans-boundary aquifers is an even more recent and less developed field. It was only in 2008 that the International Law Commission (ILC) adopted its non-binding Draft Articles on Transboundary Aquifers (ILC 2008; Eckstein and Sindico 2014, pp. 32–42; McIntyre 2011a, pp. 1–18) and in 2014 that the parties to the 1992 UNECE Water Convention adopted their Model Provisions on Transboundary Groundwaters, intended to support improved cooperation over shared groundwater resources.

Though relatively recently elaborated, international water law is already quite well settled around three key rules: the principle of equitable and reasonable utilisation, as enshrined in Articles 5 and 6 of the UNWC and widely regarded as the cardinal rule in the field (ILC 1994, p. 222); the duty to prevent significant transboundary harm, as enshrined in Article 7 of the UNWC; and the duty to cooperate in the management of shared waters, as enshrined in Article 8 of the UNWC. The first two of these rules are understood as creating substantive obligations, while the latter, the duty to cooperate, is commonly understood as a composite obligation comprising a comprehensive suite of procedural requirements. Generally, the effective implementation of these rules infers a range of related, ancillary normative requirements. Significantly for the effective implementation of the two key substantive rules, these include quite highly developed procedural rules which facilitate inter-state communication and fall within the rubric of the duty to cooperate. These notably include the duty to exchange information relevant to use of the watercourse, the duty to notify co-riparian states of planned projects potentially impacting a shared watercourse and, where necessary, duties to consult and negotiate with such states in a good faith effort to address their concerns (McIntyre 2010, pp. 475–97, 2013, pp. 239–65). However, there are also related substantive rules, principles and standards which further inform the applicable due diligence standards inherent in the three core rules. These notably include duties relating to the prevention, reduction and control of pollution of transboundary waters and concerning the maintenance and conservation of riverine ecosystems, such as those set out in Articles 20–23 of the UNWC. Since the initial identification and codification of prevailing state practice in the 1966 Helsinki Rules, almost all water resources agreements – global and regional framework conventions as well as river basin and boundary waters agreements – have established legal regimes based more or less on this model. Most notably, the 1997 UN Watercourses Convention, the 1992 UNECE Water Convention and the ILC's 2008 Draft Articles on Transboundary Aquifers all adopt broadly similar approaches based on these three basic rules along with a number of related, ancillary substantive and procedural requirements.

Despite this trend towards the convergence of international water law around three basic, yet broad and flexible principles, it is clear that this body of rules is continuously interacting with, and is increasingly being shaped by, other prolific, dynamic and highly pervasive fields of normativity, including international environmental law, international human rights law and international investment law (Boisson de Chazournes 2011, pp. 10 and 14; Maljean-Dubois 2011, pp. 25–54). While the central relevance of international environmental law to transboundary water management has long been self-evident (McIntyre 2007; Brels *et al.* 2008), the social protection values inherent to equitable and reasonable utilisation, including, in particular, the priority accorded under Article 10(2) of the UNWC to safeguarding 'vital human needs' related to shared water resources, have become very closely intertwined with the discourse on the human right to water ongoing in international human rights law (McCaffrey 2005, pp. 100–1). Of course, developments in international water law commonly arise in connection with major investment projects, often involving foreign private or public sector investors, having the potential to impact upon the environment of an international watercourse, upon another state's right to utilise the shared waters in question, or upon local people's access to adequate water resources or services. Thus, tensions may arise with normative frameworks established in the field of international economic law concerning the legal protection of foreign investors (Rieu-Clarke 2015, pp. 27–48) or compliance with the environmental and social safeguard policies of multilateral development banks (MDBs) or other international financial institutions (IFIs) (Salman 2009).

This chapter focuses on the conceptual origins, normative nature and legal implications of the two key substantive rules of general international water law – that is, the principle of equitable and reasonable utilisation and the duty to prevent significant transboundary harm. The general duty to cooperate which, as noted above, is largely procedural in nature, is examined elsewhere in this volume. Also, though substantive rules for the environmental protection of international watercourses and of basin ecosystems are of ever increasing relevance and importance in the practice of international water law, and are routinely set out separately in water resources agreements, these rules essentially elaborate upon the ever more significant environmental implications of the two key substantive rules. Thus, certain environmental aspects of both equitable and reasonable utilisation and the no-harm rule will also be touched upon briefly.

Equitable and reasonable utilisation

The principle of equitable and reasonable utilisation represents a compromise between two extreme and uncompromising positions regarding the right conferred upon states, by virtue of their territorial sovereignty, to utilise shared transboundary water resources found within or passing through their territory. The first position, based on the theory of 'absolute territorial sovereignty' and traditionally favoured by upstream states, supports the argument that a co-basin state may freely utilise waters within its territory without having any regard to the rights of downstream or contiguous states. Having absolute sovereignty over water resources while they are within its territory, a state may utilise or alter the quality of these waters to an unlimited extent, but accordingly has no right to demand continued flow or quality from another co-basin state. This approach is closely associated with the so-called 'Harmon Doctrine', named after the US Attorney-General who first elaborated the principle in the context of a dispute with Mexico over the waters of the Rio Grande (Birnie and Boyle 1992, p. 218; Fitzmaurice and Elias 2004, p. 12). Harmon did not recognise any general legal requirement for the US to safeguard the supply of water to Mexico, stating that the question of whether the US should 'take any action from considerations of comity' was one which 'should be decided as one of policy only, because, in my opinion, the rules, principles and precedents of international law impose no liability or obligation upon the United States' (Bruhacs 1993, p. 43). The second position, based on the theory of 'absolute territorial integrity' and traditionally invoked by downstream states, would confer a right on a co-basin state to demand the continuation of the full flow of waters of natural quality from another (upper) co-basin state, but confers no right to restrict or impair the natural flow of waters from its territory into that of any other (still lower) co-basin state. This approach is the antithesis of the Harmon Doctrine and would effectively grant a right of veto upon a downstream or contiguous state, as its prior consent would be required for any change in the regime of the international watercourse (McIntyre 2007, pp. 17–18). However, on the basis of a detailed analysis of state practice regarding these two extreme contrasting positions, McCaffrey concludes that they were principally invoked as 'tools of advocacy' rather than as legal principles considered likely to assist in the resolution of concrete disputes (McCaffrey 2001, p. 130). For example, he notes that shortly after Attorney-General Harmon made his (in)famous statement of opinion in 1895, the US concluded bilateral treaties with Mexico (in 1906) and Canada (in 1909) which are very much more consistent with principle of equitable and reasonable utilisation (ibid., pp. 93–4 and 111).

Over time, practically all states sharing transboundary basins have come to adopt a third approach, which is based on the theory of 'limited territorial sovereignty', which represents a compromise between the absolute positions outlined above. This approach recognises that both the sovereign utilisation rights of one (upstream) basin state and the right to territorial integrity of another (downstream) basin state are each restricted by a recognition of the equal and correlative rights of the other state (McIntyre 2007, p. 23). This approach is usually articulated in normative terms as the principle of 'equitable and reasonable utilisation', which entitles each co-basin state to an equitable and reasonable use of transboundary waters flowing through its territory. As it involves recognition the 'equality of right' of both upstream and downstream states, or of states causing and suffering pollution (McCaffrey 2001, pp. 329–30), equitable and reasonable utilisation coheres with the notion of the sovereign equality of states, a fundamental principle of public international law which is now authoritatively enshrined in Article 2(1) of the UN Charter. However, equality of right does not entitle each state to an equal share in the waters of a shared basin, but only to an equal right *vis-à-vis* its co-riparian neighbours to an equitable share of the uses and benefits of the watercourse having regard to all relevant factors. The principle is also based on the notion that there exists a 'community of interest' among all co-basin states, requiring a fair balancing of state interests which accommodates the needs and uses of each state. To permit flexibility, the concept of 'equitable and reasonable' use is consciously understood as normatively vague (Barrett 2003, p. 126) and is to be determined in each individual case in the light of all relevant factors (Lipper 1967, pp. 41–2) including, notably, the human, economic and social dependence of each state upon the water resources in question, as well as considerations of environmental protection. In essence, the principle of equitable and reasonable utilisation requires that, in using shared water resources, each co-basin state must have equitable and reasonable regard for the legitimate needs and interests of other co-basin states. As noted by one early commentator, '[i]t is only by an objective appreciation of the facts that it will be possible to discover the fair extent to which the various riparian states must take their reciprocal interests into consideration' (Sauser-Hall 1953, pp. 557–8; McCaffrey 2001, p. 138). This principle, which unquestionably provides the prevailing normative framework for identifying international watercourse rights and obligations today (McIntyre 2007, pp. 24–8), has its doctrinal origins in the sovereign equality of states, whereby all states sharing international watercourse have equivalent rights to the use of its waters.

Though equitable and reasonable utilisation is regarded as the pre-eminent substantive rule of international water law, a normative framework requiring the equitable balancing of the legitimate interests of basin states must inevitably involve intense procedural inter-state engagement (Bruhacs 1993, p. 159), which often can only be facilitated by the establishment of technically competent inter-state institutional machinery. Such institutions can ensure effective inter-state communication which might involve, *inter alia*, prior notification of planned projects potentially impacting upon the watercourse, routine exchange of information regarding the utilisation or condition of the shared waters, or expression of concerns on the part of any basin state. The pivotal role of institutional mechanisms in giving effect to the principle of equitable utilisation has long been recognised by the international community, with Recommendation 51 of the *Action Plan for the Human Environment* adopted at the 1972 Stockholm Conference calling for the 'creation of river basin commissions or other appropriate machinery for cooperation between interested States for water resources common to more than one jurisdiction', and setting down a number of basic principles by which the establishment of such bodies should be guided (UNCHE 1972). Although such institutional structures can take numerous different forms and have diverse remits, there are today at least 119 river basin organisations (RBOs) performing a very extensive range of coordination and joint management functions (Schmeier 2013, p. 65). Reliance on such institutional mechanisms to facilitate the inter-state cooperation necessary to achieve equitable and reasonable utilisation is often referred to as the 'common management' approach, which further underlines the existence of a community of interest among co-basin states (Birnie and Boyle 1992, pp. 223–4; McIntyre 2007, pp. 28–40).

The principle of equitable and reasonable utilisation enjoys very considerable support in the judicial deliberations of international and federal courts and tribunals, as well as in analogous approaches to the allocation of shared water resources adopted by municipal courts (McIntyre 2007, pp. 54–62). For example, in the *Gabčíkovo-Nagymaros* case before the International Court of Justice, Judge *ad hoc* Skubiszewski, in his dissenting opinion referred to the 'canon of an equitable and reasonable utilization' as an expression of 'general law' (ICJ 1997, p. 235). The principle receives almost universal support in treaty law, international codifications, declaratory soft law instruments and the general practice of states, as well as in the writings of leading publicists (McIntyre 2007, pp. 62–76). Indeed, on the basis of an extensive expert examination of the position having regard to all indicators of the existence of a customary rules, the International Law Commission concluded unequivocally 'that there is overwhelming support for the doctrine of equitable utilization as a general rule of law for the determination of the rights and obligations of States in this field' (ILC 1994, p. 222).

However, such near universal acceptance by states is due in large part to its flexibility and normative indeterminacy (Barrett 2003, p. 126; Lipper 1967, pp. 41–2), with the principle providing both a rather vague aspirational goal to guide transboundary water cooperation and the starting point for a process to investigate, identify and reconcile the needs, interests, entitlements and obligations of interdependent co-basin states (Bruhacs 1993, p. 159; McIntyre 2007, pp. 53–4). Indeed, Franck emphasises the procedural and institutional nature of the principle of equitable and reasonable utilisation, explaining that it involves 'a discursive process in which adversary interests need to be reconciled' (Franck 1995, p. 67). Indeed, he refers to it as a striking example of certain 'sophist rules' existing in international law, which have a 'multi-layered complexity', by virtue of which they enjoy a degree of elasticity, and 'usually require an effective, credible, institutionalized, and legitimate interpreter of the rule's meaning in various instances', a process which he characterises as 'institutionalized multilaterization' (ibid., pp. 75, 81–2 and 140). However, some commentators are more critical of the principle's normative indeterminacy (Ruiz-Fabri 1990, p. 839; Scobbie 2002, p. 924).

In essence, the principle of equitable and reasonable utilisation involves the allocation of rights in the uses and benefits of shared water resources on the basis of a distributive conception of equity having regard to all relevant factors. This suggests that uses and benefits will be shared in proportion to each basin state's needs (McIntyre 2007, pp. 147–51), where such needs are calculated through consideration of those factors which are accepted by the states concerned as relevant to water allocation. Therefore, the factors considered relevant to understanding each state's dependence on the shared waters, and thus to the calculation of each state's equitable and reasonable allocation of uses and benefits, are absolutely central and codified or conventional formulations of the principle usually include an accompanying indicative list of such relevant factors. Such a list was first set out in Article V(2) of the 1966 Helsinki Rules. Most notably, Article 6(1) of the UNWC now lists the following factors as relevant:

a. Geographic, hydrographic, hydrological, climatic, ecological and other factors of a natural character;
b. The social and economic needs of the watercourse states concerned;
c. The population dependent on the watercourse in each state;
d. The effects of the use or uses of the watercourses in one watercourse state on other watercourse states;
e. Existing and potential uses of the watercourse;
f. Conservation, protection, development and economy of use of the water resources of the watercourse and the costs of measures taken to that effect;
g. The availability of alternatives, of comparative value, to a particular planned or existing use.

Such a list is not intended to be exhaustive and a range of additional factors might be relevant in the particular circumstances of a particular basin, negotiation or dispute, such as any religious, cultural or local customary significance attached to the river in question or to its waters. Similarly, the conduct of the states concerned regarding a contested use or project might be relevant, including, for example, excessive delay in raising objections (McIntyre 2007, pp. 186–9).

While all key instruments emphasise the lack of a hierarchy among the relevant factors (UN 2007, Article 6(3); ILA 1966, Article V(3)), it is apparent from the practice of states that certain considerations will usually be accorded more significance than others. For example, while Article 6(3) of the UNWC provides that '[t]he weight to be given to each factor is to be determined by its importance in comparison with that of other relevant factors', Article 10(2) would appear to prioritise 'vital human needs', a key element in identifying the 'population dependent on the watercourse in each State' as a relevant factor under Article 6(2). Of course, this elevation of vital human needs is 'likely to enhance the "human right dimension" of the use of the waters of international watercourses' (Tanzi and Arcari 2001, p. 131; Tully 2003, p. 101). A statement of understanding agreed at the time of the adoption of the Convention advises that 'special attention is to be paid to providing sufficient water to sustain human life, including both drinking water and water required for production of food in order to prevent starvation' (UN General Assembly Working Group 1997, p. 3), a position consistent with the ongoing discourse in international law on the human right to water (McIntyre 2015, p. 345).

Indeed, it would appear from the practice of states in this field that what matters above all else is the dependence of each watercourse state upon the shared waters in question, in terms of either human, social or economic needs, and that the relevant factors listed above and elsewhere largely function to elucidate the true nature and extent of such dependence (Fuentes 1996). For example, though the UNWC suggests that existing and potential uses of a watercourse will in principle be considered equally, with the ILC noting that 'neither is given priority' and that 'one or both factors may be relevant in a given case' (ILC 1994, p. 233), it is likely that existing uses will be favoured as they can more easily be scrutinised in terms of their human, social, economic or environmental benefits (or adverse impacts), while the difficulties inherent in reliably considering the beneficial character (or negative impacts) of future uses are manifest (McIntyre 2007, p. 165; Lipper 1967, p. 50; Jiménez de Aréchaga 1960, pp. 335–6). Equally, the examination of factors such as efforts at conservation and economy of use of water resources by a particular state and the availability to a state of alternatives to a planned or existing use of shared waters primarily help to inform that state's true dependence upon the contested waters (McIntyre 2007, pp. 173–9). Further, though 'natural' factors, including the geography and hydrology of the basin are listed first under both the 1997 UNWC and the 1966 Helsinki Rules, there is general agreement among scholars that such factors are of only marginal significance as these do not relate directly to a state's dependence on the shared water and so could undermine the distributive nature of the equitable allocation envisaged under the principle of equitable and reasonable utilisation. For example, Tanzi and Arcari suggest that to accord any *a priori* pre-eminence to such circumstances 'would prejudice the principled equality among riparians' (Tanzi and Arcari 2001, p. 124) Similarly, Lipper argues convincingly that:

> Equality of right is the equal right of each co-riparian state to a division of the waters on the basis of its economic and social needs, consistent with the corresponding rights of its co-riparian states, and excluding from consideration factors unrelated to such needs.
>
> *Lipper 1967, p. 63*

The distributive nature of equity as applied in the particular field of international water law is highlighted by the fact that the significance attributed to the physical characteristics of the drainage basin, such as the length of the course of a river situated within each basin state, the extent of the drainage basin area lying in the territories of the basin states, or their relative contribution of water to the flow of a river, is relatively low (Fuentes 1996, pp. 395–408). This situation contrasts with the application of equitable principles in the territorial delimitation of the continental shelf, where the emphasis has been placed on the extent of each state's coastline (McIntyre 2007, pp. 137–42).

Although the principle of equitable and reasonable utilisation has its origins in inter-state arrangements for allocating co-basin states' quantum share of transboundary waters, it is now largely concerned with environmental requirements and the environmental consequences of incompatible uses. Indeed, as its seeks to balance economic, social and environmental imperatives in the use of water, equitable and reasonable utilisation is now widely understood as the means of operationalising the more nebulous concept of sustainable development in the specific context of transboundary water resources (Wouters and Rieu-Clarke 2001, p. 283; Kroes 1997, p. 83; McIntyre 2007, p. 247). Thus, it should come as no surprise that environmental protection and sustainability requirements are inherent to authoritative modern formulations of the principle and Article 6(1)(a) of the UNWC refers to 'ecological' factors, Article 6(1)(d) to the 'effects of the use or uses ... on other watercourse States', and Article 6(1)(f) to 'conservation, protection ... and economy of use of the water resources of the watercourse'. This connection is even more apparent in the UNECE Water Convention, under which the focus is squarely on environmental protection with the parties required, *inter alia*, to ensure 'ecologically sound and rational water management... [and] conservation of water resources and environmental protection' and 'where necessary, restoration of ecosystems' (UNECE 1992, Article 2(2)). The environmental aspects of the principle have tended to enjoy ever increasing emphasis in recent years (McIntyre 2007, pp. 78–85 and 359–80) and two seminally important cases relating to transboundary rivers brought before the International Court of Justice in recent years, in both of which the principle of equitable and reasonable utilisation was centrally relevant and was expressly linked to that of sustainable development, have revolved around environmental issues (ICJ 1997, 2010).

Prevention of significant harm

Whereas equitable and reasonable utilisation provides the cardinal, overarching rule of international water law, almost all international water resources agreements and codifications include a closely related obligation on watercourse states not to cause significant harm to other watercourse states. The existence of such a rule in general international law is supported by a wealth of authority in state practice (McIntyre 2007, pp. 198–221), and formulations are included in both the UNWC (UN 1997, Article 7) and the UNECE Water Convention (UNECE 1992, Article 2). This rule has been recognised as established customary international law by the arbitral tribunal in the 1941 *Trail Smelter Arbitration* and included among the general principles of international environmental law by the 1972 Stockholm Declaration on the Human Environment (UNCHE 1972, Principle 21). In the specific context of international water law, very many watercourse agreements contain provisions on the prevention and abatement of water pollution (Handl 1975, p. 171) and a representative formulation of the obligation of prevention was included in the seminally important 1966 Helsinki Rules (ILA 1966, Article X). A survey of watercourse agreements reveals a range of ancillary substantive provisions dealing with, *inter alia,* minimum flow requirements, the prevention of harmful effects, the protection of water quality and the application of clean technologies (McIntyre 2007, p. 88).

The requirement to prevent, however, is not absolute. While the obligation to prevent significant transboundary harm has obvious origins in territorial sovereignty and the doctrine of sovereign equality of states, it has also been linked to a number of legal maxims and doctrines upon which national legal frameworks are often based, including the Roman law maxim *sic utere tuo ut alienum non laedas* (so use your own [property] as not to harm that of another), the theory of abuse of rights (*abus de droit; Rechtsmissbrauch*) and the theory of good neighbourliness (*droit international de voisinage; Nachbarrecht*) (McCaffrey 2001, pp. 349–53; Tanzi and Arcari 2001, pp. 142–3). Indeed, its close parallels with national rules firmly established in several legal systems might help to explain the obligation's universal acceptance by states. McCaffrey points out that none of the above founding doctrines represents an absolute rule or prohibition and explains that:

> On the international plane, all three doctrines attempt to reconcile ostensibly conflicting rights of different states in the same territory or shared resource. They do so, in effect, by defining the scope of the rights of the states involved in such a way as to require that rights be exercised in a way that is reasonable vis-à-vis other states. To a large extent, they serve the function of moderating the effect of apparently absolute rules.
>
> *McCaffrey 2001, p. 351*

Therefore, the obligation to prevent significant transboundary harm is understood as a due diligence obligation relating to the taking of reasonable measures by states in the use and protection of shared water resources, rather than as an absolute prohibition on causing or permitting harm in all circumstances. As the ILC commentary explains, '[i]t is an obligation of conduct, not an obligation of result' (ILC 1994, p. 237). Thus, a state might lawfully fail to prevent significant transboundary harm, provided it had taken all reasonable measures to try to prevent such harm, which had occurred despite that state's reasonable efforts. Describing the general obligation of prevention as it appears in most environmental conventions, one commentator has observed that:

> It is clear that such agreements do not establish the strict obligation not to pollute (obligation of result), but only the obligation to 'endeavour' under the due diligence rule to prevent, control and reduce pollution. For this reason the breach of such obligation involves responsibility for fault (*rectius*: for lack of due diligence).
>
> *Pisillo-Mazzeschi 1991, p. 19*

Some commentators criticise the so-called 'no-harm' principle which, '[w]hen it comes to practice and deeds… is far too often disregarded' (Barrett 2003, p. 122). Barrett largely attributes this lack of practical compliance to uncertainty, both scientific uncertainty in relation to the causes of harm and legal uncertainty as to the interpretation and application of the rule (ibid., pp. 122–3). However, despite the principle's inherent flexibility and relativity, lawyers have a reasonably clear understanding of its key elements. For example, as regards the concept of 'harm', McCaffrey helpfully explains that, in addition to a diminution in the quantity or quality of water available,

> 'Harm' could also result from, e.g. pollution, obstruction of fish migration, works on one bank of a contiguous watercourse that caused erosion of the opposite bank, increased siltation due to upstream deforestation or unsound grazing practices, interference with the flow regime, channelling of a river resulting in erosion of the riverbed downstream, conduct having negative impacts on the riverine ecosystem, the bursting

of a dam, and other actions in one riparian state that have adverse effects in another, where the effects are transmitted by or sustained in relation to the watercourse.

McCaffrey 2001, pp. 348–9

The concept of harm is understood broadly and the obligation to prevent harm is not confined to one state's direct use of a watercourse that causes harm to another state's use thereof, as 'activities in one state not directly related to a watercourse (e.g. deforestation) may have harmful effects in another state (e.g. flooding)' (ibid., p. 349). Similarly, the ILA's commentary to Article X of the Helsinki Rules notes that, for the purposes of the no-harm rule, 'an injury in the territory of a State need not be connected with that State's use of the waters' (ILA 1966, p. 500). The continuing evolution of the so-called 'ecosystems approach' to the management of transboundary water resources, endorsed by Articles 20–23 of the UNWC, ought to ensure that UNWC Article 7 is construed broadly, at least in relation to any ecological or environmental damage (McIntyre 2004, p. 1, 2014, p. 88). As regards the significance threshold for harm prohibited under Article 7 of the UNWC, the ILC has explained that '[t]here must be a real impairment of use, *i.e.* a detrimental impact of some consequence upon, for example, public health, industry, property, agriculture or the environment in the affected State' (ILC 1988, p. 36). This approach formalises the so-called *de minimus* rule, which derives from the general principle of good neighbourliness and involves 'the duty to overlook small, insignificant inconveniences' (de Aréchaga 1978, p. 194).

In describing the nature of the due diligence obligation imposed upon states by Article 7 of the UNWC, the ILC's commentary to the 1994 Draft Articles refers approvingly to the definition of due diligence provided in the 1872 *Alabama Claims Arbitration*, which describes the concept as 'a diligence proportioned to the magnitude of the subject and to the dignity and strength of the power which is exercising it' and as 'such care as governments ordinarily employ in their domestic concerns' (ILC 1994, pp. 236–7; Moore 1898, pp. 572–3 and 612). In the specific context of international watercourses, the 'magnitude of the subject' might be expected to refer to the nature of the relevant activity and suggests, therefore, that where it is inherently dangerous 'the care required would be so great as to approach strict liability; a virtual guarantee that such a harmful event would not occur' (McCaffrey 2001, p. 373). The 'appropriate measures to prevent the causing of significant harm' required of each watercourse state under Article 7 of the UNWC can be understood to include both substantive and procedural elements, as the ICJ made clear in the *Pulp Mills* case (ICJ 2010, para. 77). Substantive due diligence might require, for example, that a basin state with the potential to cause transboundary harm should ensure the adoption and enforcement of appropriate domestic legal controls (Tanzi and Arcari 2001, p. 154), while procedural due diligence might demand that a state planning a major project likely to impact on the watercourse or on the interests of co-basin states should notify such states and, where necessary, engage in good faith consultation and negotiation regarding their outstanding concerns (McIntyre 2011b, p. 137, 2010, pp. 475–97, 2013, pp. 239–65).

The relationship between the duty to prevent significant transboundary harm and the overarching cardinal principle of equitable and reasonable utilisation has been an issue of some contention and controversy (McIntyre 2007, pp. 104–16; Nollkaemper 1996, pp. 48–54). However, on careful examination of the 1997 UNWC it becomes apparent that the no-harm rule, and other substantive rules, such as those relating to environmental and ecosystems protection, are subject to the doctrine of equitable and reasonable utilisation. In other words, the duty to prevent has in principle only a limited, though in practice probably profound, effect on the operation of the balancing of interests required under equitable and reasonable utilisation. In certain circumstances, one watercourse state's use of shared waters would have to be tolerated, even if it caused significant harm to another, where the offending use represented the equitable and reasonable allocation of benefits taking account of all relevant considerations. This hierarchy of substantive rules is apparent from the wording of Article 7(2) of the UNWC, which provides that watercourse states whose use causes significant harm 'shall ... take all appropriate measures, *having due regard for the provisions of articles 5 and 6* ... to eliminate or mitigate such harm and, where appropriate, *to discuss the question of compensation*' (emphasis added). This clearly implies that Article 7 is subordinate to the principle of equitable and reasonable utilisation (McIntyre 2007, pp. 104–16). The ILA's update on the Helsinki Rules, the 2004 Berlin Rules on Water Resources Law, inverts this relationship somewhat but endorses the same hierarchy of rules, providing that 'Basin States shall in their respective territories manage the waters of an international drainage basin in an equitable and reasonable manner *having regard for the obligation not to cause significant harm to other basin States*' (ILA 2004, Article 12(1), emphasis added).

Of course, Article 6(1) of the UNWC also lists '[t]he effects of the use or uses of the watercourses in one watercourse State on other watercourse States' as one factor among others relevant to the determination of equitable and reasonable utilisation. In the *Gabčíkovo-Nagymaros* case the ICJ strongly endorsed the principle of equitable and reasonable utilisation as the governing rule of international water law and the one on which that dispute should turn (ICJ 1997, paras. 78, 85, 147 and 150; McIntyre 1998, p. 79). The subordination of the no-harm rule possibly reflects the fact that, in many international watercourses, lower basin states tend to develop earlier and a strict prohibition on causing significant harm would effectively serve to protect existing rights and 'would therefore impede opportunities for newly developing upstream States that pursue legitimate interests for the welfare of their societies' (Nollkaemper 1996, p. 57).

However, it would be a mistake to imagine that these two key substantive rules of international water law are likely to come into conflict often, if at all. Relying on a formulation of the rule articulated by the German Staatsgerichtshof in relation to a 1927 complaint taken by the *Länder* of Württemburg and Prussia against Baden (the *Donauversinkung* case), McCaffrey, the leading scholar in this field of international law, suggests that 'the *sic utere tuo* principle is not only fully compatible with that of equitable utilization, it essentially merges with the latter principle' (McCaffrey 2001, p. 357). He explains that:

> It is thus the flexibility of the no-harm rule that makes it compatible, even if not entirely identical, with the principle of equitable utilization... rather than prohibiting the causing of harm *per se*, the law takes into account surrounding circumstances. This same process is followed in arriving at an equitable and reasonable allocation of the uses and benefits of shared freshwater resources... There is therefore no need to 'reconcile' the no-harm and equitable utilization principles. They are, in reality, two sides of the same coin.
>
> *Ibid., pp. 370–1*

There can be little doubt that the no-harm rule envisages environmental pollution and ecosystems damage as centrally relevant classes of harm to be prevented (McIntyre 2007, pp. 116–18). For example, McCaffrey recounts that, in considering the inclusion and structure of Article 7 of the UNWC, environmental protection was very much in the minds of the members of the Commission, and he explains that the environmental implications of the no-harm rule must be understood in the light of the evolving international legal frameworks for environmental protection:

> it must be recognized that much progress has been made on giving content to the notion of due diligence in international environmental law generally, and with regard

to shared natural resources, in particular. In these fields, exercising due diligence... generally means adopting and effectively enforcing legislative and administrative measures that protect other states and areas beyond the limits of national jurisdiction. The standard of protection – how stringent the measures should be – may in some cases be determined by reference to internationally agreed minimum standards in the field.

McCaffrey 2001, p. 374

Although these are not the only, or possibly even the main, causes of harm to other watercourse states, it is telling that the UNWC includes discrete substantive provisions obliging watercourse states to protect the ecosystems of international watercourses, to prevent significant harm resulting from pollution, to protect the marine environment and to ensure the sustainable development of international watercourses (UN 1997, Articles 20–24). While their precise normative status in relation to the no-harm rule and the principle of equitable and reasonable utilisation remains somewhat unclear, according to Nollkaemper, the express inclusion of such legal obligations in the Convention 'signals the evolution of normative expectations' and 'can legitimate claims and may alter the outcomes of balancing acts' (Nollkaemper 1996, p. 67). Likewise, while Tanzi and Arcari argue that their inclusion, along with the reference to 'sustainable utilisation' in Article 5(1) of the UNWC, 'further enhances the applicability of the basic water law principles to the environmental protection of the watercourse' (Tanzi and Arcari 2001, p. 177), McCaffrey observes that 'states increasingly treat pollution of international watercourses and degradation of aquatic ecosystems as a special form of harm, subject to a somewhat different regime from that applicable to allocation and utilization in general' (McCaffrey 2001, p. 364).

References

Barrett, S. (2003) *Environment and Statecraft: The Strategy of Environmental Treaty-Making*. Oxford: Oxford University Press.

Birnie, P. and Boyle, A.E. (1992) *International Law and the Environment*. Oxford: Oxford University Press.

Boisson de Chazournes, L. (2011) Features and trends in international environmental law. In: Y. Kerbrat and S. Maljean-Dubois (eds), *The Transformation of International Environmental Law*. Paris: Pedone & Hart, pp. 9–20.

Brels, S., Coates, D. and Loures, F. (2008) *Transboundary Water Resources Management: The Role of International Watercourse Agreements in Implementation of the CBD*. Montreal: Secretariat of the Convention on Biological Diversity (CBD Technical Series No. 40).

Brown Weiss, E. (2013) *International Water Law for a Water-Scarce World*. Leiden: Martinus Nijhoff.

Bruhacs, J. (1993) *The Law of Non-Navigational Uses of International Watercourses*. Dordrecht: Martinus Nijhoff.

de Aréchaga, E.J. (1978) International law in the past third of a century. *Recueil des Cours* 159, p. 9.

Eckstein, G. and Sindico, F. (2014) The Law of Transboundary Aquifers: many ways of going forward, but only one way of standing still. *Review of European, Comparative and International Environmental Law* 3(1), pp. 32–42.

Fitzmaurice, M. and Elias, O. (2004) *Watercourse Cooperation in Northern Europe: A Model for the Future*. The Hague: TMC Asser Press.

Fitzmaurice, M. and Merkouris, P. (2015) Scope of the UNECE Water Convention. In A. Tanzi, O. McIntyre, A. Kolliopoulos, A. Rieu-Clarke and R. Kinna (eds), *The UNECE Convention on the Protection and Use of Transboundary Waters and International Lakes: Its Contribution to International Water Cooperation*. Leiden: Brill/Nijhoff, pp. 101–15.

Franck, T.M. (1995) *Fairness in International Law and Institutions*. Oxford: Clarendon Press.

Fuentes, X. (1996) The criteria for the equitable utilization of international rivers. *British Yearbook of International Law* 67, p. 337

Handl, G. (1975) Balancing of interests and international liability for the pollution of international watercourses: customary principles of law revisited. *Canadian Yearbook of International Law* 13, pp. 156–94.

ICJ (International Court of Justice) (1997) *Case Concerning the Gabčíkovo-Nagymaros Project (Hungary / Slovakia). ICJ Reports* 7.

ICJ (2010) *Case Concerning Pulp Mills on the River Uruguay (Argentina v Uruguay)*, 20 April.

ILA (International Law Association) (1966) Helsinki Rules on the Uses of the Waters of International Rivers. *Report of the Fifty-Second Conference.* Helsinki, ILA, pp. 484ff.

ILA (2004) Berlin Rules on Water Resources Law. *Report of the Seventy-First Conference.* Berlin, ILA. Available from: http://www.asil.org/ilib/WaterReport2004.pdf

ILC (International Law Commission) (1988) The law of the non-navigational uses of international watercourses. *Yearbook of the International Law Commission* 2(2), pp. 22–54.

ILC (1994) Draft Articles on Non-navigational Uses on International Watercourses. *Report of the International Law Commission on the Work of its Forty-Sixth Session*, UN Gaor 49th Sess., Suppl. No. 10, UN Doc. A/49/10 (1994).

ILC (2008) Draft Articles on Transboundary Aquifers, UN Doc. A/RES/63/124 (2009). *Report of the International Law Commission on the Work of its Sixtieth Session*, UN Doc. A/63/10 (2008).

Jiménez de Aréchaga, E. (1960) International legal rules governing the use of waters of international watercourses. *Inter-American Law Review* 2, p. 329.

Kroes, M. (1997) The protection of international watercourses as sources of fresh water in the interest of future generations. In E.H.P. Brans, E.J. de Haan, J. Rinzema and A. Nollkaemper (eds), *The Scarcity of Water: Emerging Legal and Policy Responses.* The Hague: Kluwer Law International.

Lipper, J. (1967) Equitable utilization. In A.H. Garretson, R.D. Hayton and C.J. Olmstead (eds), *The Law of International Drainage Basins.* New York: Dobbs Ferry/Oceana.

Maljean-Dubois, S. (2011) The making of international law challenging environmental protection. In Y. Kerbrat and S. Maljean-Dubois (eds), *The Transformation of International Environmental Law.* Paris: Pedone & Hart, pp. 25–54.

McCaffrey, S.C. (2001) *The Law of International Watercourses: Non-Navigational Uses.* Oxford: Oxford University Press.

McCaffrey, S.C. (2005) The human right to water. In E. Brown Weiss, L. Boisson de Chazournes and N. Bernasconi-Osterwalder, *Fresh Water and International Economic Law.* Oxford: Oxford University Press, p. 93.

McIntyre, O. (1998) Environmental protection of international rivers. *Journal of Environmental Law* 10, p. 79.

McIntyre, O. (2004) The emergence of an 'ecosystems approach' to the protection of international watercourses under international law. *Review of European Community and International Environmental Law* 13(1), pp. 1–14.

McIntyre, O. (2007) *Environmental Protection of International Watercourses Under International Law.* Farnham: Ashgate.

McIntyre, O. (2010) The proceduralization and growing maturity of international water law: *Case Concerning Pulp Mills on the River Uruguay (Argentina v Uruguay). Journal of Environmental Law* 22(3), pp. 475–97.

McIntyre, O. (2011a) International water resources law and the International Law Commission Draft Articles on Transboundary Aquifers: a missed opportunity for cross-fertilisation?. *International Community Law Review* 13, pp. 1–18.

McIntyre, O. (2011b) The World Court's ongoing contribution to international water law: the *Pulp Mills* Case between Argentina and Uruguay. *Water Alternatives* 4(2), pp. 124–44.

McIntyre, O. (2013) The contribution of procedural rules to the environmental protection of transboundary rivers. In L. Boisson de Chazournes, C. Leb and M. Tignino (eds), *Freshwater and International Law: The Multiple Challenges.* Cheltenham: Edward Elgar, pp. 239–65.

McIntyre, O. (2014) The protection of freshwater ecosystems revisited: towards a common understanding of the 'ecosystems approach' to the protection of transboundary water resources under international law. *Review of European, Comparative and International Law* 23(1), pp. 88–95.

McIntyre, O. (2015) The UNECE Water Convention and the human right to access to water: the protocol on water and health. In A. Tanzi, O. McIntyre, A. Kolliopoulos, A. Rieu-Clarke and R. Kinna (eds), *The UNECE Convention on the Protection and Use of Transboundary Watercourses and International Lakes: Its Contribution to International Water Cooperation.* Leiden: Brill Nijhoff, pp. 345–66.

Moore, J.B. (1898) *History and Digest of the International Arbitrations to which the United States has been a Party, Vol. 1.* Washington, DC: US government.

Nollaemper, A. (1996) The contribution of the International Law Commission to international water law: does it reverse the flight from substance? *Netherland Yearbook of International Law* 27, pp. 39–73.

PCIJ (Permanent Court of International Justice) (1929) *Territorial Jurisdiction of the International Commission of the River Oder* case, Judgment No. 16 (10 September), *PCIJ Series A,* No. 23, 5–46.

Pisillo-Mazzwschi, R. (1991) Forms of international responsibility for environmental harm. In F. Francioni and T. Scovazzi (eds), *International Responsibility for Environmental Harm*. London: Graham & Trotman, p. 15.

Rieu-Clarke, A. (2007) The role and relevance of the UN Convention on the Law of the Non-Navigational Uses of International Watercourses to the EU and its Member States. *British Yearbook of International Law* 78, pp. 389–428.

Rieu-Clarke, A. (2015) Transboundary hydropower projects seen through the lens of three international legal regimes – foreign investment, environmental protection and human rights. *International Journal of Water Governance* 3(1), pp. 27–48.

Ruiz-Fabri, H. (1990) Règles Coutumières Générales er Droit International Fluvial. *Annuaire Francaise de Droit International* 36, p. 819.

Salman, S.M.A. (2009) *The World Bank Policy for Projects on International Waterways: An Historical and Legal Analysis*. Leiden: Martinus Nijhoff.

Sauser-Hall, G. (1953) L'Utilisation Industrielle des Fleuves Internationaux. *Recueil des Cours* 83(1953/II), pp. 465–601.

Schmeier, S. (2013) *Governing International Watercourses: River Basin Organizations and the Sustainable Governance of Internationally Shared Rivers and Lakes*. Abingdon: Routledge.

Scobbie, I. (2002) Tom Franck's fairness. *European Journal of International Law* 13, p. 909.

Tanzi, A. and Arcari, M. (2001) *The United Nations Convention on the Law of International Watercourses*. The Hague: Kluwer Law International.

Teclaff, L.A. (1967) *The River Basin in History and Law*. The Hague. Martinus Nijhoff.

Teclaff, L.A. (1991) Fiat or custom: the checkered development of international water law, *Natural Resources Journal* 31, pp. 45–73.

Tully, S.R. (2003) The contribution of human rights to freshwater resource management. *Yearbook of International Environmental Law* 14, p. 101.

UNCHE (1972) *Report of the United Nations Conference on the Human Environment*. Stockholm: UN Publication Sales No. E.73.II.A.14.

UN (United Nations) (1997) United Nations Convention on the Non-navigational Uses of International Watercourses (New York, 21 May). *International Legal Materials* 36, p. 700.

UNECE (United Nations Economic Commission for Europe) (1992) Convention on the Protection of Transboundary Watercourses and International Lakes. *International Legal Materials* 312, p. 1312.

UNECE (2004) UNECE Water Convention MoP Decision III/I, 'Amendment to Articles 25 and 26 of the Convention', UN Doc. ECE/MP/WAT/14 (12 January), adopted 28 November 2003, entered into force 6 February 2013.

UNECE (2014) Model Provisions on Transboundary Groundwaters, UN Doc. ECE/MP.WAT/40.

UN (United Nations) General Assembly Working Group (1997) *Report of the Working Group to the General Assembly*, UN Doc. A/C.6/51/SR.57.

Wouters, P.K. and Rieu-Clarke, A. (2001) The role of international water law in promoting sustainable development. *Water Law* 12, p. 281.

18

THE SIGNIFICANCE OF THE DUTY TO COOPERATE FOR TRANSBOUNDARY WATER RESOURCES MANAGEMENT UNDER INTERNATIONAL WATER LAW

Christina Leb[*]

Introduction

The general duty to cooperate gained a strong reconfirmation as one of the core principles of the law of transboundary water resources[1] with the entry into force of the United Nations Convention on the Law of Non-Navigational Uses of International Watercourses (UN Watercourses Convention 1997) on 17 August 2014. In the process of identification of general principles of international water law, the recognition of this duty as a general principle happened later in time when compared with the crystallization of other principles, such as the principle of equitable and reasonable utilization. The recognition as a general principle confirms the role of cooperation as an important factor that facilitates the implementation of other principles of international water law and, more generally, the achievement of the objectives of sustainable development and protection of freshwater resources that flow in transboundary rivers, connected aquifers and their surrounding ecosystems.

The principle of cooperation is inherent in the United Nations system established after the international society of states emerged from two devastating wars. The Charter of the United Nations (UN Charter 1945) is a foundation stone of the international law of cooperation (Verdross and Simma 1984, p. 310). In addition to maintaining international peace and security, the United Nations purpose is to 'achieve international cooperation in solving international problems of an economic, social, cultural, or humanitarian character' (Art. 1 UN Charter 1945). This Article, in combination with the additional principles and rules of cooperation outlined in Article 2, Articles 55 and 56[2] UN Charter (1945), has been quoted as the primary treaty source from which a general principle of cooperation can be derived in international law (Sahović 1972, p. 287). As a legal principle it underlies and explains the existence of specific rights and obligations of cooperation that exist as rules supporting the implementation of the principle (Fitzmaurice 1957, p. 7). The International Law Commission (ILC), in drafting the 1994 Draft

Articles on the law of non-navigational uses of international watercourses (ILC 1994, pp. 89–135) that later formed the basis for the UNWC, adhered to the spirit and purpose of the system it is embedded in[3] when eventually including the general duty to cooperate as one of the general principles of international water law.

Key considerations on the general duty to cooperate

Today, Article 8 (1) UN Watercourses Convention (1997) has become the main reference norm for the general duty to cooperate: 'Watercourse States shall cooperate on the basis of sovereign equality, territorial integrity, mutual benefit and good faith in order to attain optimal utilization and adequate protection of an international watercourse.'

This norm recalls the key principles of sovereign equality and the application of good faith in fulfilling international legal obligations upon which the system of the United Nations (Art. 2 UN Charter) and international law more generally rest. It also recalls the essential element of mutual benefit which is typically required for cooperation to emerge. Cooperation is the process through which two or more parties work together to achieve a common purpose that produces additional mutual benefits that are not achievable with unilateral action alone (Zartman 2008, p. 5). The common purpose of cooperation in the context of international water law is the attainment of optimal utilization and adequate protection of international watercourses, where optimal utilization means the generation of maximum benefits to all watercourse states, while 'achieving the greatest possible satisfaction of all their needs' and minimization of harm (ILC 1994, p. 97). In its commentary on the Draft Articles (1994) the ILC explains that optimal utilization does not imply that the technologically most efficient or monetarily most valuable use has priority, or that a state that can make the most efficient use has a superior claim on the development of an international watercourse. Nor does it mean that short-term gain should come in the way of long-term loss. The general duty to cooperate seeks to achieve sustainable utilization and takes account of the interests of all states that share the freshwater resources of an international watercourse.

This substantive element of the general duty to cooperate, the objective of achieving optimal and sustainable utilization for the benefit of all concerned watercourse states, is also one of the core objectives of the principle of equitable and reasonable utilization, another general principle of international water law.[4] This principle was, at least until the adoption of the UN Watercourses Convention (1997), considered one – if not the – cornerstone principle of the law of transboundary water resources. Until the finalization of the Draft Articles (1994) and the adoption of the UN Watercourses Convention (1997) by the General Assembly, the Helsinki Rules on the Use of the Waters of International Rivers (1966) of the International Law Association (ILA) were regarded the most comprehensive codification effort of international water law. The Helsinki Rules (1966) have been accredited with identifying and establishing the equitable and reasonable use principles as a core principle of international water law. Chapter 2 of the Helsinki Rules (1966) outlines rights and obligations linked to the equitable utilization of the waters of an international drainage basin (ILA 1967, pp. 486–94). However, an interesting aspect of the Helsinki Rules (1966) is that while they spell out the rights and obligations of individual states with respect to equitable and reasonable utilization they remain relatively mute as to the duties of the individual states in implementing this principle with respect to each other. Furthermore, the principle is framed more in terms of rights and entitlements than in terms of positive obligations for action towards achieving equity and reasonableness of use. Article IV of the Helsinki Rules (1966) states an entitlement of each basin state, 'within its territory, to a reasonable and equitable share'. Article VII provides that 'a basin State may not be denied the present reasonable use of the waters of an international drainage basin to reserve for a co-basin State a future use of such waters'.

The core normative content of the Helsinki Rules (1966) has been retained in articles establishing the applicability of the principle to the management of specific watercourses that have been included in international water agreements since then, including the UN Watercourses Convention (1997; see also Art. 3 (7–8) SADC Revised Protocol 1995; Art. 5 Lake Victoria Protocol 2003). In addition, the UN Watercourses Convention (1997) introduced an important element in the formulation of the principle that guides watercourse states in the implementation of the principle. Paragraph 2 of Article 6 UN Watercourses Convention (1997) acknowledges the fundamental role of cooperation for the achievement of equitable and reasonable utilization. According to the principle, equity and reasonableness of use is to be assessed by watercourse states based on multiple criteria including geographic, hydrographic, hydrological, climatic and ecological factors; social and economic needs of riparian states; the population dependent on the shared water resources; effects of uses on the different riparian states; existing and potential uses; conservation, protection and economic efficiency of use of the shared resources; and potential alternative uses that are comparable to a planned or existing use (Art.V Helsinki Rules 1966; Art. 6 UN Watercourses Convention 1997; Art. 3(8) SADC Revised Protocol 2000). The assessment of these different factors requires comprehensive knowledge about the conditions of and water uses within the entire basin. This is knowledge an individual state that shares a watercourse with one or more countries may not have. Information from other riparian states about the status of a basin and riparian needs will likely be required. Article 6(2) UN Watercourses Convention (1997) recognizes this fact and provides that, for the purpose of achieving equitable and reasonable utilization while taking the interests of other concerned watercourse states into account, some level of cooperation is required. Paragraph 2 prescribes that concerned watercourse states 'shall, when the need arises, enter into consultations in a spirit of cooperation'.

Consultation and provision of information on basin conditions and riparian needs will, in most cases, be required to assess whether a use is equitable and reasonable. These cooperative acts play an important role in the ability of states to achieve the objective of the principle of equitable and reasonable utilization. Given the important role cooperation plays in implementing this and other principles of international water law it is interesting that the formulation of the general duty to cooperate and its identification as a self-standing principle of international water law happened relatively late in the process of codification of the law of transboundary water resources.

Genesis of the general duty to cooperate

In 1970, the General Assembly mandated the ILC to codify the law of non-navigational uses of transboundary water resources (UNGA 1970). The process of codification up to the finalization of the '1994 Draft articles on the law of the non-navigational uses of international watercourses and commentaries thereto and resolution on transboundary confined groundwater' illustrates the progressive steps towards the recognition of the general duty to cooperate as a distinct principle of international water law.

The ILC based its work on state practice, existing agreements and the work of non-governmental expert bodies. The practice of states in managing and developing shared water resources unilaterally or cooperatively is rich. It can be elusive and hard to identify unless formalized into treaties and other forms of agreements, such as declarations of heads of states and governments, inter-ministerial memoranda of understanding or minutes of government-to-government meetings. The codification efforts of expert bodies with respect to general principles, customary rules and progressive development of law have been largely based on these arrangements, as they provide written evidence of state practice. The ILC drew on state practice, including international water treaties; in addition it also benefited from the work of the Institute of International Law (IIL) and the ILA. These two expert bodies had produced earlier texts codifying key principles of the law of transboundary water resources and continued to work on this topic in parallel with the ILC. A comparison of the writings of the IIL, ILA and ILC provides insights into process of identifying the general duty to cooperate as a general principle of international water law.

IIL Declarations concerning the law of transboundary water resources that predate the 1970 codification mandate accorded to the ILC are the Declaration on the International Regulation regarding Uses of International Watercourses for Purposes other than Navigation (1911) (Wehberg 1957, pp. 81ff.) and the Articles on Utilization of Non-Maritime International Waters (except navigation) (1961), also called Salzburg Resolution (IIL 1961, pp. 381–4). The 1911 Declaration still framed the rules of use of international watercourses in terms of obligations of abstention rather than cooperation. In the 1960s, when the negative impacts of economic activity on the freshwater resources flowing in international rivers and lakes became more widespread and, *vice versa*, impacts of natural flow conditions increasingly harmed economic activity, codification efforts slowly started to focus on cooperation that would be required between states to mitigate harmful impacts of use. While the Helsinki Rules (1966) (ILA 1967, pp. 477–531) of the ILA limited cooperation obligations to specific issue areas,[5] cooperation formed the underlying tenor of the IIL Salzburg Resolution (1961). Its preamble recognizes the interdependence of states that share transboundary water resources and their common interest in the maximum utilization and rational exploitation of these resources and its main body of norms provides for cooperation between states through notification, consultation and negotiation in situations where planned works or uses of water would significantly impact on the ability of other states to use the same waters (Arts. 5–8 1961 IIL Salzburg Resolution). Still, the Resolution did not include a general rule or principle of cooperation. A general duty to cooperate was first included in Article 2 ILA Supplementary Rules applicable to Flood Control (1974): 'Basin States shall cooperate in measures of flood control in a spirit of good neighborliness, having due regard to their interests and well-being as co-basin States' (ILA 1974, p. 47).

The documents that subsequently spelled out a cooperation duty in more general terms were the IIL and ILA resolutions of the late 1970s and early 1980s that focused on pollution. The IIL included such a general duty to cooperate with respect to pollution of rivers and lakes in its IIL Athens Resolution (1979) (IIL 1979, pp. 265–7),[6] which also provides for specific obligations to implement this general cooperation duty: including regular exchange of data, coordination of research and monitoring programmes, notification, consultation and provision of technical and financial aid to developing countries (Art. VI IIL Athens Resolution). Article 4 of the ILA Montreal Rules on Water Pollution in an International Drainage Basin (1982) includes a general rule on cooperation that is similar to the one included in the 1979 IIL document. The Article provides that '[i]n order to give full effect to the provisions of these Articles, States shall cooperate with the other States concerned' (ILA 1982, pp. 535–48).[7] In its commentary on this Article, the ILA was the first to argue for the recognition of cooperation as a fundamental principle (ILA 1982, pp. 539f.).

In parallel and shortly after the inclusion of a general cooperation duty in the IIL and ILA documents on pollution, cooperation also started to appear more specifically and prominently in the articles on non-navigational uses of international watercourses drafted by the ILC. The concept of the duty of equitable participation in the protection and management of the shared resources was introduced to the Draft Articles in 1982 (ILC 1982, p. 85) and shortly thereafter, the revised and expanded draft of 1984 included an entire, new chapter that focused solely on cooperation (ILC 1984, p. 113). This Chapter III on Cooperation and Management in Regard to International Watercourse Systems introduced, for the first time in the drafting work of the ILC, an Article stipulating a general principle of cooperation and it identified the specific rights and obligations that would be needed to implement this principle, including rights and obligations related to consultation, negotiation and the prior notification of planned measures. The formulation of the general duty to cooperate was later pulled forward in the text and rightly included into the section on General Principles (ILC 1987, pp. 70–95), where it remains and appears today as Article 8 UN Watercourses Convention (1997) and the specific cooperation obligations can now be found throughout the main text of the prevention as appropriate in the articles of Part III on Planned Measures, Part IV on Protection, Preservation and Management, Part V on Harmful Conditions and Emergency Situations, Part VI in the articles on indirect procedures, data and information vital to national defence and security and dispute settlement. Similar to the Salzburg Resolution (1961), the underlying tenor of the UN Watercourses Convention (1997) is very much one of cooperation.

Implementing the general principle of cooperation

The general duty to cooperate is formulated in Article 8(1) of the UN Watercourses Convention (1997) and in other treaties it has been incorporated (for example Art. 2(6) UNECE Water Convention 1992) as an abstract rule which requires further specification. While the rule states the basic fundamental principles and elements cooperation needs to be based on, as well as its objective, it remains abstract as to which steps the process of cooperation needs to consist of to be meaningful. As with other principles of international law, the general duty to cooperate needs to be made operational through more specific rights and obligations in order to unfold its true significance to 'attain optimal utilization and adequate protection of an international watercourse' (Art. 8(1) UN Watercourses Convention 1997). Some of these specific rights and obligations have, through consistent state practice, become rules of customary international law, such as the obligation to notify concerned states of planned measures that may cause significant harm. Others, more specific ones, can be identified based on an analysis of and their inclusion in treaties that govern the use of the waters in various individual basins.

In addition to the general duty to cooperate, the UN Watercourses Convention (1997) codified the core body of the specific rights and obligations concerning cooperation. They can be classed in three broad categories: 1) rights and obligations that primarily serve information needs; 2) those that facilitate the identification of mutual benefits among states; and, finally, 3) specific rights and obligations that help realizing mutual benefits through joint action.

Meeting information needs

As outlined above, information is not only necessary to assess whether a planned use can be considered equitable and reasonable and thus for the implementation of the principle of equitable and reasonable utilization. The provision of information and regular information and data exchange – for example, on the general status of the hydrologic system – is necessary to pursue the objective of the principle of cooperation of achieving the greatest possible satisfaction of the needs of basin riparian countries while minimizing harm.

Specific obligations connected to the provision of information relate to unilateral and mutual information sharing either on a regular or *ad hoc* basis. They comprise obligations of regular information and data exchange, provision of information upon request, the notification of planned measures and the notification of emergencies.[8]

Even though the obligation of regular exchange of information and data was included in Part II UN Watercourses Convention (1997) on General Principles, in state practice, the norm has not yet become a rule of customary law. It takes place only based on provisions included in international treaties or other arrangements. The UN Watercourses Convention (1997) recognizes that regular exchange of data and information is a specific obligation that implements the general duty to cooperate.[9] During the process of elaboration of the Draft Articles (1994) the rule was first included among the obligations implementing the general principle of cooperation (ILC 1983, paras. 102, 151), and was moved only later to Part II, where it now appears as Article 9 UN Watercourses Convention (1997). The fact that about 39 per cent of international water agreements concluded between 1900 and 2012 include a rule on the regular exchange of data and information (Leb 2013, p. 117) is evidence that states consider information sharing essential to cooperation and the coordinated management and development of transboundary freshwater resources. Often information exchange is limited to data for specific purposes such as flood control (see Memoranda of Understanding between China and India); some treaties provide for sharing of water data on flow, discharge and quality more broadly.[10]

The corollary of the obligation to share and regularly exchange is the right to request data and information. Where this involves monetary efforts on the side of the state collecting the data, treaties and also Article 9(2) UN Watercourses Convention (1997) usually make compliance conditional on the reimbursement of these costs. The quality of the ILC's drafting work that went into the UN Watercourses Convention (1997) becomes apparent in Article 9(3). To close the circle, it imposes a due diligence obligation on states to collect and process data and information 'in a manner which facilitates its utilization by other watercourse States' and thus provides for the *sine qua non* condition to make data and information sharing a possibility in the first place.

Regular exchange of data and information assists in both the prevention and mitigation of harm, such as through regular sharing of flow data during flood seasons, as well as in promoting enhanced management and development of water resources and sustainable use through improved knowledge of available water balance. In contrast, the objective of notification obligations, the unilateral *ad hoc* provision of information independent of a request, typically is the prevention of harm. As specific obligations closely linked also to the implementation of the principle of good neighbourliness and the obligation to prevent significant harm, the customary character of notification obligations is widely recognized (Bourne 1972, p. 218; McCaffrey 2007, p. 473). Notification obligations have been included in a large number of treaties and regularly refer to two kinds of situations triggering the obligation: emergencies due to natural causes or human conduct or planned measures that may cause significant harm (Convention concerning Water Economy Questions relating to the Drava 1954; Agreement between Greece and Yugoslavia concerning Hydro-economic Questions 1959; slightly more limited: Treaty relating to Utilization of Waters of the Colorado and Tijuana Rivers, and of the Rio Grande 1944; Indus Waters Treaty 1960; Convention on the Protection of the Rhine against Chemical Pollution 1976; Mekong Cooperation Agreement 1995).

The required timing of the act of notification, unless clearly spelled out in a legal instrument, can become contentious in the practical application of the rule. This is less so in the first case; given the nature of the triggering event, the state on whose territory a natural or man-made emergency occurs shall notify potentially affected states in the most expeditious manner. In the case of planned measures, this issue has been up for debate during the drafting process of the ILC Draft Articles (1994) (Rieu-Clarke 2014, p. 108) and remains open to interpretation in the final text of the UN Watercourses Convention (1997), as well as in many other treaties that require 'timely' (Art. 12 UN Watercourses Convention 1997; Art. 20 Niger Basin Water Charter 2008; Art. 24 Senegal Water Charter 2002), 'prompt' notification or notification 'in good time' or 'as early as possible' (Art. 3 Espoo Convention 1991). This means that in cases where the timing is not further specified, notification should happen 'sufficiently early in the planning stages to permit meaningful consultations and negotiations… if such prove necessary' (ILC 1994, p. 111). Or, as more generally stated by the International Court of Justice (ICJ) for all applications, the act of notification should happen at such a point in time as is required in relation to its intended objective (ICJ 2010, paras. 104f).

It is important that the notification reaches those that are potentially affected, because, where appropriate, it then triggers subsequent cooperative action, either to prevent or mitigate harm, or to identify mutual benefits. In case of emergencies, cooperation among affected states to prevent, mitigate and eliminate harmful effects as much as possible can be required by treaty instruments such as in Article 28(3) UN Watercourses Convention (1997). In case of planned measures, the notification has to be accompanied by adequate technical information so that potentially affected states are in a position to assess whether the measures planned by the notifying state significantly impact rights or interests (ICJ 2010, para. 104). In certain situations, the notification of planned measures with potentially harmful transboundary impact can facilitate the identification of additional mutual benefits; for instance, an upstream water retention infrastructure that reduces downstream flow can be operated to mitigate flood risk and provide water supply downstream in case of droughts. In its judgment on the *Dispute regarding Navigational and Related Rights (Costa Rica v Nicaragua)*, the ICJ clarified that notification is also required when awareness of a measure is required by the affected state to ensure compliance (ICJ 2009, paras. 91–7). Nicaragua, as the holder of sovereign authority over the River San Juan, is required to notify Costa Rica of the adoption of regulation affecting the latter state's right of freedom of navigation on the river to ensure 'better application of the regulation and the more effective pursuit of its purposes' (ICJ 2009, para. 96).

Notification is a unilateral act and in many cases it is the first step towards a process of cooperation among two or more states. The ensuing process may result in coordinated action – for instance, such as mutual assistance after a dam bursts to mitigate or eliminate harmful impact after emergencies. It can lead to a process of negotiation with a view to achieving an agreement on the generation of additional mutual benefits that can be achieved through planned measures, or it may remain at a level of information exchange through consultations.

Identifying mutual benefits

The UN Watercourses Convention (1997) is one of the treaties which outlines, in considerable detail, procedural steps to be taken after notification of planned measures has caused reaction by the notified state, including rules concerning specific renewable response periods of six months (Arts. 11–19). In case of a response, 'the notifying State and the State making the communication shall enter into consultations and, if necessary, negotiations with a view to arriving at an equitable resolution of the situation' (Art. 17(1)).

The UN Watercourses Convention (1997) makes a fine distinction between the legal consequences of notification of planned measures. On the one hand, there is the requirement for consultations; this is a process of bilateral or multilateral exchange of information. Where this process takes place in response to notification of planned measures, the process will focus on concerns over interests and rights of the respondent that may be impacted. In this case, the consultation process is linked to a legal duty of the notifying state to take the interests of the notified state into account (Okowa 1996, p. 302; Bourne 1972, p. 227). On the other hand, the UN Watercourses Convention (1997), demands negotiations when necessary. The process of negotiations hardly differs from consultations. The difference lies in the objective; in addition to information exchange, negotiations include the objective of achieving an agreement – though, agreement is not a legally required outcome (PCIJ 1931, p. 116). The requirement of consent would need to be explicitly stipulated in a treaty or legal agreement between concerned states. Prior consent is, for instance, required before implementation of certain measures in the Senegal River Basin (Art. 24 Senegal Water Charter 2002).

Obligations to negotiate and consult exist not only in relation to planned measures (see for example Arts. 6 and 17 UN Watercourses Convention 1997; Arts. XII (5), XIII (6) and Annex A Columbia Basin Treaty 1961; Art. 11 Danube Protection Convention 1994). They permeate international water law and are linked to the implementation of various principles and rules of international water law. In particular the UN Watercourses Convention (1997), which captures the need for cooperation on transboundary watercourses, refers to these obligations in multiple instances. Already mentioned above, information exchange through consultations can become necessary in the implementation of the principle of equitable and reasonable utilization. The information exchange that occurs through consultations allows states to voice their interests and facilitates the identification of opportunities for equitable use and the generation of benefits for all involved parties. It is through consultations and negotiations that states agree on international water treaties from which they derive benefits – for example, the Treaty relating to cooperative development of the water resources on the Columbia River Basin (1961) that led to significant flood control and hydropower benefits for both Canada and the United States; the Itaipú Treaty (1973) agreed between Brazil and Paraguay on joint hydro-electricity generation, and the Convention on Cooperation for the Protection and Sustainable Use of the Danube River (1994) which led to significant water quality improvements benefiting the numerous riparian countries of the river.

Additionally, consultation duties also appear with respect to the obligation to prevent significant harm. Article 7(2) UN Watercourses Convention (1997) provides that where significant harm occurred states shall try to reverse or mitigate the harm they caused 'in consultation with the affected State' and 'discuss the question of compensation', where appropriate (Art. 6(3)). In addition, many international treaties and international water agreements require negotiation between states as a first step to prevent or achieve settlement of disputes between states for the mutual benefit of continued peaceful relations (see, for example, Art. 33 UN Watercourses Convention 1997; Art. 24 Danube Protection Convention 1994).

Realizing mutual benefits

Where mutual benefits have been identified, states have agreed to subject themselves to specific obligations to achieve favourable outcomes through coordination and cooperation in the management and development of shared resources. For example, in Europe, where water quality issues have become a major concern affecting the well-being of people and ecosystems they depend upon, states have agreed to multiple obligations with respect to water quality improvements and monitoring. The Convention on the Protection of the Rhine against Chemical Pollution (1976) provides a framework for specific obligations for eliminate certain substances listed in Annex I of the Convention and to reduce the use of others (Annex II substances). The UNECE Convention on the Protection and Use of Transboundary Watercourses and International Lakes (1992) requires states to adopt legal and administrative measures to control pollution, including through setting emission limits for discharges from point sources (Art. 3). In the Lake Ohrid Agreement (2004), Albania and Macedonia agreed to work individually and in cooperation to 'Assure equal and integrated protection as well as sustainable development for Lake Ohrid and its watershed.'

Other opportunities to achieve mutual benefits have been seized by states in agreeing to rules governing joint investment in, or coordinate operation of, hydraulic infrastructure. On the Paraná River and also in the Senegal Basin, states agreed to joint investment to reap hydropower, flow regulation and irrigation benefits. While in the case of the Itaipú dam, the infrastructure is built on a contiguous stretch of the river shared between Paraguay and Brazil, the case of the Senegal is peculiar because, in this case, countries agreed to joint investments in infrastructure works that are located entirely in the territory of one or the other of the participating countries. In the Convention concluded between Mali, Mauretania and Senegal concerning the Legal Status of Common Works (1978) the states agreed to the future construction of regulating infrastructure, including the Diama and Manantali dams which regulate the flow of and prevent saline water intrusion into the Senegal River (Art. 3). In other basins and sub-basins, states agree to coordinate flow regulation provided through infrastructure management to provide water when needed downstream. In Central Asia, the Kyrgyz Republic agreed to operate the regulation facilities in the basin (dams, reservoirs, canals) that are located in its territory to 'achieve mutual benefit on a fair and equitable basis' (Art. 1 Agreement on the Use of Water Management Facilities of Intergovernmental Status on the Rivers Chu and Talas 2000). Kazakhstan agreed to the Kyrgyz right to be compensated for the operation and maintenance expenses incurred for its benefit.

The obligations of pollution control, ecosystem protection and coordinated infrastructure development are reflected as obligations requiring coordination and cooperation, where appropriate, in the UN Watercourses Convention (1997; see Arts. 20, 21, 25, 26). The inclusion of these obligations in this instrument with intended global reach reflects both that these good practice activities 'attain optimal utilization and adequate protection' (Art. 8(1)) and that these activities form part of usual state practice on the management and development of shared watercourses.

In order to achieve these cooperation outcomes, the UN Watercourses Convention (1997) recommends the establishment of joint mechanisms (Art. 8(2)), whereas other treaties such as UNECE Watercourses Convention (1992, Art. 9(2)) require their establishment. Joint mechanisms – joint commissions, river basin organizations or technical committees – are useful vehicles for coordinated management since they provide platforms for regular interaction, information exchange and decision-making among basin states. For example, taking the earlier example of the Chu Talas basin, it is a joint commission established by the two states that determines the detailed infrastructure operation regime and rates at which Kazakhstan compensates the Kyrgyz Republic for the operation and maintenance of the upstream infrastructure. It is very common that states pursue coordination through joint mechanisms. More than two-thirds of international water agreements concluded since 1900 establish some sort of joint mechanism, or refer specific tasks to a joint entity, with an even higher regional average in Africa, where this is the case for more than 90 per cent of the treaties concluded after 1960 (Leb 2013, p. 262).

Significance of cooperation for the management and development of transboundary water resources

The fact that specific cooperation obligations can be found in each and every Part of the UN Watercourses Convention (1997), which provides a regulatory framework for the sustainable management and development of international watercourses, is reflective of the importance of coordination between riparian states sharing tranboundary water resources. Water gives life to ecosystems and people and is an important input factor to economic and social activity. The water resources flowing in transboundary surface and groundwater bodies interconnect the territories of these states and make them interdependent. The use of significant amounts of water in one country usually affects one or more of its co-riparian states – in particular, in cases where the use is consumptive (abstraction or pollution) or where appreciable temporal or spatial diversions of the resource take place. The impacts may be negative or positive. Cooperation can be beneficial for concerned countries in both cases; to mitigate potential harm and/or to maximize benefits that can be achieved through joint or coordinated action.

As mentioned above, water quality deterioration has given rise to significant basin-wide and regional cooperation efforts in Europe, where countries have used treaties and other legal instruments to rehabilitate the continent's major watercourses (UNECE Water Convention 1992; Danube Protection Convention 1994; Rhine Protection Convention 1999; EU Water Framework Directive 2000). In the La Plata Basin, states have used international treaty instruments to avoid negative impact and harness the benefits from water flow regulation. To avoid negative impacts on navigation Argentina, Brazil and Paraguay concluded in 1979 a Tripartite Agreement on Corpus and Itaipú determining the operating level for the reservoir at Corpus and the Itaipú power plant. In a similar approach, Canada and the United States optimize cascade operations and cooperate to maximize mutual benefits for flood control and hydropower generation through temporal water retention in the Columbia River Basin (Columbia Basin Treaty 1961).

The cases cited in this chapter represent only a small number of examples of the more than 220 international water agreements concluded between countries since 1900 in cooperation on the management and development of transboundary water system. They highlight that cooperation through acts such as notification of planned uses and consultation to exchange information on the concerned states interests in the shared water resources is a common process adopted by states to achieve sustainable management of their shared water resources that is accepted as equitable by them.

Conclusion

Transboundary water bodies constitute integrated hydrologic units that are best managed by taking the basin-as-a-whole into account. There are multiple ways in which states have addressed and continue to address their hydrologic interdependence in a cooperative manner. International water law and the general principles applicable to transboundary water resources management reflect the richness of state practice that has developed over time. The United Nations system recognizes the role of cooperation as a means for peaceful settlement and prevention of conflict that occurs due to competing interests, including over a strategic resources that are essential for countries' economies and the survival and well-being of their populations. The same holds true for the role of cooperation in the management and development of the World's water resources.

The general duty to cooperate is part and parcel of the set of general principles that complement each other's effect towards achieving sustainable management of shared water resources in a peaceful manner. The specific obligations that operationalize the principle of cooperation provide the means to implement the principle of equitable and reasonable utilization, and to facilitate the prevention of significant harm and the protection of ecosystems. The general duty to cooperate and its related specific obligations and rights provide the grout for the regulatory structure applicable to the beneficial use of transboundary freshwater resources.

The regulatory framework of principles, rights and obligations of international water law, which builds on past state practice, will be increasingly tested in the future as many transboundary basins will likely experience growing water stress due to population growth and economic development, with impacts compounded in some regions due to the effects of climate change. Cooperation and coordination among riparian states will remain vital to maintain peaceful and rational development of the world's freshwater resources. Acknowledging this challenge, the international society of states has set targets in this direction as part of Goal 6 of the Sustainable Development Goals, including on the implementation of integrated water resources management at all levels, and through transboundary cooperation as appropriate.

Notes

* This chapter is based on an in depth study of the topic presented in Leb 2013. The views represented here are those of the author alone and in no way reflect the position of the World Bank, its Board of Executive Directors or any of its member countries.

1 The terms 'the law of transboundary water resources' and 'international water law' are used in this chapter interchangeably to describe the principles and norms of international law applicable to the management and development of transboundary freshwater resources that flow in international rivers, lakes and aquifers.

2 Art. 55: 'With a view to the creation of conditions of stability and well-being which are necessary for peaceful and friendly relations among nations based on respect for the principle of equal rights and self-determination of peoples, the United Nations shall promote: (a) higher standards of living, full employment, and conditions of economic and social progress and development; (b) solutions of international, economic, social, health, and related problems; and international cultural and educational cooperation; and (c) universal respect for, and observance of, human rights and fundamental freedoms for all.' Art. 56: 'All Members pledge themselves to take joint and separate action in cooperation with the Organization for the achievement of the purposes set forth in Article 55.'

3 The ILC was established by the General Assembly (GA) in 1947 to carry out the GA mandate under Article 13(1) (a) of the Charter of the United Nations (UNC) to 'initiate studies and make recommendations for the purpose of... encouraging the progressive development of international law and its codification'.

4 See Chapter 17 in this volume, by O. McIntyre.

5 Cooperation obligations included in Articles XI, XX, XXIV, XXV and XXIX–XXXI concern the specific issue areas of pollution, navigation for humanitarian purposes, dispute prevention and timber floating.

6 Article IV Athens Resolution: '*Afin de se conformer aux obligations énoncées..., les Etats utiliseront notamment les moyens suivants:... b) sur le plan international, l'exercice d'une coopération de bonne foi avec les autres Etats intéressés.*'

7 Article 4: 'In order to give full effect to the provisions of these Articles, States shall cooperate with the other States concerned.'

8 See, for example, Art. 5 ILA Montreal Rules on Water Pollution in an International Drainage Basins, which differentiates between these three reasons to provide information: 'Basin States shall: (a) inform the other States concerned regularly of all relevant and reasonably available data, both qualitative and quantitative, on the pollution of waters of the basin, its causes, its nature, the damage resulting from it, and the preventive procedures; (b) notify the other States concerned in due time of any activities envisaged in their own territories that may involve a significant threat of, or increase in, water pollution in the territories of those other States; and (c) promptly inform States that might be affected, of any sudden change of circumstances that may cause or increase water pollution in the territories of those other States.'

9 Article 9(1) UN Watercourses Convention: '*Pursuant to article 8* [*the general duty to cooperate*], watercourse States shall on a regular basis exchange readily available data and information on the condition of the watercourse' (emphasis added).

10 According to Article VI, Indus Waters Treaty, the Treaty Parties are under an obligation to exchange on a monthly basis gauge and discharge data relating to Indus tributaries, data on daily extraction and release from reservoirs and daily data related to flow volume in canals. See also Arts. 6, 11 and 13 UNECE Water Convention (1992).

References

Agreement between Greece and Yugoslavia concerning Hydro-economic Questions (1959) 18 June, 363 *UNTS* 135.

Bourne, C. (1972) Procedure in the development of international drainage basins: the duty to consult and to negotiate. *Canadian Yearbook of International Law* 10, pp. 212–34.

Columbia Basin Treaty (Treaty between Canada and the United States of America relating to co-operative development of the water resources of the Columbia River Basin) (1961) Washington, 17 January, 542 UNTS 246.

Convention concerning Water Economy Questions relating to the Drava (1954) 25 May, 227 *UNTS* 128.

Convention on the Protection of the Rhine from Chemical Pollution (1976) 3 December [online]. Available from: http://iea.uoregon.edu/pages/view_treaty.php?t=1976-Chemical-1963-RhineCommission.EN. txt&par=view_treaty_html

Danube Protection Convention (Convention on Cooperation for the Protection and Sustainable Use of the Danube River) (1994) 29 June, ECOLEX TRE-001207, 1997 *EUTSer* 68, OJ 1997 L342/19.

Espoo Convention (Convention on Environmental Impact Assessment in a Transboundary Context) (1991) 26 February, 1989 *UNTS* 309.

EU Water Framework Directive (2000) 23 October, Directive 2000/60/EC, OJ 2000 *L 327/1*.

Fitzmaurice, G. (1957) The general principles of international law, *Recueul des cours de l'Académie dedroit de La Haye* 92, pp. 1–227.

IIL (Institute for International Law) (1961) *Annuaire de l'Institut de Droit International – Session de Salzbourg, September 1961* 49(II). Basel:Verlag für Recht und Gesellschaft AG.

IIL (1979) *Annuaire de l'Institut de Droit International, Session d'Athènes 1979*. Paris: S. Karger.

ICJ (International Court of Justice) (2009) *Dispute regarding Navigational and Related Rights (Costa Rica v Nicaragua)*, judgment, ICJ Reports 213.

ICJ (2010) *Case concerning Pulp Mills on the River Uruguay, (Argentina v Uruguay)*, judgment, ICJ Reports 14.

ILA (International Law Association) (1967) *Report of the Fifty-second Conference held at Helsinki, August 14th to August 20th, 1966*. London: ILA.

ILA (1974) *Report of the Fifty-Fifth Conference, New York 1972*. London: ILA.

ILA (1982) *Report of the Sixtieth Conference, Montreal*. London: ILA.

ILC (International Law Commission) (1982) Third report on the law of the non-navigational uses of international watercourses, by Mr Stephen M. Schwebel, Special Rapporteur. *Yearbook of the International Law Commission* II(1), pp. 65–191.

ILC (1983) First report on the law of the non-navigational uses of international watercourses, by Mr J. Evensen, Special Rapporteur. *Yearbook of the International Law Commission* II(1), pp. 155–95.

ILC (1984) Second report on the law of non-navigational uses of international watercourses, by Mr J. Evensen, Special Rapporteur. *Yearbook of the International Law Commission* II(1), pp. 101–27.

ILC (1987) Summary records of the meetings of the thirty-ninth session 4 May–17 July 1987. *Yearbook of the International Law Commission* I, pp. 70–95.

ILC (1994) Draft articles on the law of the non-navigational uses of international watercourses and commentaries thereto and resolution on transboundary confined groundwater. *Yearbook of the International Law Commission* II(2), pp. 89–135.

Indus Water Treaty (1960) 19 September, 419 UNTS 125.

Lake Victoria Protocol (Protocol for Sustainable Development of Lake Victoria Basin) (2003) Arusha, 29 November, FAOLEX, no. LEX-FAOC041042.

Leb, C. (2013) *Cooperation in the Law of Transboundary Water Resources*, Cambridge: Cambridge University Press.

McCaffrey, S. (2007) *The Law of International Watercourses: Non-Navigational Uses*, 2nd edn. New York: Oxford University Press.

Mekong Cooperation Agreement (Agreement on the Cooperation for the Sustainable Development of the Mekong River Basin) (1995) 5 April, 2069 *UNTS* 3.

Memorandum of Understanding between the Ministry of Water Resources, the Republic of India and the Ministry of Water Resources, the People's Republic of China on Strengthening Cooperation on Transborder Rivers (2013) 23 October [online]. Available from: http://mea.gov.in/bilateral-documents. htm?dtl/22368

Niger Basin Charter (Charte de l'eau du bassin du Niger) (2008) 30 April [online]. Available from: http:// www.abn.ne/

Okowa, P. (1996) Procedural obligations in international environmental agreements. *British Yearbook of International Law* 67, pp. 275–336.

PIJ (Permanent Court of International Justice) (1931) *AO Railway Traffic between Lithuania and Poland*, PCIJ, Ser. A/B, No. 42.

Rhine Protection Convention (Convention on the Protection of the Rhine) (1999) 12 April, OJ 2000 L289/31.

Rieu-Clarke, A. (2014) Notification and planned measures concerning international watercourses: learning lessons from the *Pulp Mills* and *Kishenganga* Cases. *Yearbook of International Law* 24(1), pp. 102–30.

SADC Revised Protocol (Revised Protocol on Shared Watercourses in the Southern African Development Community) (2000) 7 August [online]. Available from: http://www.sadc.int/, ECOLEX TRE-001360

Sahovič, M. (1972) Codification of the Law of Friendly Relations between States. *Recueul des cours de l'Académie dedroit de La Haye* 137, pp. 243–310.

Senegal Water Charter (Charte des eaux du fleuve Sénégal) (2002) 18 May. In L. Boisson de Chazournes, R. Desgagné, M. Mbengue and C. Romano (2005), *Protection internationale de l'environnement*, 2nd edn. Paris: Pedone, pp. 301–14.

Treaty relating to the Utilization of Waters of the Colorado and Tijuana Rivers, and of the Rio Grande (1944) 14 November [online]. Available from: https://www.usbr.gov/lc/region/pao/pdfiles/mextrety.pdf

Tripartite Agreement on Paraná River Projects, President Stroessner City (1979) 19 October [online]. Available from: http://www.internationalwaterlaw.org/documents/regionaldocs/parana1.html

UN Charter (Charter of the United Nations) (1945) 26 June, 1 *UNTS* XVI.

UN Watercourses Convention (United Nations Convention on the Law of Non-Navigational Uses of International Watercourses) (1997) 21 May, UN Doc. A/RES/51/299.

UNECE Water Convention (UNECE Convention on the Protection and Use of Transboundary Watercourses and International Lakes) (1992) 17 March, 31 ILM 1312.

UNGA (United Nations General Assembly) (1970) GA Res 2669 (XXV), 8 December, UN Doc. A/RES/2699(XXV).

United Nations (n.d.) Sustainable Development Goals [online]. Available from: http://www.un.org/sustainabledevelopment/sustainable-development-goals/

Verdross, A. and Simma, B. (1984) *Universelles Völkerrecht*, 3rd edn. Berlin: Druckner & Humblot.

Wehberg, H. (1957) *Tableau Général des Résolutions (1873–1956)*. Basel: Verlag für Recht und Gesellschaft AG.

Zartman, I. W. (2008) Concept: cooperation. *PIN Points* 30, pp. 5–7.

19

JOINT INSTITUTIONAL ARRANGEMENTS FOR GOVERNING SHARED WATER RESOURCES

A comparative analysis of state practice

Susanne Schmeier

Introduction

Water conflicts are more and more often making the headlines in media around the world. Titles such as 'Water wars between countries could just be around the corner' and 'The world will soon be at war over water' (*Guardian* 2012; *Newsweek* 2015) would suggest an increasing risk of violent conflict over shared water resources. And looming disagreements – sometimes leading to conflicts[1] – in some of the world's shared basins seem to underscore these fears – for instance, when the Uzbek President threatens upstream Tajikistan over the Rogun Hydropower Project, or the former Egyptian President Mohamed Morsi states that one would be willing to confront any threats to Egypt's water security caused by Ethiopia's Grand Ethiopian Renaissance Dam (GERD). Climate change is only expected to exacerbate these challenges.

In spite of these gloomy examples of water conflict in shared basins, cooperation prevails. In the overwhelming majority of internationally shared river and lake basins (276 in total), conflicts – in their diverse forms – are rare and, if at all, occur far below the threshold of war (Wolf *et al.* 2003; DeStefano *et al.* 2010). Instead, states have in many cases engaged in establishing cooperative arrangements for preventing conflicts over shared water resources. They have acknowledged the benefits cooperation can provide in shared basins and have committed to long-term cooperation through international water treaties (IWT) and established institutionalized mechanisms for implementing these commitments – most often in the form of River Basin Organizations (RBOs).

The remainder of this chapter provides an overview of institutional arrangements, RBOs, for governing shared watercourses. It focuses, first, on why institutionalized cooperation emerges and, second, on how RBOs are designed and how their design differs across institutions and regions, before, third, turning to the question whether and how RBOs matter for governing shared water resources in an effective manner. Finally, the chapter provides an outlook towards the newly emerging challenges riparian states to most transboundary basins face and how they have started addressing them through their RBOs.

Institutionalized cooperation through RBOs

While empirical evidence from state practice shows that states in many cases tend to choose cooperation over conflict and often create institutions for securing this cooperation over time, a number of questions remain in relation to why and under which conditions states decide to establish RBOs. This section addresses these questions by summarizing findings from both scholarly research and empirical evidence.

Before embarking on the analysis of state practice with regards to institutionalizing cooperation through RBOs, it is important to note that the term 'river basin organization' or 'RBO' is used in a rather broad sense by both academic scholars and policy-makers. Often, a large variety of institutions at very different levels of institutionalization are referred to as RBOs. In the strict sense, however, RBOs are 'institutionalized forms of cooperation that are based on binding international agreements, covering the geographically defined area of an international river or lake basin, characterized by principles, norms, rules and governance mechanisms' (Schmeier *et al.* 2015, p. 4).[2] While it is important to have a clear and straightforward definition of RBOs for academic purposes (ibid., p. 3), this chapter adopts a broad approach to RBOs in order to reflect international state practice on institutionalized cooperation in a comprehensive manner. This allows for a wider range of insights into state practice relating to RBOs.

Worldwide, there are 123 RBOs covering 121 basins (see Figure 19.1) (TFDD n.d.; Schmeier 2013). In some cases, one river or lake basin is covered by more than one RBO – most often with different mandates and functions. In the Zambezi River Basin, both the Zambezi Watercourse Commission (ZAMCOM) and the Zambezi River Authority (ZRA) have been established by (different constellations of) riparian states to address particular challenges in water resources governance. In other cases, one RBO covers more than one basin. The International Joint Commission (ICJ) between Canada and the US, for instance, covers 11 different transboundary rivers between the two countries.

Data on the regional distribution of RBOs shows that states have established RBOs in only a subset of the world's 276 transboundary basins (Schmeier 2013, p. 65). A significant number of

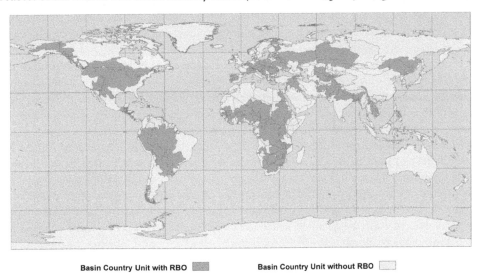

Basin Country Unit with RBO   Basin Country Unit without RBO

Transboundary river basin country units from transboundary freshwater dispute database http://www.transboundary.orst.edu/
river basin organization data from Dr Susanne Schmeier's RBO database http://www.transboundarywaters.orst.edu/research/RBO/index.html
Cartographer: Jennifer C. Veilleux © 2014 Oregon state university

Figure 19.1   RBOs and basins around the world

basins hence remains without any institutionalized cooperation mechanism. This includes some basins that are under significant pressure of water resources development, face environmental deterioration and struggle with foregone socio-economic development opportunities due to water resources insecurity – such as the Salween and Irrawaddy river basins shared by China, Myanmar and Thailand, or the Chiloango River shared by Angola, the Democratic Republic of Congo and the Republic of Congo. Most of the basins without any institutionalized cooperation mechanism are, however, rather small and of limited importance to their riparian states, which have therefore not (yet) perceived the establishment of RBOs as necessary for addressing collective action problems in these basins. Overall, state practice thus shows that riparian states to transboundary basins most often choose to cooperate through institutions in river basins of significant importance to them.

There is also some discrepancy across regions. Some regions show a significantly higher coverage of rivers and lakes by RBOs then others. In Africa and Europe as well as North America, a relatively high share of rivers and lakes are covered by RBOs, while especially in the Middle East and in Asia (and to a lesser extent in Latin America) more transboundary basins remain without any form of institutionalized cooperation (ibid., p. 66). This leads to the question of why and under which conditions states decide to establish RBOs in some basins but refrain from doing so in others. This question has attracted scholarly attention, especially since the 2000s. Empirical research has refuted the thesis that states are likely to go to war over water – or at least have been until now. Rather, it is widely maintained that states chose cooperation over conflict. Many scholars have therefore focused on factors explaining the emergence of cooperation – most often in the form of international treaties (Elhance 2000; Marty 2001; Wolf *et al.* 2003; Dinar 2007; Tir and Ackermann 2009; Brochmann and Hensel 2011) and with less of a focus on RBOs.

This research shows that there are two main factors explaining RBO establishment: the nature of the problem at stake and the constellation of actors in the respective basin (Schmeier 2013, p. 14). The nature of the problem riparian states face in a shared basin can be very different, ranging from water scarcity and related allocation problems to the management of fisheries, or from the prevention of floods to water quality concerns. First of all, a collective action problem in a basin needs to be acknowledged as such in order to lead to a decision-making process within each riparian state as to whether to address it in a cooperative or in a unilateral manner. Depending on the type of problem, states' interest in solving it in a cooperative manner then varies (Mostert 2003; Wolf *et al.* 2003; Dinar 2009). Some problems clearly lend themselves more easily to cooperation without major negotiation or bargaining needs as the benefits of cooperation are obvious, relatively evenly distributed and not intertwined in zero-sum games. Environmental protection, if perceived as a collective action problem in a certain basin, provides such benefits for all riparians without creating major losses or costs for others. The establishment of institutions addressing environmental problems in shared basins has therefore made significant progress in many areas. Similarly, states often perceive cooperation over shared water resources as helpful when joint benefits can be generated by using water resources for economic development – especially if individual capacities of each state are limited and cooperation promises gains that could not be achieved through unilateral action. Sometimes, benefits are even more equally distributed – with navigation being the most obvious example as both upstream and downstream states benefit from free navigation on all stretches of a shared river. Consequently, institutions governing navigation on shared watercourses have been among the first established by riparian states (e.g. the Central Commission for Navigation on the Rhine (CCNR), established in 1815, and the International Commission on the Danube, a predecessor of the present Danube Commission (DC), established in 1921).

Other problems, however, are often more complex and result in different perceptions about gains and losses among different riparian states. The benefits one riparian state can gain from cooperation are often perceived as losses by another riparian state, creating zero-sum-game situations in which compromise is hard to achieve. This is the case even more so if water resources are perceived as a national interest issue and of importance to national socio-economic development ambitions or political resources. The latter scenario can lead to a securitization of water. Water quantity and allocation problems – especially in regions facing water scarcity – are the most obvious example for such types of problems. And basins with high water scarcity and significant pressure on water resources, such as the Jordan, Euphrates–Tigris or Harirud river basins, show that institutions have been extremely difficult or even impossible to establish.

In addition to the nature of the problem at stake in a given basin, the constellation of riparian states also influences whether states decide to establish RBOs. Overall, peaceful relations, a high degree of overall regional cooperation (or even integration) and the absence of a (malign) hegemon are conducive to the establishment of RBOs (Bernauer 1997; Wolf *et al.* 2003; Song and Whittington 2004; Schmeier 2013). This is most obvious with European basins, where the history of cooperation and the high level of economic and political integration between riparian states made the establishment of RBOs relatively easy compared to other regions. Conflicts or limited regional cooperation in sectors other than water, however, tend to impede or obstruct the establishment of institutionalized cooperation mechanisms. In the Euphrates–Tigris River Basin, for instance, regional conflict and instability as well as a low level of cooperation between riparian countries has impeded the establishment of institutionalized cooperation mechanisms. In this particular case, the existence of an upstream hegemon with no interest in cooperation has further impeded the establishment of an RBO or any type of cooperation arrangement.

These factors determining whether states decide to establish RBOs for their transboundary basins also explain the aforementioned differences in geographical coverage. In Europe, it is the overall high level of regional cooperation and integration – including binding frameworks such as the UN Economic Commission or Europe (UNECE) Water Convention and the EU Water Framework Directive (EU WFD) – that have led to the establishment of RBOs in many transboundary basins. In addition, problems driving cooperation in European basins were – especially in the early years of cooperation – most often related to water quality and environmental protection and restorations and thus problems that are less likely to be perceived as relevant to national security considerations or likely to produce unevenly distributed gains and losses than water quantity and allocation problems.

In Africa, on the one hand, the first wave of RBO establishment in the 1960s and 1970s was characterized by riparian states' aim to fast-track economic development by developing water resources in a cooperative manner and hence a common perception that cooperation would produce joint benefits. RBOs aiming at water resources exploitation through infrastructure development, such as the Autorité du Bassin du Niger (ABN) or the Lake Chad Basin Commission (LCBC), emerged in this context. The second wave of RBO establishment in the 1990s and 2000s was driven by other factors, namely commitments riparian states made in the context of overall regional cooperation (such as in the Southern African Development Community (SADC), where the 1995 SADC Protocol on Shared Watercourses and its revised version of 2000 require the establishment of RBOs for transboundary watercourses – including the Orange–Senqu River Commission (ORASECOM), established in 2000, and the Limpopo Watercourse Commission (LIMCOM), established in 2003) and the pressure from the international development community, which massively supported the establishment of RBOs in many regions. RBOs set up with significant donor support include the Lake Tanganyika Authority (LTA), established in 2003, and the Volta Basin Authority (VBA), established in 2005.

In the Middle East and North Africa (MENA) region, on the other hand, many factors persist that make the establishment of RBOs more difficult. Political instability and conflicts, limited regional cooperation and a lack of regional economic or political integration, but also a particularly complex problem – in most cases water scarcity and disputes over water allocation – have also affected water cooperation. Where, however, cooperation over water has been established by riparians, these cooperation mechanisms – such as the Joint Water Commission (JWC) between Israel and Jordan – have made important contributions to mitigating water conflicts but also improving relations between countries beyond the water sector (Wolf 1995; Allan 2002).

The design of RBOs

State practice does not only differ with regards to whether riparian states establish RBOs, but also in the design of these RBOs. A great variation can be observed in how RBOs look, how they function and the governance mechanisms that they provide to their members and other relevant actors. This section focuses on the different design features of RBOs and maps them across different basins. It also highlights different factors influencing the design of RBOs. It is thereby differentiated between the design of an RBO – that is, its structural set-up – and the governance mechanisms it provides.

Some RBOs bring together all riparian states within a basin, or several basins, making them inclusive RBOs, while others cover a subset of riparians and thus a sub-part of the basin only. The membership structure of an RBO is one of its key design features. A slight majority of RBOs in the world are inclusive (or at least quasi-inclusive – that is, including all relevant riparian states to the basin but not necessarily riparians that share a very limited part of the basin only, such as, for instance, Albania and Poland in the Danube River Basin and thus in the International Commission for the Protection of the Danube River (ICPDR), sharing 0.03 per cent and 0.09 per cent of the basin respectively). Worldwide, 47 RBOs can be qualified as truly non-inclusive (Schmeier 2013, p. 82). They do not include one or more major riparian states to the basin. Examples include the Mekong River Basin, where the Mekong River Commission (MRC) brings together the four downstream riparians only and hence leaves out both China and Myanmar, together covering slightly more than 20 per cent of the basin's surface and inhabiting one quarter of the basin's population in their respective basin parts.

Various scholars have argued that the non-inclusiveness of RBOs can cause a number of problems for the governance of shared water resources (Mostert 2003; Zawahri and McLaughlin Mitchell 2009; Kliot et al. 2011). Given the complexity and the interconnectedness of water resources in a basin, they argue that the exclusion of basin states from governance initiatives leads to sub-optimal water resources management or even conflicts. The exclusion of upstream states, for instance, can render cooperation arrangements between downstream states largely obsolete if upstream water use affects agreements concluded between downstream states and related governance activities. Increasing water abstraction by Afghanistan (combined with the effects of global climate change), for instance, will significantly affect the – albeit hardly functional – Dostluk Commission (DOCO) on the Harirud River between downstream Iran and Turkmenistan. Also, the non-inclusiveness of RBOs can lead to situations in which different RBOs overlap while governing different parts of one basin. In the Jordan River Basin, for instance, only bilateral institutions have been established between Israel and Jordan, Israel and Palestine, and Jordan and the Palestinian Territories respectively.

For a number of reasons, some RBOs have extended their membership over time. The Commission Internationale du Bassins Congo-Oubangui-Sangha (CICOS), for instance, was originally established by Cameroon, the Central African Republic, the Republic of Congo and the Democratic Republic of Congo only, but has been enlarged towards Gabon and recently also towards Angola. Further enlargement towards Tanzania and other basin states is foreseen. Similarly, the MRC and Myanmar have discussed Myanmar's potential membership in the MRC since Myanmar's recent political opening towards the international community.

RBOs also vary with regards to the types of problems that they seek to address, the mandate they have been assigned and the functions they fulfil. In basins where water resources are scarce or contested, RBOs are often charged with overseeing and implementing water allocation agreements. The Permanent Indus Commission (PIC), for instance, was set up to oversee the 1960 Indus Waters Treaty and ensure data and information exchange between India and Pakistan on water quantity. Likewise, the JWC between Israel and Jordan is responsible for overseeing water-sharing arrangements between the two countries as agreed upon in the 1994 Israel–Jordan Peace Treaty. Other RBOs focus on water quality as member states have identified water pollution as a key concern that needed cooperation. The International Commission for the Protection of the Rhine (ICPR), for instance, was set up in order to help riparian states in jointly addressing severe water quality problems along the river that unilateral action was not able to address. Similarly, the International Commission for the Protection of Lake Geneva (ICPG) was established to ensure continuous monitoring of the water quality of Lake Geneva between France and Switzerland and to carry out required investigations on the nature, the extent and the sources of pollution in order to address it in an effective manner. Yet other RBOs address issues relating to all types of problems occurring between riparian states in shared basins. This includes fisheries management (for instance, in the case of the Lake Victoria Fisheries Organization (LVFO) or the Great Lakes Fisheries Commission (GLFC) and the Pacific Salmon Commission (PSC), both between Canada and the US), as well as hydropower development (e.g. in the case of the Komati Basin Water Authority (KOBWA) or the ZRA) or irrigation and agriculture (in the case of the Joint Irrigation Authority (JIA), an earlier institution between Namibia and South Africa).

In addition to the topics an RBO works on, one can see differences in state practice with regards to the number of functions the RBO has been assigned with. A significant number of RBOs worldwide (27 in total; Schmeier 2013, p. 85) focus on one single issue only. This includes navigation in the case of the CCNR or the International Commission for Boating on the Lake Constance (ICBL), fisheries in case of the Franco-Swiss Consultative Commission on Fishing in the Lake Geneva (FSCC) or the LVFO or hydropower development and generation in the case of the KOBWA or the ZRA. Other RBOs (26) focus on a large array of often interrelated issues. The ABN, for instance, has been assigned through the 1980 Convention creating the ABN to address issues of integrated basin development and water resources management including energy, agriculture, animal husbandry, fisheries, forestry, transport, communication and industry. Other RBOs established in Africa in earlier years – such as the LCBC or the Organisation pour la Mise en Valeur du Fleuve Sénégal (OMVS) – have similarly broad multi-issue functional scopes. This often results from riparian states' intention to jointly foster economic development in the basin through creating an RBO. The majority of RBOs, however, covers few but interrelated issues, including, for instance, water quality, environmental protection and climate change adaptation in the case of the ICPR and many other European RBOs.

Beyond the functional scope, it can also be distinguished between those RBOs mandated to coordinate river basin governance and development activities in their respective member countries (coordination-oriented RBOs) and RBOs that implement specific activities themselves and are hence more implementation-oriented (Schmeier 2010, p. 10). Typical examples for the coordination-oriented RBO type include the ICPR and the Great Lakes Commission (GLC) between Canada and the US, whereas the OMVS is a model of implementation-oriented RBOs. Both types are simply illustrative of different ends of a continuum along which RBOs can be

Susanne Schmeier

categorized and along which some RBOs tend to move. The MRC, for instance, is currently shifting from an implementation focus to a more coordinating role in the context of its organizational reform process aiming at decentralizing some river basin management functions to member states according to the subsidiarity principle.

Other design features along which state practice in transboundary basins varies but which will not be elaborated in detail in this chapter include: the degree of institutionalization and formality (most often expressed in the legal personality of the RBO and its power *vis-à-vis* its member states) (refer to Dombrowsky 2007, p. 110; UNECE 2009, p. 10; Kliot *et al.* 2011, p. 323); the organizational structure and set-up of RBOs – that is, the number of organizational bodies and the size of an RBO as well as the role of the secretariat (UNECE 2009, p. 39; Schmeier 2010; Schmeier 2013, p. 47 and 91); and the financing of RBOs (GIZ 2014).

River basin governance mechanisms

In order to govern a shared basin and hence meet the objectives defined by member states when creating an RBO, RBOs provide a number of governance mechanisms for coming to joint decisions, exchanging data and information, monitoring the basin, addressing disputes among riparians or involving basin stakeholders. In most cases, these governance mechanisms – where present – are defined in the underlying treaties or the respective RBO's rules and procedures or similar documents. This section addresses the mechanisms for three key functions RBOs provide – decision-making, data and information management, and dispute resolution.

Joint decision-making is the basis for long-term cooperation in shared basins. In order to ensure that states are able to come to joint decisions, RBOs provide negotiation forums by organizing and holding meetings of the RBO's respective governance bodies. In these meetings, decisions are taken on the basis of (in many RBOs) pre-defined decision-making mechanisms. In most cases, these are based on the consensus principle (Schmeier 2013, p. 99). Majority-based decision-making is extremely rare. They can be found in the DC and the ICPDR, where they are not actually used in practice as member states prefer negotiated solutions that take all members on board. Majority-based decision-making can also be found in some bilateral institutions where each member state is represented by a number of commissioners who take decisions on the basis of individual votes leading to certain majority constellations (e.g. the GLC between Canada and the US or the Comision Technica de Mixta de Salto Grande (CTMS) between Argentine and Uruguay).

Decisions need to be based on data and information. Taking decisions on the governance, the development or the protection of water resources requires data and information on the state of the basin, the pressures it is facing and the results of previously implemented measures. Moreover, ensuring long-term cooperation in shared basins requires the sharing of information in order to create trust among states – especially with regards to the potential impacts of development projects in one state on another. Many – though by far not all – RBOs have therefore established data and information gathering, analysis and exchange capacities. They usually consist of a general commitment to data and information sharing (most often in the legal document underlying the RBO) and the operationalization of this commitment in a data-sharing protocol (such as, for instance, the 2001 MRC Procedures for Data and Information Exchange and Sharing, or the 2010 Permanent Okavango River Basin Water Commission (OKACOM) Protocol on Hydrological Data Sharing), as well as technical means for actually gathering and analyzing the data before sharing it among the member states (such as data basis, decision-support systems, etc.).

State practice shows that some member states have chosen to manage data nationally and exchange data bilaterally with the RBO only coordinating the exchange, whereas others have

266

chosen to entrust the RBO with significant data acquisition, analysis and dissemination functions. Especially RBOs in regions where national capacity for acquiring and analysing data is limited due to technological or financial limits, RBOs can play a crucial role not only in ensuring that the basin and the challenges it faces are better understood before taking decisions on its use and development, but also in improving their member states' capacity in data and information management. Most African RBOs therefore rely on institutionalized data and information management mechanisms. These are often promoted by international donors. In this context, some RBOs have engaged in establishing their own monitoring networks and stations (in the Niger River Basin, for instance, more than 100 hydrometric stations were established in nine countries to form a data-gathering system for the ABN since 2005).

As disputes can persist or (re-)emerge once RBOs have been established, riparian states to some shared basins have equipped the respective RBOs with dispute resolution mechanisms. These dispute resolution mechanisms can prove crucial for addressing disagreements that emerge between (subsets) of riparian states in a pre-defined and orderly manner, avoiding their escalation to conflicts and the related negative consequences for the basin (Blumstein and Schmeier forthcoming). Examples from the Syr Darya Basin where Tajikistan and Uzbekistan dispute the construction of the Rogun Dam and from the Nile River Basin, where Ethiopia and Egypt have been engaging in a major conflict over the GERD illustrate this clearly.

The number of RBOs with well-defined and clear dispute resolution mechanisms is, however, limited (with only 63 RBOs; Schmeier 2013, p. 106; TFDD n.d.). Interestingly, many European RBOs – generally considered as well-developed – do not possess any defined and institutionalized dispute resolution mechanisms. Nonetheless, many disagreements that occurred in past years – for instance, between the Czech Republic and Germany over barrages constructed in the upstream part of the Elbe River Basin or between the Netherlands and upstream Rhine riparians over the opening of sluice gates for salmon migration – have not turned into proper conflicts but were instead solved in a cooperative manner.

RBOs having clearly defined dispute resolution mechanisms usually rely on structured bilateral negotiations, institutionalized mechanisms within the RBO or the predetermined referral to a third party for arbitration or adjudication – or any combination thereof. Often, a combination of two mechanisms is chosen, creating a two-step process to dispute resolution. This process usually consists of bilateral negotiations first and then referral to an RBO-related body or a third party such as the Permanent Court of Arbitration (PCA) or the ICJ, as well as regional bodies such as the SADC Tribunal (for RBOs in the SADC region).

Do RBOs matter? Insights from state practice on RBO effectiveness

In many basins – including those where RBOs exist – water resources governance challenges continue to exist or (re-)emerge over time. In the Mekong River Basin, for instance, Laos' construction of the Xayaburi Hydropower Project and many other hydropower projects on the mainstream and the Mekong's many tributaries is putting the basin's environment and the people depending on it at risk. Moreover, it has led to major disagreements with downstream Cambodia and Vietnam, fearing the negative impacts on their water resources development aspirations and their countries as a whole (Rieu-Clarke 2014; Blumstein and Schmeier forthcoming). Likewise, the LCBC has not been able to halt or counter the continuous degradation of the Lake Chad and its impacts on riparian populations as well as overall political stability in the basin. And even in the Rhine River Basin, a basin generally perceived as well governed, various newly emerging problems – ranging from fish migration to flood protection or climate change adaptation – have constantly challenged the ICPR.

Consequently, scholars, policy-makers, civil society representatives and media alike have criti-cized different RBOs for achieving little beyond 'stockpiling of reports and action plans' (Lautze *et al.* 2005, p. 26), seeing their relevance challenged (*The Diplomat* 2016) and being mere 'paper tigers' (Bernauer 1997, p. 159). This has led some scholars – as well as policy-makers – to conclude that 'the history of international river management is littered with cases in which integrated river management schemes have largely failed' (Bernauer 2002, p. 12). Other RBOs, in contrast, have been evaluated as very valuable for addressing the respective problems in the basin as they have provided effective support to riparian states in achieving their joint objectives. The mere existence of RBOs is hence insufficient for ensuring the long-term sustain-able and cooperative governance of shared water resources. Instead, the effectiveness of RBOs matters. However, research has so far not addressed the effectiveness of RBOs in great detail (for some exceptions addressing RBO effectiveness in a comprehensive manner, refer to Berardo and Gerlak 2012; Schmeier 2013; McLaughlin Mitchell and Zawahri 2015; Wingqvist and Nilsson 2015). This concerns both the definition of RBO effectiveness and the identification of factors determining whether and to what extent an RBO is effective.

What is RBO effectiveness?

Expectations of what an RBO has to achieve vary significantly across member states, the public and the international community. Member states incorporate their expectations into the man-date they assign the RBO with and the objectives they define in the RBO's underlying legal documents. Here, as well, state practice varies. Some member states expect their RBO to solve a particular water resources governance problem – such as the allocation of scarce resources in the case of the JWC between Israel and Jordan or the PIC between India and Pakistan. In other cases, member states want the RBO to develop and implement long-term approaches to improv-ing the basin's environment and ensuring long-term environmentally sound water resources. The Commissions for the Protection of the Moselle and the Sarre (CIPMS), for instance, have been set up to address pollution problems on the rivers. And in yet other basins, states entrust the RBO with promoting the overall economic development of the basin on the basis of, but beyond, water resources. Prominent examples include the ABN, mandated to ensure an inte-grated development of the Niger Basin, or the Permanent Intergovernmental Co-Ordination Committee (CIC) between Argentina, Brazil, Bolivia, Paraguay and Uruguay, established to enable harmonic and balanced development and optimum use of natural resources.

Although the mandate and the objectives of an RBO and hence also the assessment of its effectiveness vary, goal-attainment can be an important measure for RBO effectiveness. RBOs themselves, especially, need clearly defined goals and a monitoring system for tracking progress towards them.[3] And such progress towards the goals could indeed be achieved by a number of RBOs. In the Danube River Basin, for instance, the ICPDR's work on water quality has yielded considerable results. The chemical status of the river has significantly improved in the last 15 years and improvements in the biological status of the river and, in particular, with regards to hydro-morphological alterations, have also been considerable (ICPDR 2015). This has not only improved the overall environmental state of the Danube River Basin, but also the Black Sea, which was previously highly affected by pollutants intrusion from the Danube River. In other cases, however, it became clear over time that the – often very ambitious – goals of RBOs had not been achieved. In the Senegal River Basin, for instance, the two main projects of the OMVS – the Manantali and Diama Dams – were not only completed with major delay, but did also not yield the expected results (Schmeier 2013, p. 234). Extension of irrigated agriculture reached only one third of the originally intended goal and was, moreover, unevenly distributed across the countries participating in the project. Likewise, expected improvements of the navigability of the river could not be achieved.

As it has been emphasized by various scholars, goal-attainment alone, however, might be an insufficient measure for the effectiveness of international institutions in general (Peterson 1997; Young and Levy 1999) and RBOs in particular (Schmeier 2013). Goals and objectives once defined by riparian states might not hold over time as challenges in the basin change and, more importantly, they might not meet the standards of sustainable water resources development. In the case of the Senegal River Basin, for instance, strict focus of the OMVS's objectives has led to severe negative environmental and social effects. Similarly, the Lesotho Highlands Water Project (LHWP) – executed by the Lesotho Highlands Water Commission (LHWC) between Lesotho and South Africa – has had significant negative environmental and social impacts while achieving its goal of supplying South Africa with water and both countries with hydroelectric power (Klaphake and Scheumann 2006, p. 10).

What makes an RBO effective?

The factors determining RBO effectiveness are not well understood so far. While the analysis of the contribution of RBOs to attaining their goals and to ensuring sustainable development of the basin is complex already, factors determining the extent to which an RBO is successful in doing so are manifold, interdependent and thus even more complex.

Generally, it can be differentiated between exogenous factors – similar to those explaining the establishment of institutionalized cooperation in the first place – and endogenous factors relating to the RBO itself (Schmeier 2013, p. 31). Particularly challenging problems in a shared basin will make it more difficult for an RBO to address them in an effective manner. This is particularly obvious with water quantity and allocation problems, often perceived by riparian states as zero-sum games where gains for one state necessarily decrease the benefits of others. RBOs addressing this type of problem often also need to engage in mitigating water conflicts between their members. They have done so more or less successfully in different basins. In the Aral Sea Basin, the International Fund for Saving the Aral Sea (IFAS) and the related Interstate Commission for Water Coordination in Central Asia (ICWC) continue to struggle with setting up effective water-sharing arrangements for scarce resources or at least mitigate ongoing conflicts over water resources. The PIC, however, has been successful in overseeing the implementation of water-sharing arrangements from the 1960 Indus Waters Treaty and in mitigating disputes among India and Pakistan that have emerged regularly on this issue.

Another common impediment to effective water resources governance in shared basins is the constellation of actors. On the one hand, the presence of a powerful state with little to no interest in cooperation due to perceived unilateral benefits of non-cooperation makes effective basin governance particularly difficult. On the other hand, a favourable constellation of actors – such as the equal distribution of power and cooperation interests among member states that cooperate already on issues other than water – significantly facilitates the work of an RBO and hence increases its effectiveness. One example is the work of the International Commission for the Protection of Lake Constance (IGKB), which is often regarded as one of the earliest and most successful examples of international cooperation on a shared watercourse as member states have jointly implemented a range of measures for improving the lake's water quality and its overall environment and have, indeed, been successful in re-establishing the lake's natural state from before-pollution times.

While exogenous conditions certainly influence the likelihood of cooperation and its success, RBOs themselves can make a major difference. Otherwise, how could RBOs operating under highly complex circumstances such as a high number of member states with very different economic capacities and interests and a complex set of challenges in the basin – such as the ICPDR – be very effective? And what would explain the lack of effective river basin management by RBOs operating under relatively benign basin conditions such as a limited number of basins states with a high interest in cooperation and an overall agreement of joint cooperation goals – found, for instance, in the Senegal River Basin?

The design of an RBO and the governance mechanisms it provides for addressing challenges in the basin are thus crucial for effectiveness. Well-designed RBOs allow for water resources governance effectiveness even under malign exogenous conditions (ibid., p. 271). Among the many different design features of RBOs, some have proven to be particularly important: joint decision-making on the basis of sound data and information, the implementation of international water law principles through specific governance mechanisms and procedures to mitigate and resolve disputes among riparian states.

RBOs that facilitate joint decision-making in a structured manner – in particular through well-established and regular meetings of member states – are able to ensure a continuous dialogue between members on the state and the development of the basin, establishing a joint vision for the basin and a continuous perception of joint benefits. The recently adopted river basin management plan for the Congo River Basin, for instance, was developed in a cooperative process among member states with a high degree of stakeholder involvement. This has considerably increased member states' ownership in CICOS as they increasingly acknowledge the added value of institutionalized cooperation over their shared resources. This is the basis for the long-term effectiveness of an RBO.

Effective decision-making requires the availability of sound data and information. RBOs that gather data on the state of the basin and the pressures it faces in a consistent and regular manner – for instance, in the form of a state of the basin report or other monitoring mechanisms – find it easier to identify which challenges need to be addressed in order to improve the state of the basin and maintain the long-term sustainability of water resources. Moreover, they are able to trace the impacts of measures implemented by the RBO and its member states and hence adjust basin management plans accordingly. In the case of the ICPDR, for instance, activities in the 1990s focused largely on the chemical status of the river and monitored it accordingly. In recent years, however, data has shown that while the chemical status of the river has improved considerably, other parameters – especially concerning hydromorphological alterations – remain problematic. Accordingly, the ICPDR has put increasing emphasis on activities aiming at improving, for instance, fish passage facilities along the river.

Such evidence-based decision-making also ensures compliance with and implementation of international water law principles. Significant transboundary harm, for instance, can only be avoided if countries jointly decide on the development and the protection of shared water resources and sufficient data on the state of the basin and the pressures it is and will be facing is available. In the case of the Mekong River Basin, the unilateral development of hydropower projects by Laos – in spite the availability and the use of a mechanism for prior notification aiming at preventing significant transboundary harm (Rieu-Clarke 2014) – has not only led to major disagreement between riparian states, but has also considerably decreased MRC's leverage in the basin and hence affected its long-term effectiveness in governing the basin.

As disputes can occur in spite of well-functioning cooperation processes, RBOs tend to be effective especially along the political stability dimension of effectiveness if they can ensure that disputes are mitigated in a well-structured manner. However, RBOs that are not able to mitigate conflicts among their members in a structured manner necessarily find it difficult (if not impossible) to develop and implement any successful basin management activities. In the case of the Nile River Basin, for instance, the dispute over GERD (and water allocation in the basin more generally) led Egypt and (for a certain time) Sudan suspend their membership in NBI, making effective basin-wide governance impossible (Blumstein and Schmeier forthcoming).

Conclusions and outlook

This chapter has provided evidence from state practice that states choose cooperation over conflict on shared watercourses by establishing institutionalized cooperation mechanisms in the form of RBOs. It has, however, also shown that the existence of RBOs is insufficient for ensuring cooperative and sustainable water resources governance. The factors that actually matter for effective basin governance are increasingly well understood and policy action has been taken in some basins to improve the effectiveness of RBOs.[4] Recently, however, a number of new challenges have emerged that need to be addressed through cooperative governance mechanisms.

Climate change will increase the variability in precipitation and hence in water availability and, moreover, lead to an increased number of extreme events in many basins. So far, riparian states to a basin are often insufficiently equipped with mechanisms to address these climate-related risks. Moreover, existing cooperative arrangements will most likely be affected by the impacts of global climate change. Water allocation agreements that define water quotas on the basis of fixed amounts will become inefficient as water availability decreases overall or becomes more unevenly distributed intra- or inter-annually (Drieschova *et al.* 2008). Some RBOs have therefore engaged in better understanding the impacts of global climate change on the respective basins and in developing adaptation strategies at the basin level. The ICPR, for instance, has initiated a study on the impacts of climate change on the Rhine River's discharge and other parameters (e.g. water temperature). Based on these findings, a climate change adaptation strategy was developed that provides the basis for joint action among member states.

Driven by international commitments to the Agenda 2030 and the sustainable development goals (SDGs) as well as national development priorities, recent years have seen an increase in plans for water infrastructure projects, especially large-scale irrigation schemes and hydropower. While needed to reach the ambitious SDGs in the water sector and beyond and improve the livelihoods of populations, these projects come with a risk of exacerbating water conflicts. This is, in particular, the case if infrastructure projects are developed on transboundary rivers – which is very often the case given the high number of rivers suitable for large-scale infrastructure developments being transboundary in nature – and if these projects have detrimental environmental or socio-economic effects and distribute costs and benefits unevenly.

In order not to lead to conflicts, riparian states require – perhaps maybe more than ever before – effective governance mechanisms that ensure the operationalization of international water law principles. Most importantly, they include the principle of equitable and reasonable utilization and the obligation not to cause transboundary harm as well as the procedural principles of prior notification and transboundary environmental impact assessments. In many basins, however, these mechanisms do not yet exist at all. A prominent example is Tajikistan's Rogun Dam, which has not been notified to downstream states in a timely manner due to a lack of institutionalized mechanisms for doing so between riparian states.[5] Nor has information on the project and its impact been shared sufficiently with downstream and hence potentially affected states. This led to severe opposition, especially by Uzbekistan, fearing the potential consequences on its water availability for agricultural use. In other basins, such mechanisms have been included in the respective water treaties but have not been operationalized sufficiently. This makes their application to concrete cases difficult if not impossible. In yet other basins, existing mechanisms have not been proven to be sufficiently effective in addressing disputes that have emerged. The MRC's Procedures for Notification, Prior Consultation and Agreement (PNPCA) are the most prominent ones. In this case, the process worked well and ensured notification and the exchange of information in the first place, but was stuck once a decision had to be taken and interests of MRC member states diverged considerably, impeding any resolution within the RBO (Rieu-Clarke 2014). As a consequence, Laos continued its dam development against the express interest of downstream riparians (Blumstein and Schmeier forthcoming). In order to cope with future dam development in a cooperative sustainable manner, a revision of the PNPCA seems unavoidable.

These remaining challenges, as well as state practice in shared river and lake basins around the world more generally, show that cooperation requires institutionalization and the provision of governance mechanisms in order to be stable and successful over longer periods of time and, in particular, in times of (re-)emerging conflicts. RBOs are thus absolutely crucial for governing shared water resources. The mere existence of RBOs itself is, however, insufficient as the great variability in RBO effectiveness indicates. It is therefore important to initiate or support processes that allow RBOs to reform according to newly emerging needs and hence to keep up with changes in the basin – especially in times of rapid water infrastructure development. Only if the institutional capacity in a basin – and hence RBO effectiveness – is high enough to absorb change can conflicts be avoided or at least mitigated (Wolf *et al.* 2003, p. 43) and hence sustainable development ensured.

Notes

1 It should be noted here that conflicts do not necessarily mean full-fledged violence. Instead, conflicts can come in the form of minor disagreements and their verbal expression (e.g. at the diplomatic or media level), as well as in the form of violence below or even beyond the threshold of war – and any level in-between (Wolf *et al.* 2003). Moreover, conflict and cooperation tend to occur simultaneously in many basins, creating a conflict-cooperation continuum on which interactions between states take place (Sadoff and Grey, 2002; Zeitoun and Mirumachi, 2008).

2 Following this definition, a number of institutions established to govern shared watercourses would not qualify as 'real' RBOs. Most prominently, the Nile Basin Initiative (NBI), for instance, lacks a legally binding basis in the form of an international agreement (as the Cooperative Framework Agreement (CFA) has so far only been signed by six and ratified by three states and is hence not yet in force). In other cases, states have intended to establish an RBO but have never truly institutionalized it to an extent that it would have carried out any basin governance activities. This is, for instance, the case for the Helmand River Delta Commission (HRDC), which has never actively managed water resources shared by Afghanistan and Iran since its establishment in 1950.

3 Some RBOs have therefore established sophisticated monitoring systems, tracking the implementation of certain river basin management activities member states have decided to pursue as well as their contributions to the respective objectives. The ICPR, for instance, has accompanied its ambitious 'Salmon 2020' project, aiming at reintroducing salmon to the river, with a comprehensive monitoring system tracking salmon along the river in order to establish reliable data on its population but also on the effectiveness of measures taken to support it (such as fish ladders on all major transverse structures along the river).

4 A number of RBOs have recently engaged in organizational reform processes, aiming at strengthening the respective RBO and adapting it to the basin governance challenges it faces. This includes, the LCBC, aiming at improving human capacity through an extensive staffing reform, as well as the MRC, decentralizing river basin management functions to member states in order to increase effectiveness (and decrease costs) by applying the subsidiarity principle.

5 The World Bank's engagement in the process, including an environmental and social impact assessment and consultations in all Central Asian states, came relatively late in the process, when the conflict had already escalated. As a consequence, Uzbekistan refused to participate in this process.

References

Allan, T. (2002) Hydro-peace in the Middle East: why no water wars? A case study of the Jordan Basin. *School of Advanced International Studies Journal* 22(2), pp. 255–72.

Berardo, R. and Gerlak, A. (2012) Conflict and cooperation along international rivers: crafting a model of institutional effectiveness. *Global Environmental Politics* 12(2), pp. 101–20.

Bernauer, T. (1997) Managing international rivers. In O. Young (ed.), *Global Governance: Drawing Insights from the Environmental Experience*. Cambridge, MA: MIT Press, pp. 155–95.

Bernauer, T. (2002) Explaining success and failure in international river management. *Aquatic Science* 64(1), pp. 1–19.

Blumstein, S. and Schmeier, S. (forthcoming) Disputes over international watercourses: can river basin organizations make a difference?. In A. Dinar, *Management of Transboundary Water Resources under Scarcity: A Multi Disciplinary Approach*. New Jersey: World Scientific.

Brochmann, M. and Hensel, P. (2011) The effectiveness of negotiations over international river claims. *International Studies Quarterly* 55(3), pp. 859–82.

DeStefano, L., Edwards, P., DeSilva, L. and Wolf, A. (2010) Tracking cooperation and conflict in international river basins: historic and recent trends. *Water Policy* 12(6), pp. 871–84.

Dinar, S. (2007) *International Water Treaties: Transboundary River Negotiations and Cooperation*, London: Routledge.

Dinar, S. (2009) Scarcity and cooperation along international rivers. *Global Environmental Politics* 9(1), pp. 109–35.

Drieschova, E., Giordano, M. and Fischhendler, I. (2008) Governance mechanisms to address flow variability in water treaties. *Global Environmental Change* 18, pp. 285–95.

Dombrowsky, I. (2007) *Conflict, Cooperation and Institutions in International Water Management: An Economic Analysis*. Cheltenham: Elgar.

Elhance, A. (2000) Hydropolitics: grounds for despair, reasons for hope. *International Negotiation* 5(2), pp. 201–22.

GIZ (2014) *Financial Sustainability of International River Basin Organizations. Final Report*, Eschborn: Deutsche Gesellschaft für internationale Zusammenarbeit (GIZ) [online]. Available from: https://www.giz.de/fachexpertise/downloads/Wasser_GIZoriginal.pdf (accessed 2 April 2016).

ICPDR (2015) *The Danube River Basin District Management Plan. Part A – Basin-wide Overview. Update 2015*. Vienna: ICPDR [online]. Available from: https://www.icpdr.org/main/sites/default/files/nodes/documents/drbmp-update2015.pdf (accessed 2 April 2016).

Klaphake, A. and Scheumann, W. (2006) *Understanding Transboundary Water Cooperation: Evidence from Africa*. Berlin: Institute for Landscape Architecture and Environmental Planning: Working Paper 14/2006.

Kliot, N., Shmueli, D. and Shamir, U. (2011) Institutions for management of transboundary water resources: their nature, characteristics and shortcomings. *Water Policy* 3(3), pp. 229–55.

Lautze, J., Giordano, M. and Borghese, M. (2005) Driving forces behind African transboundary water law: internal, external, and implications. International Workshop on African Water Laws: Plural Legislative Frameworks for Rural Water Management in Africa, 26–28 January, Johannesburg, South Africa.

Marty, F. (2001) *Managing International Rivers: Problems, Politics and Institutions*. Berlin: Peter Lang.

McLaughlin Mitchell, S. and Zawahri, N. (2015) The effectiveness of treaty design in addressing water disputes. *Journal of Peace Research* 52(2), pp. 187–200.

Mostert, E. (2003) *Conflict and Cooperation in International Freshwater Management: A Global Review*. UNESCO IHP Technical Documents in Hydrology No. 19, Paris: UNESCO.

Newsweek (2015) The world will soon be at war over water. *Newsweek*, 24 April [online]. Available from: http://europe.newsweek.com/world-will-soon-be-war-over-water-324328?rm=eu (accessed 2 April 2014).

Peterson, M. (1997) International organizations and the implementation of environmental regimes. In O. Young (ed.), *Global Governance: Drawing Insights from the Environmental Experience*. Cambridge, MA: MIT Press, pp. 115–51.

Rieu-Clarke, A. (2014) Notification and consultation procedures under the Mekong Agreement: insights from the Xayaburi Controversy. *Asian Journal of International Law*, DOI: 10.1017/S2044251314000022.

Sadoff, C. and Grey, D. (2002) Beyond the river: the benefits of cooperation on international rivers. *Water Policy* 4(5), pp. 389–403.

Schmeier, S. (2010) The organizational structure of river basin organizations: lessons learned and recommendations for the Mekong River Commission. MRC Technical Paper, April, Vientiane.

Schmeier, S. (2013) *Governing International Watercourses: River Basin Organizations and the Sustainable Governance of Internationally Shared Rivers and Lakes*. London: Routledge.

Schmeier, S., Gerlak, A. and Blumstein, S. (2015) Clearing the muddy waters of shared watercourses governance: conceptualizing international river basin organizations. *International Environmental Agreements*, DOI: 10.1007/s10784-015-9287-4.

Song, J. and Whittington, D. (2004) Why have some countries on international rivers been successful negotiating treaties? A global perspective. *Water Resources Research* 40(5), pp. 1–18.

The Diplomat (2015) Why the Mekong River Commission may be in Peril. *The Diplomat*, 10 October [online]. Available from: http://thediplomat.com/2015/10/why-the-mekong-river-commission-may-be-in-peril/ (accessed 2 April 2016).

Guardian (2012) Water wars between countries could be just around the corner, Davey warns. *Guardian*, 22 March [online]. Available from: http://www.theguardian.com/environment/2012/mar/22/water-wars-countries-davey-warns (accessed 2 April 2016).

Tir, J. and Ackermann, J. (2009) Politics of formalized river cooperation. *Journal of Peace Research* 46(5), pp. 623–40.

TFDD (Transboundary Freshwater Dispute Database) (n.d.) *International River Basin Organization Database* [online]. Available from: http://www.transboundarywaters.orst.edu/research/RBO/ (accessed 2 April 2016).

UNECE (2009) *River Basin Commissions and other Institutions for Transboundary Water Cooperation*, Geneva: UNECE Convention on the Protection and Use of Transboundary Watercourses and International Lakes [online]. Available from: http://www.unece.org/?id=11628 (accessed 2 April 2016).

Wingqvist, G. and Nilsson, A. (2015) *Effectiveness of River Basin Organisations: An Institutional Review of Three African RBOs*. Stockholm: Sida's Helpdesk for Environment and Climate Change.

Wolf, A. (1995) *Hydropolitics along the Jordan River: Scarce Water and its Impact on the Arab–Israeli Conflict*. Tokyo: UN University Press.

Wolf, A., Yoffe, S. and Giordano, M. (2003) International waters: identifying basins at risk. *Water Policy* 5(1), pp. 29–60.

Young, O. and Levy, M. (1999) The effectiveness of international environmental regimes. In O. Young (ed.), *The Effectiveness of International Regimes: Causal Connections and Behavioral Mechanisms*. Cambridge, MA: MIT Press, pp. 1–32.

Zeitoun, M. and Mirumachi, N. (2008) Transboundary water interaction I: reconsidering conflict and cooperation. *International Environmental Agreements* 8(4), pp. 297–316.

20

STRENGTHENING THE IMPLEMENTATION OF TRANSBOUNDARY WATER AGREEMENTS

Insights from the UNECE Water Convention Implementation Committee[1]

Nataliya Nikiforova

Introduction and objective of the mechanism to support implementation and compliance

The 1992 United Nations Economic Commission for Europe Convention on the Protection and Use of Transboundary Watercourses and International Lakes (UNECE Water Convention 1992) is a universal legal framework for protecting and ensuring the quantity, quality and sustainable use of shared water resources by facilitating transboundary cooperation. In fact, as much as its subjective scope of application was originally confined to the pan-European region, following the entry into force of its 2003 amendments in February 2013, the Convention is now open to accession by all United Nations Member States.

The core legal obligation under the Convention is the one of harm prevention. On the one hand, it is carefully crafted in due diligence terms – that is, as a duty to take all appropriate measures to 'prevent, control and reduce any transboundary impact', as opposed to an absolute obligation of prevention of transboundary impact (Tanzi *et al.* 2015). On the other hand, the obligation in point incorporates the equitable and reasonable utilization principle (Article 2, paragraph 2 (c)), as well as the sustainable development principle (Article 2, paragraph 5 (c)). Furthermore, in the same Article 2 (paragraph 6), the Convention complements the prevention obligation in hand with the principle of cooperation as a core means to achieve prevention, through bilateral and multilateral cooperation agreements and the obligation for riparians to set up joint water bodies (Article 9).

The objective assessment of compliance considering the due diligence nature of the core substantive obligation in hand required on a case-specific basis analysis of the subjective degree of the capacity of a Party to, indeed, undertake all appropriate measures to prevent, control and reduce transboundary harm. Against this background, the UNECE Water Convention has focused, since its entry into force in 1996, on supporting and facilitating implementation by

Parties as well as the Convention's application by countries that are not yet Parties through providing an intergovernmental platform for the development of transboundary cooperation and developing a broad range of tools to support implementation.

The establishment of the Implementation Committee at the sixth session of the Convention's Meeting of the Parties (Rome, 28–30 November 2012), with the aim to facilitate, promote and safeguard the implementation and application of and compliance with the Convention, represents a major milestone in the development of its institutional structure which is expected to further enhance the institutional cooperation within the Convention (Tanzi and Contartese 2015).

According to the decision setting up the mechanism in hand (UNECE 2013a), the latter was envisaged to be simple, non-confrontational, non-adversarial, transparent, supportive and cooperative in nature, building on the distinctive collaborative spirit of the Convention, with the emphasis being placed on difficulties in implementation rather than on compliance or breach of obligations under the Convention which would be addressed under Article 22 of the Convention (Tanzi 2015b; Lammers 2014).

At the same time, while the procedure to support implementation and compliance should be without prejudice to Article 22 of the Convention on the settlement of disputes, Tanzi and Contartese (2015, p. 322) argued that: 'The Implementation Committee, and Article 22 on dispute settlement, in combination with the procedural provisions on cooperation, provide for a highly integrated framework for the prevention, management and settlement of water disputes in line with the general trend.'

Background and adoption of the mechanism to support implementation and compliance

As regards efforts to support accession as well as implementation of and compliance with the UNECE Water Convention, the establishment of the Implementation Committee was preceded, from the material law standpoint, by the *Guide to Implementing the Water Convention* (UNECE 2013b) adopted at the fifth session of the Convention's Meeting of the Parties (Geneva, 10–12 November 2009). The *Guide* was then updated before its publication in 2013 in order to reflect the new realities of the Convention that became a global instrument open for accession to all United Nations Member States in early 2013.

The preparation of the *Guide*, in response to requests for clarification of the legal, technical and economic implications of accession, was carried out by the Convention's Legal Board and the Working Group on Integrated Water Resources Management, with a broad participation of legal and water experts from Parties and non-Parties to the Convention as well as non-governmental organizations and academia. This practical guidance constitutes a comprehensive commentary to the Convention's provisions, providing explanations of legal, procedural, administrative, technical and practical aspects of its requirements and illustrates them with examples from Parties and other States.

While the Convention did not explicitly provide for the mechanism under consideration, the possibility for its establishment was considered by the Legal Board during the elaboration of the above mentioned *Guide* as a means to strengthen the dispute prevention and management role assigned to the Meeting of the Parties. The Chair of the Legal Board therefore, in his note to the fifth session of the Meeting of the Parties, presented the proposal for a decision to pave the way for establishing such mechanism under the Convention, as follows (UNECE 2009, p. 3):

> At the present stage of the Convention's evolution, the establishment of a mechanism
> to support implementation and compliance based on the experience of similar

mechanisms under other conventions and on the work carried out so far under the Convention, as well as the Guide to implementing the Convention, would seem to be a natural step forward.

The Meeting of the Parties endorsed this proposal entrusting the Legal Board with the preparation of the draft regulatory framework concerning the functions, composition and powers of a possible implementation review mechanism to be submitted for consideration of the Meeting of the Parties at its following session. The preparatory study and related negotiations within the Legal Board, in combination with the Working Group on Integrated Water Resources Management, took three years concluding with the submission of the draft decision on the matter, which was adopted without amendments at the sixth session of the Meeting of the Parties in 2012, including the core rules of procedure that would allow this body to be operative immediately after its establishment.

Referring to this momentous decision, Bernardini (2015, p. 47) observed that:

> The creation of the Committee is a bold decision by the Meeting of the Parties, undoubtedly constituting the latest major evolution of the Convention from the institutional point of view. Even though it is too early to appraise the impact of the Committee, its potential to provide a unique forum for the prevention of disputes and conflicts is certainly astounding. It can thus be hoped that countries will take advantage of such opportunity.

Membership

According to the decision establishing the mechanism (UNECE 2013a, p. 2), the Implementation Committee is composed of nine members. The Committee membership has a number of distinctive characteristics that allow for the appropriateness of its expertise, as well as its independency and impartiality.

First, the core rules of procedure of the Implementation Committee (UNECE 2013a, p. 9) point out that: 'Each member of the Committee shall serve in his or her personal capacity and, with respect to any matter that is under consideration by the Committee, act in an independent and impartial manner and avoid any real or apparent conflict of interest.' Consequently, the Committee members are not allowed to represent governments or organizations in meetings of other bodies of the Convention, except for technical expert meetings.

The Committee members are therefore independent experts serving in their personal capacity and not as state representatives. This was a controversial issue during the Legal Board deliberations, as a number of countries were in favour of the Committee members representing state interests. While this is understandable taking into account the political dimension and relevance of transboundary issues, it is noteworthy that the Parties finally opted unanimously for the independence and impartiality of the Committee members who are also supposed to serve objectively, in the best interest of the Convention.

According to the compromise eventually found, the candidates to the Committee membership are to be nominated by individual Parties to be finally elected by the Meeting of the Parties. It is to be noted that the candidates are not necessarily nationals of the nominating state, nor of any other Party to the Convention. In this connection, it is remarkable that one of the current members of the Committee, whose candidature was suggested by one of the Parties of the Convention based on his prestige and recognized expertise, is a national of a country that is not Party to the Convention – that is, the United States of America.

It is even more remarkable that Parties may – even though they are not obliged to – take into consideration proposals for candidates made by Signatories or by non-governmental organizations (NGOs) qualified or having an interest in the fields to which the Convention relates. The possibility of involving the latter provides a significant entry point for the participation of civil society in the selection process of the membership of the Committee.

The following criteria are to be considered by the Meeting of the Parties in the election of the Committee members:

- *Experience and expertise*: the Meeting of the Parties is to consider whether the candidates have sufficient experience and recognized expertise in the fields related to the Convention, including, but not limited to, water management. The aim is also to have a balanced representation between legal, and scientific and technical expertise. The more diverse the experience and expertise of the Committee the more valuable is the advisory potential for, and the capacity in handling the implementation difficulties of, the Parties to the Convention.
- *Geographical distribution*: in the election of the Committee members, consideration would also be given to the geographical distribution of membership. In selecting candidates, the Meeting of the Parties should aim at a membership that represents different geographical subregions covered by the Convention.

As regards the term of office, the Meeting of the Parties, at its sixth session, elected five members for a full term of office and four members for a half term of office. A full term of office commences at the end of an ordinary session of the Meeting of the Parties and runs until its second ordinary session thereafter. Subsequently, at its seventh session, three out of the four members whose term had expired were re-elected for a full term and one more member was elected for a full term (UNECE 2015b).

If a member of the Committee can no longer perform his or her duties for any reason, the Bureau of the Meeting of the Parties would appoint another member fulfilling the above criteria to serve for the remainder of the term. Members would normally not serve for more than two consecutive terms, unless the Meeting of the Parties decides otherwise.

A Committee member may also find himself or herself facing a direct or indirect conflict of interest with respect to an individual matter under consideration by the Committee (UNECE 2013a, p. 9). In any such circumstance, the member in question should bring the conflict of interest to the attention of the Committee before consideration of that particular matter, or as soon as he or she becomes aware of it. The Committee itself may also become aware of a possible conflict of interest of its members and would then take an appropriate decision. If the conflict of interest is confirmed, the Committee member involved would not attend the parts of the meetings concerning the matter in question. It is noteworthy that being a citizen of the state whose implementation is to be discussed is not in itself a reason for conflict of interest.

The Committee elects its own chair and vice-chair for one term, with the possibility of re-election.

The decisions of the Committee should be adopted by consensus or, at least, every effort should be made towards achieving it. If all efforts have been exhausted and no agreement has been reached, decisions will be adopted by a three-quarter majority of the members present and voting or by a majority of five members, whichever is the greater number.

Functions and powers of the Implementation Committee

In accordance with its objective of facilitating, promoting and safeguarding the implementation and application of and compliance with the Convention, the Committee's functions concern difficulties that the Parties may face in implementing and complying with the Convention's provisions, distinguishing between specific and general issues.

Specific issues concerning difficulties in implementation or application

The action by the Committee whereby it would consider and examine specific implementation and compliance issues may be triggered in different ways that will be analysed in detail below. The following are the specific functions of the Committee:

- considering any request for advice relating to specific issues concerning difficulties in implementation or application by one or more State Parties;
- considering any submission by a State Party relating to specific issues concerning difficulties in implementation and compliance by another Party;
- considering undertaking its own initiative concerning specific difficulties in implementation or application by one or more State Parties it has become aware of;
- examining, at the request of the Meeting of the Parties, specific issues of implementation of and compliance with the Convention.

Advisory procedure

The advisory procedure is the innovative feature of the Water Convention's Implementation Committee compared to implementation and compliance mechanisms established under other multilateral environmental agreements. Under this procedure, the role of the Committee would be limited to considering a request for advice and providing such advice to the Party, or Parties, that have resorted to the procedure in question. It is to be emphasized that the initiation of such a procedure should not be regarded as a presumption of a situation of non-compliance.

The rationale of this non-confrontational and non-adversarial procedure lies in the political dimension and complexity of transboundary water issues coupled with the presumption that riparian states concerned would otherwise, in principle, be reluctant to admit difficulties in compliance with the Convention and to bring a self-submission before the Committee.

Two different scenarios are foreseen under the advisory procedure in the decision establishing the Committee (UNECE 2013a, p. 4):

- A Party requesting advice from the Committee about its own difficulties in implementing the Convention – as stated above, the existence of the advisory procedure is likely to attract Parties that are indeed facing difficulties with the Convention's implementation but would not want to be regarded as involved in a compliance procedure.
- A Party or Parties requesting advice from the Committee about its or their efforts to implement or apply the Convention vis-à-vis each other, other Parties and/or non-Parties – an example would be a number of riparians jointly approaching the Committee to request advice and support in developing a basin treaty.

The advisory procedure may, therefore, open the door for the participation of non-Parties to the mechanism under consideration following an invitation by the Committee. However, the participation by non-Parties would be subject to their consent. If a non-Party decides not to participate in the advisory procedure conducted in relation to a State Party and it is still considered to be potentially concerned by it, it would be kept informed about the continuation of the procedure. A similar approach would apply with respect to the Parties to the Convention that are not the requesting Parties, but are considered to be concerned by a particular matter.

The content of the possible advice provided to the Parties to the procedure would depend on the kind of difficulty submitted by the interested Party, or identified by the Committee. The different kinds of advice or facilitation of assistance that may be afforded by the Committee are outlined in the implementation procedure as follows:

- suggesting or recommending that domestic regulatory regimes be set up or strengthened and relevant domestic resources be mobilized as appropriate;
- assistance in establishing transboundary water cooperation agreements and arrangements for strengthening cooperation and sustainable management of transboundary waters;
- facilitating technical and financial assistance, including information and technology transfer, and capacity-building; and
- assistance in seeking support from specialized agencies and other competent bodies, as appropriate.

The Committee may also decide to request the Party or Parties concerned to develop an action plan to facilitate implementation of the Convention within a time frame to be agreed upon by the Committee and the concerned Parties and assist them to that effect, as appropriate.

The Committee could consider inviting the Party or Parties concerned to submit progress reports on the efforts that they are undertaking to implement its obligations under the Convention, with no possibility, unlike in its other functions, to recommend to the Meeting of the Parties to take any additional measures.

Submission by Parties

The self-submission foreseen by the implementation procedure implies that the submission is brought before the Committee by a Party that concludes that, despite its best endeavours, it is or may be unable to comply fully with the Convention.

Another possibility is for a Party, or a group of Parties, that consider themselves to be, or likely to become, affected by another Party's difficulties in implementing and/or complying with the Convention. It should be noted, however, that before addressing a submission and the relevant supporting information in writing to the secretariat for its subsequent transfer to the Committee, the Party that intends to make a submission should inform the Party whose implementation and/or compliance is in question. In the same manner, the secretariat should send a copy of the submission to the latter Party within two weeks of receiving it.

The latter procedure triggers an active involvement of the Party that is considered to experience difficulties with Convention's implementation as it is required to submit a reply with corroborating information to the secretariat within a three-month deadline which may be extended to up to six months should the circumstances of a particular case so require. The reply would be transmitted through the secretariat to the submitting Party or a group of Parties expecting its or their reaction. Based on this exchange, the Committee would then consider the matter as soon as practicable.

Committee initiative

The implementation procedure may also be triggered at the initiative of the Committee where it becomes aware of possible difficulties in implementation by a Party of or a possible non-compliance by a Party with the Convention.

The possibility for a Committee of the kind in question to take action upon its own initiative dates from the establishment of the Implementation Committee of the 1991 Convention on Environmental Impact Assessment in a Transboundary Context (Lammers 2014) but has not necessarily become a general rule for all the implementation and compliance mechanisms. For example, it is not envisaged in the compliance procedure under the Protocol on Water and Health to the UNECE Water Convention. Against this background, the introduction of a self-trigger mechanism allowing the Committee to be proactive in addressing the difficulties experienced by Parties is to be considered as a particularly progressive development in the direction of third-party implementation facilitation process.

The question is how may the Committee become aware of such difficulties? The implementation procedure (UNECE 2013a) clearly specifies that such awareness may arise 'including from information received from the public'. Such a possibility is yet another compromise formula reached in the negotiations within the Convention's Legal Board where the possibility of affording the public the power of communication to the Committee was also initially considered. While it was finally decided not to introduce communications from public, having the possibility of providing information to the Committee on the possible difficulties of a Party or Parties to implement the Convention became the compromise formula that would nonetheless involve a significant element of public participation in the process (Vykhryst 2015).

Next to generic information-gathering tools, one is to note the recent introduction of reporting by the Meeting of the Parties to the Convention at its seventh session (Budapest, 17–19 November 2015). Reporting under the Convention will start with a pilot exercise in 2016–17. The reporting template consists of three sections focusing on transboundary water management at national level (national legislation, economic, financial and technical measures), followed by questions for each transboundary basin or a group of basins (status and content of agreements, status and activities of joint bodies, other questions concerning different articles of the Convention), as well as final questions and comments. As stressed in the relevant decision, reporting would allow for the identification of emerging issues and difficulties in the implementation of the Convention and would therefore inform the work of the Implementation Committee (UNECE 2016). As soon as the Committee becomes aware of a situation which may potentially trigger its initiative, it may decide to request the Party concerned to provide further information on the matter. Upon receipt of information from the Party concerned, the Committee would need to determine whether undertaking its initiative would be appropriate and legitimate in that particular case. Such decision could not be left at the sole discretion of the Committee which is to base its determination on a number of criteria, namely, *inter alia*:

- the source of the information, by which the Committee has become aware of possible difficulties in the implementation by a Party of or possible non-compliance by a Party with the Convention, is known and not anonymous;
- the information is the basis for a reasonable assumption of possible difficulties in implementation or possible non-compliance;
- the information relates to the implementation of the Convention;
- an appropriate amount of time and resources are available to the Committee to consider the matter.

Examination of specific issues at the request of the Meeting of the Parties

The Committee should also examine, at the request of the Meeting of the Parties, specific issues of implementation of and compliance with the Convention. Presumably, such request would also need to contain additional instructions, taking into account the specific circumstances of each case, to be followed by the Committee as no further guidance is provided for in the decision establishing the Committee.

Information gathering and consultation

As a means to perform its functions, the mechanism to support implementation and compliance foresees that the Committee may undertake a number of activities to gather more accurate and detailed information beyond that provided by the Party concerned, or seek advice from the Convention's governing bodies and consult experts, as appropriate (UNECE 2013a, pp. 11–12).

In particular, the Committee may request further information on matters under its consideration, undertake, with the consent of the Party concerned, information gathering in the territory of that Party; gather any information it deems appropriate, including from the secretariat and the knowledge of the Committee members; invite the Parties and non-Parties concerned to attend its meetings, seek the services of experts and advisers, including governments, NGOs, academia and intergovernmental organizations as appropriate; seek the advice of the Meeting of the Parties and consult with other bodies of the Convention, as appropriate. Relevant information gathered through the above means or made available to the Committee, including from the public, should be duly taken into account by the Committee as long as the reliability of the source and the interests and motivations of its provider have been considered.

At first glance, the information-gathering function may be seen as complementary to other Committee's functions mentioned above. However, this has been confronted to a certain degree by the Committee's recent practice whereby acquiring information has been an ongoing self-standing activity of the Committee even if it could potentially lead to other types of actions: prior to its second meeting (Geneva, 12 December 2013), the Committee received information provided by a member of public expressing concerns regarding difficulties in transboundary water cooperation between two Parties and two non-Parties to the Convention. Without prejudice to any future decision and within the framework of its information-gathering function, the Committee decided to request the Parties concerned to provide their views on the matter. The information-gathering activity has been since ongoing further to the Committee's deliberations at its third and fourth meetings (Bologna, 15 May 2014 and London, 4 December 2014, respectively). Currently, following its fifth meeting (Vienna, 5–6 May 2015), the Committee is still in the information-gathering phase, with a view to considering the appropriate course of action to take on the matter in hand (UNECE 2013c, 2014a, 2014b, 2015a).

The possibility that the Committee could be activated by *referrals by the secretariat* had been considered by the Legal Board during its deliberations. This option was eventually discarded, partly in light of the fact that in similar implementation and compliance mechanisms it has been rarely used (Lammers 2014). Most importantly, the Legal Board grounded its decision on the point in hand in order to preserve the neutrality of the role of the secretariat vis-à-vis States Parties.

Measures to facilitate and support implementation and compliance and to address cases of non-compliance

The measures that the Committee may decide upon are the same as those that were mentioned above in relation to the advisory procedure, with the addition of the possibility for the Committee to address the Meeting of the Parties with a recommendation to take additional measures. Such measures are as follows (UNECE 2013a, pp. 7–8):

- Providing advice and facilitating assistance to individual Parties and groups of Parties in order to facilitate their implementation of and/or compliance with the Convention, which may include:
    - ° suggesting or recommending that domestic regulatory regimes be set up or strengthened and relevant domestic resources be mobilized as appropriate;
    - ° providing assistance in establishing transboundary water cooperation agreements and arrangements for strengthening cooperation and sustainable management of transboundary waters;
    - ° facilitating technical and financial assistance, including information and technology transfer, and capacity-building; and
    - ° assistance in seeking support from specialized agencies and other competent bodies, as appropriate.
- Requesting and assisting, as appropriate, the Party or Parties concerned to develop an action plan to facilitate implementation of and compliance with the Convention within a time frame to be agreed upon by the Committee and the Party or Parties concerned;
- Inviting the Party concerned to submit progress reports to the Committee on the efforts that it is making to comply with its obligations under the Convention;
- Recommending to the Meeting of the Parties that it takes a number of measures.

The Committee may also address the Meeting of the Parties to the Convention, as its superior governing body, if it considers that additional measures would be needed to ensure the implementation and application of and compliance with the Convention. The Committee would usually address the Meeting of the Parties in a report containing specific recommendations submitted to one of its ordinary sessions upon consideration of which the Meeting of the Parties may decide upon one or more of the following measures:

- to take the measures also available to the Committee as outlined above and other non-confrontational, non-judicial and consultative measures as appropriate;
- to recommend that Parties provide financial and technical assistance, training and other capacity-building measures and facilitate technology transfer;
- to facilitate financial assistance and provide technical assistance, technology transfer, training and other capacity-building measures, subject to financial approval, including, when appropriate, seeking support from specialized agencies and other competent bodies;
- to issue a statement of concern;
- to issue declarations of non-compliance;
- to issue cautions;
- to suspend, in accordance with the applicable rules of international law concerning the suspension of the operation of a treaty, the special rights and privileges accorded to the Party.

Presumably, measures taken by the Meeting of the Parties may add value to the Committee activities, due to its greater outreach, resources and possibility of providing support to capacity-building as well as the political authority. The implementation procedure outlines a number of criteria to be taken into account by the Meeting of the Parties when considering the recommendations the Committee: its decision should depend on the circumstances of each given case and should take into account the cause, type, degree and frequency of the difficulties with implementation and/or of non-compliance by a Party or group of Parties. It is to be expected, based on the distinctive collaborative spirit of the Convention and its focus on support to implementation that the four last measures from the list above would be rather exceptional.

The Committee would be responsible for monitoring the consequences of implementation of action pursuant to its measures or the measures taken by the Meeting of the Parties.

General issues of implementation and compliance

In addition to the above mentioned functions, the Committee should also carry out any other functions that may be assigned to it by the Meeting of the Parties, including examination of general issues of implementation and compliance that may be of interest to all Parties, and report to the Meeting of the Parties accordingly (UNECE 2013a, p. 4). These reports, submitted to each ordinary session of the Meeting of the Parties, would include information on the Committee's activities and the reasoning for its decisions, as well as any appropriate recommendations

In particular, the Committee has recently submitted its first report containing a decision on general issues of implementation to the seventh session of the Meeting of the Parties (UNECE 2015c, p. 8). The decision specifically recalled that 'transboundary cooperation is a key principle of the Convention, as it supports the achievement of the Convention's object and purpose', stressing, however, that: 'Cooperation *per se* is not the only objective of the Convention and that the principles of reasonable and equitable use and of prevention, control and reduction of transboundary impact are no less important.'

The following rules and considerations, *inter alia*, are also applicable to all Committee procedures (UNECE 2013a, pp. 4, 6 and 7):

- Entitlement to participate
  The Party or Parties concerned, as well as the members of public submitting relevant information, would be entitled to participate in the Committee discussions with respect to a particular case, and to comment on the finalization of findings and measures by the Committee. The same entitlement applies to the Parties and/or non-Parties that the Committee considers to be potentially concerned, if they have expressed their consent to participate in the procedure. Only the members of the Committee, however, would take part in the preparation and adoption of its findings and measures.
- Confidentiality
  As a general rule, no information held by the Committee would be kept confidential. Nevertheless, if the information has been explicitly provided in confidence, the Committee and any person involved in its work would be responsible for ensuring its confidentiality. In the same manner, the meetings of the Committee would be held in public and would be open to observers, unless the Committee decides otherwise.
- Availability of resources
  The Committee will carry out its functions in accordance with the time and resources available to it, with reference to both financial and human resources in terms of involvement of the Committee members and secretariat support. This clause may become particularly important in future in case the workload of the Committee would increase, judging by the experience of, e.g., the Compliance Committee of the UNECE Convention on Access to Information, Public Participation in Decision-making and Access to Justice in Environmental Matters (UNECE Aarhus Convention 1998) which, following its establishment in 2003,

experienced a major increase in the number of submissions and cases, starting from 2011 and it is expected that its workload would be increasing further.

Future plans and prospects

In its report to the sixth session of the Meeting of the Parties to the Convention (UNECE 2015c), the Committee encouraged Parties and other stakeholders to seek its assistance, support and facilitation to address difficulties in implementing and complying with the Convention and in order to prevent water-related disputes.

The Committee also decided to further promote its mandate and main functional duties as well as continue raising awareness on the principles of international water law and provisions of the UNECE Water Convention. One may expect the Committee to engage in both general and targeted promotion activities, including regional and sub-regional events on international water law, involving NGOs and civil society through communication to the broader public, and also encouraging the participation by Committee members in meetings beyond the programme of work of the Convention.

At the same time, the recently adopted regular reporting mechanism will undoubtedly provide the Committee with an objective source of information regarding the Convention´s implementation by its Parties which the Committee needs to implement its core functions. Indeed, reporting is recognized as essential for reviewing and enhancing national implementation of the Convention by Parties and possibly also by countries that are not yet Parties.

In sum, as stressed by Tanzi and Contartese (2015, p. 324):

> It is arguable that, if appropriately resorted to by Parties and/or functioning upon its own initiative, also on the basis of a reporting system, the Implementation Committee could assist Parties in appropriately complying with the Convention, including with their duties of cooperation, hence, furthering dispute prevention among Riparian States.

Note

1 The chapter has benefited from comments by Professor Attila Tanzi, Full Professor of International Law, Chair of the Implementation Committee of the UNECE Water Convention.

References

Bernardini F. (2015) The normative and institutional evolution of the Convention. In A. Tanzi, O. McIntyre, A. Kolliopoulos, A. Rieu-Clarke and R. Kinna (eds), *The UNECE Convention on the Protection and Use of Transboundary Watercourses and International Lakes: Its Contribution to International Water Cooperation*. Leiden: Brill Nijhoff, ch. 3, pp. 46–7.

Convention on the Protection and Use of Transboundary Watercourses and International Lakes (1992) 17 March (entered into force on 6 October 1996). Treaty Series. Treaties and international agreements registered or filed and recorded with the secretariat of the United Nations, vol. 1936, No. 33207, p. 269. Available from: https://treaties.un.org/Pages/showDetails.aspx?objid=0800000280044685 [accessed on 22 March 2016].

Convention on Access to Information, Public Participation in Decision-making and Access to Justice in Environmental Matters (1998) 25 June (entered into force on 30 October 2001). Treaty Series. Treaties and international agreements registered or filed and recorded with the secretariat of the United Nations, vol. 2161, No. 37770, p. 447. Available from: https://treaties.un.org/doc/Publication/UNTS/Volume%202161/v2161.pdf [accessed on 22 March 2016].

Lammers, J.G. (2014) The Helsinki Water Convention: a new implementation mechanism and committee. *Environmental Policy and Law* 44(1–2), pp. 117–24.

Tanzi A. (2015a) *The Economic Commission for Europe Water Convention and the United Nations Watercourses Convention: An Analysis of Their Harmonized Contribution to International Water Law*. Geneva: United Nations.

Tanzi A. (2015b) *Dispute Settlement in UNECE Water Convention* [video online]. Available from: https://www.youtube.com/watch?v=-qlXInsjWHA [accessed 4 January 2016].

Tanzi A. and Contartese C. (2015) Dispute prevention, dispute settlement and implementation facilitation in international water law: the added value of the establishment of an implementation mechanism under the Water Convention. In A. Tanzi, O. McIntyre, A. Kolliopoulos, A. Rieu-Clarke and R. Kinna (eds), *The UNECE Convention on the Protection and Use of Transboundary Watercourses and International Lakes: Its Contribution to International Water Cooperation*. Leiden: Brill Nijhoff, ch. 22.

Tanzi A., Kolliopoulos A. and Nikiforova N. (2015) Normative features of the UNECE Water Convention. In A. Tanzi, O. McIntyre, A. Kolliopoulos, A. Rieu-Clarke and R. Kinna (eds), *The UNECE Convention on the Protection and Use of Transboundary Watercourses and International Lakes: Its Contribution to International Water Cooperation*. Leiden: Brill Nijhoff, ch. 9.

UNECE (United Nations Economic Commission for Europe) (2009) *Facilitating and Supporting Implementation and Compliance: A Needed Step in the Convention's Evolution* [pdf]. Geneva: United Nations. Available from: http://www.unece.org/fileadmin/DAM/env/documents/2009/Wat/mp_wat/ECE_MP_WAT_2009_3_e.pdf [accessed on 4 January 2016].

UNECE (2013a) *Report of the Meeting of the Parties on its Sixth Session. Decisions and vision for the future of the Convention (ECE/MP.WAT/37/Add.2). Decision VI/1. Support to implementation and compliance. Annex 1: Mechanism to support implementation and compliance and annex 2: Core rules of procedure of the Implementation Committee* [pdf]. Geneva: United Nations. Available from: http://www.unece.org/fileadmin/DAM/env/water/mop_6_Rome/Official_documents/ECE_MP.WAT_37_Add.2_ENG.PDF [accessed on 4 January 2016].

UNECE (2013b) *Guide to Implementing the Water Convention*. Geneva: United Nations, pp. 98–100.

UNECE (2013c) *Report of the Implementation Committee on its Second Meeting* [pdf]. Geneva: United Nations. Available from: http://www.unece.org/fileadmin/DAM/env/water/meetings/Implementation_Committee/1st_meeting/Documents/ECE_MP.WAT_IC_2013_2_report_ENG.pdf [accessed on 4 January 2016].

UNECE (2014a) *Report of the Implementation Committee on its Third Meeting* [pdf]. Geneva: United Nations. Available from: http://www.unece.org/env/water/3rd_implementation_committee_2014.html#/ [accessed on 4 January 2016].

UNECE (2014b) *Report of the Implementation Committee on its Fourth Meeting* [pdf]. Geneva: United Nations. Available from: http://www.unece.org/fileadmin/DAM/env/documents/2014/WAT/12Dec_4_IC_London/ECE_MP.WAT_IC_2014_4_report.pdf [accessed on 4 January 2016].

UNECE (2015a) *Report of the Implementation Committee on its Fifth Meeting* [pdf]. Geneva: United Nations. Available from: http://www.unece.org/fileadmin/DAM/env/documents/2015/WAT/05May_5-6_IC/ece.mp.wat.ic.2015.2_report_E.pdf [accessed on 4 January 2016].

UNECE (2015b) *Election of the Members of the Implementation Committee* [online]. Available from: http://www.unece.org/env/water/mop7.html#/ [accessed on 4 January 2016].

UNECE (2015c) *Report of the Implementation Committee to the Meeting of the Parties.* [pdf] Geneva: United Nations. Available at: http://www.unece.org/fileadmin/DAM/env/documents/2015/WAT/11Nov_17-19_MOP7_Budapest/ECE_MP.WAT_2015_5_IC_report_ENG.pdf [accessed on 4 January 2016].

UNECE (2016) *Report of the Meeting of the Parties on its Seventh Session* [online]. Geneva: United Nations. Available from: http://www.unece.org/fileadmin/DAM/env/documents/2015/WAT/11Nov_17-19_MOP7_Budapest/ece.mp.wat.49_for_submission_FINAL.pdf [accessed on 21 August 2016].

Vykhryst S. (2015) Public information and participation under the Water Convention. In A. Tanzi, O. McIntyre, A. Kolliopoulos, A. Rieu-Clarke and R. Kinna (eds), *The UNECE Convention on the Protection and Use of Transboundary Watercourses and International Lakes: Its Contribution to International Water Cooperation*. Leiden: Brill Nijhoff, ch. 18.

21

THE ROLE OF NON-STATE ACTORS IN THE DEVELOPMENT AND IMPLEMENTATION OF INTERNATIONAL WATER LAW

Komlan Sangbana

Introduction

Since the adoption of the *Rio Declaration on Environment and Development* in 1992 and the Aarhus Convention on Access to Information, Public Participation in Decision-Making and Access to Justice in Environmental Matters in 1998 (hereinafter the Aarhus Convention), non-state actors have played an increasingly prominent role in environmental issues (Boisson de Chazournes 2013; Louka and Krchnak 2005; Razzaque 2009). This humanisation in the governance of environmental resources is reflected in the area of shared freshwater resources through the development of legal rules and mechanisms for integrating individuals and communities into the management and protection of transboundary freshwater. Indeed, projects related to transboundary water resources can have an impact on the living conditions and health of local populations. Unsurprisingly, individuals and communities likely to be affected by such projects are increasingly demanding that they be consulted prior to the implementation of a project and that transparency in the decision-making processes is enhanced (Tignino 2010).

The recognition of non-state actors in the governance of these resources has its basis in the concept of public participation, which is one of the principles governing the management and protection of shared waters in several agreements regarding water sharing (Senegal River 2002, Article 13; Victoria Basin 2003, Article 3(1), 4(2)(h), 12(2), 22; Niger Basin 2008, Chapter VII; Great Lakes 2012, Article 2(4)(k); Chad Basin 2012, Article 73). In view of this broad recognition of the importance of involving the public in the governance of transboundary freshwater, we might ask what impact these actors actually have on the elaboration and application of norms for the management and protection of international freshwater. In this respect, this chapter aims to underline the role of non-state actors in the development of rules governing shared freshwater and the mechanisms that promote the contribution of non-states actors. The role of these actors can be observed through the prism of various practices. This chapter, however, will focus in particular on the practice of institutional mechanisms put in place by freshwater agreements. Indeed, river basin institutions increasingly facilitate greater public involvement in the process of norm development for the protection of transboundary water resources. This process, which may be qualified as a participatory approach, is justified by the dominant paradigm of integrated water resources management (IWRM). IWRM implies the involvement of all stakeholders and interested parties (Sangbana 2011). The practice of basin institutions is also an interesting area of study because, on the one hand, it evidences the contribution of non-state actors in the development of norms governing the management and protection of freshwater and, on the other, it demonstrates how these actors contribute to the implementation of these norms.

Contribution of non-state actors to the development of norms governing the management and protection of freshwater

Contribution through observer status

Observer status is subject to specific and varied criteria both in terms of how it may be acquired and exercised. The criteria for granting observer status are used to identify actors that may participate in the works of the organs of the basin institution. The analysis of the criteria developed in each basin institution considered reveals a variety of actors. This disparity of approach is related to the mandate conferred on each organisation. For structures with essentially technical competences, preference is given to non-governmental organisations (NGOs) with relevant expertise for the accomplishment of the basin institution's mission. This is the case with the International Commission for the Protection of the Rhine (ICPR). The Rules of Procedure of the ICPR only admits observer NGOs that have specific technical or scientific competences that align with the objectives of the Convention (ICPR 2010, point 8.1.b). The application must therefore include a description of the skills and experience that the NGO can offer to the Commission and indicate the reasons why it believes that its contribution would be useful to the work of the Commission (ibid., point 8.2.a.b). The International Commission for the Protection of the Danube River (ICPDR) likewise reproduces these criteria in the Guidelines for Participants with Consultative Status and for Observers to the ICPDR (ICPDR 2005, point 3.2 and 3.3.a). It, however, further requires that the applicant organisation be present in at least three states bordering the Danube (Mandl 2015, p. 51). The anticipated perspective is clearly that of regional organisations. Local user associations are not entitled to benefit from the observer status. Their participation is organised in each Member State of the ICPDR.

In the case of basin organisations with general expertise for basin development, priority is given to entities representing the interests of local populations and communities. This is the case before the Permanent Water Commission of the Organisation for the development of the Senegal River (OMVS). According to Article 23 of the Senegal River Water Charter (2002), observer status with the Permanent Water Commission may be granted to representatives of river users and of NGOs. This participation is, however, framed in terms of the selection procedure of these entities. Indeed, 'observer status' is granted by the decision of the Council of Ministers on the proposal of the OMVS High Commissioner (Article 23). Additionally, entities are screened by the national authorities of each country and proposed to the High Commissioner of the OMVS. To participate, these entities must be directly concerned with the case before the Permanent Water Commission. The immediate consequence of this participation mechanism is that the granting of observer status is efficient.

Observer status confers on its beneficiaries the right to participate in meetings of the organisations into which they are admitted. This right of participation notably includes a right of initiative and the right to intervene. The right of initiative is reflected in the ability, for entities with observer status, to submit documents and proposals to be discussed during the sessions in advance. ICPDR accords such observers the right 'to submit relevant papers and proposals to the International Commission, which are distributed by the Secretariat and may be discussed at the meeting' (ICPDR 2005, point 3.1.d; also ICPR 2010, point 3.1.d). This right is particularly important since it enables these organisations to persuade Member States to take action on emerging issues relating to the protection of water resources. Organisations enjoying observer status are therefore no longer only involved in the decision-making process, they can also partici-pate in the process from the very beginning. This practice is common at the ICPDR. For instance, to fight against the negative externalities of hydropower development in the Danube, NGOs have introduced draft principles to the Expert Group on River Basin Management of the Commission. The first project, entitled 'Danube NGO joint position concerning hydropower development in the Alpine, Carpathian, Danube and Western Balkan regions' was introduced by three NGOs, namely the World Wide Fund for Nature (WWF), the International Association for Danube Research (IAD) and the European Anglers Alliance (EAA). It identifies, first, the prin-ciples governing the planning of hydroelectric development projects in national policies and, second, those related to the implementation of the project. Among others, the duty to carry out an environmental impact assessment (EIA), the use of best practices and available technologies and the involvement of local communities are all included within these principles. The second project, 'Guiding principles of hydropower development', submitted by the Danube Environmental Forum, suggests principles to guide development of sustainable hydropower, such as the principle to protect biodiversity and ecosystems and the polluter-pays principle. Both projects have allowed the conduct of work for the adoption of the Guiding Principles on Sustainable Hydropower in June 2013.

The question as to whether the delegates of Member States are required to review the docu-ments and proposals submitted to them by NGOs may be resolved by having recourse to their right of initiative. While the practice is not explicit on the issue, the answer must be in the nega-tive. Even though the right of initiative grants real influence to NGOs in shaping the regulation, the decision to review and incorporate the proposals remains the responsibility of Member States' delegates. The right of initiative does not imply an obligation upon states to decide in accordance with the initiative introduced. The ICPR confirms this interpretation by stating in its Rules of Procedure that even the review of information documents and observers' proposals 'is left to the discretion of the participants in relevant meetings' (ICPR 2010, point 8.8).

The right to intervene appears as the capacity for entities with observer status to intervene during debates to give their points of view and opinions. The Guidelines of the ICPDR stipulate that the observers have the right 'to participate in meetings organised in the framework of the Convention, in which they are entitled to participate, with the possibility to express their posi-tion and views and to have them reflected in the relevant documents' (ICPDR 2005, point 3.1.c). This provision confers real power on these observers. It requires that the instruments to be adopted reflect the views or opinions given by these actors. The participation of these entities therefore makes perfect sense. They may influence the final recommendations and contribute to more effective protection of transboundary freshwater. Nevertheless, it is necessary to add that the expansion of this right to intervene is not the same everywhere. For instance, the Rules and Procedures of the Permanent Commission of the Okavango River Basin Commission considers an observer as, 'a person that... has been allowed by the Co-Chairperson of the Commission at a particular meeting... to listen to the discussions at the meeting that may be of the interest to the observer or the institution that is represented by the observer' (OKAKOM 2010, point 4.7.1). From this definition of the quality of the observer, the Rules and Procedures mentioned above stipulate: 'observers shall not be permitted to participate in the discussions at a Commission meeting unless specifically requested to do so by the Co-Chairperson' (ibid., point 4.7.3).

The right to participate in the discussions is therefore restrained; observers essentially have the function 'to inform the Commission or to clarify some issues raised during the discussions at the meeting in connection with issues relevant to his or her presence in the meeting' (ibid., point 4.7.2).

Contribution through the channel of public hearing

Public consultation is being increasingly promoted to allow the involvement of non-state actors in the policy development process and regulations for river basin. They appear as a procedure that allows basin institutions to collect opinions and proposals from the local population and communities on issues relating to the management and protection of transboundary water resources. Their specificity lies in the fact that it targets the participation of the largest number of people affected. The participation of the greatest number of stakeholders increases the legitimacy of the measures to be adopted.

Public hearings could concern general programme and policies relating to the basin management. The practice of the International Joint Commission (IJC) illustrates this trend. The 2012 Great Lakes Water Quality Agreement includes among the responsibilities of the Commission:

> consulting on a regular basis with the Public about issues related to the quality of the Waters of the Great Lakes, and about options for restoring and protecting these waters, while providing the Public with the opportunity to raise concerns, and tender advice and recommendations to the Commission and the Parties.
>
> *Article 7(1)(g)*

By using 'public', the Agreement meant 'individuals and organizations such as public interest groups, researchers and research institutions, and businesses and other non-governmental entities' (Article 1(f)). It is in this perspective that the Agreement provides for the holding of a Public Forum on the Great Lakes in the year of its inception and, subsequently, every three years (Article 5(1)). The Public Forum on the Great Lakes is notably an opportunity for 'the Parties to discuss and receive Public comments on the state of the lakes and binational priorities for science and action to inform future priorities and actions' (Article 5(1)(a)).

The involvement of local communities through public consultations may also be observed in the practice of African basin organisations. The practice within the Organisation for the Development of the Senegal River (OMVS) illustrates this point. Local coordinating committees, which exist in the context of OMVS National Cells in each Member State, serve as excellent platforms for the participation of local populations because of their geographic area of competence (Sow and Touré 2008). By bringing together, among others, professional association groups and local NGOs' representatives (Bedredine and Fabre 2010), these committees have facilitated such public consultations during the development of the Water Development and Management Master Plan (SDAGE) of the Senegal River.

Similarly, the 'Basin Wide Forum' set up in the framework of the Permanent Commission of the Waters of the Okavango River Basin facilitates the holding of these public consultations. This particular structure brings together groups such as tribal government representatives, community institutions, such as village development communities, and the technical committees of the villages, as well as representatives of groups of water users – particularly farmers, fishermen and craftsmen (OKAKOM n.d.). One of the essential functions of the Forum is to provide the parties with socio-economic and environmental landscape information to facilitate the adoption of adequate standards of protection of the basin (ibid.).

Long ignored in the legal framework of the Mekong River Commission (MRC) (Schmeier 2013), the importance of the involvement of non-state actors in decision-making has been recognised in the Public Participation Strategy of the MRC, adopted in 2003, in these terms: '[S]takeholder involvement in decision-making about sustainable development is fundamental to achieving feasible, equitable and lasting solutions' (MRC 2009). This statement has found application for the first time in the context of the elaboration of the development plan for the basin in 2008. The development of this plan led to public consultations in its phase 2 (MRC 2008a). Similarly, the adoption of the 'Initiative on Sustainable Hydropower' was accompanied by public consultations which assembled at least 200 non-state actors who gave commentaries on the programmes (MRC 2008b).

Public consultations are also held in several European basin organisations, such as the ICPR and the ICPDR (Mandl 2015). In particular, they are part of the development of plans for flood risk management.

Public hearings may also be realised in the context of an EIA. These hearings permit consultation with local populations about the risks of a planned project identified in an EIA. In the Pulp Mills case, the ICJ considered the EIA as 'a requirement under general international law' (Pulp Mills 2010, paragraph 204). The Court also referred to the consultation with local populations during an EIA, recognising that public hearings were held in Argentina and Uruguay before the execution of the project on the Uruguay River (ibid., paragraph 217–219). The implementation of an EIA requires, in general, the consultation of a largest number of people affected. This point is underlined by the Espoo Convention on Environmental Impact Assessment in a Transboundary Context (1991), which states that:

> [t]he Party of origin shall provide, in accordance with the provisions of this Convention, an opportunity to the public in the areas likely to be affected to participate in relevant environmental impact assessment procedures regarding proposed activities and shall ensure that the opportunity provided to the public of the affected Party is equivalent to that provided to the public of the Party of origin.
>
> *Article 2(6)*

> Similarly, the Convention on the Sustainable Management of Lake Tanganyika specifies that 'individuals and communities living within the Lake Basin… have the right to participate at the appropriate level, in decision-making processes that affect the Lake Basin or their livelihoods, including participation in the procedure for assessing the environmental impacts of projects or activities that are likely to result in adverse impacts.
>
> *Article 17*

Thus EIA appears as a means to involve non-state actors in the process of developing the regime of protection and management of transboundary freshwater.

Contribution through expert status

It is increasingly common for some institutions to involve non-state actors in the development of norms through the granting of expert status. Experts may be academics or highly qualified researchers in water management. This participation shows a new form of involvement of non-state actors in the management of transboundary water resources. However, the degree of involvement in the development of norms varies among institutions.

The most common approach is that of invitation. This leaves the institution with the opportunity to judge whether to call upon external experts or not. The ICPR and the International

Commission of the Meuse adopt this approach. These commissions may decide to consult external experts and invite them to their meetings (e.g. Convention on the Protection of the Rhine (1999), Article 14(5); International Agreement on the River Maas (2006), Article 6(5)). Usually, in these cases, their intervention is restricted to specific themes and their role is to provide elements of insight on matters falling within their field of expertise.

The participation of external experts may also be institutionalised. As part of the IJC, the Great Lakes Water Quality Agreement empowers the Great Lakes Science Advisory Board to advise the Commission and Great Lakes Water Quality Board (Article 8(1) and (2)). Under Article 8(4) on the Advisory Board, the latter also comprises representatives of the parties such as 'managers of Great Lakes research programs and recognized experts on Great Lakes water quality problems and related matters'. The website dedicated to the organ specifies its functions (IJC n.d.). One can notice a distinction, thereby, between government research managers and non-governmental research managers. In addition, it also distinguishes between governmental experts and non-governmental scientific experts. The balance between government representatives and non-governmental representatives is also an objective expressly pursued by the Commission, as mentioned on its website. Therefore, under the IJC framework, external actors as well as national experts are fully involved in the development of standards against pollution. This remarkable approach strengthens the participation of non-state actors in the process of developing standards against pollution. However, such participation remains under the supervision of the IJC. According to the Great Lakes Water Quality Agreement (2012), Council members are actually appointed by the Commission, subject to consultation with the relevant government (Article 8 (2)).

The contribution of non-state actors in the implementation of norms governing the management and protection of freshwater

Non-state actors as a conduit of implementation

Public participation contributes to the protection of transboundary water resources. Non-state actors appear in this context as full participants in the implementation of these standards for the governance of shared water resources. The contribution of non-state actors in this area is particularly noticeable in the context of the implementation of norms for the prevention, reduction and control of pollution.

Non-state actors can facilitate the application of standards against pollution by providing the institutions tasked with the management of these resources the additional information they need to assess the impact of utilisation on the quality of the resource and the aquatic ecosystem. As a result, basin institutions are assisted in adopting the most appropriate preventive and corrective measures. This is highlighted by the public forum organised in the context of the management of the Great Lakes between the United States and Canada. This Forum gives an opportunity to non-state actors – such as public interest groups, researchers and research institutions – to provide information concerning the quality of water. Generally speaking, six months after each Forum, in consultation with the Executive Committee of the Great Lakes, the parties establish binational scientific programmes and priority actions needed to address current and future threats to the Great Lakes' water quality. According to the Great Lakes Water Quality Agreement (2012), these measures are defined based on an assessment of the state of the Great Lakes, on the comments received during the Public Forum of the Great Lakes and the recommendations of the Commission (Article 5(2)(c)).

In other cases, non-state actors are more directly involved in the application of the norms. For instance, they can be involved in the monitoring of freshwater quality. The Niger Basin Authority

(NBA) practice illustrates this point. Indeed, the establishment of a quality water monitoring system across the Niger Basin is one of the major projects in the framework of the fight against pollution. In practice, this operational mechanism will mean the installation of the network for measuring the water quality in the Niger River Basin. The monitoring system includes two categories of networks, namely conventional networks for measuring the water quality and the alternative networks for measuring water quality. Unlike the conventional network based on the technical expertise of scientists, the alternative network relies on the contribution of local communities in monitoring water quality. Concretely, NBA provides measuring kits to the concerned populations to enable them to monitor the water quality of the Niger. It must be said that this involvement, aside from the collection of information, helps to raise public awareness on the need to preserve water quality.

Another example of direct involvement is provided by the practice within the OMVS (Sangbana 2015). The organisation particularly refers to users' associations for the carrying out of certain activities within the framework of the fair and coordinated management of the Senegal River Basin water resources, especially through the maintenance of the hydraulic axes, erosion control or the construction of drinkable water supply systems. This process was marked by innovative actions such as the cleaning of access points for livestock watering, the dredging of ditches and runoff channels, the removal of typha, reforestation and production of stony cords etc. (OMVS n.d.). As in the case of the Niger, these initiatives aim to empower local communities and populations. The need for empowerment of local populations in the conservation of transboundary freshwater is a major issue in the protection of transboundary freshwater. This is especially the case in contexts where polluting practices relate mostly to discharges of wastewater and pesticides directly into the watercourse. In this context, public involvement here also aims to educate in the sense of raising awareness among these populations about the importance of water resources and ecosystems.

Involvement of non-state actors in monitoring of implementation

Non-states actors are increasingly involved in follow-up and control activities. This involvement is twofold. The public can either be an external source of information for monitoring and control, or a direct monitoring mechanism for the implementation of certain standards.

The use of the public as an external source of information appears as a main component of prior consultation procedures of planned measures. The Xayaburi project is a good illustration of this. A typical case is the construction by Laos of a large hydroelectric dam on the main course of the Mekong River (MRC 2011a). In accordance with Article 5 of the Agreement on the Mekong Basin of 1995, which requires prior consultation in the case of proposed measures and procedures, as well as the Mekong Commission's Guidelines, the Laotian authorities submitted documents concerning the Xayaburi project (MRC 2011a).

Several working groups were formed to support the process of prior consultation, including the MRC Prior Consultation Joint Committee Working Group, composed of four representatives of each Member State. This Group was charged with conducting the entire process. At its first session on 26 October 2010, members of the Working Group agreed on the need to include a series of stakeholder consultations in the process of prior consultation. Stakeholder consultations are an important milestone in the MRC prior consultation process because they had not been envisaged previously (Rieu Clarke 2015). Public consultations were conducted between January and February in Cambodia (10 and 28 February 2011), in Thailand (10, 12, 22 January and 16 February 2011) and in Vietnam (14 January and 22 February 2011) (MRC 2011b, p. 10). The Laotian authorities did not consider it necessary to conduct these consultations, arguing that they had already carried them out during the preparation of the impact survey.

The consultations were conducted at two levels in the three states: on the one hand, between local communities living downstream from the dam project who may be affected and, on the other, between a wide range of stakeholders at the national level. The actors consulted were able to identify several negative impacts of the project on the river and requested more detailed studies on the consequences of the project (MRC 2011b, pp. 17–18). On the basis of these consultations, the pre-consultation process was not conclusive. Indeed, in accordance with the wishes expressed by the groups that were consulted, Cambodia, Thailand, Vietnam requested the extension of the period of six months provided for in the Procedures on Notification and Pre-consultation of the Mekong Commission. This position was challenged by Laos, which decided to unilaterally proceed with the construction of the dam (Rieu Clarke 2015). This case shows how the public can contribute by emphasising the failures of states in the implementation of international standards and thereby provide assessment criteria for the basin institution in the exercise of its control activities.

In other cases, the public is involved in the review of the implementation of norms by states. This is the case for the IJC. The 2012 Great Lakes Water Quality Agreement requires that parties: 'shall publicly report, in the Progress Report of the Parties, state of the Great Lakes Report and Lakewide Action and Management Plans, on the progress in achieving the General Objectives, Lake Ecosystem Objectives and Substance Objectives' (Article 3(4)). Article 5(2)(e) of the Agreement provides for parties to prepare:

> a binational Progress Report of the Parties to document actions relating to this Agreement, taken domestically and binationally. The first such report shall be provided to the Public and the Commission before the second Great Lakes Public Forum, and subsequent reports shall be provided before each subsequent Great Lakes Public Forum.

Non-state actors are thus seen as being tasked with monitoring, in addition to the Commission. In order to facilitate the exercise of this prerogative, the Commission makes available to the public all the advice and recommendations made to parties, subject to the authorisation of the parties (Article 7(4)).

The possibility for the public to review the application of these standards helps to strengthen the decisions made by the IJC. Moreover, in the case of the IJC, the Great Lakes Water Quality Agreement states that the evaluation of the progress report provided by the Commission after the review of the Parties Progress Report must include a summary of public comments on it (Article 7(1)(k)(ii)). Parties are required to consider this evaluation report when adopting relevant measures (Article 5(4)). This form of highly developed environmental democracy is quite unique. Public participation in decision-making appears here not only as the right to participate in the process leading to the adoption of the standard, but also as the right to conduct monitoring of the application of this standard.

Conclusion

As has been shown, non-state actors participate actively in the development and application of international water law. The variety of mechanisms developed in the framework of basin institutions facilitate the participation of these actors in the governance of transboundary freshwater resources. This trend generates an improvement in the content of these standards. Indeed, the participation of non-state actors contributes as much to the legitimacy as to the effectiveness of the measures adopted in the framework of basin institutions since these measures have the support of local communities. This emerging trend highlights the importance of going beyond the interstate context to ensure better management of shared water resources.

References

Bedredine F. and Fabre, G. (2010) Exemple de gestion concertée et solidaire d'un bassin fluvial partagé (Guinée, Mali, Mauritanie, Sénégal) [online]. Available from: http://www.riob.org/spip.php?page=mot-pays&id_mot=85&lang=fr (accessed 7 January 2016).

Boisson de Chazournes, L. (2013) *Fresh Water in International Law*. Oxford: Oxford University Press.

Chad Basin Water Charter (2012) 30 April.

Convention on Access to Information, Public Participation in Decision-Making and Access to Justice in Environmental Matters (1998) 25 June, United Nations Treaty Series, 2161 (1998) 447.

Convention on Cooperation for the Protection and Sustainable Use of the Danube River (1994) 29 June [online]. Available from: https://www.icpdr.org/main/icpdr/danube-river-protection-convention

Convention on the Protection of the Rhine (1999) 12 April [online]. Available from: http://www.iksr.org/fileadmin/user_upload/Dokumente_en/convention_on_tthe_protection_of__the_rhine.pdf

Convention on the Sustainable Management of Lake Tanganyika (2003) 12 June [online]. Available from: http://www.ecolex.org/server2.php/libcat/docs/TRE/Full/En/TRE-001482.pdf

Espoo Convention on Environmental Impact Assessment in a Transboundary Context (1991) 25 February, UNTS, 1989 (1991) 309.

Great Lakes Water Quality Agreement (2012) 7 September [online]. Available from: http://www.ijc.org/fr_/Great_Lakes_Water_Quality

ICPDR (2005) Guidelines for Participants with Consultative Status and for Observers to the ICPDR [online]. Available from: http://www.icpdr.org/main/icpdr/observers

ICPDR (2013) Guiding Principles on Sustainable Hydropower Development in the Danube Basin [online]. Available from: http://www.icpdr.org/main/activities-projects/hydropower

ICPR (2010) Rules of Procedure and Financial Regulations of the ICPR [online]. Available from: http://www.iksr.org

IJC (n.d.) Website: http://ijc.org/en_/sab/Great_Lakes_Science_Advisory_Board

International Agreement on the River Maas 1st December 2006, [Online] Available from http://www.cipm-icbm.be/files/files/FR1.pdf.

Louka E. and Krchnak K.M. (2005) Improving water governance through increased public access to information and participation. *Sustainable Development Law and Policy* 5(1), pp. 34–9.

Mandl, B. (2015) Public participation in the International Commission for the Protection of the Danube River. In M. Tignino and K. Sangbana, *Public Participation and Water Resources Management: Where Do We Stand in International Law ?*. Paris: UNESCO, pp. 48–54.

MRC (2008a) Stakeholder Consultation on MRC's Basin Development Plan Phase 2 (BDP2) and its Inception Report. Consultation Proceeding. 12–13 March, Vientiane, Laos.

MRC (2008b) Regional Multi-Stakeholder Consultation on the MRC Hydropower Programme Consultation Proceedings. 25–27 November, Vientiane, Laos.

MRC (2009) Public Participation Strategy, Draft Report. June. Un-published.

MRC (2011a) Prior Consultation Project Review Report: Stakeholder Consultations Related to the Proposed Xayaburi Dam Project. 24 March [online]. Available from: http://www.mrcmekong.org/assets/Publications/Reports/PC-Proj-Review-Report-Xaiyaburi-24-3-11.pdf

MRC (2011b) Project Review Report, Stakeholder Consultations Related to the Proposed Xayaburi Dam Project, volume 2. 24 March [online]. Available from: http://www.mrcmekong.org/assets/Consultations/2010-Xayaburi/2011-03-24-Report-on-Stakeholder-Consultation-on-Xayaburi.pdf

Niger Basin Water Charter (2008) 30 April [online] Available from: http://www.abn.ne/index.php?option=com_content&view=frontpage&lang=fr

OKAKOM (2010) Rules and Procedures of the Permanent Okanvango River Basin Water Commission [online]. Available from: http://www.okacom.org/site-documents/key-documents

OKAKOM (n.d.) Website: http://www.okacom.org/

OMVS (n.d.) Website: http://www.portail-omvs.org/participation-du-public/associations/association-usagers-adu

Protocol for Sustainable Development of Lake Victoria Basin (2003) 29 November [online]. Available from: http://www.lvbcom.org/

Pulp Mills on the River Uruguay (Argentina v Uruguay) (2010) Judgment, ICJ Reports 2010, p. 14.

Razzaque, J. (2009) Public participation in water governance. In J.W. Dellapenna and J. Gupta, *The Evolution of the Law and Politics of Water*. Dordrecht: Springer, pp. 353–71.

Rieu Clarke, A. (2015) Transboundary hydropower projects on the mainstream of the lower Mekong River: the case of public participation and its national implications for basin states. In M. Tignino and

K. Sangbana, *Public Participation and Water Resources Management: Where Do We Stand in International Law?*. Paris: UNESCO, pp. 91–7.

Sangbana, K. (2011) La gestion intégrée des ressources en eaux partagées et les organismes de bassin en Afrique les cas de l'Autorité du Bassin du Niger et de l'Autorité de la Volta. In SFDI, *L'eau en droit international*. Pedone: Paris, pp. 235–44.

Sangbana, K. (2015) La participation du public dans le cadre de l'Organisation pour la mise en valeur du fleuve Sénégal. In M. Tignino and K. Sangbana, *Public Participation and Water Resources Management: Where Do We Stand in International Law?*. Paris: UNESCO, pp. 77–83.

Schmeier, S. (2013) *Governing International Watercourses. River Basin Organizations and the Sustainable Governance of Internationally Shared Rivers and Lake*. New York: Routledge.

Senegal River Water Charter (2002) 28 May [online]. Available from: http://www.portail-omvs.org

Sow, M. and Touré, B. (2008) État des lieux du cadre juridique et institutionnel dans la région d'intervention en Mauritanie, Doc. OMVS/GEF/BFS.

Tignino, M. (2010) Les contours du principe de la participation du public et la protection des ressources en eau transfrontières [online]. Available from: http://vertigo.revues.org/9750

22

HYDRO-HEGEMONS AND INTERNATIONAL WATER LAW

Rebecca L. Farnum, Stephanie Hawkins and Mia Tamarin

Introduction: power and international law

A chapter about power and hegemony may seem out of place in a law and policy handbook. The law is, after all, supposed to be free from politics, blind to its supplicants, an equal arbiter of justice. However, while certain mechanisms within legal frameworks strive to be apolitical, it cannot be ignored that the systems creating and upholding those frameworks are far from apolitical themselves. It is people who write and enforce law; it is states that produce and implement international law. International law is thus a product of society, influenced by power and politics.

In his opening to *The Politics of International Law*, Christian Reus-Smit (2004, p. 3) demonstrates the social construction of international law, showing how states and other actors use politics to determine 'not only "who gets what when and how", but also who will be accepted as a legitimate actor and what will pass as rightful conduct'. International relations occur 'within a framework of rules and norms' (ibid.), but those very rules and norms are the product of states' behaviours and wishes. In a system created by its own users, politics and power cannot be divorced from the product or its application. Analysis of those politics and the way power plays out on the global stage is thus an integral part of understanding international law. This chapter relies on three related assumptions about the relationship between power and law in order to examine the more specific connections between hydro-hegemons, freshwater and international water law:

1. the political processes that create international law are shaped in large part by power relations between states;
2. legal norms and institutions often serve to reinforce existing power dynamics;
3. legal norms and institutions can be leveraged for or against extant hegemonic orders.

International relations are inherently political, shaped by domestic pressures, representatives' rapport and global events. While international relations and international law are frequently separated as fields of study, they are inextricably bound, with relations creating law and law informing relations (Reus-Smit 2014). While all states are, in principle, equal, there exist 'firsts among equals' in global politics. The five permanent seats on the United Nations (UN) Security Council are a prime example. The Security Council is one of the few international institutions that can produce legally binding resolutions, yet only permanent members wield veto power, giving those states more power over the production of international law. This power imbalance reflects political processes and directly impacts international law's content, scope and applicability.

Ideally, 'law is the protector of the weak' (Frederick Schiller). Too often, though, it seems that '[t]he function of the law is not to provide justice or to preserve freedom. The function of the law is to keep those who hold power, in power' (Spence 1996, p. 90). The international legal system has come a long way since the 1648 Treaty of Westphalia and 1945 founding of the UN, but the world system is still technically anarchic. Under the Westphalian model of social organisation, states have theoretical sovereignty over their territories and there is no greater power than the nation-state. A citizen of a country is bound to that country's laws whether or not she wishes to be. In contrast, states are not answerable to a force higher than themselves: the collective international community of states. International law emanating from treaties and rulings from the International Court of Justice (ICJ) are applicable and enforceable only with states' voluntary compliance (see Crawford 2012). This requirement of consent is a core principle of international law. As this chapter demonstrates, principles of sovereignty and consent embody formal equality in the face of considerable inequality. The political nature of international legal creation, along with the system's imperial origins, gives rise to international legal norms and systems that generally benefit and reflect the worldview and values of the powerful. International law thus becomes another avenue through which powerful actors wield their influence – and acquire more.

The fact that law is created through political processes dominated by powerful players does not seem to bode well for the ability of international law to rein in domineering states or protect weaker states from bullying hegemons. However, the international legal requirements of notification and environmental impact assessment (EIA) for major projects, the ways in which activists make claims based on legal rights to campaign for vulnerable populations and the growing power of the ICJ despite its potential to limit state sovereignty all suggest that the law can be used as a tool in counter-hegemonic efforts.

This chapter builds from these ideas to explore how the Framework of Hydro-Hegemony (FHH) intersects with international law. Particular attention is given to geographical context; the principles of equitable and reasonable use, no significant harm and sovereignty; the fragmentation of international law; and procedural matters. The chapter will conclude by comparing lessons from across the given examples and identifying weaknesses in international water law and hydro-hegemony that this analysis reveals.

Power analysis, international relations and freshwater: The Framework of Hydro-Hegemony

Hegemony is a concept from politics and international relations referring to leadership and rule. A hegemon is an actor or group of actors with authority in their sphere. The famous theorist of hegemony, Antonio Gramsci, argued that Mussolini's Fascist regime in Italy was upheld not only through the state's brute force but also through cultural institutions, with the ruling class maintaining power through the reproduction of 'common sense' ideas. 'Hegemony' goes beyond domination (leadership built on brute force) to become leadership upheld through authority and consent.

Hydro-hegemony refers to hegemony at the river basin level, wherein one state, or a bloc of states, has more control over water flows and usage than other riparians. The FHH was developed in order to understand 'who gets how much water, how and why' (Zeitoun and Warner 2006, p. 435). The FHH is rooted in international relations literature on power analysis, hegemony theories and security studies. Much of that literature assumes a simplistic dualism of conflict and cooperation leading to either absolute control or equal co-management. In contrast, hydro-hegemony theorists argue that the outcomes of transboundary water distribution result from varying configurations of complex political interplays between interested actors. To analyse the nuances of these interplays, the FHH adapts Lukes' (1974) theory of power.

Building from Lukes, political and social scientists suggest three 'faces' of how power is operationalised: material, bargaining and ideational. The most obvious of these forms is material power, the capacity of an actor to tangibly achieve their interests through physical or economic force. States with greater abilities to extract or divert water from a basin and the military capability to destroy unwanted infrastructure on the river have more material power.

The second dimension, bargaining power, revolves around the ability to control the 'rules of the game' (Zeitoun and Warner 2006, p. 442), influencing the agenda and what is and is not on the negotiating table. The chairperson of a river basin organisation (RBO) wields bargaining power as she determines what will and will not be discussed during a meeting. Without recognition from the chairperson, a state may not be able to bring proposals to the RBO for consideration, much less implementation.

The third dimension of power is the most difficult to concretely grasp, and also probably the most difficult to counteract. Through ideational power, hegemons influence ideas and assumptions – not merely their own, but also other actors'. The third dimension is the capacity to create, uphold and destroy narratives, perceptions and knowledge. The core organisation of the world by nation-states is a hegemonic concept backed up by ideational power. That states should be the primary actors governing transboundary basins is 'common sense' in the current global order. The ideational power of the state system is demonstrated by how few people actively consider or work towards alternatives.

The last two dimensions of power – bargaining and ideational – are sometimes referred to as 'soft power' (see Nye 2004). In contrast, the first, material face is often called 'hard power'. Both soft and hard powers play a role in determining water distribution outcomes and reinforcing or countering hydro-hegemony within basins.

Power is seen in the FHH as the 'prime determinant enabling the successful execution of the water resource control' (Zeitoun and Warner 2006, p. 451). To illustrate the significance of differing kinds of power and analyse where and how power is held and employed within river basin relations, the FHH suggests four 'pillars' of power useful in evaluating how states exert influence over shared waters (Figure 22.1). In addition to Lukes' (1974) three faces of power, the FHH includes geography (the position of a state in relation to a watercourse) as an influential force in hydropolitics. The relationship between international law and these four pillars will now be explored.

International law as soft power and a tool of (counter-)hegemony

Hydro-hegemony and international law are powerfully connected, with law frequently used as a tool to both reproduce and resist hydro-hegemonic realities. Hydro-hegemons have greater levels of authority and influence to ensure that processes and outcomes of water distribution manifest in their favour. These hydro-hegemons exist, to some degree, in virtually every transboundary water relationship. It is no surprise that stronger players 'win the game' more frequently. Weaker players in transboundary water relations are typically constrained in their actions by the hydro-hegemon's interest, as hegemons wield their power through a variety of compliance-producing mechanisms ('carrots') and authoritarian strategies ('sticks'). Even so, non-hydro-hegemonic riparian states hold the potential to push against both the carrots and sticks.

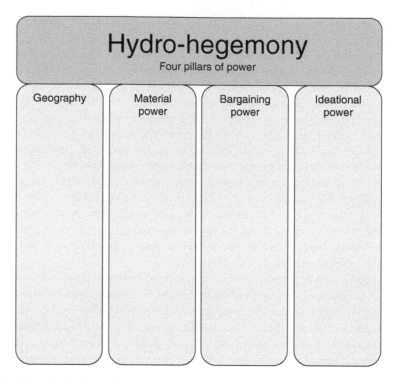

Figure 22.1 The revised pillars of hydro–hegemony
Source: Cascão and Zeitoun 2010, p. 32

Cascão's (2008) work on 'counter-hegemony' explores how non-hegemonic states might resist hegemonic control. Various studies and theorisations suggest that soft forms of power are particularly useful tools for non-hegemons (Zeitoun *et al* 2011; Cascão and Zeitoun 2010). It is primarily through this emphasis on soft power that issues of international law clearly come into questions of hydro-hegemony.

The ability to create, write and influence international law straddles bargaining and ideational powers. As a repository and creator of ideas, international law is a tool and actor in ideational power; as a repository of international 'rules', it is a tool and actor in bargaining power. This power is held and utilised by academics and lawyers via the drafting of articles; activists, corporations and civil society leaders in their campaigns; and states themselves through their participation in treaty making and in organisations such as the UN. Appeals to international law, including claims about either the 'rightness' or 'wrongness' of a riparian's actions, are a component of bargaining power. A study of 165 territorial disputes since 1945 found that actors with strong legal claims were more than twice as likely to seek negotiations before using force (Huth *et al.* 2011). This suggests a different kind of 'battle' using bargaining power in 'lawfare' in addition to (or instead of) material power in warfare (Kennedy 2012).

International law reflects and reproduces global discourses on issues, influencing domestic and foreign policy. It shapes and perpetuates norms of behaviour. As such, international law wields significant soft power – and thus influences hydro-hegemonic relations – even when it does not carry with it a strong global police force with hard power (Daoudy 2008). It can be argued, however, that there is also a material element to the intersection between international law and hydro-hegemony: the unequal capacity of states to employ lawyers. The ambiguity of much of international law leaves room for 'duelling experts' to determine the winner of a legal dispute over international waters. In such cases, countries with greater fiscal resources and legal capacity are likely to triumph.

Soft power is far from the only issue in hydro-hegemony, and those with the soundest legal arguments will not always prevail over an otherwise dominating power. However, international law is a significant source of soft power, and soft power is relevant to hydro-hegemonic relations and analyses. The next sections consider how some of the core principles and procedures of international law are being leveraged both by hydro-hegemons to further consolidate their position and by non-hydro-hegemons to counter the *status quo*.

Geographical context and its influence on hydro-hegemony

The original FHH included three pillars, with all three faces of power in one pillar, and 'Riparian position' and 'Exploitation potential' receiving dedicated attention in the others. In the revised conceptualisation (Figure 22.1), power is nuanced as each face gets its own pillar and 'Geography' gets the fourth.

Geographical factors are also considered in international water law: Article 6 of the 1997 United Nations Convention on the Law of the Non-navigational Uses of International Watercourses (UN Watercourses Convention 1997), for example, includes explicit attention to '[g]eographic, hydrographic, hydrological, climatic, ecological and other factors of a natural character', when weighing up the factors that states might take into account in determining what constitutes an equitable and reasonable use of an international watercourse. This section examines the influence of geographical considerations, most obviously riparian position for river basins and exploitation potential for aquifers, on hydro-hegemonic relations. As will be demonstrated, geography creates an important context for the way in which states draw upon the principles of international water law.

Geography and river basins

The geographical factor most pertinent to hydro-hegemony in river basins is riparian position, whether a state is upstream or downstream of its neighbours sharing the river. Upstream states can theoretically use all of the water from what would otherwise be a shared resource. In the 1890s, Mexico complained of United States (US) practices wastefully diverting water from the Rio Grande to the detriment of downstream users. At the time, the US Attorney General asserted *absolute* territorial sovereignty (the Harmon Doctrine), claiming the US had no obligation under international law to restrict the use of territorial waters.

Today, international water law has evolved such that the principles of equitable and reasonable utilisation (ERU) and no significant harm (NSH) signify *limited* territorial sovereignty. Through this theory of riparian rights, states may use water from a common source provided their use does not unreasonably interfere with other riparian states' uses. Limited territorial sovereignty could be said to be a counter-hegemonic tool for downstream riparians like Mexico, allowing them to counter the upstream hydro-hegemon's claims to absolute sovereignty.

However, not all hydro-hegemons are upstream. In the Ganges–Brahmaputra–Meghna River Basin, India is the clear hydro-hegemon even as Nepal is the uppermost riparian. Nepal and Bhutan share water but not a border, with India cutting between them. This significantly limits negotiation potential between the two weaker parties. This also emphasises the need for careful consideration of legal language: 'shared river'/'transboundary river' and 'transboundary'/ 'transgovernmental' are not necessarily synonymous pairs in all basins, and can affect the applicable scope of legal rules. For example, on the one hand, the term 'transboundary' frames international problems as occurring only at the border, undermining the regional issues inherent in the use of international watercourses, as well as the 'genuinely global dimension of local eco-system health' (Conca 2006, p. 16). The term 'shared', on the other hand, can be interpreted as requiring allocation of volumes of water, and indeed the recent bilateral Agreement between Jordan and Saudi Arabia relating to the Management and Utilisation of the Groundwaters in the Al-Sag/Al-Disi Layer (2015) – that neglects to allocate water volumes – avoids this language in favour of the term 'joint'. The language that denotes the geographical scope of international legal frameworks is intrinsically linked to power, since it prescribes who is included in, and excluded from, international negotiations over freshwater.

The case of the Ganges–Brahmaputra–Meghna River Basin provides an example of the consequences of such geographical framing, as well as demonstrating that the ideas of ERU and NSH (see 'Equitable and reasonable use, no significant harm and sovereignty' below), so useful for some non-hegemons, can be used by a hydro-hegemon to further cement its position. In the water-sharing Ganges Treaty (1996) between India and Bangladesh, claims to 'do no harm' and 'reasonable use' are made, but disingenuously so, in ways that do not truly limit India's hydro-hegemonic potential (Hanasz 2014). Nepal, the upstream riparian, is not included in the agreement. This emphasises that geographic context is an element rather than a determining factor in hydro-hegemony.

Geography and aquifers

The original FHH was developed to analyse power relations between states at the basin level, with the river as the primary unit of water under consideration. As a consequence of being 'out of sight, therefore out of mind', groundwater and aquifers have been largely ignored in hydro-hegemony theory, and have only recently gained more explicit attention in international legal instruments. For example, the International Law Commission's (ILC's) Draft Articles on the Law of Transboundary Aquifers (Draft Aquifer Articles 2008) only recently brought aquifers explicitly into the international legal framework for freshwater. This instrument also includes the ERU and NSH principles with attention to 'natural characteristics'.

Geographic issues are arguably, though not necessarily characteristically, different between surface and groundwater. The FHH's focus on surface water and riparian position has led to an underappreciation of issues pertinent to aquifers. This chapter does not go into depth about the hydrogeological and legal differences between surface water and groundwater. Nor does it explore whether there are significantly different hegemony considerations for surface water and groundwater, since the water contained in both carry similar distributive concerns. Instead, it argues that the original FHH's 'Exploitation potential' pillar can be read into the 'Geography' and 'Material power' elements of the revised framework; from this angle, hydro-hegemony specific to groundwater and aquifers can be considered.

Geographical factors like the depth of the water table, direction of flow, location of recharge and discharge zones, and amount of territory a state has over an aquifer influence the ease and cost of abstraction. The extent of this control by a state is then determined by the economic and technical capacity.

This relationship between geological ease of abstraction and technical-financial capacity is a two-way street, as pumping can alter flow directions and speeds, altering ease of abstraction in other places. Given this and the fact that flow dynamics can take different speeds and direction through various aquifer layers, state relationships over aquifers quickly become complicated. As international law over aquifers is still emerging and the subject of much debate, regional agreements are particularly important in current considerations of the relationship between geography, aquifers, hydro-hegemony and international law.

In the case of the Ceylanpınar (*Ra's al 'Ayn* in Arabic) Aquifer shared between Turkey and Syria, over-exploitation by increased use and unlicensed wells in both countries has led to increased water deficits and reduced flows to the Khabour River (Öztan and Axelrod 2011). However, as the recharge zone lies predominantly in Turkey, Turkey has more control over the aquifer's management for continued use. As the majority of the aquifer lies under Turkish territory, Turkey also has more exploitation opportunities than Syria. This uneven geographical situation and the countries' disparate economic power mean that exploitation potential is critically unequal.

Another example includes the Disi Aquifer shared by Jordan and Saudi Arabia. The aquifer has low permeability and negligible recharge, with its finite resources leading to a silent pumping race between the two states (Ferragina and Greco 2008). The recent Disi Aquifer Agreement (2015) restricts exploitation only in the relatively small buffer zone at the international border. Since Saudi Arabia has a significantly larger portion of the aquifer under its territory, Saudi Arabia's current exploitation is relatively unaffected, and the practical implications of the treaty are far from equitable.

A considerable proportion (around 94 per cent) of the groundwater exploitation happening within the Guarani Aquifer System is taking place within Brazil, the clear regional hegemon in Latin America (for more on this issue, see the Guarani Aquifer System 2009). With the principle of sovereignty underpinning the Guarani Aquifer Agreement (2010), the regional treaty maintains and promotes the hydro-hegemonic *status quo*.

In the case studies above, inequalities created largely by geographic factors demonstrate how legal rules focused on contextual situations can further or help mitigate hegemonic relationships. Geography and the hard power of technical and economic capacity thus remain relevant in hydro-hegemonic analysis and the practical implications of international water law. The next section will explore how these and other factors play out in states' use of soft power through a more detailed consideration of the principles of ERU, NSH and sovereignty.

Equitable and reasonable use, no significant harm and sovereignty: (counter-)hydro-hegemonic bargaining tools

The duty to ensure ERU and NSH when utilising a watercourse are the primary principles of international water law. They are now legally binding on state signatories to the UN Watercourses Convention (1997) and form the bedrock of multiple bi- and multi-lateral agreements. However, the ways in which these principles are used and interpreted varies widely across states.

The principle of ERU acts as the legal entitlement of riparian rights with NSH as the regulatory check, though there has been historical disagreement between states over their interaction and which should be prioritised (Wouters 1999). The two are generally seen as conflicting, with states emphasising one and understating the other according to their preferences. A lower riparian state would supposedly favour the principle of NSH to protect against the use from upstream states. Upper riparian users are likely to favour the principle of ERU, which provides more scope to make use of water without consideration for downstream users (Salman 2007).

ERU in the UN Watercourses Convention (1997, Article 5) and the ILC Draft Aquifer Articles (2008, Article 4) requires that states use and develop a watercourse taking into account all watercourse states concerned. Article 6 of the UN Watercourses Convention (1997) and Article 5 of the ILC Draft Aquifer Articles (2008) both list a series of factors to be taken into account in determining ERU, including social, economic, cultural and historical considerations. However, this list of factors is flexible. Just as states stress NSH or ERU in accordance with their preferences, they emphasise those ERU factors most helpful to their agendas. As a result, claims tend to balance each other out legally, with an absence of legal hierarchy over factors leading to maintenance of the *status quo* (Lankford 2013).

The Nile River Basin is a particularly interesting case study for hydro-hegemony and international law. Egypt has historically been the hydro-hegemon, with a great deal of control over the waters of the Nile despite its downstream position. On the one hand, by emphasising the extant uses, dependent population and socioeconomic needs factors of ERU, Egypt leverages international water law to legitimise its continued role as receiving the lion's share of the Nile's waters. Upstream states, on the other hand, can make claims to the principle of ERU's focus on natural factors and potential uses in their arguments. The equivocality of ERU thus allows for its use as a tool of both hydro-hegemony and counter-hydro-hegemony.

In addition to the *use* of these principles, involvement in their *formation* illustrates how states seek to use international water law to further their positions – and ensure other states do not gain additional bargaining power.

For example, in comments preceding the General Assembly's adoption of the ILC Draft Aquifer Articles (2008), Brazil pushed for the NSH principle to be interpreted as an obligation of 'conduct' and not one of 'result' (ibid., p. 30), further advocating that the obligation to prevent, reduce and control pollution should not result in 'undue hardship' on origin states disproportionate to the benefits of potentially harmed states (ibid., p. 37). Similarly, Turkey appealed to weaken the obligation of NSH by suggesting that the phrase 'shall try' replace 'shall take all appropriate measures' (ibid., p. 32). Portugal, a non-hydro-hegemon, expressed concern over the lack of a definition for 'significant harm' and argued that there is 'danger in leaving such subjective terms to be interpreted by States on a case-by-case basis, in accordance with their own interests of the moment. In fact, it may create an unjustified disadvantage to weaker States' (ibid., p. 21).

Sovereignty is a central international legal principle, but one that has a complex place in international water law. While the principles of ERU and NSH have effectively restricted sovereignty as a legitimate justification of unrestricted water utilisation, the ILC Draft Aquifer Articles (2008) actually emphasise the sovereignty of aquifer states. In the context of aquifer utilisation, sovereignty is now part of the 'toolbox' of bargaining power that states can use in their hydro-hegemonic and counter-hydro-hegemonic efforts (more on this divergence around the sovereignty principle is discussed in the next section).

Sovereignty as sustained in Article 3 of the ILC Draft Aquifer Articles (2008) can be used to lessen states' responsibilities towards their neighbours and has been advocated for by various hydro-hegemons, including Turkey (ibid., p. 22). Brazil expressed its support for the inclusion of sovereignty as 'fundamental' (ibid.), claiming the need 'to maintain such a balance and avoid excessive restrictions to legitimate activities' (ibid., p. 8). Non-hegemons might try to leverage the sovereignty principle for counter-hydro-hegemonic action by using it to prevent interference from other states' aquifer utilisation. However, such use could inadvertently support unrestricted resource rights, creating a claim–counter-claim culture with no objective mediator. Such a system is vulnerable to the administrative advantage of hegemons with abundant resources.

The wording contained in framework treaties is important, since regional and bi-/multi-lateral agreements governing watercourses and aquifers 'borrow' from these overarching instruments. Yet, powerful actors can direct the emphasis given to certain principles over others during their creation. Such variations may be positive or negative from the perspective of equity. The Revised Southern African Development Community Protocol on Shared Watercourses (2000), for example, draws upon the provisions of the UN Watercourses Convention (1997) in its preamble and includes ERU and NSH language very similar to the UN formulation – but adds explicit attention to poverty alleviation, sustainable development and the needs of future generations. In contrast, the ILC Draft Aquifer Articles (2008) have been recognised as influential in the creation of the Guarani Aquifer Agreement (2010), and its emphasis on sovereignty could indicate hydro-hegemonic influence in its drafting and implementation.

States' use of and engagement with the formation of these principles reveals the bargaining power inherent in international law. Turkey has a history of using international legal principles as soft power to increase its control over the transboundary waters; Iraq and Syria have likewise leveraged international water law principles in counter-hydro-hegemonic attempts to curb upstream control of the surface waters (Daoudy 2008). Similarly, states' comments on the ILC Draft Aquifer Articles (2008) illustrate how power can be wielded through the *wording* of provisions. Ambiguous and subjective terms are generally preferred by those states with the capacity to interpret them in their favour. More defined, narrow provisions are preferred by those with less power, demonstrating a possible reliance on the law to make up for their lack of influence in other spheres. Additionally, the very fact that hydro-hegemons like Brazil and Turkey are concerned with the wording of legal instruments at all, in spite of their dominant position, reveals concern over being held accountable by legal principles and the perceived legitimacy of their actions.

Hydro-hegemony in other fields of law: the fragmentation of international law

Though the discussion above focused primarily on issues specific to international water law, many other areas of international law are relevant to hydro-hegemony. Global water governance frames water, among other things, as an environmental, economic and human rights concern. Managing and using river basins and aquifers involves much more than the water itself; it also has implications for pollution, land rights, biodiversity and wetlands, for example. International environmental law, international human rights law and international economic law, among others, all govern various components of freshwater resources (Boisson de Chazournes 2013). The various regional agreements, domestic laws and private sector regulation mechanisms also intersecting with shared waters further complicate matters. While the principle of *lex specialis* can help in determining which law takes precedence when there are both general and specific rules, norms over shared water have emerged in different fields of law holding equivalent specificity or generality. The structure of international law thus leads to legal fragmentation, making it unclear which rules are applicable when.

Legal fragmentation over transboundary river basins and aquifers has implications for hydro-hegemony. As demonstrated above, states divergently prioritise the principles of ERU, NSH and sovereignty to further their own (counter-)hydro-hegemonic agendas. The fragmentation of law provides an even larger set of tools from which states can pull.

Consider, for example, the Jordan River Basin. The region is characterised by relative water scarcity and vast inequalities between populations. Israel, Jordan, Lebanon, Syria and the West Bank are all riparians, with Israel the clear hydro-hegemon, especially in regards to the occupied West Bank. Ongoing debate over the status and legality of Israel's occupation and activities in the West Bank includes disputes over the applicability of international humanitarian law and international human rights law, further fragmenting and undermining applicable international water law over the Jordan River Basin.

'Water rights' are an oft-used legal principle for counter-hydro-hegemonic action in the Jordan River Basin, with states, NGOs and communities all making claims. Water rights, formulated in slightly different ways, are frequently applied in parallel legal provisions – including international water law, international human rights law and international humanitarian law. The multiple emphases on rights to water can be a boon for advocates and campaigners: in this case, they have a number of sources from which to pull arguments, potentially giving more weight to their claims. But, in the legal realm, fragmentation often diminishes non-hegemons' chances for legal redress. With multiple mechanisms, jurisdiction may be passed between often contradictory arenas to such an extent that some legal claims essentially fall through the cracks and may legitimise legal breaches.

The advent of the law of transboundary aquifers arguably further fragments international law governing freshwater. Established international law over water, environmental and biodiversity concerns has effectively eroded sovereignty in favour of attention to commonly held vital resources. Although the ILC Draft Aquifer Articles (2008) amount only to non-binding 'soft law', its reintroduction of sovereignty to certain transboundary water resources arguably adds legitimacy to interpretations of *absolute* sovereignty (see McCaffrey 2008). Thus, hydro-hegemons are potentially once again able to use that principle as a bargaining power tool, although limited territorial sovereignty remains an established customary principle of international law. Fragmentation, read in this light, seemingly becomes a handmaiden to hydro-hegemony, as contradictory rules make space for 'interest-based' interpretation, reflective of the indeterminate structure of law itself. Advocating for systematic interpretation, contextualising treaties with relevant external language and events, may help to counter this concern, although the interpretation of those in power will tend to prevail.

While it is necessary for international legal frameworks to be flexible (after all, it is rare that 'one size fits all'), the fragmentation of law undermines its purported objectivity and certainty. Without careful attention to the potential resultant harms, international legal fragmentation will likely bolster hydro-hegemonic arrangements by providing yet more tools for hydro-hegemons to leverage in their water use.

Procedural rules and (counter-)hydro-hegemonic action

Thus far, this chapter has primarily considered substantive rules of international water law. There are, however, a number of procedural obligations also relevant to shared waters and hydro-hegemonic relations. International water law explicitly obliges a duty of cooperation on states, including prior notification and consultation during the planning of projects likely to impact a watercourse. Information sharing and the precautionary principle are also prominent. The overlap and fragmentation of international law suggests that procedural obligations from international environmental law, such as the conducting of EIAs and the involvement of public participation in decision-making, may also be germane.

Procedural rules are used by both hegemons and non-hegemons to delegitimise and justify various actions. The Grand Ethiopian Renaissance Dam (GERD) in the Nile River Basin can be seen as a counter-hydro-hegemonic action by Ethiopia, a move against decades of hydro-hegemonic 'bullying' from downstream riparian Egypt. Egypt has claimed the vast majority of the Nile's water for decades, asserting historic rights based on 1929 and 1959 colonial treaties between Egypt and Sudan.

Ethiopia's announcement of the GERD was met with loud protests and even threats from Egypt, concerned about possible reductions to downstream flow. Egypt claimed that Ethiopia had acted unilaterally without providing sufficient prior notification and called for Ethiopia to halt construction until adequate consultation and negotiations with riparians were satisfactorily concluded. Egypt, Ethiopia and Sudan were locked in months of tension. While an agreement has now been reached, Egypt's response to the GERD demonstrates how hydro-hegemons make use of procedural obligations to delegitimise actions threatening their favoured *status quo*.

Egypt's legal claims against the GERD included calls for a proper assessment of the project's impacts. EIAs are one of the strongest procedural duties in international environmental law, regarded as a necessary component of due diligence against transboundary harm. EIAs are now standard – and generally required – for any major infrastructure development, including dams, pipelines and pumps (see ICJ 2009, 2010).

Transboundary EIAs are a powerful way to shed light on the harms of, and thus potentially push against, hydro-hegemony. The procedural EIA requirement creates a space for negative impacts like reduced downstream flow and aquifer pollution to be named and mitigation measures called for. Additionally, the conducting of an EIA is a more static, strict rule, difficult to interpret subjectively. There is thus less space for a hydro-hegemon to deny a breach of the EIA requirement.

There is, however, a great deal of space for EIAs and other procedural rules to be used politically as well as legally, in ways that may further hydro-hegemony rather than countering it. For example, Israel and Jordan recently signed an agreement for a major project to desalinate water from the Gulf of Aqaba and pipe some of it north to replenish the Dead Sea.

The Red Sea–Dead Sea Conveyance project was full of counter-hydro-hegemonic potential, and was certainly packaged that way to supporters at its beginning. An early proposal written jointly by Israel, Jordan and the Palestinian Authority spoke of greater access to potable water for everyone and an economic and environmental avenue for cooperation. The World Bank administered a multi-million dollar fund from multiple donors to conduct a feasibility study (including an EIA and social impact assessment) that emphasised public participation a great deal in its design. In December 2013, Israel, Jordan and Palestine signed a memorandum of understanding to build a conduit.

Yet, at the end of all this potentially counter-hydro-hegemonic work, an agreement was signed between Israel and Jordan – without the Palestinian Authority. Instead of being an equal part of an agreement, Palestine may be able to broker a separate agreement with Israel to buy additional water. Such a result furthers Israel's hydro-hegemonic position over Palestine. This hydro-hegemonic outcome illustrates the limitations of procedural rules to guard against unjust results. Potentially worse, attention to procedural obligations (even if only lip service) may solidify future hydro-hegemonic possibilities by securing the consent of neighbouring riparians, which can, in turn, limit future claims. Though the signed agreement deviates from the original plan, Israel can point to Palestine's participation in the process to delegitimise future protests.

The use of EIAs to further (counter-)hydro-hegemonic agendas also raises issues of the politicisation of science and information asymmetry. Water systems are complex, and the likely impacts of projects and utilisation debatable. A state will hire experts who will testify in ways that fit its agenda and focus on the scientific knowledge that most supports its aims. While a non-hegemon may leverage the precautionary principle to push against a major project initiated by a hydro-hegemon, that hydro-hegemon may use its material power to buy a great deal of expert knowledge able to produce an authoritative EIA discounting neighbours' concerns. Information asymmetry and the capacity of more powerful states like Israel to generate but withhold knowledge further challenge counter-hegemonic efforts.

Procedural rules have a great deal of counter-hydro-hegemonic potential and can be leveraged as an important part of weaker states' bargaining power. But it is important to recognise that the political nature of science and procedure means that these rules can just as easily be leveraged by hydro-hegemons to further cement their positions. Also, the very use of legal structures may reinforce the legitimacy held by hydro-hegemons, as will be addressed in the next and final section.

Conclusions

Understanding the interplay between international water law, power and hydro-hegemony is vital to understanding water policy and outcomes on a global level. International water law, as well as other bodies of law, can be a source of bargaining power for states in transboundary water interactions. Legal principles, both substantive rules and procedural obligations, provide a set of norms to which states can refer in justifying or delegitimising actions. Given states' interest and involvement in the development of international law, it seems that even powerful states are concerned with issues of accountability, legitimacy and the potential bargaining power provided by international law.

As a result, law can serve as a tool for both justice and oppression. The fragmentation and indeterminacy of law (see Miéville 2004) creates avenues for the exercise of soft power by states to develop legitimising 'legal narratives' for their actions (Shehadeh 1996). Non-hegemons can leverage international law to strengthen their bargaining power; principles of international water law can help bolster counter-hydro-hegemonic advocacy campaigns. Yet, the reality of overarching hegemonic structures and extreme power asymmetries between states may limit the extent to which using law for counter-hydro-hegemony is effective: legally, as states with greater resources generally win the 'duel of experts', and practically, as being in the legal right is no guarantee of success or protection.

Examining international water law and hydro-hegemony together reveals the weakness of their state-centred outlook. The imperial Westphalian system is the framework in place for global organisation, with the state the central actor. However, as the Red Sea–Dead Sea case study demonstrates, semi-state actors, indigenous peoples, local and municipal governments, donors, corporations, international and non-governmental organisations, individuals and ecosystems are growing in prominence and have agency in shaping outcomes. Trends such as public participation and international human rights mechanisms show the evolution of international law in recognising these non-state actors, but states remain the key actors with legal personality. Hydro-hegemony theory, too, has been highly state-centric. States are the unit of analysis with virtually no attention to which citizens or groups within the state wield and benefit from the state's hegemony.

The world is changing, and state-centric analyses can limit understanding of broader influences. Decision-making that impacts transboundary waters occurs on all levels, not only the state level. In many places, local and municipal actions have greater impacts than national activities. Major development projects are nearly always backed by international organisations such as the Asian Development Bank or International Monetary Fund.

Hydro-hegemony theory must move beyond its realist assumptions of international relations theory if it is to be of continued use. Similarly, international water law must embrace developments and evolve if it is to continue being relevant. International law and hydro-hegemony scholarship and practice must incorporate both more micro- and macro-level actors. When individuals, corporations and semi-states are not included in international legal frameworks and analysis, their place in maintaining, or being subject to, oppressive hydro-hegemonic arrangements remain unacknowledged and unaddressed.

In addition to emphasising the importance of non-state actors to international water law and hydro-hegemony, this chapter's consideration of aquifers as well as river basins reinforces the importance of understanding geographical context as well as economic, social and political factors. Hydro-hegemony theory should, accordingly, develop a more nuanced appreciation of how riparian position and exploitation potential influence hydro-hegemonic outcomes, and of the material power differences that arise between various catchment types.

Lastly, but perhaps most importantly, using hydro-hegemony as a lens to study international water law reveals the structural ways in which international law serves to further hegemonic arrangements. There is a violent relation in law itself: *which laws apply, when, by whom, to what end?* (Kennedy 2012, p. 164). It is precisely these questions, applied to transboundary water distribution, that hydro-hegemony seeks to answer. The international legal system is political; subject to the same power relations that create hydro-hegemonic realities and inequalities. Non-hegemons making use of international water law as a form of bargaining power in counter-hydro-hegemony must consider the unequal nature of that law, and the ways in which their use of it reinforces wider power dynamics, in their efforts.

References

Agreement between the Government of the Heshemite Kingdom of Jordan and the Government of the Kingdom of Saudi Arabia for the Management and Utilization of the Ground Waters in the Al-Sag/Al-Disi Layer (2015) 30 April [online]. Available from: http://www.internationalwaterlaw.org/documents/regionaldocs/Disi_Aquifer_Agreement-English2015.pdf [unofficial translation].

Boisson de Chazournes, L. (2013) *Fresh Water in International Law*. Oxford: Oxford University Press.

Cascão, A. (2008) Ethiopia: challenges to Egyptian hegemony in the Nile basin. *Water Policy* 10(2), pp. 13–28.

Cascão, A. and Zeitoun, M. (2010) Power, hegemony and critical hydropolitics. In A. Earle, A. Jägerskog and J. Ojendal (eds), *Transboundary Water Management: Principles and Practice*. London: Earthscan, pp. 27–42.

Conca, K. (2006) *Governing Water: Contentious Transnational Politics and Global Institution Building*. Cambridge: MIT Press.

Convention on the Law of the Non-navigational Uses of International Watercourses (1997) 21 May (entered into force 17 August 2014). *International Legal Materials* 36 (1997) 700.

Crawford, J. (2012) *Brownlie's Principles of Public International Law*, 8th edn. Oxford: Oxford University Press.

Daoudy, M. (2008) Hydro-hegemony and international water law: laying claims to water rights. *Water Policy* 10, pp. 89–102.

Ferragina, E. and Greco, F. (2008) The Disi project: an internal/external analysis. *Water International* 33(4), pp. 451–63.

Guarani Aquifer System (2009) Guarani Aquifer Strategic Action Programme [online]. Available from: http:// iwlearn.net/iw-projects/974/reports/strategic-action-program/view

Guarani Aquifer Agreement (2010) 2 August [online]. Available from: http://www.internationalwaterlaw.org/documents/regionaldocs/Guarani_Aquifer_Agreement-English.pdf [unofficial translation].

Hanasz, P. (2014) Sharing waters vs sharing rivers: the 1996 Ganges Treaty. *Global Water Forum* [online]. Available from: http://www.globalwaterforum.org/2014/07/28/sharing-waters-vs-sharing-rivers-the-1996-ganges-treaty

Huth, P.K., Croco, S.E. and Appel, B.J. (2011) Does international law promote the peaceful settlement of international disputes? Evidence from the study of territorial conflicts since 1945. *American Political Science Review* 105(2), pp. 415–36.

ICJ (2009) 13 July. *Dispute Regarding Navigational and Related Rights* (Costa Rica *v* Nicaragua), ICJ Rep. 213.

ICJ (2010) 20 April. *Pulp Mills on the River Uruguay (Argentina v Uruguay)*, ICJ Rep. 14.

ILC (2008) Shared natural resources: comments and observations by governments on the draft articles on the law of transboundary aquifers. *International Law Commission Sixtieth Session*, UN Doc. A/CN.4/595.

ILC Draft Articles on Transboundary Aquifers (2008) *Official Records of the General Assembly*, 63rd Session, Supp. No. 10, UN Doc. A/63/10.

Kennedy, D. (2012) Lawfare and warfare. In J. Crawford and M. Koskenniemi (eds), *Cambridge Companion to International Law*. Cambridge: Cambridge University Press.

Lankford, B. (2013) Does Article 6 (Factors Relevant to Equitable and Reasonable Utilization) in the UN Watercourses Convention misdirect riparian countries? *Water International* 38(2), pp. 130–45.

Lukes, S. (1974) *Power: A Radical View*. London: Macmillan.

McCaffrey, S.C. (2008) Comments on the International Law Commission's Draft Articles on the Law of Transboundary Aquifers 2006. *Social Science Research Network*. Available from: http://papers.ssrn.com/sol3/papers.cfm?abstract_id=1114988

Miéville, C. (2004) The Commodity-Form theory of International Law: an introduction. *Leiden Journal of International Law* 17(2), pp. 271–302.

Nye, J.S. (2004) *Soft Power: The Means to Success in World Politics*. New York: Public Affairs.

Öztan, M. and Axelrod, M. (2011) Sustainable transboundary groundwater management under shifting political scenarios: the Ceylanpinar Aquifer and Turkey–Syria relations. *Water International* 36(5), pp. 671–85.

Reus-Smit, C. (2004) The politics of international law. In C. Reus-Smit (ed.), *The Politics of International Law*. Cambridge: Cambridge University Press.

Revised Southern African Development Community Protocol on Shared Watercourses (2000) 7 August (entered into force on 22 September 2003). *International Legal Materials* 40 (2001) 321.

Salman, S.M.A. (2007) The Helsinki Rules, the UN Watercourses Convention and the Berlin Rules: perspectives on international water law. *International Journal of Water Resources Development* 23(4), pp. 625–40.

Shehadeh, R. (1996) The weight of legal history: constraints and hopes in the search for a sovereign legal language. In E. Cotran and C. Mallat (eds), *The Arab–Israeli Accords: Legal Perspectives*. Boston: Kluwer Law International.

Spence, G. (1996) *From Freedom to Slavery: The Rebirth of Tyranny in America*. New York: St Martin's Press.

Treaty between the Government of the People's Republic of Bangladesh and the Government of the Republic of India on Sharing of the Ganges/Ganges Waters at Farakka (1996) (entered into force 12 December 1996). *International Legal Materials* 36 (1996) 519.

Wouters, P. (1999) The legal response to international water scarcity and water conflicts: the UN Watercourses Convention and beyond. *German Yearbook of International Law* 42, p. 293.

Zeitoun, M. and Warner, J. (2006) Hydro-hegemony: a framework for analysis of trans-boundary water conflicts. *Water Policy* 8(5), pp. 435–60.

Zeitoun, M., Mirumachi, N. and Warner, J. (2011) Transboundary water interaction II: the influence of 'soft' power. *International Environmental Agreements: Politics, Law and Economics* 11(2), pp. 159–78.

PART 3

Cross-cutting issues

23

WATER RESOURCES AND INTERNATIONAL INVESTMENT LAW

Ana María Daza-Clark

Introduction

Water and business are intimately connected. Few economic sectors could develop without access to fresh water resources. By way of example, the energy sector requires access to water resources for both hydroelectric and thermoelectric generation, neither of which would be possible without this valuable resource.

Increased variability of the hydrological cycle, unpredictable water events and the effects of climate change could mean harder competition for water, among different socio-economic sectors. Competition may arise between mining and farming, or between irrigation and environmental sustainability and access to water. To address such challenges, policy makers and water managers may need to implement adjustment measures, such as reallocation of water permits or the adoption of stricter pollution controls – ultimately, affecting prior uses, among which there may be some granted to both national and foreign investors.

The manner in which we study the relationship between business and water resources may fall within the scope of different areas of law, such as environmental law, human rights law, mining law and energy law. International investment law is one such area of law whose scope of regulation may overlap with the regulation and management of water resources. The regulation of water resources is not specifically governed under international investment agreements (IIAs), which include bilateral investment treaties (BITs) and free trade agreements (FTAs). These agreements serve the purpose to promote and protect foreign investment in each contracting party, they tell us nothing about water resources. IIAs do, however, state that contracting parties 'shall admit investment in accordance with its domestic laws and regulations' (Mexico – United Kingdom BIT 2007). This means that foreign investors are bound by the national order of the host state, related to the management and use of water resources. In order to explore the linkages between the law that governs foreign investment and water resources management, the second section of this chapter provides a brief introduction to the general structure of IIAs. The third section, through a discussion of relevant investment cases, classifies water uses by foreign investors, into three groups: 1) as a tradable good, 2) as an input of economic activities and 3) as a service, for the provision of water supply and sanitation. Each of these uses presents distinctive characteristics and different types of relationship between users.

International investment law and arbitration

Perhaps similar to water law and governance, international investment law is a relatively new area of international economic law. However, in contrast to many other areas of law, international investment law is rapidly developing due to a number of factors: 1) states continued efforts to liberalize, promote and attract foreign investment; 2) active negotiation and conclusion of BITs, FTAs and other types of investment treaties; and 3) a very dynamic dispute settlement mechanism, investor-state arbitration, among others. All these factors are under continued scrutiny and development, by academics and practitioners, which enables a fast-growing area of law (UNCTAD 2015).

Historically, protection of property and aliens abroad, through diplomatic means, and protection of private property, through early treaties on navigation and commerce, have all contributed to the development of modern international investment law and arbitration. In the sector of natural resources governance, among which water resources may be included, international disputes began quite early. For instance, the Russian Communist revolution (1917) and the Mexican agrarian revolution (1938) proclaimed state ownership and control over their natural resources. In such circumstances, the expropriation of foreign property was not accompanied by payment of compensation in favour of the affected owners. The underlying justification for such nationalizations was the sovereignty of states over their natural resources (Subedi 2008, p. 16).

Mexico rejected the conditions for payment of compensation demanded by the US. In the exchange of diplomatic correspondence between the two countries, Cordell Hull, the American Secretary of State, put forward what is currently the most widespread formulation of the standard of compensation. Indeed, the 'Hull formula' of prompt adequate and effective compensation has been incorporated into a large number of IIAs as a requirement for a lawful act of expropriation by host states.

Eventually, the discussion over the treatment of aliens and the protection of their assets was taken to a new forum, the United Nations General Assembly (UNGA). After several years of consideration by the Commission on Permanent Sovereignty over Natural Resources (PSNR), the General Assembly issued Resolution 1803 of 1962 which: 1) recognized the rights of peoples and nations to permanent sovereignty over their natural wealth and resources 2) promoted the application of national legislation to foreign capital and 3) stated that expropriation and nationalization should only be carried out for public purposes and under payment of appropriate compensation (UNGA 1962).

Later in 1974 after further negotiations, the General Assembly Resolution 3201 established a New International Economic Order (NIEO) (UNGA 1974a). This Resolution ratified the principles set out in previous resolutions on PSNR and stated the need to regulate and monitor the operations of transnational corporations in the receiving country. Further, the Charter of Economic Rights and Duties of States, recognized the right of each state to nationalize and expropriate against payment of appropriate compensation to be determined by the expropriating state. Furthermore, in case of controversy, the dispute would be solved under domestic law by the national courts of the state (UNGA 1974b).

The Charter was adopted by a majority of 120 votes against six, and ten abstentions. However, it also showed the increasing imbalance and lack of agreement between major capital-exporting states and capital-importing states, which held and continue to hold, an overwhelming majority in the General Assembly.

It is interesting to observe that, parallel to the negotiations of the UN General Assembly, capital-exporting and capital-importing states had been negotiating investment treaties bilaterally (Sornarajah 2010, pp. 23–4). Such negotiations show, to some extent, the contrast between collective positions of developing countries in their international economic relations during the NIEO negotiations and the stances adopted individually at the national level in order to attract foreign investment (Sornarajah 2010).

IIAs and their scope of protection

The law of foreign investment, as Dolzer and Schreuer (2012, p. 3) suggest, 'consists of layers of general international law, of general standards of international economic law, and of distinct rules peculiar to its domain'. By December 2015 there were over 2,500 IIAs in force. Virtually all these agreements contain clauses under which the contracting parties agree to provide: stable legal frameworks to foreign investors; non-discriminatory conditions *vis-à-vis* national and other foreign investors; and payment of compensation – and other guarantees – in case of expropriation.

IIAs have often sought to contain expansive provisions on the definition of 'investment'. This definition generally encompasses a wide range of property rights and interests; 'every kind of asset' is the leading formula to a non-exhaustive definition of 'investment' (OECD 2006). Therefore, the definition of investment includes tangible and intangible assets and other property rights deemed necessary to the attainment and operation of the investment – for instance, water permits and licences, linked to investments in mining, oil extraction and energy generation.

The Energy Charter Treaty (ECT) provides one such definition in Article 1(6), which states that:

> 'Investment' means every kind of asset, owned or controlled directly or indirectly by an Investor and includes... any right conferred by law or contract or by virtue of any licences and permits granted pursuant to law to undertake any Economic Activity in the Energy Sector.
>
> *Energy Charter Treaty 1994*

In *AbitibiBowater*, a US investor in the province of Newfoundland and Labrador, sought relief under the North American Free Trade Agreement (NAFTA) for the expropriation of its water, timber and land rights by the government of the Province (*AbitibiBowater Inc. v the Government of Canada* 2009). In this case the claimant alleged that in 1905 the Province had granted the company a perpetually renewable lease of 99 years for 2000 square miles of surface, timber and water rights in the Exploits River watershed (ibid.). According to the investor, these rights fell under the definition of investment pursuant to Article 1139 of NAFTA and were thus protected under the relevant provisions of NAFTA.

Once it has been determined that the investor has property rights or property interests under the laws of the host state and that these rights are covered investments in accordance with the relevant IIA, the investor may invoke protection under the substantive provisions of the IIA, such as national treatment (NT), most favoured nation treatment (MFN), fair and equitable treatment (FET), full protection and security (FPS) and expropriation.

Relative standards of protection: MFN and NT

The first two standards of protection – NT and MFN – are both relative standards. This means that the state should treat foreign investors 'no less favourably' than its own national investors, in the case of NT. An important factor for the application of this standard is the basis for comparison – that is, the extent to which both national and foreign investors are in similar circumstances. Such factor begs the question of whether both national and foreign investors are in the same line of business or should they only operate within the same economic sector (Dolzer and Schreuer 2012, pp. 198–200)? A different treatment may be justified by the circumstances and context under which the governmental measure was adopted.

Methanex Corporation, a Canadian investor and distributor of methanol, submitted a claim to arbitration against the United States for alleged breaches of NAFTA provisions, resulting from a ban on the use or sale in California of the gasoline additive MTBE. The claimant argued that the measure sought to favour domestic producers of ethanol, another fuel additive allegedly competing with methanol, in violation of Article 1102 of NAFTA (NT). The tribunal disagreed with Methanex, stating that methanol and ethanol producers were not in similar circumstances. The ban had been applied to all, national and foreign, MTBE manufacturers in the state of California on a non-discriminatory basis (*Methanex Corporation v United States of America* 2005, Part IV).

Often governmental measures seek to address natural resources scarcity or prevent pollution. As in the case above, arbitral tribunals may consider the geographical and physical context of the governmental measure. In *Methanex* the tribunal addressed the drought problems that the state of California could face should the use of MTBE continue (ibid., Part III).

Under the MFN standard, the state should treat the foreign investor no less favourably than it treats foreign investors from third states. As a way of illustrating the application of the MFN standard, in the *Methanex* scenario, the claimant Methanex could have brought a claim under NAFTA's MFN provisions, if the US had applied the MTBE ban only against Methanex and not against other foreign investors, party or not to NAFTA. Like the NT standard, a determination under the MFN standard requires a consideration of 'like circumstances' among foreign investors.

The analysis of the MFN standard is not straightforward. The question that has confronted several arbitral tribunals is: what benefits fall within the scope of the MFN or, in other words, in which areas is MFN meant to apply? Some IIAs, for instance, have expressly excluded the application of MFN standard from dispute resolution clauses (Dolzer and Schreuer 2012, pp. 209–11).

Fair and equitable treatment (FET)

The FET standard is probably the most prominent standard in international investment law. As broadly described by Dolzer (2013, p. 12):

[A]cceptance of the standard is a response to the danger of the 'obsolescent bargain' which may threaten an investor who was welcomed by the host state before his investment, who sunk its money into the project, but who later on finds itself subject to the upper hand of the host state.

Ascertaining the FET standard involves consideration of several principles, such as transparency, due process and good faith. Lack of stability and consistency, may frustrate investor's legitimate expectations, which constitutes the backbone of the FET standard. However, the standard ought be interpreted according to the facts of the case, as stated by the arbitral tribunal in *Mondev International Ltd. v United States of America* (2002, 118).

In 2003, Tanzania undertook the process of the privatization of its water utility infrastructure, to expand and update the Dar es Salaam water and sanitation systems with funding from the World Bank and other financial institutions. Biwater Gauff, incorporated as City Water Services Limited, was granted a 40-year concession for the provision of water services. City

Water's project eventually failed, due to difficulties in meeting its contractual obligations regarding billing and collection of tariffs from consumers. Tanzanian government officials adopted a number of measures seeking to recover control of the water services company; these measures included the cancellation of the contract, the occupation of City Water's facilities, taking over the management of the company and deportation of senior managers. In response, Biwater Gauff invoked several breaches of the Tanzania – United Kingdom BIT, among which was the breach of the FET standard by the government of Tanzania.

The arbitral tribunal analyzed the arguments of the claimant against the measures adopted by the government. It found a breach of FET in a few actions and omissions by the government of Tanzania. One such breach of the Tanzanian government was the failure to manage the expectations of the users in relation to the reasonable speed of improvements in the network. Indeed, the government did not promote a peaceful termination of the contract for City Water; instead it interfered and polarized the public opinion against Biwater Gauff. In view of the tribunal, this inconsistent behaviour by the government, constituted a breach of the FET obligation under the BIT (*Biwater Gauff (Tanzania) Ltd. v United Republic of Tanzania* 2008, 627–8).

In the context of water resources management, one may ask how an adaptive approach to the management of water resources can be reconciled with the requirements of consistency and stability under the FET standard of investment law. In a recent decision on the alteration and adjustment of legal frameworks, put in place to promote renewable energy, the arbitral tribunal analyzed the element of legitimate expectations. The tribunal considered whether the respondent, Spain, provided specific commitments to investors with regard to the legal environment in place. Following its conclusion that no such commitments had been made, the tribunal then asked whether the legal framework in force could in itself generate legitimate expectations (*Charanne and Construction Investments v Spain* 2016, 493–4). The tribunal further concluded that, in the absence of specific commitments, investors could not expect that rules will not be modified; citing a previous decision in *Electrabel v Hungary* (ibid., 499–500), the tribunal agreed that:

> [w]hile the investor is promised protection against unfair changes, it is well established that the host state is entitled to maintain a reasonable degree of regulatory flexibility to respond to changing circumstances in the public interest. Consequently, the requirement of fairness must not be understood as the immutability of the legal framework, but as implying that subsequent changes should be made fairly, consistently and predictably, taking into account the circumstances of the investment.

The FET has largely evolved in the last eight to ten years. Its content is increasingly defined in treaties and jurisprudence; yet, the factual context of the case remains of utmost importance.

No expropriation without compensation

The standard of expropriation with compensation has been traditionally the most invoked standard in investment treaty arbitration, rapidly being overtaken by the FET standard, discussed above. Expropriation is not illegal *per se*; under international law, a state may expropriate under certain conditions such as: public purpose, non-discrimination, due process and payment of compensation. A common expropriation provision, included in the ECT, states:

> Investments of Investors of a Contracting Party in the Area of any other Contracting
> Party shall not be nationalised, expropriated or subjected to a measure or measures

having effect equivalent to nationalisation or expropriation (hereinafter referred to as 'Expropriation') except where such Expropriation is:

(a) for a purpose which is in the public interest;
(b) not discriminatory;
(c) carried out under due process of law; and
(d) accompanied by the payment of prompt, adequate and effective compensation.

With variations in wording, states generally include direct and indirect expropriation or measures having equivalent effect. A determination of direct expropriation is rather straightforward, as it requires the transfer of ownership from the investor to state or a third party determined by the state. In *AbitibiBowater*, discussed above, the Province of Newfoundland and Labrador issued Bill 75 16 December 2008, entitled 'An Act to Return to the Crown Certain Rights Relating to Timber and Water Use Vested in Abitibi-Consolidated and to Expropriate Assets and Lands Associated with the Generation of Electricity Enabled by those Water Use Rights'. The investor alleged this was an unlawful expropriation, since the Act did not provide for payment of compensation (*AbitibiBowater Inc. v the Government of Canada* 2009, 8–9).

Conversely, a determination of an indirect expropriation presents far more challenges. Regulatory prerogatives exercised by host states to protect the environment, protect public health and safety may often impose an economic burden on investors. Regulatory expropriation claims have not been brought against developing states only; developed states have also been challenged for adopting measures aiming to protect the general welfare. In *Chemtura Corporation v Canada* the tribunal dismissed the claimant's challenge of a Canadian phase-out regulation on the pesticide lindane. The tribunal's determination of whether the investment had been expropriated, considered the level of deprivation suffered by Chemtura. The tribunal concluded that the investor did not suffer a substantial deprivation, since its sales of lindane were just a portion of the company's overall business in Canada. The tribunal added that Canada had validly exercised its police power in a non-discriminatory manner, considering the risk of lindane on health and the environment (*Chemtura Corporation v Canada,* 2010, 363–6).

The question that follows is whether a governmental measure, aiming to protect public welfare, would be only or mainly assessed on the basis of the level of deprivation it has on the investment. The answer to this question has not always been consistent; arbitral tribunals have sometimes differed in their approach to ascertain the existence of an indirect expropriation breach. In *Methanex* the tribunal considered that non-discriminatory measures for a public purpose, in accordance with due process, is not deemed expropriatory and compensable 'unless specific commitments had been given by the regulating government' (*Methanex Corporation v United States of America* 2005, Part IV). In this case the tribunal proposes a dividing line between legitimate non-compensable regulation and compensable indirect expropriation. The way in which we assert such a dividing line will be discussed later in this chapter.

Finally, tribunals have also introduced a proportionality analysis to the determination of whether a breach of an IIA has occurred. In this case, arbitral tribunals may balance multiple potential competing interests. In *Tecmed v Mexico*, where the respondent, Mexico, denied the renewal of a permit to operate a landfill, the tribunal rejected the view that regulatory measures are excluded from the definition of expropriation. The tribunal considered, instead, the negative financial impact of the measure, in addition to whether the governmental measure was proportional to the public interest and Mexico's obligations under the BIT. The tribunal found inspiration in the jurisprudence of the European Court of Human Rights: 'There must be a reasonable relationship of proportionality between the charge or weight imposed to the foreign investor and the aim sought to be realized by any expropriatory measure' (*Tecnicas Medioambientales TECMED S.A. v The United Mexican States* 2003, 122).

The implications of water as an economic good

Principle No. 4 of the Dublin Statement states that: 'Water has an economic value in all its competing uses and should be recognized as an economic good' (ICWE 1992). The Dublin Statement recognizes: 1) access to water and sanitation is a basic right which should be provided at an affordable price, 2) providing water an economic value may avoid overuse, creating adequate incentives for efficient use and conservation and 3) misuse of water resources is directly linked to environmental damage.

Looking at water resources in light of economic principles does not have the purpose to project water resources as a tradable good. Rather this outlook may assist in clarifying the real value that water has for business. Competition for water among users in the same economic sector, or among users in different economic sectors, is increasingly likely to involve foreign investors. The following classification of water resources in the context of foreign investment identifies three different ways in which water resources relate to business and foreign investment law.

Water resources for international trade

Trading fresh water may fall under the category of trading natural resources. Academics and practitioners have posed the question of whether water resources may be defined and treated as a commodity. If so, the question that follows is: at what point does the water becomes a tradable good; when it is still in a watershed, when it has been bottled or when it has been put in a pipeline or ship for transportation (Hughes and Marceau 2013, p. 267)? The World Trade Organization (WTO), for instance, does not have specific provisions regulating the trade of water resources; hence general rules from the General Agreement on Tariffs and Trade (GATT) and other covered agreements may apply in case of a dispute. The discussion on rules of trade and its dispute settlement is outside the scope of this chapter. Instead, the focus is on the investment side of potential water trading.

In 1993, the governments of Canada, Mexico and the United States issued a Joint Statement with the purpose to correct false interpretations regarding water resources:

> The NAFTA creates no rights to the natural water resources of any Party to the Agreement.
>
> Unless water, in any form, has entered into commerce and become a good or product, it is not covered by the provisions of any trade agreement, including the NAFTA. And nothing in the NAFTA would oblige any NAFTA Party to either exploit its water for commercial use, or to begin exporting water in any form. Water in its natural state in lakes, rivers, reservoirs, aquifers, waterbasins and the like is not a good or product, is not traded, and therefore is not and has never been subject to the terms of any trade agreement.

Canada, nonetheless, had expressed concern in connection with the implications of its specific obligations under the NAFTA. In Canada's view, the relationship between bulk water removal and Chapter 11 of NAFTA (investment provisions) could affect its regulatory freedom, in relation to the NT and expropriation with compensation standards under NAFTA.

At the end of 1980 the Government of British Columbia had adopted an initiative for the export of fresh water. Several investors, both national and foreign, expressed interest in exporting water to the United States. Sun Belt, an investor from the United States applied for a water export licence to provide bulk water to California. In 1991, the government of British Columbia imposed a moratorium on new and expanded export licences. The investor, Sun Belt, alleged that this moratorium was discriminatory and sought to benefit Western Canada Water Enterprises Ltd, a Canadian exporter, breaching Canada's NT obligation under Chapter 11 NAFTA.

In 1999, Sun Belt filed a Notice of Arbitration against Canada, under NAFTA, seeking to reverse the national ban imposed on the export of fresh water by marine tankers from the Great Lakes. Sun Belt claimed an initial temporary loss of US$ 468 million, alleging that these costs could rise to US$ 1.5 billion (*Sun Belt Inc. v Her Majesty the Queen (Canada)* 1999, 4). Canada did not assert a defence under NAFTA; the government settled the dispute, under conditions which remain unknown.

Recent negotiations on the Comprehensive Economic and Trade Agreement between the European Union (EU) and Canada (CETA) include a specific provision on the 'Rights and obligations relating water', under Article 1.9:

1. The Parties recognize that water in its natural state, including water in lakes, rivers, reservoirs, aquifers and water basins, is not a good or a product. ...
2. Each Party has the right to protect and preserve its natural water resources. Nothing in this Agreement obliges a Party to permit the commercial use of water for any purpose, including its withdrawal, extraction or diversion for export in bulk.
3. If a Party permits the commercial use of a specific water source, it shall do so in a manner consistent with this Agreement.

The issue of ownership over water resources was discussed in the context of a claim to the waters of Rio Grande/Rio Bravo, by Texas farmers against Mexico, under the provisions of NAFTA. The claimants argued that Mexico had seized water owned by them, thus breaching NT, the minimum standard of treatment and expropriation obligations.

The permits had been granted by the state of Texas, after the allocation of water resources as provided for in the Treaty for the 'Utilization of the Waters of the Colorado and Tijuana Rivers and of the Rio Grande/Rio Bravo', signed on 3 February 1944 between Mexico and the United States. The tribunal articulated an important *obiter dictum* in relation to ownership over water resources:

> One owns the water in a bottle of mineral water, as one owns a can of paint. If another person takes it without permission, that is theft of one's property. But the holder of a right granted by the State of Texas to take a certain amount of water from the Rio Bravo/Rio Grande does not 'own', does not 'possess property rights in', a particular volume of water as it descends through Mexican streams and rivers towards the Rio Bravo/Rio Grande and finds its way into the right-holders irrigation pipes.
> *Bayview Irrigation District et al. v United Mexican States 2007, 116*

The dispute was not solved on the merits, as the arbitral tribunal declined jurisdiction, because none of the claimants had actual investments in Mexico; therefore, the dispute fell outside the scope of NAFTA (ibid., 122).

Water resources as an input for other economic activities

The relationship between investment law and water management is probably more fluid under this category. All economic activities require water resources in different amounts in order to be operational. Some economic activities would be simply impossible without water permits or entitlements. Access to water resources and large extensions of arable land, in certain parts of Africa, have provided incentives to invest in the irrigation of sizeable extensions of farmland in Africa for the production and export of food, biofuels mass and other agricultural products. This type of investment has been considered problematic for its significant effect on local communities (Hey 2013, pp. 299–300).

So far there have been no investment treaty disputes relating to large commercial farming, as a result of conflicts over the use of water resources. It may also be that such disputes have not been framed under claims over competition for water resources. In *Von Pezold v Zimbabwe*, an arbitral tribunal was constituted to decide a dispute between large commercial farming investors and the government of Zimbabwe (*Bernhard Friedrich Arnd Rüdiger von Pezold et al v Republic of Zimbabwe* 2012). The claimants alleged that the fast-track land reform programme breached expropriation, FPS and non-discriminatory obligations, under the German–Zimbabwe BIT. The issue of 'land grabs' was brought to the attention of the tribunal by the European Center for Constitutional and Human Rights (ECCHR), as an *amicus curiae* submission, in relation to indigenous communities' human right to traditional ownership or use of land (ibid., 22). The tribunal rejected the submission of the ECCHR, expressing the difficulties in deciding substantial human rights questions, which were outside the scope of the dispute under analysis (ibid., 60–3).

Disputes more specific to water resources may also arise in the energy and mining sectors. Changes in the quality of water resources could affect users downstream and alter the wellbeing of ecosystems.

In 2007 the city of Hamburg approved the development of a coal-fired power plant situated at the banks of the Elbe River by Swedish investor Vattenfall. The local government approved a provisional licence, despite the opposition of environmental and political groups concerned about the quality of water in the Elbe River (Bernasconi 2009, p. 1). At the time, and to issue the final approval, the city of Hamburg included additional restrictions to the new permit that sought to avoid impact on the volume of water, temperature and oxygen content (*Vattenfall AB, Vattenfall Europe AG, Vattenfall Europe Generation AG & Co. KG v Federal Republic of Germany* 2009, 37–40). The claimant observed that, without the water permit, the company would not be entitled to the emission control permit for the construction of the plant. Such modifications led Vattenfall to initiate an investment dispute against Germany for €1.4 billion in compensation (*Vattenfall AB, Vattenfall Europe AG, Vattenfall Europe Generation AG & Co. KG v Federal Republic of Germany* 2009). In 2010 Germany and Vattenfall settled the dispute, and Germany issued new water permits for the development of the project.

In the mining sector there are currently a number of ongoing investment treaty disputes, as a result of conflicts between the local communities and foreign investors. Communities and other stakeholders have expressed concern over the dangers of environmental degradation that the projects may cause, affecting their access to clean water and agricultural practices (*Bear Creek Mining Corporation v Republic of Peru* 2015). Note that such disputes have not only arisen in a developing country context; developed countries have also been subject to investment disputes. In *St Marys v Canada*, the investor St Marys alleged that, as a result of quarry opponents' lobby, the Ministry of Environment unilaterally expanded the conditions to acquire water permits (*St Marys VCNA, LLC v Government of Canada* 2011). It also claimed that the government stalled tests on the quarry site, establishing unreasonable conditions and arbitrary criteria for the approval of tests (ibid.). The quality of water resources often brings distrust against mining and other extractive industry projects. Local communities in El Salvador, Peru, Indonesia, Costa Rica and others have come forward against the issuance of permits that may endanger their livelihoods and quality of natural resources. However, when such permits have been already issued and the projects are partially developed, the situation for investors and host states could involve high transaction costs.

Water as service: supply and sanitation

Perhaps the relationship between water and investment is most visible in the context of water services and sanitation. There has been a number of high profile investment treaty disputes closely followed by the water law community.

The disputes arising out of water services and sanitation rarely relate to competing demands or potential scarcity, as in the previous cases. This type of dispute rather tends to touch upon the regulation and management of water infrastructure, setting of tariffs and compliance with contractual obligations.

In some cases, investors have not been able to live up to the commitments made in the tendering processes and concession contracts. In *Biwater Gauff*, referred to earlier in this chapter, the arbitral tribunal addressed several of the bad business judgments made by the investor which led to the dispute. Considering the contractual performance in the context of the BIT claims, the tribunal observed: 'As a result of the poor bid, coupled with numerous management and implementation difficulties, [Biwater] (and City Water) did not generate the income which had been foreseen, and accordingly the project quickly encountered substantial difficulties' (*Biwater Gauff (Tanzania) Ltd. v United Republic of Tanzania* 2008, 486).

The tribunal found that Tanzania had expropriated Biwater contractual rights, through different politically motivated actions, the usurpation of the management and the deportation of staff. Notwithstanding such a determination, the tribunal considered that the expropriation did not cause actual economic loss to the claimant because the fair market value of the investment, at the moment of the expropriation, was nil (ibid., 797).

In the case of Cochabamba, Bolivia water services concession, the tariffs associated to the concession contract were raised between 50 and 100 per cent, in some cases a quarter of monthly income (Finnegan 2002). As a result, the citizens of Cochabamba reacted against the investor and the concession contract, participating in protests, which after a period of social unrest, concluded with the abandonment of the investment by Aguas del Tunari S.A.

In some cases, substantial changes in the conditions of operation of the project may arise, affecting negatively the economic equilibrium of the contract.

Economic emergencies, war and other events could give rise to such situations. The measures adopted by Argentina to tackle the severe economic crisis of 2000 affected several concession contracts, not only in the area of water services and sanitation, but also electricity, telecommunication and the financial sector.

The defences invoked by Argentina, have put forward important issues, relevant to the relationship between water and international investment law; one of the most contentious issues is the human right to water. Academics and practitioners, in both areas of law, pose the question of whether arbitral tribunals constitute an appropriate forum to ascertain the human right to water. If so, what is the methodology to realize such a right? In *SAUR*, Argentina invoked its international obligation of protecting its citizens' human right to water as a defence for treaty breaches. The tribunal recognized Argentina's obligation to protect this fundamental right through various prerogatives associated with the regulation and enforcement of the provisions agreed upon in the concession contract. The tribunal, however, did not find that such prerogatives would conflict with Argentina's obligations to protect the investment under the IIA. While the company was under the state's regulatory power to guarantee fundamental rights, such as the right to water, this power was not absolute and should be exercised consistently with the obligation to respect the rights of the investor (*SAUR International SA v Republica Argentina* 2012, 328–32).

Developments and conclusion

Under IIAs the contracting parties have committed to provide legal stability and predictability to each other's foreign investment. It has been argued that during the negotiations of the first generation of IIAs, capital-exporting countries, which were mainly developed countries, proposed tight standards of investment protection so as to guarantee an adequate business environment for their national investors (Álvarez 2009, p. 13). Time has shown that developed countries have also found themselves in circumstances in which regulatory action, deemed to protect public welfare, could have breached investment obligations under IIAs. As discussed above, social, environmental and economic regulatory action has been evaluated by arbitral tribunals, not in relation to their inherent legitimacy, but in the context of the provisions relative to investment obligations. As noted above, arbitral tribunals have been cautious in considering, and above all, making any determination in relation to respondent states' other international obligations, such as human rights.

Several states have undertaken to modify and adapt IIA provisions, in order to secure the exercise of regulatory prerogatives in relation to the public interest. The United States, for instance, adopted such modification in its 2004 BIT Model, which was later reflected in its 2012 BIT Model. The new provisions seek to provide interpretative guidance in the application of certain standards of protection. One of the relevant guiding principles, in the context of expropriation provides: 'Except in rare circumstances, non-discriminatory regulatory actions by a Party that are designed and applied to protect legitimate public welfare objectives, such as public health, safety, and the environment, do not constitute indirect expropriations'(US Model BIT 2012, Annex B (Expropriation), (4)(b)).

As noted above, in the context of the standard of expropriation, tribunals often assessed the level of deprivation of the governmental measure over the investment, in order to determine the existence of an indirect expropriation. Under this provision, arbitral tribunals are required to interpret claims of indirect expropriation in light of the legitimate objectives that the measure may be pursuing. The measure shall be applied in a non-discriminatory manner, but tribunals are yet to determine what kind of measures may fall in the category of a 'rare circumstance'. Good faith and proportionality appear to be important elements of assessment.

These provisions have been widely introduced in a new generation of IIAs, mainly negotiated by United States and Canada with trading and investment partners such as Korea, the EU, Kuwait and Central America. Other states such as China, Uzbekistan and Tanzania have also adopted these provisions in their new IIAs. Water resources are at the heart of the environment and public health, thus regulatory measures related to water resources management and protection are likely to fall within the wider sphere of regulatory freedom, as recognized under this new generation of IIAs.

References

AbitibiBowater Inc. v the Government of Canada (2009) Notice of Intent to Submit a Claim of Arbitration under Chapter Eleven of NAFTA, 23 April.

Agreement for the Promotion and Reciprocal Protection of Investments between Mexico and the United Kingdom (2007) Available from: http://investmentpolicyhub.unctad.org/IIA/country/136/treaty/2545

Aguas del Tunari, S.A. v Republic of Bolivia ICSID Case No. ARB/02/3.

Álvarez, J.E. (2009) A BIT on custom. *New York University Journal of International Law and Politics* 42(1).

Azurix Corp. v The Argentine Republic (2006) ICSID Case No. ARB/01/12.

Bayview Irrigation District et al. v United Mexican States (2007) ICSID (Additional Facility) Case No. ARB (AF)/05/1), Award 19 June.

Bear Creek Mining Corporation v Republic of Peru (2015) ICSID Case No. ARB/14/2, Claimant's Memorial on Merits of 29 May.

Bernasconi, N. (2009) Background paper on *Vattenfall v Germany* arbitration. Winnipeg, International Institute for Sustainable Development (IISD).

Bernhard Friedrich Arnd Rüdiger von Pezold et al v Republic of Zimbabwe (2012) ICSID Case NO. ARB/10/15, Procedural Order No. 2, 26 June.

Biwater Gauff (Tanzania) Ltd. v United Republic of Tanzania (2008) ICSID Case No. ARB/05/22, Award of 24 July.

Charanne and Construction Investments v Spain (2016) Arbitraje No. 062/2012, bajo las reglas de SCC, Award of 21 January.

Chemtura Corporation v Canada, UNCITRAL (2010) (NAFTA), Award signed 2 August.

Compañiá de Aguas del Aconquija S.A. and Vivendi Universal S.A. v Argentine Republic (2007) ICSID Case No. ARB/97/3.

CETA (Comprehensive Economic and Trade Agreement) between Canada, and the European Union (n.d.) Consolidated text not yet into force.

Electrabel S.A. v Republic of Hungary (2015) ICSID Case No. ARB/07/19, Award 25 November.

Energy Charter Treaty (1994) Consolidated version.

Dolzer, R. (2013) Fair and equitable treatment: today's contours. *Santa Clara Journal of International Law* 7.

Dolzer, R. and Schreuer, C. (2012) *Principles of International Investment Law*, 2nd edn. Oxford: Oxford University Press.

Finnegan, W. (2002) Leasing the rain: the world is running out of fresh water, and the fight to control it has begun. *The New Yorker*, 8 April.

Hey, E. (2013) Virtual water, 'land grab' and international law. In L. Boisson de Charzounes, C. Leb and M. Tignino (eds), *International Law and Freshwater Water: Multiple Challenges*. Cheltenham: Edward Elgar.

Hughes, V. and Marceau, G. (2013) WTO and trade in natural resources. In L. Boisson de Charzounes, C. Leb and M. Tignino (eds), *International Law and Freshwater Water: Multiple Challenges*. Cheltenham: Edward Elgar.

ICWE (International Conference on Water and the Environment) (1992) Dublin Statement on Water and Sustainable Development, Dublin, 26–31 January.

Joint Statement by the Governments of Canada, Mexico and the United States (1993).

Methanex Corporation v United States of America (2005) Under UNCITRAL Rules, Final Award on Jurisdiction and Merits of 3 August.

Mondev International Ltd. v United States of America (2002) ICSID Case No. ARB(AF)/99/2, Award of 11 October.

OECD (2006) *International Investment Perspectives: 2006 Edition*. OECD: Paris.

Pac Rim Cayman LLC v Republic of El Salvador (2012) ICSID Case No. ARB/09/12.

SAUR International SA v Republica Argentina (2012) ICSID Case No. ARB/04/4, Decisión sobre Jurisdicción y sobre Responsabilidad, 6 June.

Sornarajah, M. (2010) *The International Law of Foreign Investment*, 3rd edn. Cambridge: Cambridge University Press.

Subedi, S. (2008) *International Investment Law: Reconciling Policy and Principle*. Oxford and Portland, OR: Hart.

Suez, Sociedad General de Aguas de Barcelona S.A., and InterAguas Servicios Integrales del Agua S.A. v The Argentine Republic (2010a) ICSID Case No. ARB/03/17, Decision on Liability 30 July.

Suez, Sociedad General de Aguas de Barcelona S.A. and Vivendi Universal, S.A. v Argentine Republic (2010b) ICSID Case No. ARB/03/19 and *AWG Group v Argentine* under UNICTRAL Rules, Decision on Liability 30 July.

Sun Belt Inc. v Her Majesty the Queen (Canada) (1999) Notice of Claim and Demand for Arbitration, 12 October.

St Marys VCNA, LLC v Government of Canada (2011) Under the UNCITRAL Rules and NAFTA, Notice of Arbitration of 14 September.

Tecnicas Medioambientales TECMED S.A. v The United Mexican States (2003) Case No. ARB (AF)/00/2, Award of 29 May.

UNCTAD (United Nations Conference on Trade Development) (2015) *World Investment Report 2015. Reforming International Investment Governance*. New York and Geneva: UNCTAD.

UNGA (1962) Resolution 1803 (XVII) of 14 December.

UNGA (1974a) Resolution 3201 (S-VI) of 1 May.

United Nations (1974b) General Assembly Resolution 3281 (XXIX) Charter of Economic Rights and Duties of States, adopted in 2315th plenary meeting, 12 December.

United States Model Bilateral Investment Treaty (2012). Available from: https://ustr.gov/sites/default/files/BIT%20text%20for%20ACIEP%20Meeting.pdf

Vattenfall AB, Vattenfall Europe AG, Vattenfall Europe Generation AG & Co. KG v Federal Republic of Germany (2009) ICSID Case No. ARB/09/6), Notice of Arbitration of 30 March.

24

WATER JUSTICE

Understanding the philosophical underpinning of decision-making in the context of international water governance

Marian J. Neal (Patrick) and Peter S. Wenz

Introduction

Water allocation is a fundamental part of water resources management. Water allocation has been described as an unavoidable conflictual process because it is a political process (Allan 2003) and it involves multiple uses and users of water. The scarcity of water resources, driven by anthropogenic and/or natural means, exacerbates the already politically sensitive process of water allocation. Issues of justice arise when resources are, or are perceived to be, in short supply. In these situations individuals or groups of people are concerned about getting their fair share and arrangements are made, or institutions created, to allocate resources (Wenz 1988).

Over the last two decades the international community's concern for shared water resources has prompted a burgeoning collection of declarations, treaties, agreements and joint basin organizations in the quest for peaceful cooperation over shared water (Giordano and Wolf 2003). There is, however, no universally accepted blueprint that will ensure that shared water resources will be allocated justly. International water law plays an important role in ensuring that states that share water resources subscribe to common principles that enable ongoing dialogue and negotiation over how water is allocated.

International water law has, however, often been criticized for its inability to solve trade-off problems or prioritize water users or uses for transboundary water resources. The expectation of water law to give the 'right' answer to these allocation challenges is undeserved since its primary purpose is to create a 'culture of communication' or provide a framework for negotiation processes and criteria that *ought* to be considered when making shared water distribution decisions (Rieu-Clarke and Rocha Loures 2013).

In this chapter we first describe what existing principles and factors are available to the discourse of international water sharing from the international water law perspective. We then describe how the issue of scale impacts on our notions of water justice. We then examine, through elimination, what existing justice and moral theories and principles or existing sources of knowledge are available to assist us making decisions about right and wrong from a philosophical perspective. Each perspective is applied in a transboundary water context. We conclude by discussing how this philosophical approach can assist us in understanding water justice in the context of international water allocation decision-making.

International water law principles

International water law provides us with the concept of *limited territorial sovereignty* as a basis upon which to build principles of water allocation and sharing between states. This doctrine stipulates that all watercourse states shall enjoy an equal right to the utilization of a shared resource, and each watercourse state must respect the sovereignty and reciprocal rights of other watercourse states. This doctrine emerged from two conflicting theories of water allocation:

- absolute territorial sovereignty; and
- absolute territorial integrity.

The theory of *absolute territorial sovereignty* allows for the unlimited use and development of an internationally shared watercourse within the national borders of each state and generally favours the upstream state. The theory of *absolute territorial integrity* entitles every riparian state to the natural flow of a river system crossing its borders and generally favours the downstream state. One of the main advantages of the concept of *limited territorial sovereignty* is that it simultaneously recognizes the rights of both upstream and downstream states without sacrificing the principle of sovereignty (UN Watercourses Convention Factsheet n.d.).

Using the concept of *limited territorial sovereignty*, the UN Watercourses Convention (UNWC) employs two primary principles:

- equitable and reasonable use; and
- no significant harm.

Article 5.1 of the Convention states that:

> watercourse states shall in their respective territories utilize an international watercourse in an equitable and reasonable manner. In particular, an international watercourse shall be used and developed by watercourse states with a view to attaining optimal and sustainable utilization thereof and benefits therefrom, taking into account the interests of the watercourse states concerned, consistent with adequate protection of the watercourse.

Article 7.1 of the Convention states that 'watercourse states shall, in utilizing an international watercourse in their territories, take all appropriate measures to prevent the causing of significant harm to other watercourse states'.

Article 6 of the Convention describes factors relevant to determining equitable and reasonable utilization. They include:

- geographic, hydrographic, hydrological, climatic, ecological and other factors of a natural character;
- the social and economic needs of the watercourse states concerned;
- the population dependent on the watercourse in each watercourse state;
- the effects of the use or uses of the watercourses in one watercourse state on other watercourse states;
- existing and potential uses of the watercourse;
- conservation, protection, development and economy of use of the water resources of the watercourse and the costs of measures taken to that effect; and
- the availability of alternatives, of comparable value, to a particular planned or existing use.

UN 1997

Water justice and negotiation

International water law in general and the UNWC in particular certainly provide a comprehensive framework for water allocation decision-making. They capture distributive, substantive and procedural justice elements which are all vital for fair and just allocation outcomes. There are, however, many claims of injustice and unfair treatment by water users within watercourse states; and there remain concerns, disputes and tensions about getting one's fair share of this communal resource or the benefits thereof. Patrick (2014) posits that this is because of scale issues – that is, because international water law deals with state actors while the recipients of water allocation decision-making are primarily actors at the subnational and local levels.

The case of the Lesotho Highlands Water Project

The Lesotho Highlands Water Project is an international water-sharing project between Lesotho and South Africa. South Africa pays royalties to Lesotho for water transferred to its economic hub; and Lesotho receives hydropower electricity for its domestic use. The treaty signed between Lesotho and South Africa in 1986 is considered by some as an example of good practice of benefit sharing (Haas *et al.* 2010). All water-related uses and development with the Orange–Senqu River Basin fall under the jurisdiction of the Orange–Senqu River Commission (ORASECOM). The ORASECOM Agreement explicitly recognizes the principles of 'equitable and reasonable use' and 'no significant harm' from the UNWC through the regional instrument the Revised Protocol on Shared Watercourses in the Southern African Development Community (SADC), which quotes the factors relevant to equitable and reasonable utilization listed in the UNWC verbatim.

However, the infrastructure involved in this interbasin transfer and hydropower scheme includes large dam development, which has had significantly negative impacts on the long-term ability of the affected local communities within Lesotho to maintain their livelihoods even though they received monetary compensation or were resettled. Those displaced households that opted for monetary compensation forfeited the opportunity to own land; as a result, the next generations will not be able to reap the benefits of such land. In this case monetary compensation was an unsustainable measure and exacerbated further food insecurity and impoverishment for both the present households and future generations (Mokorosi and Van der Zaag 2007). This example illustrates that decisions that are sometimes considered good practice at one level (in this case at the international level between South Africa and Lesotho) do not necessarily translate to positive outcomes at other levels (in this case at the local community level within Lesotho).

Patrick (2014) developed a conceptual framework, entitled the Cycles and Spirals of Justice, that explicitly takes scale into account when evaluating whether the outcomes of water allocation decision-making are considered just or unjust. The framework comprises three elements: a continuum, a cycle and a spiral. The continuum is based on the question 'justice for whom?', which challenges us to be explicit about who should be included in the decision-making process and illustrates that our notions of justice are based on our perspectives and worldviews. The cycle element illustrates our changing values as a society for different uses of water (the social, economic and environmental uses of water). These values also depend on the priorities of water users and continuously cycle based on these changing values and priorities. What was once considered just or fair in the past now might be socially or environmental unacceptable. The spiral element explicitly addresses scale problems in that it illustrates how each cycle of justice and injustice at one level is linked to other cycles of (in)justice at other levels. A continuous shifting of potentially just and unjust outcomes sets up a spiralling motion between the various levels in a complex system. It is this motion that illustrates that often what appears to be a just decision or allocation outcome at one level can create injustices at another.

The continuum, the cycle and the spiral illustrate that the notion of justice is not static and that it is socially constructed. Understanding this presupposes that our notion of justice is, in fact, pluralistic. But where does this leave us with regards to attempting to make water allocations more fair and just?

In the context of water allocation negotiations, the outcome of the negotiation depends on each negotiators' willingness to make trade-offs. The vital ingredient in non-zero-sum negotiations is trust in the process and confidence that all parties will do what they promised (Susskind and Islam 2012). This is an example of where the distributional or substantive elements of justice intersect with the procedural elements of justice. But the question is: how does one make 'good' trade-off decisions? Good is a subjective outcome and each negotiator will have a different set of underpinning rationales for defining what is a good or just decision.

During a negotiation it is common practice to refer to legitimate and reliable sources of knowledge to justify why we need, for example, water for a particular use. This is 'explicit knowledge' which has been codified in some form – for example, setting a water quality management goal based on the scientifically derived standard of water quality that is acceptable for human consumption. There is, however, within each of us, a vast storehouse of tacit knowledge based on emotions, culture, insights, intuition and experience that we subconsciously draw upon to make decisions and that we regard as philosophically right or good. A deeper understanding of our tacit knowledge and how it shapes our decision-making with regard to values related to right and wrong, and good and bad is necessary if we are to understand how to progress negotiations from rights to needs and to being interest-based. To be able to explicitly articulate the rationales we adopt to justify the water allocation and management decisions that we make, we need a philosophical understanding of the sources of knowledge upon which our tacit knowledge rests.

Understanding the rationales behind decision-making

Philosophers are much better at showing what is wrong with other people's understandings of good and bad than at presenting solutions of their own. Solutions are identified by following the implications of the reasons for rejecting other philosopher's solutions. That is the method applied here.

In this section, we consider and reject basing good and bad, right and wrong judgements about justice on:

- the word of God;
- pure reason (Kant);
- utilitarian calculations of what promotes the greatest overall happiness or preference satisfaction (Bentham *et al.*);
- cost–benefit analysis;
- contractarianism (Rawls's Theory of Justice); and
- cultural relativism.

The word of God

The suggestion that God or some religious text ultimately determines what is right and wrong has been discarded by philosophers for several reasons. First, it cannot be proven philosophically or by reason alone that there is such a being as a god who rules the universe and lays down laws for human conduct. Second, if belief in God rests on faith instead of reason, then only those with the appropriate faith would have access to information about morality, whereas morality should govern everyone's conduct. Third, because faith can be misplaced, the beliefs of the faithful about morality may be inaccurate. Related to this is the historical fact that religious texts such as the Bible have been interpreted in different eras in many different ways. For example, many people in the American South believed sincerely 150 years ago that the Bible endorsed the enslavement of black people by white people. Today, some biblically oriented thinkers believe that the Bible verse 'Be fruitful and multiply' means that no one should use artificial means of birth control, whereas other equally sincere biblically oriented thinkers reject that interpretation.

Although religious differences are often the underlying cause of conflict, it is also possible to use the 'Word of God' as an avenue for cooperation. The Jordan River is a transboundary river basin shared between Jordan, Lebanon, Palestine, Israel and Syria and is highly symbolic for Jews, Christians and Muslims. The river runs through the heart of spiritual traditions: some of the founding stories of Judaism, Christianity and Islam are set along its banks and the valley contains many sacred sites. EcoPeace-Middle East, a regional NGO, has developed a set of publications on the Jordan River outlining its value from each of the faith's teachings.

For Christians, the Jordan is a symbol of purity. Christ's baptism in the river marks His revelation as the Son of God and the beginning of His ministry on earth. For thousands of years the Jordan River brought life to one of the cultural heartlands of the Islamic world. The Jordan River is a site where the People of Israel crossed into freedom, Elijah the Prophet ascended to heaven, and Naaman was healed and revived (EcoPeace 2014).

This common value for the river, even though from different religious perspectives, is used to motivate decision-makers to find common ground for the rehabilitation of the river.

Reason

Immanuel Kant is justly famous for attempting to base morality on pure reason, using what he called the categorical imperative. According to Kant, it would be a contradiction ruled out by pure reason for a person to make herself an exception to rules that she thought should govern everyone else. Just as the rules of logic and the laws of nature apply to everyone equally, so do the rules of morality, which brought Kant to what he called the categorical imperative. You should always act in such a way as you could will that everyone else followed the same rule of action that you are following. If you are returning favours with favours, you should be able to will that everyone else do the same in similar circumstances. If you are honouring your parents, you should be able to will that everyone do the same in similar circumstances and so forth.

With this in mind, Kant claimed that certain duties were absolute because a contradiction is involved in the very thought of everyone breaking that duty. Telling the truth, for example, is a duty of this sort because lying is effective only against a background of expected truth telling. If everyone lied all the time, lies would not achieve their goal of imparting false beliefs because a lie works only when the person lied to thinks she's being told the truth. But no one would mistake a lie for a truthful statement if lying were the norm instead of the exception. So, liars must count on most people telling the truth most of the time. It is essential to lying that the liar makes herself an exception to a general observed rule against lying. Thus, lying runs afoul of the categorical imperative, as the liar counts on others obeying a rule that she is unwilling to obey. Lying is therefore always wrong.

Most people refuse to accept some consequences of this reasoning. For example, if you were secretly keeping Anne Frank in your attic during World War II and Nazi officers asked you if you'd seen a Jewish-looking girl in the neighbourhood, Kant would say that the only morally appropriate answer is 'Yes.' 'Do you know where she is now?' 'Yes.' 'Can you lead us to her?' 'Yes.' If Kant had his way, the *Diary of Anne Frank* might have been a lot shorter. In sum, the pure reason approach yields answers to some moral questions that we find unacceptable.

In order to ameliorate this situation in transboundary water negotiations, setting the 'rules of the game' is an integral part of the negotiation process. The game of negotiation takes place at two levels. At one level, negotiation addresses the substance; at another, it focuses – usually implicitly – on the procedure for dealing with the substance (Fisher and Ury 1991). In social psychology these rules are referred to as distributive justice rules. They include *inter alia* the need for consistency, accurate information, opportunity to correct decisions, representation of all affected parties – the procedural equivalent to distributive equality, interpersonal behaviour, articulation of reasons for allocation decisions, accountability and treating affected parties with respect (Brockner and Wiesenfeld 1996; Gross 2011).

Within the UNWC there are a number of articles related to procedural justice:

- Article 8 General obligation to cooperation: Watercourse states shall cooperate on the basis of sovereign equality, territorial integrity, mutual benefit and *good faith* in order to attain optimal utilization and adequate protection of an international watercourse.
- Articles 11 to 19: Planned measures and notification. Procedures are articulated in the event of a dispute between two or more parties concerning the interpretation or application of the present Convention, the parties concerned shall, in the absence of an applicable agreement between them, seek a settlement of the dispute by peaceful means.

UN 1997, emphasis added

Utilitarianism

Taking a hint from our objection to Kant's philosophy, some philosophers suggest evaluating actions by their tendency to produce good results. The problem with Kant's view seems to be that it recommends actions in some circumstances that lead to bad results – in fact, worse results than clearly available alternatives. We think it's reasonable to lie to the Nazis about Anne Frank – that is, to violate Kant's absolute duty to tell the truth, because under the circumstances more good than harm would come from telling a lie. In general, the consequentialist view of morality is that actions are right when they produce the best consequences – in other words, the most good and the least bad. Otherwise, they are wrong.

But this just replaces the problem of discovering what is right and wrong with the problem of deciding what is good and bad. What is the good that people are supposed to be promoting? The most influential answers to this question, what is good and bad, are supplied by utilitarians. In their classic formulation, they claim as a matter of common sense that pleasure (or happiness or positive experience) is the only good and that pain (or suffering or negative experience) is the only bad, and that actions are right when they produce the greatest amount of pleasure (or happiness etc.) and the least amount of pain (or suffering etc.). The pleasure of each individual should be counted equally with that of all others. In the formulation of Jeremy Bentham (1748–1832), the doctrine's first major figure, the goal of all actions should be to provide the greatest good for the greatest number, which is the greatest balance of pleasure over pain among all who are affected by the action in question.

Utilitarian theory can get quite complicated, but the major flaw is that common sense rejects some of the implications of utilitarianism just as clearly as it rejects some of the implications of Kant's theory. Suppose, for example, that a famous person is picked up for shoplifting, as was the actress Winona Ryder some years ago. If the prosecutor is a utilitarian, she should use her discretion to prosecute the case much more vigorously than she would prosecute the case of most other shoplifters because the Ryder case will receive a lot of publicity. The pain Ryder experiences at being doggedly pursued and harshly punished will yield the good result of publicizing the evils of shoplifting and impress on other people the suffering of those who are caught. This will maximize happiness in society by deterring many others from shoplifting, which will make store owners happy, keep down retail prices for paying customers and save many would-be shoplifters (those who would have got caught) from punishment. Clearly, from a utilitarian perspective, Ryder should be treated more harshly than most others who do the same illegal act. The pain of harsh punishment of most shoplifters, those who aren't already famous, will not yield as much good, through publicity, as the harsh punishment of Ryder. So, overall good consequences in society justify harsh punishment in her case but lesser punishments in other cases that are otherwise similar. But common sense morality tells us that like cases should be treated alike. Winona Ryder shouldn't be punished more than anyone else would be for the same offence. Utilitarianism recommends, in fact requires, treating Rider in a manner that common sense morality considers unjust.

Another problem with utilitarianism, and there are many, is that not everyone considers happiness to be the most important matter in life. As pharmaceuticals become more sophisticated, it may be possible to keep many people happy by keeping them permanently hooked on drugs in such a way that they have delusions of engaging in their favourite pursuits. Of course, really they will be just having hallucinations while lying in bed, but they will be happier than they could expect to be with non-hallucinatory experiences because reality so often gets in the way of a good time. If happiness (or positive experience) were the only good, people should jump at the chance of living their lives as drug addicts, supported by others. However, few people actually prefer a life on drugs, no matter how happy, to a life with experience of the real world, regardless of the jeopardy of unhappiness. Again, utilitarian reasoning leads to judgements that most people don't share.

Some utilitarians suggest substituting 'the satisfaction of preferences' for happiness as the good that people should attempt to maximize. Most often, satisfying people's preferences makes people happy – I prefer having a Prius; I get a Prius; I'm happy – so the goal of maximizing preference satisfaction recommends the same actions as the goal of maximizing happiness. However, as we just saw, sometimes people prefer to be less happy than they could be. In these cases the goals differ and the suggestion here is that we follow the goal of preference satisfaction rather than happiness in such cases of conflict. Thus, if people don't prefer to be happy having drug-induced hallucinations, utilitarianism wouldn't recommend the drug programme. This reconciles utilitarianism with common sense.

However, preference satisfaction and happiness share a common weakness – neither can be directly observed. Both are internal psychological states that we infer in other people based on verbal and other behaviour. We need to make such inferences in order to practice utilitarianism because our moral duty as utilitarians is to act so as to bring about the greatest good for the greatest number. We need to know how our actions are affecting the happiness or preference satisfaction of others in order to know which actions utilitarianism requires of us. But, assuming that we switch from happiness to preference satisfaction as our utilitarian goal, how do we estimate how satisfied people are? What behaviours should we be looking at?

Verbal behaviour is not very helpful, because there is no common, interpersonal measure of preference satisfaction to serve as a reference point for interpersonal comparisons. One person's 'very satisfied' may be the same as another person's 'somewhat satisfied', and one person's 'fighting mad' may be the same as another person's 'kind of annoyed'.

According to Lincoln (1986) three types of interests need to be met in order to achieve a sustainable outcome from a negotiation process. They are substantive, procedural and psychological. These three interests are often called the 'satisfaction triangle'. Each interest represents a different side of the triangle; ideally any conflict management process would be designed to reach the centre point within the triangle. This point, in some sense, represents an optimal satisfaction of the procedural, psychological and substantive interests of each of the parties. Psychological satisfaction is, as outlined above, difficult to conceive and measure (Delli Priscoli 2003).

Cost–benefit analysis

One way to solve this problem is to substitute 'revealed preferences' for 'preference satisfaction'. Economists consider purchasing behaviour in a free market to reveal preferences because people tend to buy goods and services that they think will lead to personal satisfaction. This economic-oriented manner of judging preference satisfaction turns utilitarianism into cost–benefit analysis (CBA), where the goal is to maximize the ability of people to purchase what they want in a free market, as this will maximize their chances of satisfaction.

There are many problems with CBA as a substitute for utilitarianism. First, many satisfactions, including some of the most important to human beings, cannot be bought, such as the satisfaction of mastering a musical instrument or a foreign language, and the satisfaction of an enduring, close relationship with another person. Second, if purchasing behaviour is the measure of preference satisfaction and the goal is to maximize the satisfaction of preferences in society as a whole, people will be given unequal weight in moral calculations. People with more money than most others will have more weight than most others in moral calculations proportionate to their greater wealth. They will be able to 'reveal' more preferences through more purchases. Society will be geared to keeping them satisfied because, instead of treating the satisfaction of each person equally, CBA treats the expenditure of each dollar equally. Because rich people have more money to spend, their preferences will be disproportionately catered to in actions based on CBA calculations. But this contradicts our common sense moral belief that all people deserve the same consideration in moral calculations. So, we're back to the drawing board looking for the basis of morality.

Since the Dublin Principles (ICWE 1992, principle 4) stated that 'Water has an economic value in all its competing uses and should be recognized as an economic good', this principle has often been used as a premise for conflict resolution. Benefit sharing is espoused to be a standard of best practice when it comes to water resources sharing in the international river basin context. This approach explicitly articulates the options for water sharing and their trade-offs with the ultimate goal of promoting the sustainable use of transboundary water resources. Emphasis is on the development of 'baskets of benefits' at a regional level. Conflicts resolution is found by moving from a zero sum outcome to identifying positive-sum outcomes or win–win solutions to benefit all water-sharing states. However, as previously stated, these benefits don't necessarily trickle down to the end-users at the local level.

Contractarianism

One suggestion is to base morality on agreements that are mutually advantageous. Think of morality as a kind of business deal or contract wherein people decide on the basis of self-interest what rules they will abide by in their relationships with one another. Everyone decides not to kill or steal from others, for example, because they do not want themselves to be victims of murder or theft.

One problem with basing morality on such an agreement is that people in any social group have different levels of power and influence. What is to prevent the more powerful from using force or bribes to get agreement on a set of rules that reinforce their already privileged position, such as rules that exempt the wealthy from taxes, as was the case at times in medieval France? Morality can't stand in the way of such injustice because, on this contractarian view, the contract *defines* morality. There is no morality before or outside the contract, so whatever the contract endorses as morally correct is for that reason morally correct. No appeal to standards outside the contract is allowed. The obvious problem is that we don't agree with this judgement. We want the contract to reflect what we already recognize to be morally appropriate.

An influential response to these problems is contained in Harvard philosopher John Rawls's *A Theory of Justice*. Rawls's response is to *imagine a hypothetical contract among people who are equal to one another*. Because the people are equal, no individual or group can take advantage of others through force or bribery and the rules that result from the contract are likely to be ones that we would consider appropriate.

However, there are three problems with this. First, if people are agreeing to the contract out of self-interest, why would it be immoral for them to break the rules of morality agreed upon in the contract whenever they can get away with it? After all, if the contract defines morality, there is no morality at the time of agreement to the contract that requires anyone to keep his word – that is, to actually act the way he has agreed in the contract to act. For this reason, the creation of morality through a hypothetical contract doesn't permit people to criticize as immoral those who fail to keep their agreement to abide by the contract. Again, this doesn't reflect what most of us consider morally correct. Second, when some people break their word without fear of criticism, others will do the same and the whole contract-created morality will crumble for failure of compliance. Finally, if the reason to go from an actual to a hypothetical contract is to ensure that the contract reflects our current moral views, morality doesn't really stem from the contract. The contract is just a way of justifying, or adding further justification to what we *already* consider morally correct. This raises again the issue of where our views of morality come from in the first place and how they are to be justified as correct, which was the original problem. The contract view hasn't helped to solve this problem.

The global distribution of water is uneven in space and variable in time, therefore the underlying water signature of each state that comprises a transboundary river basin results in the uneven distribution of water. This provides a natural advantage of good quantity, quality and/or timeliness of water to some states but not all. This advantage is not related to any claims humans might make (such as deservedness), but it does create better opportunities for some and not for others to survive and prosper. Thus a hypothetical contract among the states where all are equal is not possible in a transboundary water context. Human intervention to overcome the natural uneven distribution of water necessitates the technical manipulation of the hydrological cycle and the creation of organizations and institutions to manage this manipulation.

The idea that all parties should be regarded as equal is especially evident in situations where a state is both the upstream and downstream riparian. As previously mentioned, the upstream states generally align themselves with the principle of 'absolute territorial sovereignty', while the downstream states often espouse 'absolute territorial integrity' as their underpinning interest. India, for example, is located upstream from Pakistan and Bangladesh, but downstream from Nepal and Bhutan. It is thus in India's interest to ensure that the rationale it uses as an upstream state is compatible with the principles it endorses as a downstream state; this, in some ways, levels the playing field within the region.

Cultural relativism

As a matter of historical fact, people receive morality as children, and the morality they receive reflects their culture. This leads to the suggestion that whatever your culture says is morally correct is, for that reason, morally correct – at least for you and others who participate in your culture. This view is known as cultural relativism. Cultures define what is morally correct, so whatever a person's culture specifies as right or wrong is right or wrong for that person and for everyone else who participates in that culture. Because cultures differ on some matters of morality, what is right and wrong on these matters is relative to the cultures in question.

The problem with cultural relativism is, again, that it leads to moral judgements that most of us find unacceptable. Some cultures, for example, think it's morally required for young women to have their clitoris removed so that they will not experience pleasure in having sex, which will reduce the incidence of infidelity among these girls when they become married women. Some cultures have been overtly racist, believing, for example, that black people should be subservient to and remain economically and intellectually dependent upon white people. Most of us not only reject the application of these views to our culture, but also believe that these views are incorrect in other cultures as well. We don't believe that morality stems from, is completely dependent upon, and is relative to culture.

As with religious faiths, cultural identity and practices can also contribute to cooperation over water if there is cultural significance associated with a specific river system.

The Mekong River is shared between Cambodia, Thailand, Myanmar, Vietnam, China and Lao PDR. The river system is home to a great cultural diversity, with more than 95 different ethnic groups living in the Mekong Basin. In the lower basin, Khmer, Lao, Thai and Vietnamese people have depended upon the Mekong's resources for thousands of years and the river forms an integral part of local culture. Many of these cultural practices, especially those linked to wild-fish capture, are under threat from the development of dams along the mainstream and its tributaries. This has resulted in tensions over the joint management of the river as positions in the negotiation process become more entrenched.

Intuition

Where does this leave us? We suggest basing morality on what people 'intuit' as morally correct. That is the standard that we've used in this brief review and rejection of the views on morality just considered. Reliance on such intuitions is not the same as endorsement of cultural relativism because we can use, and just have used our intuitions to criticize cultural relativism. More generally, the difference is this: whereas cultural relativism insists that whatever a given culture says about right and wrong is right and wrong within the context of that culture, reliance upon moral intuitions allows for the possibility of internal critiques of a culture. This possibility arises because cultures are not completely isolated from one another and therefore individuals' intuitions are reasonably influenced by other cultures.

Consider, for example, someone who lived in the American South 160 years ago when slavery was part of the local culture. This southerner could use her intuition to reject slavery because she had resources that could enable her to internalize a point of view outside her immediate culture, but within a broader culture that was accessible to her. Thinkers of the French Enlightenment, such as Voltaire, had already explained that people should not be considered property. The United States' Declaration of Independence had used these enlightenment insights to declare that 'all men are created equal'. The Christian religious tradition had long included (among many other strands) the view that people are all equal in the sight of God. Abolitionists in the Northern United States were, at that time, claiming on such bases that slavery was morally wrong. Putting all this together, the southerner of 160 years ago could reason that the abolitionists were right; slavery is morally wrong.

This is an example of how cultures can evolve through moral critiques, both external and internal critiques. As worldwide communication and interaction increase, people are seldom isolated within a particular culture. Although they will always, at least initially, view other cultures with the mind set or worldview of their own cultures, the cultures of their origin, they can, nevertheless, learn from other cultures, change their views and effect change in their original culture. Thus, for example, people in cultures which consider women so intellectually inferior to men that they should be denied equal educational opportunity can learn from the accomplishments of educated women in different cultures that their initial beliefs about women's intellectual potential was mistaken. This can be combined with the results of research showing that literate women tend to have fewer and healthier children than illiterate women and that societies which reduce fertility tend to be more prosperous. Putting all this together can result in the acceptance of women's education in societies where previously such education was considered unnecessary or evil.

Relying in this way on moral intuitions for our basic information about right and wrong is unsatisfying to some thinkers because it always leaves open the possibility that what we accept today as morally correct may someday be frowned upon as unenlightened barbarity. So, we can never be completely secure in our moral views. People who want certainty in their lives, especially about morality, find this unsettling and unacceptable. But they shouldn't. People are fallible and relying in a non-culturally relative way on moral intuitions simply reflects our fallibility. It's always possible that we're wrong, which is why we should never stop questioning our views and should never stop giving some heed to people who differ from us.

A moral and legal code that includes all possible particulars and exceptions to what is right and wrong is not feasible because:

- human beings are intellectually incapable of foreseeing all possible future conflicts among current moral and legal principles and the proper resolutions of all those conflicts;
- new moral and legal principles may be incorporated into our moral thinking in the future and these could not be included in the code at this time; and
- any moral and legal code that tried to include all of these complexities would be too complex for people to follow, so it would not serve as a practical guide to human conduct in general and water allocation in particular.

Conclusion

A fundamental part of water justice is understanding the underlying rationales or principles people appeal to when deciding on how water ought to be allocated between users and uses. Water justice is particularly significant when changes in water governance occur due to altered allocations, reforms in the institutional rules of the game relevant to water or when there are changes in the underlying hydrological regime, due most especially to climate change (Patrick *et al.* 2014).

Not only does the context of the water governance domain constantly change along with the underlying hydrological regime, but so too does the meaning and concept of justice and its underpinning principles. They evolve as human understanding and technologies evolve (Wenz 1988). What was just in the past may not be today and what is just in one context is often not in another. This is why striving for or managing justice is not a static act. If justice is achieved at one level, it might not be at another. What is often perceived as a just outcome at one level of one scale could result in injustices at another level or scale. It is important to recognize that there exists a cycling of the definition of what is considered to be just or unjust and because of this we need to be cognisant of the relationship between justice and injustice in the decision-making process (Patrick 2014). There exists a distinct possibility that we might be unaware of the injustices that our actions at one level might have at another.

Given the plurality of philosophies that underpin what is right and wrong and good and bad, we need to recognize that law can provide us with the space to negotiate and brings all the options to the table, but cannot tell us what is the right (or only) solution to cooperation or conflict. Therefore the principles espoused in international water law will not be able to answer what water uses or users ought to be prioritized over others, but it does provide a platform to explicitly understand what factors ought to be taken into account in our water-sharing decisions.

In particular, where international agreements are concerned, we should take special care to guard the rights and promote the well-being of local communities because, as the case of the Lesotho Highlands Water Project illustrates, government negotiators are prone to prioritize national over local interests, which can lead to unjust outcomes. Justice is served, as Aristotle said, by giving each his due. He also said that what is right is often the mean between two extremes, and this applies to justice as the extremes are giving a stakeholder more than is due and giving that stakeholder less than is due. Ideally, the goal is to hit the mark by giving each exactly what is due. When there is a tendency to diverge from justice by giving one party more than is due, Aristotle thought, decision-makers should aim at giving more than is due to the other party, as that will compensate for the tendency to favour the first party. In the context of local versus national priorities in water-sharing agreements, if there is a tendency to favour national over local interests, decision-makers should deliberately favour local interests somewhat to compensate for a built-in tendency to favour national interests.

The same reasoning applies to the interests of poor people, to the interests of future generations, to the interests of nonhuman species, and to the health of ecosystems. Each of these is likely to have a less powerful voice in international negotiations and is therefore in jeopardy of having its interests underserved by international agreements. So, negotiators should aim at giving such interests more than they are due in order to end up giving them what they are due.

Still, there is no easily applied formula to arrive at decisions that accord justice to all parties – local and national, rich and poor, human and nonhuman. A plurality of values are in play, and no one theory of justice or ethics provides a key to properly weighing these values so as to give all stakeholders their due. We must rely on fallible human judgement without the benefit of calculative certainty available in mathematics. However, the plurality of principles articulated in international water law, both procedural and substantive, combined with good will and the advice to favour parties most likely to be underserved, should go a long way towards avoiding or at least ameliorating injustice in the context of international water governance.

References

Allan, J.A. (2003) IWRM/IWRAM: a new sanctioned discourse? Occasional Paper 50. SOAS Water Issues Study Group, School of Oriental and African Studies/King's College London, University of London.

Brockner, J. and Wiesenfeld, B.M. (1996) An integrative framework for explaining reactions to decisions: interactive effects of outcomes and procedures. *Psychological Bulletin* 120, pp. 189–208.

Delli Priscoli, J. (2003) Participation, consensus building and conflict management training course (Tools for achieving PCCP). UNESCO-IHP, Paris.

EcoPeace (2014) River out of Eden: Water, Ecology and the Jordan River in the Christain, Islam and Jewish Tradition. Available from: http://foeme.org/www/?module=publications&project_id=23

Fisher, R. and Ury, W. (1991) *Getting to Yes: Negotiating an Agreement Without Giving In*, 2nd edn. London: Random House Business.

Giordano, M.A. and Wolf, A.T. (2003) Sharing waters: post Rio international water management. *Natural Resources Forum* 27, pp. 163–71.

Gross, C. (2011) Why justice is important. In D. Connell and R.Q. Grafton (eds), *Basin Futures: Water Reform in the Murray–Darling Basin*. Canberra: ANU E Press.

Haas, J.M., Mazzei, L. and O'Leary, D.T. (2010) Lesotho Highlands Water Project: communication practices for governance and sustainability improvement. *World Bank Working Papers*. Washington, DC: World Bank.

ICWE (International Conference on Water and the Environment) (1992) Dublin Statement on Water and Sustainable Development, Dublin, 26–31 January.

Lincoln, W.F. (1986) *The Course in Collaborative Negotiations*. Tacoma, WA: National Center Associates, Inc.

Mokorosi, P.S. and van der Zaag, P. (2007) Can local people also gain from benefit sharing in water resources management? Experiences from dam development in the Orange-Senqu River Basin. *Physics and Chemistry of the Earth* 32, pp. 1322–9.

Patrick, M J. (2014) The cycles and spirals of justice in water allocation decision-making. *Water International* 39, pp. 63–80.

Patrick, M.J., Syme, G.J. and Horwitz, P. (2014) How reframing a water management issue across scales and levels impacts on perceptions of justice and injustice. *Journal of Hydrology* 519.

Rieu-Clarke, A. and Rocha Loures, F. (2013) Introduction. In F. Rocha Loures and A. Rieu-Clarke (eds), *The UN Watercourses Convention in Force: Strengthening International Law for Transboundary Water Management*. London and New York: Routledge.

Susskind, L.E. and Islam, S. (2012) Water diplomacy: creating value and building trust in transboundary water negotiations. *Science and Diplomacy* 1.

UN (United Nations) (1997) Convention on the Law of the Non-navigational Uses of International Watercourses on the 21 May 1997 (entered into force 17 August 2014). *International Legal Materials* 36 (1997), 700.

UN Watercourses Convention Factsheet (n.d.) Theories of Resource Allocation: User Guide Fact Sheet Series Number 10. Available from: http://www.unwatercoursesconvention.org/documents/UNWC-Fact-Sheet-10-Theories-of-Resource-Allocation.pdf

Wenz, P.S. (1988) *Environmental Justice*. New York: State University of New York Press.

25

ADAPTIVE WATER GOVERNANCE

A theoretical approach reflected in the Mauri and Desaguadero River Basin adaptation plan

Juan Carlos Sánchez Ramírez, Paula Pacheco Mollinedo and Natalia Aguilar Porras[*]

Introduction

It has been estimated that, by 2025, 1.8 billion people will be living in countries or regions with absolute water scarcity (less than 500 m^3 per year per capita), and two-thirds of the world's population could be living under water-stressed conditions (WWAP 2012).

The hydrological cycle, the continuous natural mechanism to transport water over the Earth through different physical stages, is likely to experience some of the major impacts from predicted climate changes due to its direct connection to hydrological components, such as the rise in the temperature of the water and the air (IPCC 2001). Human beings, highly dependent on water availability to fulfil their basic needs (including biological, recreational, spiritual, agricultural, energetic and industrial), will be highly impacted as the water cycle will be affected by climate change and other drivers. Furthermore, ecosystems dependent on water availability will thus alter their provisioning of services for human sustenance.

Climate change not only raises concerns for human livelihoods and ecosystem sustainability, it also introduces a new array of preconditions under which we cannot keep applying the same rules and norms for decision making and governing natural resources that have been used for former circumstances. This whole new environmental stage demands from us no less than commitment into the study of new ways of governance for our threatened natural resources. New strategies for coping with climate change are and will be necessary as it is expected for rainfall patterns and global temperatures to change, bringing more frequent droughts and floods, and, alongside them, more frequent and severe storms, which will affect both quantity and quality of water ecosystems.

In this context, much can be learned from the 'ecosystems approach' promoted and developed under the Convention on Biological Diversity (1992). This strategy for the integrated management of land, water and living resources that promotes conservation and sustainable use in an equitable way can be complementary to the paradigm of IWRM. Building on both concepts, 'adaptive water governance' gains traction as a framework for ensuring that water governance regimes are ready to tackle climate variability in a given basin. The first section of this chapter will examine the theoretical components of 'adaptive water governance'. The second section of this chapter, will examine how those elements unfold in the Bolivian context.

Ecosystems and climate change

Ecosystems

While moving water through different forms all over the Earth, the hydrological cycle also con-nects water with broader biophysical environments: atmospheric, marine, terrestrial, aquatic and subterranean. By doing this, the hydrological cycle provides water for the ecosystems present on those environments (every flora and fauna species has its thermic thresholds and water needs within which it can develop (Saladié Borraz 2013)) and, at the same time, those ecosystems influ-ence and drive the water cycle (Barchiesi *et al.* 2014).

As an example of this relationship, Barchiesi *et al.* (ibid.) reflect on the forest ecosystems. They explain that tree canopies in the forests intercept precipitation and, through evaporation and tran-spiration from foliage, they reduce groundwater and streamflow. Forests ecosystems (natural and human-established) are one of the major users of water. However, activities that demand new forms of land cover, such as agriculture and grazing, add pressure to the use of these ecosystems and cause the partial removal of tree cover. This reduction accelerates water discharge, thus increasing down-stream water flow, while also increasing the risk of floods during rainy seasons. Less tree cover can also contribute to drought conditions because daytime temperatures reduce water storage of soils.

Water cycle and ecosystems are interconnected through physical and biological processes, like precipitation and evapotranspiration by vegetation, respectively. Changes in one drive conse-quences in the next one. Barchiesi *et al.* (ibid.) affirm that the same happens in the water- and land-use relationship; land use depends on water appropriation, and the quality of fresh water systems is directly affected by land use. That is why ecosystems are distributed over the landscape depending on drainage basins and water divides, which is determined by the topography of the land (ibid.).

Ecosystem services and climate change

One of the most popular and simplest definitions of ecosystems services is the one given by the Millennium Ecosystem Assessment (2005) in their *Ecosystems and Human Well-Being* synthesis report: ecosystems services are the benefits that people obtain from ecosystems.

According to the Millennium Ecosystem Assessment (ibid.), human beings are fully depen-dent on Earth's ecosystems due to the services that they provide. Particularly, wetland ecosystems, including rivers, lakes, marshes, rice fields and coastal areas, provide many services that contribute to human well-being and poverty alleviation, among those, fish supply and water availability.

Healthy ecosystems are described as more capable of providing services because they are more resilient than degraded ones, which means that their constitution makes them more flexible, adaptable and able to absorb disturbances – such as those posed by climate change. Resilience has been traditionally conceived as the ability to return to an equilibrium state following perturbation (Holling and Gunderson 2002). Perturbation is an event that disrupts the number, diversity, inter-relations, feedback mechanisms and adaptive capacities of the components of a complex system (Craig and Ruhl 2014).

Notwithstanding, the need to maintain ecosystem functionality so it can provide adequate services for human living has been always a challenge. Many of the uses we make of ecosystem services put them under great pressure (IUCN 2000), and climate change is likely to exacerbate that pressure while jeopardizing both the quality and quantity of water.

Altogether, the impacts of climate change on water quality and quantity are countless. As an example, in terms of the quantity, the United States Environmental Protection Agency (EPA)

(EPA 2016) warns that less snowpack in the mountains and earlier snowmelt mean that less water likely will be available during the summer months, when demand is highest, making it more difficult for water managers to satisfy water demands throughout the course of the year. Moreover, it indicates that freshwater resources along the coasts face risks from sea level rise. As the sea rises, saltwater moves into freshwater areas. This may cause public utilities to find potable water from other sources, including an increase in the need for desalination (or removal of salt from the water) for some coastal freshwater aquifers used as drinking water supplies. Additionally, as most services and industrial sectors count on water quotas, if climate change results in lower stream-flows or changes in the timing of streamflows, it will reduce the amount of hydroelectricity that can be produced. Lower water flows would also reduce the amount of water available to cool fossil-fuel and nuclear power plants. Agriculture and livestock also depend on water. Heavy rain-fall and flooding can damage crops, increase soil erosion and delay planting; areas that experience more frequent droughts will have less water available for crops and livestock (ibid.).

Water quality is likely to be degraded by higher water temperatures and CO_2 concentrations (IPCC 2007a; Lettenmaier *et al.* 2008). EPA (2016) also remarks that, in some areas, increases in runoff, flooding, or sea level rise are a concern. These effects can reduce the quality of water and can damage the infrastructure that we use to treat, transport and deliver water. This problem goes along with others, such as increasing ocean acidification due to absorption of CO_2 emissions, which reduces ocean's pH and changes the chemistry of the water, threatening marine ecosystems.

Changes in flow regimes or water quality due to activities upstream, or due to climate change, will disturb coastal ecosystems and provoke devastation impacts on estuaries, wetlands and the marine environment. In turn, coral reefs, mangroves, seagrasses and nearshore terrestrial ecosystems such as lagoons, which are highly interconnected, will be disturbed (Silvestri and Kershaw 2010).

Integrated actions for adaptive water governance

New challenges for water governance

It can be deduced from the preceding paragraphs that studies of climate change impacts on water are setting new preconditions for water systems and, therefore, there is an increasing necessity for new suitable management approaches. The IPCC came to this conclusion by stating, in its 2007 Fourth Assessment Report that:

> [t]raditionally, hydrological design rules have been based on the assumption of station-ary hydrology, tantamount to the principle that the past is the key to the future. This assumption is no longer valid. The current procedures for designing water-related infrastructures therefore have to be revised.
>
> *IPCC 2007b, 3.6.1*

As hydrologists have observed, water should be managed in a holistic, more integrated way (Richter *et al.* 2003), addressing the entire social-ecological system (Cosens and Williams 2012). In the words of Savenije and Van der Zaag, 'there is an intimate relationship between groundwa-ter and surface water, between coastal water and fresh water, etc.' Consequently, Savenije and Van der Zaag go on to suggest that: 'Regulating one system and not the others may not achieve the desired results' (2008, p. 290).

Together with the need for a holistic approach consistent with water's natural cycle, some of the challenges that climate change has posed for water governance relate to high levels of uncertainty. Rapid and sometimes irreversible changes in the state of resources and ecosystems increase the need for inter-sectorial and inter-institutional coordination (including among levels of government). Such change also heightens the need to actively involve and build the adaptive capacity of numerous and diverse stakeholders on multiple levels, and better recognize the role that freshwater ecosystem health and sustainability play in fostering adaptive capacity, as well as environmental and social resilience (Barchiesi *et al.* 2014, p. 21).

IWRM and ecosystem-based approach

The notion of managing water resources as integrated units has been formulated by the international community for a little more than two decades. The idea was exposed during the preparatory conferences for the Rio Summit in 1992, mainly through the inputs of the International Conference on Water and the Environment (ICWE 1992), which was held in Dublin, Ireland, from 26–31 January. At this conference, the Dublin Principles were launched, along with the key concepts for an integrated water management. Already, by early 2000, delegations from 130 countries had managed to agree on the concept of IWRM during the Second World Water Forum held in The Hague, Netherlands (Second World Water Forum and Ministerial Conference 2000).

According to the Global Water Partnership (GWP 2000), IWRM is 'a process which promotes the coordinated development and management of water, land and related resources, in order to maximize the resultant economic and social welfare in an equitable manner without compromising the sustainability of vital ecosystems'. IWRM, therefore, integrates the concept of sustainable development, which incorporates consideration of natural aspects of water systems, equity and participation. Savenije and Van der Zaag (2008) suggest that IWRM acknowledges the entire water cycle with all its natural aspects, as well as the interests of the water users in the different sectors of a society (or an entire region); hence it addresses both the natural and the human dimensions of water.

Moreover, the ecological dimension of IWRM is emphasized by linking it to ecosystem-based strategies and insights. The rationale behind this complementarity is to enhance the essential role of ecosystems and the crucial role they have for society. In light of climate change, and taking into consideration elements from IWRM but also from the ecosystems approach in the governance field, researchers and analysts are starting to further the notion of adaptive water governance (Cosens and Williams 2012). This concept can serve as useful guidance for drafting or adapting 'rules and norms' of a given water governance regime to the new pressing challenges of climate change.

Ecosystem-based approaches recognize that healthy ecosystems have an important role to play in helping people to adapt, and uses knowledge in biodiversity to place them and their services at the centre of governance approaches. In this way, it fosters sustainable management, conservation and restoration of those ecosystems, increasing their resilience and ensuring their provision of services to society and the environment. For example, conservation of forests around watersheds contributes to the hydrological balance and, therefore, to maintain water quality. This is achieved through minimizing soil erosion on-site, reducing sediment in water bodies (wetlands, ponds and lakes, streams and rivers), and trapping/filtering other water pollutants in forest litter and underwood. Also, forestry activities neither involve the use of fertilizers, pesticides or fossil fuels, nor do they result in outfalls from domestic sewage or industrial processes; therefore, they contribute indirectly to water quality (Sánchez and Roberts 2014, p. 7). Additionally, as water and land have reciprocal effects (land use depends on water appropriation, while the quality of freshwater system is directly affected by land use), a true ecosystem approach to water planning and management will take into account all land and water uses at the catchment or basin level, as well as related groundwater and coastal ecosystems (Sánchez and Roberts 2014).

Within this perspective, acknowledging all possible biophysical components of aquatic ecosystems that can eventually affect directly or indirectly their functionality can help in developing more integrated adaptive strategies.

Milestones for adaptive water governance

With attention to the identified ongoing governance challenges for water, it is possible to distinguish three main pillars or milestones under which we can cluster the different governance concerns and the responses to them based on the IWRM and ecosystem approach concepts. These essential governance needs are: the incorporation of multilevel governance aspects; effectively addressing participation and engagement of public and stakeholders; and ensuring flexibility in all laws, policies, management practices and institutional mechanisms (Troel and Swanson 2014). Each of these variables will be explained in detail below.

Multilevel governance

Issues related to water resources are present at different levels of governance, the international, the regional, the basin, the national, the provincial or district and the local level. Multilevel cross-cutting governance adds further complexity to the design of these new ways of management, but, in return, it provides major integration and therefore resilience to the water resources and ecosystem service users. In other words, when thinking about different decisions on water resource management at different levels, considering different physical stages of water and locations, such as watersheds, sub-catchments and river basins, the necessity of coordination appears more evident.

Partly because of the intrinsic multi-sectorial nature of water management (health, environment, agriculture, energy, culture, social, economic etc.), and partly because of territorial divisions along the basins, fragmentation remains a problem among water governance attempts around the world (see Chapter 15, this volume). Adaptive governance, therefore, aims to challenge those water-related sectors to develop more effective mechanisms for horizontal integration, understood as the inter-institutional or inter-sectorial coordination and cooperation; but also vertical integration is needed in order to ensure coordination among different levels of water governance. Vulnerabilities, as well as positive capacities to address these requirements, vary widely within communities and localities. While enhancing local participation can be a powerful tool to underpin vertical integration, a lack of capacities, information, political commitment or strategic planning can constitute major impediments towards such integration. An important factor to achieve effective vertical integration is the principle of subsidiarity, which requires making decisions at the lowest appropriate level (Savenije and Van der Zaag 2008).

Outside the national borders, water governance also plays an essential role. A third of the world's river basins are shared by two or more riparian states; this happens because 'Natural freshwater basin boundaries do not usually coincide with man-made borders, more than 500 international freshwater rivers, lakes, and aquifers traverse the frontiers of as many as 148 countries' (Drieschova and Eckstein 2014, p. 51).

Acknowledging the cross-border character and multilevel perspective of water is fundamental for better managing water resources within a context of climate change. Thereafter, strong communication and coordination are also requisite steps in order to engage all stakeholders within the basin. IWRM and the ecosystem approach come together with the basic idea of addressing the river basin as the logical and natural unit and replace fragmented management of water. However, many water policies do not consider enough situations that can derive from this changing reality, not taking account of circumstances that can derive from phenomena that take place outside their political or territorial limits. Riparian states, at all levels, should anticipate these events and strengthen, or in some cases create, communication channels and even dispute settlement mechanisms at a transboundary level.

Participation

The subsidiarity principle requires the implementation of effective participation of stakeholders at all levels (Nordic Freshwater Iniative 1991). Higher levels of decision making need to be informed of interests and decisions taken at the local levels. Likewise, decisions made at higher governance levels should take into account the opinions and positions of those at local levels. This constant bargaining across levels leads to better implementation of decisions, by means of an institutionalized active exchange of updated information regarding concerns, strategies, plans and knowledge. These are all fundamental concerns for adaptive water governance.

According to the Dublin principles (ICWE 1992), water resources development and management should be based on a participatory approach, involving all relevant stakeholders. Besides, water has been commonly considered as a matter of public interest, which justifies and encourages the use of public participation mechanisms at all governance levels in water-related decisions or plans. The effectiveness of such mechanisms is crucial for the participatory legitimacy of the aforementioned decisions and plans (Cosens and Williams 2012). As stated by the Convention on Access to Information, Public Participation in Decision-Making and Access to Justice in Environmental Matters (1998), in the field of the environment, improved access to information and public participation in decision making enhances the quality and the implementation of decisions (United Nations Conference on Environment and Development 1992).

While it is very difficult to achieve, active public participation constitutes another cross-cutting component in adaptive governance. Adaptation requires, as pointed out by Sánchez and Roberts 2014,

> effective and ongoing engagement of local individuals, communities and formal and informal institutions to ensure that these policies, plans and activities reflect the actual and evolving vulnerabilities, adaptive capacities, coping mechanisms, needs, and priorities of local populations, and that they are in line with national water management and development policies.
>
> *p. 28*

Increasing awareness is also a desirable effect from participation, but insufficient (Arnstein 1969); public meetings and even the provision of information on environmental impact fail in providing enough deliberative legitimacy. As stated by McKinney and Harmon (2004), adaptive governance requires more than public comment, it requires meaningful public comment; this is a two-way flow of information in which governmental agencies work not only to provide information from their own expertise, but to also incorporate local knowledge and work towards a greater role for public input in decision making. When engaged with the plans in a more effective manner, individuals develop ownership over the decisions and, thus, contribute to the monitoring of compliance (Buchy and Hoverman 2000).

Additionally, participation should take place from the beginning. Many participatory approaches make the mistake of involving stakeholders too late, making it more difficult for them to adapt to the process and limiting their options, as well as contributions, to pre-given decisions or schemes. Also, involving stakeholders with different social, academic, cultural experience and backgrounds might represent a big challenge if the preparation and empowerment of stakeholders, in order to provide a knowledge and capacity for negotiation and discussion, does not take place previously.

Flexibility

Uncertainty is probably the most representative and known challenge posed by climate change to water governance structures. Many water tools and instruments all around the globe are based on stationary standards and take into account different ranges of probability to set the concordant management responses. Many of these instruments set quotas within transnational basin agreements, making them inflexible towards climatic changes. Climate, as UNECE (2015) describes, is a fundamental driver of so many environmental and economic processes, most of which have not previously accounted for any kind of change in water availability, much less for dramatic shifts resulting from new climatic conditions.

As uncertainty about the timing, scale, intensity and character of the potential impacts are the new assumption for future adaptive governance tools, new proposals have to recognize the likelihood that water flow might vary largely between and within seasons, therefore, calling for more flexible provisions (Craig 2010).

Climate change threats will keep limiting water availability and water demand for economic development, and population growth will increase stress on finite resources – thus conflicts, especially among riparian countries which exert asymmetrical power activities, are likely. The inclusion of an accessible and efficient communication system is fundamental, especially since a more adaptive approach for future agreements appears to be shifting from listing and establishing rigid water rights and quotas, as a mean for preventing misunderstandings and conflicts, to institutionalizing communication (Garmestani and Benson 2013).

Hopefully, while conflicts are inevitable, the possibility of more cooperative arrangements will also emerge. The recent entering into force of UN Watercourses Convention (1997) provides a flexible framework to which riparian countries can join without affecting the validity of their previous agreements but acquiring the protection of international standards such as the right to a reasonable and equitable share, and duty of not causing significant harm to others.

In transboundary scales, a common platform of understanding is likely to help in legitimating actions taken under adaptive water governance while avoiding conflicts among signatory countries of those agreements. International consensus on the most basic principles can lead to a more effective implementation of specific bilateral or multilateral agreements in moments when efficient responses are needed the most.

The construction of the strategic plan for climate change resilience of Mauri and Desaguadero River Basin in Bolivia: an overview of the process through the lens of adaptive governance

In the Bolivian central plateau, the effects of climate change translate into temperature increase, changes in rainfall regime and more extreme meteorological phenomena that constantly affect human livelihoods and natural systems (Agua Sustentable 2016a). Both the Mauri and Desaguadero are bi-national rivers, which are situated in the plateau. Mauri, born in Peru, is the main tributary of Desaguadero, which starts at Lake Titicaca; and both are part of the Titicaca, Desaguadero, Poopó and Salares (TDPS) system, which drains the waters of the Bolivian central plateau, covering the southern end of the Department of La Paz and northeast of the Department of Oruro (Agua Sustentable 2016b). The basin has an area of 43,782 km² and it is part of the Department of Oruro and the Department of La Paz (Agua Sustentable 2016b).

The methodology for the development of the plan is based on social learning processes and previous experiences supported by the expertise of the Bolivian NGO Agua Sustentable, whose work supports the sustainable management of water and environment at national and international levels. The construction of the plan had three main stages: research, agreement and planning.

During the first stage, research, scientific meteorological, hydrological, social, economic, institutional and biological information was collected. Historical patterns, climate variability and influence of climate change in the hydrologic cycle at the system level (TDPS system) were analysed and combined with social, political and economic data, which included information about the uses, rights and policies at the local and national level. Different ways of water management under different water availability scenarios were considered to determine how that availability might impact the main production activities of the communities around the basin. Traditional and local knowledge provided by community members on how to tackle climatic hazards was also taken into account.

The second stage, agreement, has been conceived as social public deliberative spaces; roundtables, where civil society, social organizations, departmental and municipal institutions, national state institutions and other actors get together and are informed by the information generated in the first stage. The agreement stage considers also advocacy with key stakeholders and decision makers at multiple levels.

Finally, the third stage, planning, which is meant to draw from the previous stage, prioritizes the adaptation measures and develops institutional recommendations together with the translation of the scientific information for decision makers.

The process of developing a Resilience Plan for Climate Change was built following the milestones of adaptive water governance. The three strategic lines are explained below.

As defined previously, issues related to water resources have to be managed at different levels of governance. The initiatives of strengthening resilience need to be integrated at all coordination levels, and decisions on climate change adaptation have to consider the multisectoral point of view.

Figure 25.1 depicts a pyramid representing the overall bottom-up approach for the implementation of the plan. As illustrated, the plan has its basis in a wide range of local actions that aims to foster capacity building at this level to subsequently generate suitable inputs to influence decision making in the upper levels while enriching national plans.

Results from the second stage, as mentioned, are meant to influence budgetary decisions at the lower levels, while improved national plans advance in gaining acceptability for international adaptation funds.

Horizontal integration was achieved as the inter-institutional and intersectoral cooperation succeeded in generating multidisciplinary studies, including social and hydrological scopes, while using participatory research in order to involve local stakeholders. The main goal was to warrant that the local needs assessed in the studies should be considered at high levels of decision making. Vertical integration took a major role during the roundtables. Indeed, participants were stakeholders ranging from local communities to the Bi-National Authority of the Titicaca Lake and the Plurinational Authority of Mother Earth.

The nature of the two rivers is already complex. The basin is shared by two regional governments and two countries. Therefore, in the construction of solutions for water management, integration of those different actors is essential, with an effective mechanism of communication.

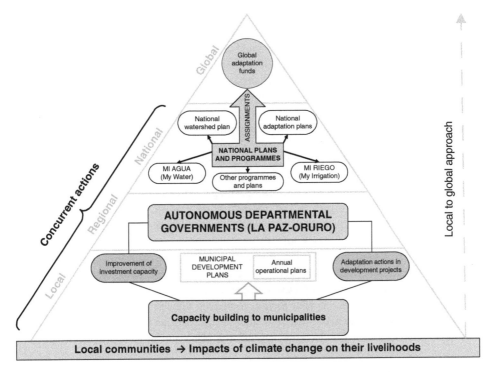

Figure 25.1 Pyramid
Source: Agua Sustentable Atlas 2016b

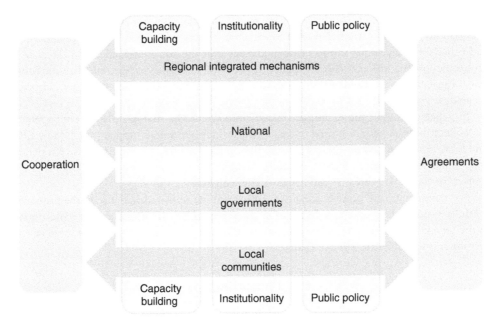

Figure 25.2 Communication mechanism
Source: Pacheco and Sánchez 2012

As discussed, in multilevel systems communication is crucial for integration of the different levels. Notwithstanding, involving, and sometimes even simply finding, all relevant stakeholders on water-related issues might have turned into a major challenge. Despite the difficulties, the project managed to involve seven groups of participants in the agreement for the construction of the adaptation plan: central government, departmental government, municipal government, social organizations, work unions, NGOs and international cooperation agencies and universities. The percentage of members from each category was balanced, and high levels of female participation were achieved. Those different groups of actors can be clustered in the four different levels depicted above.

As seen, integrated capacity building was applied to maximize the participation of all attendees in the tables. Plus, workshops, meetings, interviews, fieldtrips and fairs were held all along the basin in order to promote the participation of communities, institutions, local organizations and municipalities in the project. Furthermore, 122 women took part in a special leader's programme, where they attended to workshops, meetings and national and local gatherings, even outside the country.

Ensuring participation from the initial stages, as well as the inclusion of the community's ancestral knowledge and traditional capacities as part of the relevant information, was a successful decision because it led to a more effective engagement.

A comprehensive document, in table format, containing the inputs from the roundtables was set up as a basis for the construction of the plan, allowing participants to provide information, which later serves as an input for the decisions. Overall, the plan's aim of setting a participatory-based agenda that will potentially integrate part of the municipalities' budgets gives participants a desirable decision-making power.

New adaptive water governance plans should ensure a certain level of flexibility to respond not only to climate, but also to social and political changes. To enhance it, apart from the communication structure, the project includes research on a series of technical, geographical, hydrological and climatology studies, including the social management of water. The generation of these data was used to build climatological scenarios in which water availability might change. An economic analysis was carried out to determine potential increases in the production costs and other socio-economic impacts that might occur at the basin level, and then calculate the economic loss that the area might expect in the future. That analysis was linked to the institutional capacities of the municipalities and how their GDP reduction can affect their adaptation capacities.

To generate possible future climatic scenarios to which the strategic plan needs to respond, the collected information was used to create two scenarios based on models proposed by IPCC, one more optimistic and one pessimistic in order to foresee possible changes in the temperature and rainfall in the region. Expected floods, droughts, frosts, hailstorms, flash floods and their impact for society are also analyzed based on the scenarios.

The consideration of possible scenarios will contribute to the adaptability of the plan and to the follow-up of the programmed actions in the region. After this experience, decision makers in the region can count on more information to help them make better plans and more effective investments to respond to climate change.

Conclusion

Climate change and its multiple hazards to humanity have made us recognize the necessity to review and, when necessary, adjust all national and international instruments relating to the governance of water. Reflected within the hydrological cycle, water dynamics and their connection to ecosystems and livelihoods must be considered in a holistic way in order to adhere to adaptive water governance thinking.

IWRM has emerged to fulfil the need for integration of the different spatial scales, sectors and users into water management and to more fully encompass the hydrological cycle when addressing water-related decisions. Likewise, the ecosystem approach has evolved to complement adaptive water strategies with biological components, which are needed to enhance resilience of ecosystems and, with that, ensure the perpetuity in their role of providing services for nature and humanity.

The different milestones for adaptive water governance can be drawn from the concerns and responses posed by IWRM and ecosystem-based perspectives towards climate change adaptation. Multilevel governance, participation, flexibility and integrated approaches as ecosystems' are necessarily linked, and their overall effectiveness relies highly on each other's implementation.

Demonstrating and spreading the benefits of basin-wide governance, and the role of ecosystems in fostering water quality and quantity, will keep shaping policies and other instruments towards a more suitable framework for water. Furthermore, adoption of international instruments can be useful in providing clarification about common international water law principles that complement and enrich IWRM and ecosystem-based approaches. A common platform of understanding is likely to help in legitimizing actions taken under adaptive water governance, while avoiding conflicts among signatory countries of those agreements. International consensus on the most basic principles, as those contained under the UN Watercourse Convention, can lead to a more effective implementation of specific bilateral or multilateral agreements in moments when efficient responses are required.

Finally, the milestones for adaptive water governance can be seen as orientation principles for the development of adaptation plans. The analysis of the adaptation plan for Mauri and Desaguadero has been used as an example of how the usage of those principles in the improvement of planning for future adversities such as climate change can be verified.

Note

* The authors would like to acknowledge the support given by Agua Sustentable in the provision of relevant data and information for this chapter.

References

Agua Sustentable (2016a) Agua Sustentable Non Governmental Organization. Leaflet of the project. Retrieved from its website www.aguasustentable.org on 16 May 16 2016; and from https://issuu.com/aguasustentable/docs/mauri_desaguadero_triptico

Agua Sustentable (2016b) Draft document of the project [online]. Available from: http://mauri.cambioclimaticoyagua.org.bo/imagenes-plan/documentos/Plan%20de%20Adaptacion%20Mauri-Desaguadero.pdf

Arnstein, S.R. (1969) A ladder of citizen participation. *Journal of the American Institute of Planners*, 35:4, p. 216-224, DOI: 10.1080/01944366908977225 [online]. Available from: http://dx.doi.org/10.1080/01944366908977225

Barchiesi, S., Welling, R., Dalton, J. and Smith, M. (2014) Sustaining ecosystems through better water management for climate change adaptation. In J.C. Sánchez and J. Roberts (eds), *Transboundary Water Governance: Adaptation to Climate Change*. Gland, Switzerland: IUCN.

Buchy, M. and Hoverman, U.S. (2000) Understanding public participation in forest planning: a review. *Forest Policy and Economics* 1(1), May, pp. 15–25 [online]. Available from: http://www.sciencedirect.com/science/article/pii/S138993410000006X

Convention on Biological Diversity (1992) 22 May (entered into force 29 December 1993) [online]. Available from: https://www.cbd.int/convention/text/

Convention on Access to Information, Public Participation in Decision-Making and Access to Justice in Environmental Matters (1998) Aarhus, Denmark, 25 June [online]. Available from: http://www.unece. org/fileadmin/DAM/env/pp/documents/cep43e.pdf

Cosens, B.A. and Williams, M.K. (2012) Resilience and water governance: adaptive governance in the Columbia River basin. *Ecology and Society* 17(4), p. 3 [online]. Available from: http://dx.doi.org/10.5751/ ES-04986-170403

Craig, R.K. (2010) Stationarity is dead: long live transformation: five principles for climate change adaptation law. *Harvard Environmental Law Review* 31, pp. 9–75.

Craig R.K. and Ruhl, J.B. (2014) Designing administrative law for adaptive management. *Vanderbilt Law Review 1*. Vanderbilt University, 67(1), January.

Drieschova, A. and Eckstein, G. (2014) Cooperative transboundary mechanisms. In J.C. Sánchez and J. Roberts (eds), *Transboundary Water Governance: Adaptation to Climate Change*. Gland, Switzerland: IUCN, p. 51.

EPA (United States Environmental Protection Agency) (2016) Retrieved from its official website on 13 May 2016. Based on US Global Change Research Program. T. R. Karl, J. M. Melillo and T. C. Peterson (eds) (2009), *Global Climate Change Impacts in the United States*. New York: Cambridge University Press [online]. Available from https://www3.epa.gov/climatechange/impacts/water.html and https://downloads.globalchange.gov/usimpacts/pdfs/climate-impacts-report.pdf

Garmestani, A.S. and Benson, M.H. (2013) A framework for resilience-based governance of social–ecological systems. *Ecology and Society* 18(1), p. 9. http://dx.doi.org/10.5751/ES-05180-180109

GWP (Global Water Partnership) (2000) *Integrated Water Resources Management*. Technical Advisory Committee, Background Paper No. 4. Stockholm, Sweden.

Holling, C.S. and Gunderson, L.H. (2002) Resilience and adaptive cycles. In L.H. Gunderson and C.S. Holling (eds), *Panarchy: Understanding Transformation in Human and Natural Systems*. Washington, DC: Island Press.

ICWE (International Conference on Water and the Environment) (1992) Development Issues for the 21st Century. Adopted in Dublin, Ireland, from 25–31 January [online]. Available from: http://www.ircwash. org/sites/default/files/71-ICWE92-9739.pdf

IPCC (Intergovernmental Panel on Climate Change) (2001) *Climate Change 2001 Impacts, Adaptation and Vulnerability*. Cambridge University Press [online]. Available from: http://www.ipcc.ch/ipccreports/tar/ wg2/pdf/wg2TARfrontmatter.pdf

IPCC (2007a) Ecosystems, their properties, goods, and services. *Climate Change 2007: Impacts, Adaptation and Vulnerability*. Contribution of Working Group II to the Fourth Assessment Report of the IPCC. Cambridge: Cambridge University Press, p. 233.

IPCC (2007b) Freshwater resources and their management. *Climate Change 2007: Impacts, Adaptation and Vulnerability*. Contribution of Working Group II to the Fourth Assessment Report of the IPCC. Cambridge: Cambridge University Press, p. 196.

IUCN (International Union for the Conservation of Nature) (2000) *Vision for Water and Nature: A World Strategy for Conservation and Sustainable Management of Water Resources in the 21st Century*. Gland, Switzerland, and Cambridge, UK: IUCN.

Lettenmaier, D., Major, D., Poff, L. and Running, S. (2008) Water resources. In *The Effects of Climate Change on Agriculture, Land Resources, Water Resources, and Biodiversity in the United States*. A Report by the US Climate Change Science Program and the Subcommittee on Global Change Research. Washington, DC, pp. 133 and 142 [online]. Available from: http://www.usda.gov/oce/climate_change/SAP4_3/ CCSPFinalReport.pdf

McKinney, M. and Harmon, W. (2004) *The Western Confluence: A Guide to Governing Natural Resources*. Washington, DC: Island Press.

Millennium Ecosystem Assessment (2005) *Ecosystems and Human Well-Being: Wetlands and Water Synthesis*. Washington, DC: World Resources Institute [online]. Available from: http://www.millenniumassessment.org/documents/document.358.aspx.pdf

Nordic Freshwater Iniative (1991) Copenhagen report: Implementation mechanisms for integrated water resources development and management: report from Copenhagen Informal Consultation, 11–14 November [online]. Available from: http://www.ircwash.org/resources/copenhagen-report-implementation-mechanisms-integrated-water-resources-development-and

Pacheco, P., Sánchez, J.C. and Tattle, Q. (2012) Data and information management in transboundary water governance. *Journal of Water Law* 23(2), p. 66 [online]. Available from: https://www.researchgate.net/ publication/266387353_Data_and_Information_Management_in_Transboundary_water_Governance

Richter, B.D., Mathews, R., Harrison, D.L. and Wigington, R. (2003) Ecologically sustainable water management: managing river flows for ecological integrity. *Ecological Applications* 13, pp. 206–24.

Saladié Borraz, O. (2013) El cambio climático es una realidad y no adaptarse supone un riesgo. In S. Borras Pentinat and P. Villavicencio Calzadilla (2013) *Retos y Realidades de la Adaptación al Cambio Climático. Perspectivas Técnico Jurídicas*. Spain: Thomson Reuters Aranzadi (free translation).

Sánchez, J.C. and Roberts, J. (eds) (2014) *Transboundary Water Governance: Adaptation to Climate Change*. Gland, Switzerland: IUCN.

Savenije, H.H.G. and van der Zaag, P. (2008) integrated water resources management: concepts and issues. *Physics and Chemistry of the Earth* 33, pp. 290–7.

Second World Water Forum and Ministerial Conference (2000) *The Hague: From Vision to Action: Final Report*. The Hague: World Water Council and Netherlands Ministry of Foreign Affairs. 17–22 March.

Silvestri, S. and Kershaw, F. (2010) *Framing the Flow: Innovative Approaches to Understand, Protect, and Value Ecosystems Services Across Linked Habitats*. Cambridge: UNEP World Conservation Monitoring Center, p. 5.

Troel, J. and Swanson, G. (2014) Adaptive water governance and the principles of international water law. In J.C. Sánchez and J. Roberts (eds), *Transboundary Water Governance: Adaptation to Climate Change*. Gland, Switzerland: IUCN, p. 31.

UNECE (United Nations Commission for Europe) (2015) *Water and Climate Change Adaptation in Transboundary Basins: Lessons Learned and Good Practices* [online]. Available from: https://www.unece.org/fileadmin/DAM/env/water/publications/WAT_Good_practices/ece.mp.wat.45.pdf, p. 14.

United Nations (1997) Convention on the Law of the Non-Navigational Uses of International Watercourses [online]. Available from: http://legal.un.org/ilc/texts/instruments/english/conventions/8_3_1997.pdf

United Nations Conference on Environment and Development (1992) *Agenda 21, Rio Declaration, Forest Principles*. New York: UN.

WWAP (World Water Assessment Programme) (2012) *The United Nations World Water Development Report 4: Managing Water under Uncertainty and Risk*. Paris: UNESCO, p. 541 [online]. Available from: http://unesdoc.unesco.org/images/0021/002156/215644e.pdf

26

A BRIEF HISTORY OF GLOBAL WATER GOVERNANCE

Joshua Newton

What is global water governance?

Even with a recent proliferation of actions at the global level related to water, it took some time before organisations and researchers started to recognise that little work had been done on what global water governance was or its potential impact on the world's water resources. Global water governance, however, is already happening (Gupta 2011), and although there is 'no global, comprehensive, intergovernmental structure for water, there is a very dynamic process of advancing international understanding and cooperation on water for sustainable development' (WEHAB 2002, p. 25).

Ken Conca states that the 'world's water is indeed subject to deeply and increasingly transnational forms of governance'. Rules, roles and practice have become further embedded in water-related policy decisions. 'If those acts include such things as the framing of policy, the setting of standards and the mobilization and allocation of resources, then water is indeed subject to governance that is increasingly, though certainly not exclusively, global' (Conca 2006, p. 5).

Claudia Pahl-Wostl, Joyeeta Gupta and Daniel Petry define global water governance as 'the development and implementation of norms, principles, rules, incentives, informative tools and infrastructure to promote a change in behaviour of actors at the global level in the area of water governance' (Pahl-Wostl *et al.* 2008, p. 422).

Robert Varady, Katherine Meehan and Emily McGovern, who work extensively on global water initiatives, a subset of global water governance, argue that the structure of global water governance is 'not characterized by an even or smooth distribution of power, decision-making or policy across space'. They continue to explain that this type of governance happens in specific places such as the location of UN agencies/international organisations or the World Water Fora through 'networks of knowledge transfer and communication' (Varady *et al.* 2008a, p. 6).

Pahl-Wostl *et al.* expand on this view to state that global water governance is actually a form of 'mobius-web' typology of governance (Pahl-Wost *et al.* 2008) that James Rosenau describes as a 'web-like process that neither begins nor culminates at any level or any point in time' (Rosenau 2000, p. 13), where a 'number of top-down, bottom-up, network, and side-by-side governance elements exist in parallel' (Pahl-Wostl *et al.* 2008, p. 427). Conca reinforces the idea that this form of global governance for water is not a 'neat, uncontested set of water norms' (Conca 2006, p. 5). Gupta calls this system 'less subject to control, less predictable, less in line with good governance principles, but more flexible, more reflective of power structures, more adaptive, more efficient and, at times, more effective' (Gupta 2011, p. 8).

Within this web-like structure Pahl-Wostl *et al.* believe there are six elements (Pahl-Wostl *et al.* 2008):

1. International law;
2. Global intergovernmental agencies;
3. Regional bodies;
4. Non-state actors/stakeholders;
5. Private sector; and
6. Global communities of scientists and water professionals.

Other authors call the global water governance regime 'not fully legitimized' (Mukhtarov 2007, p. 2), 'diffuse' and charge that it reflects 'fuzzy governance, one that scarcely qualifies to be called a global water governance regime' (Dellapenna and Gupta 2009, p. 9). It is a regime that is 'fragmented,' has no leadership (Pahl-Wostl *et al.* 2008) or coordination (Pahl-Wostl 2012) nor a 'natural center of gravity' with 'competing actors and interests and… no real consensus process to deal with water science' (Gupta *et al.* 2013, p. 9). Water is 'a rising issue of global governance characterized by comparatively young and immature structures and processes that have slowly evolved over the past two decades' (Pahl-Wostl *et al.* 2008, p. 427), but where still no global policy framework exists (Dellapenna and Gupta 2009).

A look back at global water governance

Activities related to water at the global level have been gaining momentum over the past century. It could be said that there has been exponential growth of the number of events and organisations that address global water issues. The following section will describe these events that loosely make up part of the current piecemeal, both informal and formal, global water governance regime.

The first water-related intergovernmental meeting took place in Paris in 1851, known as the International Sanitary Conference, which began a series of 14 conferences on the subject that occurred at irregular intervals until 1938 and the beginning of World War II. These conferences provided the spirit behind which the World Health Organization (WHO) was founded in 1948. While several initiatives were put forth to the conference on the spread of cholera and other communicable diseases, the conferences were fairly unsuccessful until the Seventh Conference in 1892, when a treaty was approved to establish maritime quarantine regulations (Harvard University n.d.).

In 1873, the precursor to the World Meteorological Organization (WMO), the International Meteorological Organization, was founded after the first international meeting on meteorology in Brussels in 1853, which was designed to share weather information between states (Rodda 1995).

Following World War I, and directly related to the initiatives of the League of Nations, water started to appear on the international agenda, specifically with regards to navigation and hydropower.

The first convention on water was the *Convention and Statute on the Regime of Navigable Waterways of International Concern*, signed in Barcelona in 1921 (League of Nations 1921), which set minimum standards for navigation (Malla 2005) and stated that: 'Each riparian State is bound, on the one hand, to refrain from all measures likely to prejudice the navigability of the waterway'

(League of Nations 1921). While this Convention and Statute, which was signed by 40 states (UN Treaty Collection n.d.), was groundbreaking in being the first such treaty, it failed, because the convention did not have a truly global perspective in that it did not 'effectively combine the different approaches to the principle of freedom of navigation that had emerged on different continents' (Boisson de Chazournes 2009, p. 2).

Shortly thereafter, the first convention on non-navigable uses was signed, which was the *Convention Relating to the Development of Hydraulic Power Affecting More than One State, and Protocol and Signature* in 1923 (League of Nations 1923). This convention encouraged states to take into consideration the interests of other riparian states when carrying out hydropower projects on shared rivers (Boisson de Chazournes 2009) and promoted cooperation where possible (League of Nations 1923).

It was also around this time that international professional and scientific societies and organisations related to water began to emerge, most of which are still in existence today. The first was the International Association of Hydrological Sciences (IAHS) in 1922, an organisation that is dedicated to expanding the knowledge of the science of hydrology.[1] In 1928, the International Commission on Large Dams (ICOLD)[2] was founded, followed, in 1935, by the International Association of Hydraulic Engineering Research (IAHR).[3]

In realisation that many of the world's issues cross national borders, following World War II, the UN was created to reduce the amount of conflict by improving the well-being of the world's citizens. This was when global activity on water issues truly commenced. Given the post-war boom in development in industrialised countries, this period was also marked by large, ambitious water-related projects such as dams, tidal barrages, hydroelectric plants, irrigations schemes, tidal barrages, river diversions, the draining of wetlands and inter-basin transfers (Varady *et al.* 2008b).

Following the proliferation of UN agencies, UNESCO's International Hydrological Decade, from 1965–74, was 'man's first concerted attempt to take stock of his diminishing available resources of fresh water and to co-ordinate world-wide research on ways of making better use of them' (Nace 1969). A comprehensive inventory of the world's water resources was finished in 1978, allowing, for the first time, examination of the state of the world's freshwater availability (Varady *et al.* 2008a).

At the same time that international organisations started to focus on water, organisations that concentrated on international law were starting to turn their attention to international water issues. While the Institute of International Law (IIL), a non-governmental organisation, put forward the Madrid Declaration in 1911 prohibiting activities that might hurt other riparian nations within shared river basins,[4] the real push came from the International Law Association (ILA) in 1956 and 1958 with its Dubrovnik Statement and the more refined New York Resolution, respectively (Salman 2007). The New York Resolution introduced the terminology that is still used today emphasising that each co-riparian state is entitled to a reasonable and equitable share in the beneficial uses of the waters of the drainage basins (ILA 1958). This was the same year that the United Nations was conducting work on 'integrated river basin management' (United Nations 1958), which culminated in UN General Assembly (UNGA) Resolution 1401 (XIV) in 1959 that initiated preliminary studies on the legal problems relating to the use of international rivers (United Nations 1959). The ILA's work continued over the next eight years, finishing at a meeting in Helsinki in 1966, whose outcome was the *The Helsinki Rules on the Uses of the Waters of International Rivers* (ILA 1966). It was in these rules that the principle of 'equitable utilization' was established and became the 'guiding rule for the work of the ILA in the field of international rivers' (Salman 2007). Also of note within the *Helsinki Rules* was the first mention of groundwater in any global legal instrument (ILA 1966).

By the late 1960s, many countries, both developed and developing, were starting to feel the impact of humans on the environment, most notably in terms of pollution, and they needed to be addressed urgently. On 3 December 1968, the UNGA decided, in Resolution 2398 (XXIII), to convene a United Nations Conference on the Human Environment (United Nations 1972a), which was to be held on 5–16 June 1972 in Stockholm, Sweden.

Overall, the Stockholm Conference was a huge success. The goal of the conference was to raise awareness about the environment and, because of the preparatory process, this was achieved before the meeting took place. The Stockholm Conference also legitimised the environment as a national and international concern, was instrumental in building national and international instruments in the field of environment, established a framework for treaty making, was the first UN event where civil society participated and had an impact on the outcome and became a model for UN conferences (Engfeldt 2009).

It was about this time as well that the UNGA, in 1970, asked the International Law Commission (ILC) to study the topic of international watercourses. In 1971 they started working on a draft Convention that would take almost three decades to finish (Salman 2007).

In the same year, the Convention on Wetlands of International Importance especially as Waterfowl Habitat (also known as the Ramsar Convention) was signed. It has 168 contracting parties, and protects over 2,100 wetlands covering over 205 million hectares.[5] Even though not a United Nations convention, Ramsar primarily focuses on the protection, conservation and wise use of wetlands for waterfowl (Ramsar Conference 1971), although this has expanded to include all uses of wetlands (Iza 2004).

In 1971, the Committee on Natural Resources of the United Nations, in its first session, started to discuss the possibility of convening a major water conference, initially with the idea to share experiences in water management between countries. In 1975, Resolution 3513 was passed in the UNGA to organise such a meeting (Biswas 1988).

The United Nations Water Conference, held in Mar del Plata, Argentina, from 14–25 March 1977, was the first global water meeting of its kind, with participants from a high policy-making level (WMO–UNESCO 1991), and helped focus attention exclusively on water for the first time (Salman 2003). The outcome was the 60-page Mar del Plata Action Plan (MPAP), full of recommendations ranging from assessment of water resources to training and research, and culminating with ten resolutions.

Although cooperation had existed between water-related UN bodies and agencies since the 1950s through the Committee on Natural Resources and the Administrative Sub-Committee on Coordination (Mageed and White 1995), Mar del Plata caused a more formal approach to be taken, most notably in the form of the UN Intersecretariat Group for Water Resources (ISGWR) in 1979. At the time, there were 24 UN entities that were part of the ISGWR (Rodda 1995), which resided under the Administrative Committee on Coordination (ACC). The purpose of the ISGWR was to 1) cooperate in the monitoring of the progress of the MPAP, 2) promote cooperation over water-related activities within the UN system and 3) assist in coordinating activities at the country and regional levels (Mageed and White 1995).

One of the main outcomes of the UN Water Conference was that 'governments commit to provide all people with water of safe quality and adequate quantity and basic sanitary facilities by 1990', as the Conference stated that 'all peoples, whatever their stage of development and their social and economic conditions, have the right to have access to drinking water in quantities and of a quality equal to their basic needs'. The main mechanism of this was the International Drinking Water Supply and Sanitation Decade (hereafter Decade) (1981–90), which was proposed first at the United Nations Conference on Human Settlements in 1976 and reiterated in the MPAP (United Nations 1978).

At the end of the Decade a meeting was convened in New Delhi in 1990,[6] evoking the theme: 'Some for all rather than more for some', which was the first truly global meeting on water and sanitation (Beyer 1991), and seen as a follow-up to the Decade. One-third of the world's population still lacked water and sanitation, 'these two most basic requirements for health and dignity' (United Nations 1990b). The end of the Decade also produced a resolution by the United Nations that created the Water Supply and Sanitation Collaborative Council (WSSCC), which was charged with completing the work that was unfinished by the 1981–1990 Decade (United Nations 1990a).

A ten-year review of the 1972 Stockholm Conference by the United Nations Environment Programme (UNEP), which was created following the conference, resulted in the independent World Commission on Environment and Development (WCED), otherwise known as the Brundtland Commission, which, while it put sustainable development on the global agenda, water, even though mentioned throughout the report, was not a major or even minor focus of the Commission (WCED 1987), which many people attribute to a 'water blindness' that took place during the decade of the 1980s (FAO 2000; Biswas 1988), also known as the 'lost decade' for international water policy (Scheumann and Klaphake 2001).

The 1990s was when water truly started to gain attention and momentum at the global level. Following New Delhi, the next major global conference on water was the International Conference on Water and the Environment (ICWE), held in Dublin, Ireland, in the beginning of 1992. The purpose of ICWE was to prepare for the primary input from the water community to the United Nations Conference on Environment and Development (UNCED), otherwise known as the Earth Summit, in Rio di Janeiro, Brazil, which was to take place later that year.

ICWE was hosted by the Irish Government and organised by the United Nations ACC-ISGWR, whose chair at the time was WMO, which was administratively responsible and housed the ICWE Secretariat in Geneva. The United Nations was the convening entity, but it was not a normal UN meeting in that the participants were government-nominated experts as the participants did not want to pre-empt their governments' positions at the preparatory meetings to UNCED. This meant that the outcome of the meeting, which was officially not intergovernmental, did not have to be taken into account during the UNCED process (Young *et al.* 2004).

The major outcome of Dublin was what is known as the Dublin Principles, which have been heavily referenced ever since the conference. In brief, they are (ICWE 1992):

1. freshwater is a finite and vulnerable resource, essential to sustain life, development and the environment;
2. water development and management should be based on a participatory approach, involving users, planners and policy-makers at all levels;
3. women play a central part in the provision, management and safeguarding of water; and
4. water has an economic value in all its competing uses, and should be recognised as an economic good.

The Dublin Principles 'can be judged as one of the clearest, most comprehensive and far-reaching statements of water management up to today' (Scheumann and Klaphake 2001, p. 6) and have had a large impact on shaping water governance (Gupta 2013). Even though the concept of integrated water resources management (IWRM) had been discussed for many years, Dublin was the first time that this concept was adopted by the international community and codified in the first two principles (Savenije and Hoekstra 2003). According to the Global Water Partnership (GWP), IWRM is 'a process which promotes the co-ordinated development and management of water, land and related resources, in order to maximize the resultant economic and social welfare in an equitable manner without compromising the sustainability of vital ecosystems' (GWP 2000, p. 22). IWRM continues to be the primary theoretical management paradigm for water.

Twenty years after Stockholm and a few short months after Dublin, the Earth Summit took place in Rio in June 1992, capping off a three-year negotiation process whose aim it was to create 'a blueprint for action to achieve sustainable development worldwide' (United Nations n.d.c). During the previous 20 years, the polarising interests and concerns related to sustainable development of developing and developed countries had increased a significant amount, which was one of the major obstacles heading into UNCED (Engfeldt 2009).

Outside the Rio Declaration and the three Rio Conventions,[7] the primary outcome of the Earth Summit was Agenda 21, a 'handbook for action' that not only had specific implementation language, but also provided cost estimates for such implementation (ibid.). While water was not one of the main focuses of the negotiations (Scheumann and Klaphake 2001), and received not a single mention in the Rio Declaration (UNCED 1992b), a full chapter (UNCED 1992a) was spent outlining water-related activities for the coming decades, 'by and large, an elaboration of the Mar del Plata Action Plan' (Salman 2003, p. 493).

On the whole, the most controversial water topics were left out of Chapter 18, namely the mention of large infrastructure (dams), privatisation and pricing (Conca 2006). In addition, there was weak language for transboundary water issues. The issue of water as an economic good was revisited and the wording from the Dublin Principles was softened, due to pressure by a number of developing countries (Scheumann and Klaphake 2001) to include mention of water as both an economic *and* social good (UNCED 1992a). Overall, Agenda 21 was not successful in creating a strategic approach to international water policy (Kibaroğlu 2002) and failed to ascribe a great deal of urgency or priority to water issues (Grover and Biswas 1993), but still remains a cornerstone in global water governance as will be made clear later in this chapter.

Shortly after the Earth Summit, the UNGA passed a resolution, based on a recommendation of UNCED, to observe a World Water Day, held every year on 22 March, and which invites countries to devote a day to public awareness with regard to water and the recommendations coming from Agenda 21 (United Nations 1992). Each year thousands of events worldwide are organised by local, national, regional and international bodies, from all sectors, promoting general water issues and a theme selected by UN-Water.[8] During the same period, the predecessor of UN-Water (see below) was changed from the ISGWR to the ACC Sub-Committee on Water Resources (ACC SWR) (Rodda 1995).

The year 1991 marked the start of a tradition in the water community, with the first ever World Water Week. This annual event, held in Stockholm, was originally organised by the city of Stockholm, but through the Week's development, the Stockholm International Water Institute (SIWI) was established and took over its organisation. World Water Week is important because it has been an informal platform now for over two decades for the exchange of ideas, fostering new thinking and linking best practices, scientific understanding and decision-making (SIWI 2010).

In 1992, at a regional level, but with repercussions at the global level, the Member States of the United Nations Economic Commission for Europe (UNECE) agreed upon the Convention on the Protection and Use of Transboundary Watercourses and International Lakes (UNECE Water Convention), which entered into force in 1996. Not only has the UNECE Water Convention provided crucial building blocks' for customary international water law (Tanzi 2000), but, as of October 2015, with a ratification of an amendment to the Convention from the Ukraine, the parties to the UNECE Water Convention agreed to open the convention to states that were not from the UNECE region, thereby setting the groundwork for making it a global convention (UNECE 2015).

The UNECE Water Convention has not only added value to the issue of transboundary waters, but has also provided guidance on other areas such as climate, ecosystems and through its Protocol on Water and Health, which aims to protect 'human health and well being by better water management, including the protection of water ecosystems, and by preventing, controlling and reducing water-related diseases' (UNECE n.d.).

In the 'follow-up' section of the Dublin Statement, the proposal was made to consider a 'world water council' in which all stakeholders could participate in (ICWE 1992). At the 1994 International Water Resources Association's World Water Congress, a special session agreed to create the World Water Council. This World Water Council was to serve as an umbrella organisation uniting various stakeholders to be a think tank with global reach in a position to alert political leaders to give water a priority that had not existed before (Coulomb 2011). In 1996, the World Water Council became a reality, with its headquarters established in Marseille, France.

During the same period a second initiative was brewing in the corridors of the World Bank and UNDP. Upon the invitation of those two organisations, contributions were sent to develop the Global Water Partnership (GWP), which was founded also in 1996 with the concept to develop the conceptual framework of IWRM and to establish regional Technical Advisory Committees to promote IWRM in the regions. In 2002, GWP officially became an intergovernmental organisation in Sweden.[9]

While GWP is known for its championing of IWRM, the World Water Council is most recognised for organising of the World Water Forum every three years, in partnership with a host country. With no such meeting place in the United Nations for water, the Fora have filled a gap where not only national governments, but also other stakeholders can come together to discuss the most pressing issues within the water community.

The Fora have grown in size and scope since the first World Water Forum that took place in Marrakech in 1997. That first meeting was to raise awareness about water issues to global leaders (Coulomb 2011). The idea behind the second Forum, in the Hague in 2000, was to put into practice, 'From Vision to Action', the momentum gained in the first Forum, with a strong focus on governance. This Forum also produced the first Ministerial Statement, a staple for all Fora to come.

Successive Fora produced new ideas and activities. The third Forum witnessed the release of the first ever *World Water Development Report* (WWAP 2003),[10] developed by the World Water Assessment Programme (WWAP), as well as 'Financing water for all' (Winpenny 2003) also known as the Camdessus Report. The forth World Water Forum saw the introduction of the Parliamentarian and Local Authorities' Processes, where legislators and mayors from around the world went to discuss water in Mexico City. The fifth World Water Forum saw a first Heads of State meeting, as well as the *Istanbul Water Guide* (5th World Water Forum 2009), a compilation of 150 recommendations, which supported the Ministerial Statement. The sixth Forum in Marseille in 2012 focused on solutions, while the seventh, in South Korea, concentrated on the follow-up to the implementation of those solutions in practice.

While the WWC and GWP were being conceptualised and established, other global events and meetings were taking place around the world with regard to water. In 1994, after 23 years of work, the ILC finally adopted articles for a draft convention on international watercourses, which they then submitted to the UNGA (ILC 1994). The Sixth Committee of the UN (Legal Committee) spent three years deliberating before the General Assembly adopted the Watercourses Convention on 21 May 1997. And on 17 August 2015, the Watercourses Convention entered into force with its 35th ratification, that of Vietnam.

Several water meetings took place in 1998, including the International Conference on Water and Sustainable Development, organised by the French Government, where over 120 ministers and high-level officials attended, including French President Jacques Chirac, to elaborate strategies for improving water resources conservation and management to guarantee improved provision of drinking water supply, sanitation and irrigation (IISD 1998).

Later that year, the Sixth Session of the UN Commission on Sustainable Development (UNCSD), which was a body set up by the UNGA to 'ensure effective follow-up' to the Earth Summit,[11] met to review the progress made on 'Strategic Approaches to Freshwater Management'. Among other recommendations, UNCSD encouraged governments to improve IWRM and invited the ACC SWR to be more transparent in their actions, to enhance coordination within the UN system and accelerate implementation of Chapter 18 (UNCSD 1997/8).

The momentum of the 1990s kept going right into the early 2000s for global activity related to water. Within the first three years of the new millennium, four major events took place that helped shape how we govern water today: the World Commission on Dams (WCD), the Millennium Declaration and the International Conference on Freshwater and the World Summit on Sustainable Development (Rio+10).

During the latter part of the twentieth century, opposition to large dams started to grow because of the impact on people, river basins and ecosystems. At first, this opposition was limited to the local areas that were affected by large dams (World Commission on Dams 2000), but then coalesced into a well-networked worldwide movement in the mid-1990s. The World Bank, financers of large infrastructure projects across the globe became the main focal point of the protestor's movement. To respond to the 'growing paralysis of dam-building efforts around the world', the World Bank and the World Conservation Union (IUCN) convened a multi-stakeholder meeting in 1997 that formally created the World Commission on Dams, a two-year, 12-member commission. The WCD presented their findings in 2000, recommending a set of 26 guidelines for 'good practice' in the implementation of dam projects (Conca 2006).

A few months before the release of the WCD findings, the United Nations Millennium Summit was held in New York in 2000 at its 55th General Assembly, where the world's nations released the 'Millennium Declaration', a set of eight goals, the millennium development goals (MDGs), with corresponding targets that are indicators for improving the lives of people in developing countries.

Target 7.C under *Goal 7: Ensure Environmental Sustainability* was set for water to 'Halve, by 2015, the proportion of the population without sustainable access to safe drinking water and basic sanitation'.[12] The inclusion of this target is important, because water (and sanitation) is directly linked to human health, overall economic development and equity (Lenton *et al.* 2005). While not a binding agreement it has been 'a vital instrument which led to a new global focus, led to the formulation of national policies and priorities, stimulated increased knowledge and capacity, and resulted in increased funding streams for investments in water and sanitation' (UNDP 2013).

Much like Dublin was in preparation for the Earth Summit, the German Government convened the International Conference on Freshwater in 2001 to prepare for the World Summit on Sustainable Development (WSSD) in Bonn by building a bridge between all the water activities that were occurring outside the United Nations and UNCSD process leading to Johannesburg (Lane 2009). The outcome of the Bonn conference was called the 'Bonn Keys' (International Conference on Freshwater 2001), which highlighted meeting water security needs for the poor, decentralisation, new partnerships, cooperative agreements in shared river basins and better governance arrangements.

The WSSD in Johannesburg, South Africa, in 2002, a follow-up meeting within the UNCSD process to the Earth Summit ten years earlier, was important for water for two major reasons. First, a target on sanitation was added to the MDGs, also with the objective to halve 'the proportion of the population without access to basic sanitation by 2015' (UNCSD 2002, p. 4). The sanitation target was seen as missing from the first version of the MDGs and inextricably linked to drinking water supply, as had been the case during the International Drinking Water Supply and Sanitation Decade.

The second reason was that a new target was set, although not part of the MDGs, to 'Develop integrated water resources management and water efficiency plans by 2005' (UNCSD 2002, p. 15). This gave additional support to developing countries to implement IWRM and reinforced the focus of such organisations as the GWP. While the target was never met, with still much work to go (UN-Water 2012), integrated approaches to water resources management are still seen as a key to achieving sustainable development (United Nations 2012a).

In December 2000, the UNGA had declared 2003 the 'International Year of Freshwater', which was supposed to drive UN activity to 'increase awareness of the importance of freshwater and to promote action at the local, national, regional and international levels' (United Nations 2010b).

In 2001, ECOSOC renamed the ACC the Chief Executives Board (CEB) for Coordination and all subsidiary bodies of the ACC ceased to exist at that time (United Nations 2013). This meant that the ACC SWR was effectively abolished. It was suggested that an *ad hoc* approach would be the best way to move forward to address inter-agency coordination (UN-Oceans 2008).

Some of the water programme directors of the UN agencies did indeed continue to meet following the ACC SWR's demise. Through the effort of several individuals, the idea was floated to create a more formalised coordination entity within the UN to organise the activities of the agencies doing water work with the UN system and following Johannesburg, the CEB confirmed the creation of UN-Water, an inter-agency mechanism to follow up the water-related decisions of the WSSD (United Nations 2003b).

In 2004, after eight years of work, the ILA approved the Berlin Rules on water resources (ILA 2004), a 'revised version of earlier conventions, including the Helsinki Rules' (Sivakumar 2011). The major difference between the Berlin Rules and the Helsinki Rules is that the Berlin Rules, also non-binding in nature, do not just consider international drainage basins, but national ones as well (Salman 2007), 'integrating domestic and international water law' (Dellapenna and Gupta 2008, p. 445). Another difference is that environmental issues are taken into account more explicitly as are the role of stakeholders in decision-making and suggestions for the remedies of damages (Gupta 2009). From the outset of the Berlin Rules, the probability that states would use them was drawn into question because of how modern some of the legal concepts were (Dellapenna and Gupta 2008); to this day that question remains.

Also in 2004, UN Secretary-General Kofi Annan made another significant move to raise the profile of water within the UN system and established the United Nations Secretary-General's Advisory Board on Water and Sanitation (UNSGAB), which was an 'independent body… to give him advice as well as to galvanize action on water and sanitation issues'.[13] The ten years following its inception, UNSGAB has been chaired by successive world leaders that have been able to bring the issue of water and sanitation to levels in the global political arena where they otherwise might not have reached. UNSGAB has also produced the three Hashimoto Action Plans, named for UNSGAB's first Chair, Mr Ryutaro Hashimoto, the former Prime Minister of Japan.[14]

After a UN resolution on 23 December 2003, sponsored by the Government of Tajikistan, the United Nations proclaimed that the decade of 2005–15 would be the International Decade for Action – Water for Life where the goals of the Decade would be to reinvigorate the efforts towards meeting the water-related Agenda 21 and Millennium Declaration goals (United Nations 2003a).

In 2008, the ILC submitted to the UNGA the draft articles codifying the law of transboundary aquifers after eight years of work, which was endorsed at the assembly on 11 December 2008 (United Nations 2009). While the final form of the draft articles is not certain, as this will eventually be decided by the Member States, the resolution encourages the countries that share aquifers with other nations to 'make appropriate bilateral or regional arrangements for the proper management of their transboundary aquifers, taking into account the provisions of the draft articles that are annexed in the resolution' (ibid.).

A significant moment in the global water community happened on 28 July 2010 when the United Nations recognised the human right to water sanitation in Resolution A/RES/64/292, sponsored by Bolivia, which passed with 122 Member States in favour, none against and 41 abstentions (United Nations 2010a).

The human right to water and sanitation did not begin there, however. Water was absent from the Universal Declaration of Human Rights because it was considered an abundant resource and those that were suffering from lack of safe access to water and sanitation were not at the negotiating table because of colonialism still in effect in the late 1940s (de Albuquerque and Roaf 2010). Several international human rights treaties since then have explicitly mentioned water.[15] But, it was not until 2002, when the 29th Session of the Committee on Economic, Social and Cultural Rights adopted General Comment No. 15 on the right to water, that the momentum started to build towards recognition by the UN of the right to water and sanitation (OHCHR 2003). Following the United Nations General Assembly Resolution in 2010, the Human Rights Council in 2011 passed a resolution taking the human right to water and sanitation to a new level by asking countries to ensure financing for the sustainable delivery of water and sanitation services (United Nations Human Rights Council 2011).

Also in 2010, a new global initiative was started because it was felt that there were gaps in terms of 'policy, planning, financing, information, technical assistance that are impeding global progress' on sustainable access to sanitation and drinking water. Sanitation and Water for All[16] was established linking developing countries, donors, multilateral agencies, civil society and other development partners to work towards this goal (Sanitation and Water for All 2010).

To mark the 20th anniversary of the Earth Summit in Rio in 1992, the United Nations decided that it would convene the UN Conference on Sustainable Development, also known as Rio+20, yet again in Rio, Brazil, in June 2012. While water almost did not make it into the final document, *The Future We Want*, of Rio+20 due to issues countries had with the transboundary waters, it was later saved and made some headway in terms of its prominence in the global sustainable development agenda, recognising in Paragraph 119 that 'water is at the core of sustainable development' (United Nations 2012b).

Pursuant to UNGA Resolution 67/203, the Member States of the UN decided to formally abolish the Commission on Sustainable Development in favour of a high-level intergovernmental forum that would be negotiated through the General Assembly (United Nations 2013). The last session of the UNCSD, its 20th, took place in September 2013, handing over its work to the newly formed forum. This is part of the ongoing process of looking beyond the MDGs ended in 2015. For water, this was significant in that the outcome of the High-Level Panel of Eminent Persons on the Post-2015 Development Agenda, formed by Secretary-General Ban Ki-Moon after Rio+20 (United Nations 2013), and consultations with stakeholders (UNDP 2013) showed that water should feature prominently in a Post-2015 Development Agenda.

After long deliberations of the Open Working Group and several rounds of negotiations between Member States of the UN, along with other parallel processes, on 25 September 2015, the United Nations General Assembly approved the 2030 Agenda for Sustainable Development, the successor of the MDGs. Within the text of the agenda are the sustainable development goals (SDGs, also known as the Global Goals), within which water has a standalone goal with six targets (see Conclusion, this volume, for a complete overview of the SDGs).

The years 2014 and 2015 proved to be quite active within the global water arena, as was seen by the entry into force of the United Nations Watercourses Convention, the seventh World Water

Forum, the adoption of a standalone water goal within the SDGs and activities related to water during the 21st Conference of the Parties of the United Nations Framework Convention on Climate Change (UNFCCC) in Paris. What also was remarkable about global water governance in 2015 was the efforts of the Organisation for Economic Co-operation and Development (OECD).

Starting in preparation for the sixth World Water Forum in Marseille in 2012, the OECD's Water Governance Initiative, with a stakeholder group of over 100 member organisations, developed the OECD Principles on Water Governance over the course of three very active years of work. These 12 Principles fall under the broader categories of effectiveness, efficiency and trust and engagement, 'provide a framework to understand whether water governance systems are performing optimally and to help adjust them where necessary' (OECD 2015). On 4 June 2015, Ministers from OECD's 34 member countries backed the Principles on Water Governance, which launches the next step in identifying indicators to monitor and evaluate the future implementation of the Principles at the national level. While OECD countries do not represent the entirety of the world's nations, successful implementation with OECD countries of such principles will undoubtedly spread the application to non-OECD countries.

The most recent significant movements to take place within the global water governance space were the formation of two high-level panels. The Global High-Level Panel on Water and Peace was launched in Geneva in late 2015, with the mandate to 'develop a set of proposals aimed at strengthening the global architecture to prevent and resolve water-related conflicts, and facilitate the use of water as an important factor of building peace' (Geneva Water Hub 2015).

The second emerged from the World Economic Forum's Annual Meeting in Davos in January 2016, where the United Nations and World Bank have launched a High-Level Panel on Water. Comprised of ten heads of state and two special representatives, the panel sets out to 'motivate effective action and advocate on financing and implementation' (UN-Water 2016).

With water rising on the political agenda nationally and internationally alike, the future for global water governance appears to be on a trajectory of continued development and transformation. More initiatives are added each year and discussions are starting about whether to create a more formal mechanism for water within the United Nations system. What can be said is that it has not been a simple road to arrive at where the world is today with regards to global water governance and it will continue to be a complicated path due to the complicated nature of water, its disregard for political boundaries and tendency to evoke people's strongest sentiments. But, it's a path that can still be shaped for a more sustainable future.

Notes

1 See http://www.iahs.info
2 See http://www.icold-cigb.org
3 See http://www.iahr.org. Now known as the International Association for Hydro-Environment Engineering and Research.
4 The Madrid Declaration was subsequently replaced by the IIL's Salzburg Resolution in 1961 (IIL 1911, 1961).
5 See http://www.ramsar.org
6 UNDP Global Consultation on Safe Water and Sanitation for the 1990s, 10–14 September 1990.
7 1)UN Framework Convention on Climate Change (UNFCCC), 2) UN Convention to Combat Desertification (UNCCD) and 3) Convention on Biological Diversity (CBD).
8 See UN-Water n.d.
9 See GWP n.d.
10 The World Water Development Report was released every three years on World Water Day during the World Water Forum until the fourth edition in 2012. From then, a variety of reports will be produced by WWAP at regular intervals.

11 See UNCSD n.d.
12 See United Nations n.d.a.
13 See http://www.unsgab.org
14 See https://sustainabledevelopment.un.org/topics/water/unsgab
15 See Convention on the Elimination of All Forms of Discrimination against Women (CEDAW), Convention on the Rights of the Child (CRC) and Convention on the Rights of Persons with Disabilities (CRPD).
16 See http://www.sanitationandwaterforall.org

References

5th World Water Forum (2009) *Istanbul Water Guide*. Istanbul, Turkey, 22 March.

Beyer, Martin G. (1991) The global consultation on safe water and sanitation for the 1990s. *Natural Resources Forum* 15(2).

Biswas, Asit K. (1988) United Nations Water Conference action plan. *International Journal of Water Resources Development* 4(3).

Boisson de Chazournes, Laurence. (2009) Freshwater and international law: the interplay between universal, regional and basin perspectives. World Water Assessment Programme Side publication series, *World Water Development Report* 3.

Conca, Ken (2006) *Governing Water: Contentious Transnational Politics and Global Institution Building*. Cambridge, MA: MIT Press.

Coulomb, Réne (2011) *The World Water Council: From its Origins through the World Water Forum in the Hague*. Paris: Editions Johanet.

de Albuquerque, Catarina and Roaf, Virginia (2010) *On the Right Track: Good Practices in Realising the Rights to Water and Sanitation*. Lisbon: Entidade Reguladora de Serviços de Águas e Residuos.

Dellapenna, Joseph W. and Gupta, Joyeeta (2009) The evolution of global water law. In Joseph W. Dellapenna and Joyeeta Gupta (eds), *The Evolution of the Law and Politics of Water*. Dordrecht: Springer Science + Business Media.

Engfeldt, Lars-Göran (2009) *From Stockholm to Johannesburg and Beyond: The Evolution of the International System for Sustainable Development Governance and its Implications*. Stockholm: Swedish Ministry of Foreign Affairs.

FAO (2000) *New Dimensions of Water Security: Water, Society and Ecosystem Services in the 21st Century*. Land and Water Division. Rome: FAO.

Geneva Water Hub (2015) Global High-Level Panel on Water and Peace – Secretariat. Available from: https://www.genevawaterhub.org/projects/global-high-level-panel-water-peace-secretariat [accessed 8 May 2016].

Grover, Brian and Biswas, Asit K. (1993) It's time for a world water council. *Water International* 18(2).

Gupta, Joyeeta (2009) Glocal water governance: controversies and choices. In G.J. Alaerts and N.L. Dickinson (eds), *Water for a Changing World: Developing Local Knowledge and Capacity*. London: Taylor & Francis.

Gupta, Joyeeta (2011) An essay on global water governance research challenges. In Michael R. van der Valk and Penelope Keenan (eds), *Principles of Good Governance at Different Water Governance Levels*. Papers presented at a workshop held 22 March, Delft, The Netherlands.

Gupta, Joyeeta (2013) Global water governance. In Robert Falkner (ed.), *The Handbook of Global Climate and Environmental Policy*. Handbooks of Global Policy Series. West Sussex: Wiley-Blackwell.

Gupta, Joyeeta *et al.* (2013) Policymakers' reflections on water governance issues. *Ecology and Society* 18(1), Art. 35.

GWP (Global Water Partnership) (n.d.) *History*. Available from: http://www.gwp.org/en/About-GWP/History [accessed 21 October 2013].

GWP (2000) *Integrated Water Resources Management*. Global Water Partnership Technical Committee Background Papers No. 4. Denmark: Global Water Partnership.

Harvard University (n.d.) *International Sanitary Conferences*. Available from: http://ocp.hul.harvard.edu/contagion/sanitaryconferences.html [accessed 22 October 2013].

IISD (1998) Summary Report of the International Conference on Water and Sustainable Development. *Sustainable Development* 13(4), 22 March.

IISD (2013) Summary of the Third Session of the UN General Assembly Open Working Group on Sustainable Development Goals: 22–24 May 2013. *Earth Negotiations Bulletin* 32(3).

ILA (1958) *Resolution on the Use of the Waters of International Rivers*. Report of the 48th Conference, New York.

ILA (1966) *The Helsinki Rules on the Uses of the Waters of International Rivers.* Report of the 52nd Conference, Helsinki, August.

ILA (2004) *Berlin Conference (2004) – Water Resources Law.*

ILC (1994) *Draft Articles of the Law of Non-navigational Uses of International Watercourses and Commentaries Thereto and Resolution on Transboundary Confined Groundwater.* Report of the forth–sixth session of the International Law Commission.

IIL (Institute of International Law) (1911) *International Regulation Regarding the Use of International Watercourses For Purposes other than Navigation.* Declaration of Madrid, 20 April.

IIL (1961) *Resolution on the Use of International Non-Maritime Waters.* Salzburg, 11 September.

International Conference on Freshwater (2001) *Conference Report: Water – A Key to Sustainable Development.* 3–7 December. Bonn: Lemmens Verlags & Mediengesellschaft mbH.

ICWE (International Conference on Water and Environment) (1992) The Dublin Statement on Water and Sustainable Development. 26–31 January. Available from: http://www.wmo.int/pages/prog/hwrp/documents/english/icwedece.html

Iza, Alejandro (ed.) (2004) *International Water Governance: Conservation of Freshwater Ecosystems: Volume I – International Agreements Compilation and Analysis.* IUCN International Law Programme, IUCN Environmental Policy and Law Paper No. 55. Gland, Switzerland: IUCN.

Kibaroğlu, Ayşegül (2002) *Building a Regime for the Waters of the Euphrates–Tigris River Basin.* International and National Water Law and Policy Series. London: Kluwer Law International.

Lane, Jon (2009) Global water conferences: a personal reflection. In Asit K. Biswas and Cecilia Tortajada (eds), *Impacts of Megaconferences on the Water Sector.* Berlin and Heidelberg: Springer-Verlag.

League of Nations (1921) *Convention and Statute on the Regime of Navigable Waterways of International Concern.* Treaty Series, Vol. VII, Barcelona, 20 April.

League of Nations (1923) *Convention Relating to the Development of Hydraulic Power Affecting More than One State, and Protocol and Signature.* Treaty Series XXXVI, Geneva, 9 December.

Lee, Terrence (1992) Water management since the adoption of the Mar del Plata Action Plan: Lessons for the 1990s. *Natural Resources Forum* 16(3).

Lenton, Roberto, Wright, Albert M. and Lewis, Kirsten (2005) Health, dignity, and development: what will it take?. UN Millennium Project Task Force on Water and Sanitation. London: Earthscan.

Mageed, Yahia A. and White, Gilbert F. (1995) Critical analysis of existing institutional arrangement. *International Journal of Water Resources Development* 11(2).

Malla, Katak B. (2005) *The Legal Regime of International Watercourses: Progress and Paradigms Regarding Uses and Environmental Protection.* Stockholm: Stockholm University.

Mukhtarov, Farhad G. (2007) Global water governance and the concept of legitimacy. Proceedings of the GRSC/GARNET International Conference on 'Pathways to Legitimacy'. University of Warwick, 17–19 September.

Nace, Raymond L. (1969) *The International Hydrological Decade: Water and Man; A World View.* UNESCO. Paris: Imprimeries Oberthur.

OECD (2015) *OECD Principles on Water Governance.* 4 June.

OHCHR (2003) *General Comment No. 15: The Right to Water (Arts. 11 and 12 of the Covenant).* E/C.12/2002/11, Twenty-ninth session of the Committee on Economic, Social and Cultural Rights, 20 January.

Pahl-Wostl, Claudia (2012) Governance and water needs issues. Presentation, Natural and Social Capital (NASCap). Winnipeg: International Institute for Sustainable Development.

Pahl-Wostl, Claudia, Gupta, Joyeeta and Petry, Daniel (2008) Governance and the global water system: a theoretical exploration. *Global Governance* 14.

Ramsar Conference (1971) *The Final Act of the International Conference on Conservation of Wetlands and Waterfowl.* Ramsar, Iran, 30 January–3 February.

Rodda, John C. (1995) Whither world water?. *Water Resources Bulletin* 31(1).

Rosenau, James (2000) The governance of fragmentation: neither a world republic nor a global interstate system. *Studia Diplomatica* LIII(5).

Salman, Salman M.A. (2003) From Marrakech through The Hague to Kyoto: has the global debate on water reached a dead end?'. *Water International* 28(4).

Salman, Salman M.A. (2004) From Marrakech through The Hague to Kyoto: has the global debate on water reached a dead end? – Part two. *Water International* 29(1).

Salman, Salman M.A. (2007) The Helsinki Rules, the UN Watercourses Convention and the Berlin Rules: perspectives on international water law. *International Journal of Water Resources Development* 23(4).

Sanitation and Water for All (2010) Sanitation and water for all: a global framework for action. Draft Concept Note, 1 March.

Savenije, Hubert H.G. and Hoekstra, Arjen, Y. (2003) Water resources management. In *Knowledge for Sustainable Development: An Insight into the Encyclopedia of Life Support Systems*, Volume II. Paris: UNESCO/EOLSS.

Scheumann, Waltina and Klaphake, Axel (2001) *Freshwater Resources and Transboundary Rivers on the International Agenda: From UNCED to Rio+10*. Bonn: Deutsches Institut für Entwicklungspolitik.

Sivakumar, Bellie (2011) Water crisis: from conflict to cooperation – an overview. *Hydrological Sciences Journal* 56(4).

SIWI (2010) *World Water Week Celebrates Twenty Years*. Stockholm: Stockholm International Water Institute.

Tanzi, Attila (2000) The Relationship between the 1992 UNECE Convention on the Protection and Use of Transboundary Watercourses and International Lakes and the 1997 Convention on the Law of Non-Navigational Uses of International Watercourses. Report of the UNECE Task Force on Legal Administrative Aspects, Geneva, February.

UN Treaty Collection (n.d.) *Convention and Statute on the Regime of Navigable Waterways of International Concern*. Available from: https://treaties.un.org/pages/LONViewDetails.aspx?src=LON&id=558&chapter=30&clang=_en [accessed 22 October 2013].

UN-Oceans (2008) *The Demise of the ACC Sub-Committee on Oceans and Coastal Areas and ICP Proposal for a New Mechanism for Coordination*.

UN-Water (2012) *Status Report on the Application of Integrated Approaches to Water Resources Management*. United Nations Environment Programme.

UN-Water (n.d.) *World Water Day*. Available from: http://www.unwater.org/wwd [accessed 22 October 2013].

UN-Water (2016) *UN and World Bank announce members of joint high-level panel*. 22 April. Available from: http://www.unwater.org/news-events/news-details/en/c/411728/

UNCED (1992a) *Agenda 21 – Chapter 18*: Protection of the quality and supply of freshwater resources: application of integrated approaches to the development, management and use of water resources. 3–14 June. Available from: http://www.un-documents.net/a21-18.htm

UNCED (1992b) Rio Declaration on Environment and Development. United Nations Conference on Environment and Development, Rio de Janeiro, Brazil, 3–14 June.

UNCSD (1997/8) *Report of the Sixth Session*. Economic and Social Council, Official Records, Supplement No. 9, 22 December 1997 and 20 April–1 May 1998.

UNCSD (2002) *Plan of Implementation of the World Summit on Sustainable Development*. 26 August–4 September. Available from: http://www.un.org/esa/sustdev/documents/WSSD_POI_PD/English/WSSD_PlanImpl.pdf

UNCSD (n.d.) Commission on Sustainable Development (CSD). Available from: http://sustainabledevelopment.un.org/csd.html [accessed 22 October 2013].

UNECE (2015) *Ukraine paves the way for globalising the Water Convention*. 13 October. Available from: http://www.unece.org/info/media/presscurrent-press-h/environment/2015/ukraine-paves-the-way-for-globalising-the-water-convention/doc.html

UNECE (n.d.) *Introduction: About the Protocol on Water and Health*. Available from: http://www.unece.org/env/water/pwh_text/text_protocol.html [accessed 22 October 2013].

United Nations (1958) *Integrated River Basin Development: A Report by a Panel of Experts*. Sales No. 58.II.B.3.

United Nations (1959) *Resolution 14/1401 – Preliminary studies on the legal problems relating to the utilization and use of international rivers*. Fourteenth session of the United Nations General Assembly, 21 November.

United Nations (1972a) *Declaration of the United Nations Conference on the Human Environment*. United Nations Audiovisual Library of International Law. Available from: http://legal.un.org/avl/pdf/ha/dunche/dunche_ph_e.pdf

United Nations (1972b) *Report of the United Nations Conference on the Human Environment*. A/CONF.48/14/Rev.1, Stockholm, 5–16 June.

United Nations (1978) *Water Development and Management: Proceedings of the United Nations Water Conference*. Water Development, Supply and Management Series. Oxford: Pergamon Press.

United Nations (1990a) *International Drinking Water Supply and Sanitation Decade*. A/RES/45/181, 71st Plenary Meeting, United Nations General Assembly, 21 December.

United Nations (1990b) *New Delhi Statement: Some for all rather than more for some*. A/C.2/45/3, Forth–fifth session, United Nations General Assembly, 11 October.

United Nations (1992) *Observance of World Day for Water*. A/RES/47/193, 93rd Plenary Meeting, United Nations General Assembly, 22 December.

United Nations (2003a) *Messages from Lake Biwa and Yodo River Basin*. A/RES/58/217, 78th Plenary Meeting, United Nations General Assembly, 23 December.

United Nations (2003b) *Summary of conclusions of the United Nations System Chief Executives Board for Coordination at its second regular session of 2003.* CEB/2003/2, UN Headquarters, 31 October–1 November. 5 December 2003.

United Nations (2009) *Resolution 63/24 – The law of transboundary aquifers.* Sixty-third session of the United Nations General Assembly, 15 January.

United Nations (2010a) *General Assembly Adopts Resolution Recognizing Access to Clean Water, Sanitation as a Human Right.* GA/10967, Sixty-fourth session of the United Nations General Assembly, 108th Meeting, 29 July. Available from: http://www.un.org/News/Press/docs/2010/ga10967.doc.htm

United Nations (2010b) *Resolution 65/154 – International Year of Water Cooperation, 2013.* Sixty-fifth session of the United Nations General Assembly, 20 December.

United Nations (2012a) *Resolution 67/203 – Implementation of Agenda 21, the Programme for the Further Implementation of Agenda 21 and the Outcomes of the World Summit on Sustainable Development and of the United Nations Conference on Sustainable Development.* Sixty-seventh session of the United Nations General Assembly, adopted 21 December. 27 February 2013.

United Nations (2012b) *The Future We Want,* Outcome of the Rio+20 United Nations Conference on Sustainable Development. A/CONF.216/L.1,19 June.

United Nations (2013) *A New Global Partnership: Eradicate Poverty and Transform Economies through Sustainable Development.* The Report of the High-Level Panel of Eminent Persons on the Post-2015 Development Agenda. New York: United Nations.

United Nations (n.d.a) *Goal 7: Ensure Environmental Sustainability.* Available from: http://www.un.org/millenniumgoals/environ.shtml [accessed 22 October 2013].

United Nations (n.d.b) *Reports of the Administrative Committee on Coordination (ACC).* Available from: http://www.un.org/esa/documents/acc.htm [accessed 22 October 2013].

United Nations (n.d.c) *UNCED Conference.* Available from: http://www.un.org/geninfo/bp/enviro/html [accessed 22 October 2013].

United Nations Human Rights Council (2011) *Resolution 18/1 – The Human Right to Safe Drinking Water and Sanitation.* 18th Session of the Human Rights Council, 12 October.

UNDP (2013) *A Million Voices: The World We Want.* Available from: http://www.undp.org/content/undp/en/home/librarypage/mdg/a-million-voices–the-world-we-want.html

Varady, Robert G., Meehan, Katharine and McGovern, Emily (2008a) Charting the emergence of 'global water initiatives' in world water governance. *Physics and Chemistry of the Earth* 34(3).

Varady, Robert G., Meehan, Katharine, Rodda, John, McGovern, Emily and Iles-Shih, Matthew (2008b) Strengthening global water initiatives. *Environment* 50(2).

WCED (1987) *Our Common Future.* New York: Oxford University Press.

WEHAB (2002) *A Framework for Action on Water and Sanitation.* WEHAB Working Group, World Summit on Sustainable Development, Johannesburg, South Africa, August.

Winpenny, James (2003) Financing water for all: report of the world panel on financing water infrastructure. World Water Council, March.

WMO–UNESCO (1991) *Report on Water Resources Assessment: Progress in the Implementation of the Mar del Plata Action Plan and a Strategy for the 1990s.* Paris: WMO–UNESCO.

World Commission on Dams (2000) *Dams and Development: A New Framework for Decision-making.* The Report of the World Commission on Dams. London: Earthscan, November.

WWAP (2003) *The United Nations World Water Development Report 1: Water for People, Water for Life.* Barcelona: UNESCO/Berghahn.

Young, Gordon J., Dooge, James C.I. and Rodda, John C. (2004) *Global Water Resource Issues.* Cambridge: Cambridge University Press.

27

THE SUSTAINABLE DEVELOPMENT GOALS IN RELATION TO WATER MANAGEMENT

What role for legal frameworks?

Anna Schulz

The challenges we face

Sustainable development of water resources is widely seen as critical to address the increasing range of water-related challenges faced by humanity. The most recent *World Water Development Report* (*WWDR*) stresses that water is critical to sustainable development across the three pillars of economic development (for agriculture and industrial production), social development (access to drinking water and sanitation is a human right) and environmental protection (water provides a wide range of ecosystems services) (WWAP 2015, pp. 2–3). However, the *WWDR* outlines that, due to a range of unsustainable practices, water resources and therefore human development is at risk. The report outlines predictions that, under a business as usual scenario, by 2030 there will be a 40 per cent global water deficit, in part due to population growth, rapid urbanization and increased demand by the agriculture and energy sectors, all of which take place in the context of climate change, which is likely to increase the variability of precipitation (WWAP 2015, p. 11).

In addition, the year 2015 was a critical turning point in the global effort to achieve sustainable development. The MDGs, the first set of time-bound development goals, were designed to expire in 2015 and while they did galvanize the world to action all of the goals were not met. In response, the international community negotiated a new set of SDGs to both finish the work of the MDGs and expand on their scope and reach. It is therefore an ideal time to reflect on the role of sustainable development in water management at the international and national levels. This chapter will: examine the evolution of sustainable development, including specific elements related to water; discuss the international law of sustainable development and sustainable development in international water law; address sustainable development in national legal frameworks; and examine the implications of the new SDG on water.

Sustainable development

The concept of sustainable development emerged to address both the world's development chal-lenges and the rampant environmental degradation that was associated with traditional growth-oriented development models. The term 'sustainability' existed for a considerable period of time. The first use of the term was the German *nachhaltigkeit*, by Hans Carl von Carlowitz, in 1713. However, it did not gain momentum as a principle at the international level until the early 1980s.

The UN Conference on the Human Environment (UNCHE), held in Stockholm, Sweden, from 5–16 June 1972, first recognized the connection between human development and the environment. The preamble in part reads 'the protection and improvement of the human envi-ronment is a major issue which affects the well-being of peoples and economic development throughout the world; it is the urgent desire of the peoples of the whole world and the duty of governments' (UNCHE 1972, ¶ 2). In addition, the *Stockholm Declaration* begins to enshrine the principles that underpin sustainable development, including the obligation to protect and improve the environment for present and future generations (principles 1 and 2); consideration of the environment in economic decision-making (principle 4); and the sovereign right of states to use their natural resources provided they do not cause damage to the environment of other states (principle 21) (UNCHE 1972).

The term 'sustainable development' was first used in formal publication at the international level in the IUCN's *World Conservation Strategy: Living Resource Conservation for Sustainable Development*, which defines development as sustainable if it 'take[s] account of social and ecological factors, as well as economic ones; of the living and non-living resource base; and of the long term as well as the short term advantages and disadvantages of alternative actions' (IUCN 1980, ¶ 3).

In 1982 the UN General Assembly (UNGA) adopted the World Charter for Nature, which reiterated many of the principles of the 1972 *Stockholm Declaration* and in paragraph 4 of its general principles made the first use of the term 'sustainable' in a UN outcome document, provided that,

> Ecosystems and organisms, as well as the land, marine and atmospheric resources that are utilized by man, shall be managed to achieve and maintain optimum *sustainable* productivity, but not in such a way as to endanger the integrity of those other ecosys-tems or species with which they coexist.
>
> *UNGA 1982, ¶ 4, emphasis added*

Following shortly on, in 1983 the UNGA established the World Commission on Environment and Development (WCED) (UNGA 1983), also called the Brundtland Commission after Commission Chair Gro Harlem Brundtland, to 'propose long-term environmental strategies for achieving sustainable development to the year 2000 and beyond' (ibid., ¶ 8). The conclusion of the WCED, 'Our Common Future' or the Brundtland Report, was released in 1987 and is widely credited as the foundation on which all further development of the principle of sustain-able development is based. The report defines sustainable development as:

> development that meets the needs of the present without compromising the ability of future generations to meet their own needs. It contains within it two key concepts: the concept of 'needs', in particular the essential needs of the world's poor, to which the overriding priority should be given; and the idea of limitations imposed by the state of technology and social organization on the environment's ability to meet present and future needs.
>
> *WCED 1987, ch. 2, ¶ 1*

The report also began to coalesce in its organization around the three pillars of sustainable development – social, environmental and economic – and called for an international conference to further develop the objectives of sustainable development.

The 1992 UN Conference on Environment and Development (UNCED), also called the 'Earth Summit', held in Rio de Janeiro, Brazil, from 3–14 June 1992, was a direct response to the work of the Brundtland Commission, and resulted in adoption of the *Rio Declaration on Environment and Development* and Agenda 21 by 178 governments. The *Rio Declaration* does not redefine sustainable development, but rather begins the elaboration of principles that contribute to achieving sustainable development. These include:

- the sovereign right of States to use their resources provided it does not harm the interests of other States (Principle 2);
- the right to development must equitably meet the needs of present and future generations (Principle 3);
- common but differentiated responsibility (CBDR) (Principle 7);
- right to participate in decision-making processes (Principle 10);
- precautionary principle (Principle 15);
- polluter pays (Principle 16);
- environmental impact assessment (Principle 17); and
- notification of transboundary environmental effects (Principle 19).

UNCED 1992a

The principles contained in the *Rio Declaration* are elaborated in Agenda 21, which aims to operationalize the principles in various sectors.

Agenda 21 is a massive document, comprising over 700 pages, including chapters on: poverty eradication; consumption patterns; demographic dynamics; human health; human settlement; integration of environment and development decision-making; atmosphere; land resources; deforestation; desertification and drought; mountain development; agriculture and rural development; biological diversity; biotechnology; oceans and seas, coastal areas and use of their living resources; water resources; toxic chemicals; solid waste and sewage-related issues; radioactive wastes; the role of Major Groups; women; children and youth; indigenous peoples; NGOs; local authorities; workers and trade unions; business and industry; the scientific and technological community; farmers; financial resources; transfer of technology and capacity-building; science; education; public awareness and training; capacity-building; institutional arrangements; legal instruments and mechanisms; and information for decision-making.

Calling it 'the beginning of a new global partnership for sustainable development', in Agenda 21 governments outlined ambitious programmes and strategies to achieve sustainable development and created institutions such as the UN Commission on Sustainable Development (UN CSD) to assess progress towards their achievement (UNCED 1992b).

Following the Earth Summit, in 1997 a special session of the UNGA, also called Earth Summit +5, took place in New York, from 23–28 June, to assess progress on the implementation of Agenda 21. The meeting resulted in a special UNGA resolution adopting a 'Programme for the Further Implementation of Agenda 21' (UNGA 1997). The assessment of implementation found uneven progress in achieving the objectives of Agenda 21 and noted the continued deterioration of the environment (ibid., ¶ 8–10). The resolution stressed that trends remain unsustainable. The resolution also represented the crystallization of the three pillars of sustainable development stating 'economic development, social development and environmental protection are interdependent and mutually reinforcing components of sustainable development' (ibid., ¶ 23).

In 2000, the 55th session of the UNGA, designated 'The Millennium Assembly of the UN', convened a Millennium Summit, in New York, from 6–8 September, at which 189 governments signed the Millennium Declaration. The Declaration in part launched a partnership to reduce extreme poverty, including a series of targets to be achieved by 2015 that came to be known as the MDGs. The eight MDGs include: eradicate extreme poverty and hunger; achieve universal primary education; promote gender equality and empower women; reduce child mortality; improve maternal health; combat HIV/AIDS, malaria and other diseases; ensure environmental sustainability; and develop a global partnership for development (UNGA 2000).

At the World Summit on Sustainable Development (WSSD), also referred to as Earth Summit +10, in Johannesburg, South Africa, from 26 August–4 September 2002, the UN continued the process of assessing and further developing the global agenda on sustainable development, adopting the *Johannesburg Declaration on Sustainable Development* and the *Johannesburg Plan of Implementation (JPOI)*, both of which aim to further the implementation of Agenda 21 and the MDGs (WSSD 2002a, 2002b).

As the twentieth anniversary of the Earth Summit approached in 2012, a preparatory process was launched for the UN Conference on Sustainable Development (UNCSD or Rio+20). The process took place in the context of growing realization that the global effort to achieve the MDGs resulted in uneven progress, while some of the MDGs had already been achieved, it became clear that others were unlikely to be reached by 2015 (UNDG 2012). Rio+20, held from 20–22 June 2012, in Rio de Janeiro, Brazil, resulted in adoption of *The Future We Want* (UNGA 2012). Many commentators were disappointed with the outcome of Rio+20, characterizing it as a 'lowest common denominator' agreement and, rather than advancing the sustainable development agenda, of being a simple regurgitation or even a walking back from existing principles (Fitzmaurice *et al.* 2014, pp. 1–3; Hendry 2014, p. 4).

However, the conference did make some critical steps forward, such as launching a process towards the post-2015 development agenda and the formulation of SDGs to replace the MDGs (UNGA 2012, Section B). The outcome also strengthened the role of the UN Environment Programme (UNEP) and ECOSOC (Economic and Social Council of the UN) in monitoring implementation of sustainable development, including through the establishment of the UNEP Governing Council (UNGA 2012, ¶ 88), which has been operationalized by the UNGA through the annual UN Environment Assembly (UNEA), and replacement of the widely criticized UN CSD (Fitzmaurice *et al.* 2014, p. 3) with the annual High-Level Political Forum on Sustainable Development (HLPF) (UNGA 2012, ¶ 84–6).

The process launched in 2012 resulted in the adoption in 2015 at the UN Sustainable Development Summit, held in New York from 25–27 September, of 'Transforming Our World: the 2030 Agenda for Sustainable Development' (UNGA 2015). The resolution announced 17 SDGs and 169 associated targets, including on: no poverty; zero hunger; good health and wellbeing; quality education; gender equality; clean water and sanitation; affordable and clean energy; decent work and economic growth; industry, innovation and infrastructure; reduced inequalities; sustainable cities and communities; responsible consumption and production; climate action; life below water; life on land; peace, justice and strong institutions; and partnerships for the goals.

Sustainable development and water

Water has been recognized, since the very beginning, as an important element of the still evolving principle of sustainable development. For instance, by Earth Summit +5 the recognition that 'water resources are essential for satisfying basic human needs, health and food production, and the preservation of ecosystems, as well as for economic and social development in general' was included in the resolution (UNGA 1997, ¶ 34).

Yet such recognition developed over time and water-related issues were often treated as cross-cutting, rather than addressed as standalone issues. While addressing water as a cross-cutting issue reflects the importance of water to many of the aspects of sustainable development and their frequent interdependencies, it often led to piecemeal approaches to water-related sustainable development issues (WWAP 2015, pp. 93–4).

Water has, however, been included in even the nascent attempts to address the linkages between development and the environment. The *Stockholm Declaration* calls for water to be managed for the benefit of present and future generations (UNCHE 1972, Principle 2). The World Charter for Nature emphasized that water should be reused or recycled (UNGA 1983, ¶ 9(c)). The Brundtland Report treated water as a cross-cutting issue (WCED 1987).

The Earth Summit in 1992 left water out of the *Rio Declaration* altogether, however, in Agenda 21 water received an entire chapter related to the 'Protection of the quality and supply of freshwater resources: application of integrated approaches to the development, management and use of water resources' (UNCED 1992b, chapter 18). In addition, drought was included in 'Managing fragile ecosystems: combating desertification and drought' (chapter 12), sanitation issues were addressed in 'Environmentally sound management of solid wastes and sewage-related issues' (chapter 21) and water was treated as a cross-cutting issue in other chapters of Agenda 21 such as those on: protecting and promoting human health conditions; promoting sustainable human settlement development; promoting sustainable agriculture and rural development; and managing fragile ecosystems: sustainable mountain development, among others.

The MDGs, in Goal 7 on environmental sustainability, included the first time-bound target (7c) related to water resources, particularly to 'halve the proportion of people who are unable to reach or to afford safe drinking water and basic sanitation' by 2015 (Way 2015, p. 58). The target on safe drinking water was adopted in the Millennium Declaration (UNGA 2000, ¶ 19.1), while the target on basic sanitation was added at the WSSD (WSSD 2002b, ¶ 8), it was also a cross-cutting element of the rest of the MDGs (Hendry 2014, 4; WWAP 2015, 94).

There is no doubt that the MDGs galvanized political will and action at the international, national and local levels. Progress towards the MDGs was regularly assessed, including through annual MDG Reports launched in 2004. The MDG Report 2015 marks the finish of the MDGs highlighting uneven progress made towards achieving Target 7c. On access to improved drinking water, at the global level the target to halve the proportion of people without access was reached in 2010, increasing from 76 per cent to 91 per cent from 1990–2015. However, analysis at the regional level shows that Sub-Saharan Africa failed to meet the target, with the Report highlighting that half of the population without access to improved sources live in that region (Way 2015, p. 58). On basic sanitation, the target was missed at the global level with an increase in access to basic sanitation between 1990 and 2015 from only 54 per cent to 68 per cent. Regionally, the Caucasus and Central Asia, Eastern Asia, Northern Africa and Western Asia met the target to halve lack of access to basic sanitation. The Report notes that the drinking water target was achieved by 147 countries, the sanitation target was achieved by 95 and both targets were achieved by 77 (ibid.).

One of the criticisms of the MDGs was the lack of a standalone goal on water and the absence of different aspects of water-related issues, such as allocation, water quality and conservation of aquatic ecosystems. Moreover, the water target focused solely on water, sanitation and hygiene (WASH) issues (WWAP 2015, p. 94; ENB 2013, p. 5). Another criticism, explicitly addressed in the process of developing the SDGs, was that the MDGs were perceived to be the result of a top-down process, which hampered implementation (WWAP 2015, p. 94).

Intensive work by UN-Water and other stakeholders, in advance of the adoption of the SDGs, resulted in the formulation of a dedicated goal on water, which served as the basis for water-related discussions under the UNGA Open Working Group (OWG) on Sustainable Development during 2014 and the post-2015 intergovernmental negotiations in 2015. The result in the final SDGs was the adoption of SDG 6 to 'ensure availability and sustainable management of water and sanitation for all' (UNGA 2015, Goal 6). The Goal includes six substantive and two operational targets:

- By 2030, achieve universal and equitable access to safe and affordable drinking water for all;
- By 2030, achieve access to adequate and equitable sanitation and hygiene for all and end open defecation, paying special attention to the needs of women and girls and those in vulnerable situations;
- By 2030, improve water quality by reducing pollution, eliminate dumping and minimizing release of hazardous chemicals and materials, halving the proportion of untreated wastewater and substantially increasing recycling and safe reuse globally;
- By 2030, substantially increase water-use efficiency across all sectors and ensure sustainable withdrawals and supply of freshwater to address water scarcity and substantially reduce the number of people suffering from water scarcity;
- By 2030, implement integrated water resources management at all levels, including through transboundary cooperation as appropriate;
- By 2020, protect and restore water-related ecosystems, including mountains, forests, wetlands, rivers, aquifers and lakes;
- By 2030, expand international cooperation and capacity-building support to developing countries in water- and sanitation-related activities and programmes, including water harvesting, desalination, water efficiency, wastewater treatment, recycling and reuse technologies; and
- Support and strengthen the participation of local communities in improving water and sanitation management.

UNGA 2015, Goal 6

Each of the targets will be supported by a series of indicators still under development and the HLPF is mandated to oversee and review progress in achieving the SDGs at the international level (UNGA 2015, ¶ 47–8, 72–7).

The materials presented above comprise the history of sustainable development at the global policy level and in the specific field of water resources. The objective of this chapter is, however, to explain the relationship between, and role of, legal frameworks in the advancement of water management through the SDGs.

Law and sustainable development

It is critical to note that efforts at the global level have focused on generating international *policy* on sustainable development rather than generating legally-binding outcomes that would constitute international law. Such processes result in what is often termed *soft law*, including the declarations, resolutions, plans and goals, adopted by the UNGA and the various UN conferences on sustainable development. As Cordonier Segger and Khalfan note, soft law 'can constitute evidence of emerging customary international law, act as tools to interpret treaties and custom, and serve as templates for generating the precise texts of treaty law' (2004, p. 3).

Legal frameworks in the case of water and sustainable development have to function at four levels – international, regional, national and local – and achieving the SDGs will require action at all four levels, however, this work will focus primarily at the international and national levels. The international level is of particular importance in relation to water, due to the large number of transboundary water resources. At the international level it is necessary to look at both the international law of sustainable development and its associated principles, as well as principles of international water law that fulfil the objectives of and could possibly serve as models for the development of the broader international law of sustainable development (Rieu-Clarke 2005).

The national level is also essential as national legislation plays a critical role in translating international policy into national policy processes. The section on national level legal frameworks will examine the types of legal frameworks that can be used to undertake the sustainable development of water resources, despite the fact that these will likely vary widely due to national circumstances.

International law of sustainable development

As noted above, efforts by governments at the international level on sustainable development largely constitute soft law; given this, is sustainable development a principle of international law?

The recognition that international law in the field of sustainable development requires further development predates the principle itself. Principle 22 of the 1972 *Stockholm Declaration* says 'States shall co-operate to develop further the international law regarding liability and compensation for victims of pollution and other environmental damage caused by activities within the jurisdiction or control of such States to areas beyond their jurisdiction.' Principle 24 also emphasizes:

> International matters concerning the protection and improvement of the environment should be handled in a co-operative spirit by all countries, big and small, on an equal footing. Co-operation through multilateral or bilateral arrangements or other appropriate means is essential to effectively control, prevent, reduce and eliminate adverse environmental effects resulting from activities conducted in all spheres, in such a way that due account is taken of the sovereignty and interests of all States.
>
> *UNCHE 1972, ¶ 22*

The 1992 *Rio Declaration*, in its Principle 27, emphasizes 'States and people shall cooperate... in the further development of international law in the field of sustainable development' (UNCED 1992a, ¶ 27).

In the Earth Summit +5 in 1997 the UNGA Resolution did laud the uptake of legal principles including CBDR, the precautionary principle, polluter pays and environmental impact assessment (EIA) at the national and international levels in both treaties and agreements and in national legal systems. However, the Resolution also stresses that 'While some progress has been made in implementing the UNCED commitments through a variety of international legal instruments, much remains to be done to embody the Rio principles more firmly in law and practice' (UNGA 1997, ¶ 14).

The *Johannesburg Declaration* reaffirms commitment to the principles of international law related to sustainable development, but calls to progressively develop international law related to sustainable development were no longer included (UNGA 2002a, ¶ 32), nor is it addressed in the *JPOI*.

By the adoption of *The Future We Want* at Rio+20 in 2012, language related to the international law of sustainable development had firmly transitioned from calls for the further development of international principles to reference to achieving various objectives 'consistent', 'in conformity' or 'in accordance' with international law (UNCSD 2012, ¶ 27, 29, 58, 158, 168, 170, 177, 246).

The Resolution adopted at the UN Sustainable Development Summit 2015, that launched the SDGs, states that in implementing the SDGs 'we reaffirm our commitment to international law and emphasise that the Agenda is to be implemented in a manner that is consistent with the rights and obligations of States under international law' (UNGA 2015, ¶ 18).

There has been a marked shift in the language between early recognition even up to 1997 of the nascent character of the international law of sustainable development and the need for its further development and the current language which indicates that existing international law should be applied. Despite this shift, many legal scholars still emphasize that the status of various principles of the international law of sustainable development and even the principle of sustainable development itself still constitute emerging principles of international law. The International Law Association (ILA) in 2012 elaborated guiding principles to their 2002 New Delhi Declaration, noting a 'genuine reluctance to formalise a distinctive legal status' for the principle of sustainable development (ILA 2012, ¶ 1). In addition, the ILA observed that while the principles of EIA and CBDR are well established in treaty law, others remain less well established (ILA 2012, ¶ 5, 10).

The principle of sustainable development

The principle of sustainable development can be viewed at two levels: first, it is a principle in its own right and, second, it is itself a collection of subsidiary principles (Cordonier Segger and Khalfan 2004, pp. 46–7). On the first point, the principle of sustainable development is still an emerging principle in international law, albeit one that is increasingly accepted. One of the challenges posed by the principle is that it remains very broad and ill defined (Handl 1995, p. 37; Cordonier Segger and Khalfan 2004, p. 3). As Sands notes, there are three legal implications of principles in international law: 1) they can be used by practitioners of international law as a means of interpreting rules; 2) they are used in the negotiation and further elaboration during the negotiation of new instruments; and 3) they can affect the way procedural rules are applied (Sand 1995, pp. 56–7).

Sands emphasizes that the principle of sustainable development 'means for international law the treatment of economic development, environment and human rights in an integrated and interdependent manner' (Sands 1995, p. 53; Cordonier Segger and Khalfan 2004, pp. 46–7). This corresponds to the interpretation that, as a principle, sustainable development is a lens through which other rules should be judged.

Since the inception of the principle of sustainable development, its application in international law has grown. The principle of sustainable development is present in a number of international treaties, including the ASEAN Agreement on the Conservation of Nature and Natural Resources (ASEAN 1985), UN Framework Convention on Climate Change (UNFCCC 1992), the Convention on Biological Diversity (CBD 1992), the North American Free Trade Agreement (NAFTA 1993), the UN Economic Commission for Europe (UNECE) Convention on the Transboundary Effects of Industrial Accidents (UNECE 1992), the UN Convention to Combat Desertification in Those Countries Experiencing Serious Drought and/or Desertification, Particularly in Africa (UNCCD 1994), the Marrakesh Agreement to Establish the World Trade Organization (Marrakesh Agreement 1994, pp. 1143–4), the International Tropical Timber Agreement (ITTA 1994) and the UNECE Convention on Access to Information, Public Participation in Decision-Making and Access to Justice in Environmental Matters (Aarhus Convention 1998).

In the 1997 International Court of Justice (ICJ) judgement on the Gabčíkovo-Nagymaros Project the Court emphasized that, despite the newness of the norm, sustainable development defined as the 'need to reconcile economic development with protection of the environment' must be taken into account, even in this case where projects had been agreed by treaty before the emergence of the norm (ICJ 1997a, 78). In a separate opinion, ICJ Judge Weeramantry goes further, stressing that despite its relative novelty, the principle of sustainable development has normative value and is a recognized principle of international law (ICJ 1997b, pp. 88–9).

In the field of international water law, the 1997 UN Convention on the Law of the Non-Navigational Uses of International Watercourses in its preamble expresses conviction that the convention 'will ensure the utilization, development, conservation, management and protection of international watercourses and the promotion of optimal and sustainable utilisation thereof for present and future generations' (UN Watercourses Convention 1997, preamble). Use of the term 'sustainable development' occurred relatively early in the field of international water law, with the 1987 Agreement on the Action Plan for the Environmentally Sound Management of the Common Zambezi River System aiming to 'strengthen their regional co-operation for sustainable development' (ZACPLAN Agreement 1987).

Use of natural resources in a sustainable and equitable manner

One of the central principles of sustainable development is the use of natural resources in an equitable and sustainable manner. While this obligation rests in delicate balance with the recognized sovereign right of states to utilize their natural resources it is a fundamental principle to several strands of international law, including biodiversity, forests and water. ICJ Judge Weeramantry highlights, in his separate opinion to the Gabčíkovo-Nagymaros Project judgement, the challenge applying the principle sustainable development, where one country argues the right to development and the other the need to protect the environment, underscoring the duality of economic and environmental interests embodied in the principle (ICJ 1997b, p. 90).

The obligation to use resources in an equitable and sustainable manner is embodied in a number of international agreements. The preamble of the Convention on Biodiversity reaffirms that 'States are responsible for conserving their biological diversity and for using their biological resources in a sustainable manner' (CBD 1992, preamble).

The CDB defines sustainable use as 'the use of components of biodiversity in a way and at a rate that does not lead to the long-term decline of biological diversity, thereby maintaining its potential to meet the needs of present and future generations' (ibid., Art. 2). The UNCCD stresses that the objective of the Convention can only be achieved through 'conservation and sustainable management of land and water resources' (UNCCD 1994, Art. 2.2).

In addition to the sovereign right to use natural resources and the corresponding obligation to use natural resources in a sustainable and equitable manner, there is also the duty to avoid transboundary harm. This extends from the principle of *sic utere tuo ut alienum non laedas* from Roman law, otherwise known as the principle of no-harm. The principle is outlined in the *Stockholm Declaration*, which notes that:

> States have… the sovereign right to exploit their own resources pursuant to their own environmental policies, and the responsibility to ensure that activities within their own jurisdiction or control do not cause damage to the environment of other States or of areas beyond the limits of national jurisdiction.
>
> *UNCHE 1972, Principle 21*

No-harm forms the basis of a number of international treaties, in particular those dealing with transboundary pollution.

In the field of international water law, the 1992 UNECE Convention on the Protection and Use of Transboundary Watercourses and International Lakes recognizes the role of international cooperation in the 'sustainable use of transboundary waters' and obliges member states to make their administrative and legal systems compatible with the promotion of 'sustainable water-resources management, including the application of the ecosystems approach' (UNECE Water Convention 1992, preamble and Art. 3(i)).

The 1997 UN Watercourses Convention includes sustainable utilization as a component of the equitable and reasonable utilization of watercourses. In Article 5.1 the Convention specifies 'an international watercourse shall be used and developed by watercourse States with a view to attaining optimal and sustainable utilisation thereof and benefits from,... consistent with adequate protection of the watercourse' (UN Watercourses Convention 1997, Art. 5.1). The equitable use of international watercourses predates the emergence of the principle of sustainable development and the concept, in particular the factors used to determine equitable and reasonable utilization have been proposed as a model for the sustainable use of other natural resources (Rieu-Clarke 2005, pp. 5, 158–9).

The sustainable use of watercourses was rapidly adopted following the Earth Summit in 1992 in a number of water agreements, including: Agreement on the Establishment of a Permanent Okavango River Basin Water Commission (OKACOM Agreement 1994); Convention for the Protection and Use of the Danube River (Danube River Convention 1994); South African Development Community Protocol on Shared Watercourse Systems (SADC Protocol 1995); Agreement on the Cooperation for the Sustainable Development of the Mekong River Basin (Mekong River Agreement 1995); Convention on the Protection of the Rhine (1999); Revised SADC Protocol on Shared Water Resources (Revised SADC Protocol 2000); Tripartite Interim Agreement on the Protection and Sustainable Utilisation of the Water Resources of the Incomati and Maputo Watercourses (Tripartite Interim Agreement 2002); International Agreement on the River Meuse (2002); Framework Agreement on the Sava River Basin (2002); Convention on the Sustainable Management of Lake Tanganyika (Lake Tanganyika Convention 2003); Protocol for the Sustainable Development of Lake Victoria Basin (2003); Agreement for the Protection and Sustainable Development of Lake Ohrid and its Watershed (Lake Ohrid Agreement 2004); Cooperative Agreement for the Sustainable Development and Integrated Use of the Rio Apa Basin (Rio Apa Agreement 2006); Niger River Basin Water Charter (Niger Water Charter 2008); and Agreement on the Nile River Basin Cooperative Framework (CFA 2010).

The principle of no-harm is a well-established, if much debated, principle of international water law. Debate around the principle centres on whether the principle is or is not subordinate to the principle of equitable and reasonable utilization. If subordinate, no-harm becomes one of the factors considered in the determination of whether a specific use is equitable and reasonable; if not, the principle could, in theory, prevent most upstream development of shared watercourses, as nearly any upstream use could constitute harm (ILC 1988, p. 29). This tension is resolved in the UN Watercourses Convention by qualifying the principle, requiring states take 'appropriate measures to present the causing of significant harm', and that, where significant harm is caused, the provisions on equitable and reasonable use and its associated factors for determination are considered (UN Watercourses Convention 1997, Art. 7).

Common but differentiated responsibilities

The principle of CBDR, the idea that States have different levels of responsibility for the problems being faced, different circumstances and different capacities to address those challenges, is well established in international law. CBDR is contained in provisions from Conventions adopted before the Earth Summit, such as the 1988 Vienna Convention for the Protection of the Ozone Layer, which calls for 'taking into account the circumstances and particular requirements of developing countries' (Vienna Convention 1988, preamble). CBDR, or accounting for national circumstances and different capabilities, is also specifically referenced in: UNFCCC 1992; UNCCD 1994; and Minamata Convention on Mercury 2013. In the field of international water law, CBDR is less relevant. This is in part due to the nature of water resources. For instance, transboundary waters shared within specific regions and basin states are more likely to have similar levels of socioeconomic development.

Public participation

The UNCCD calls for 'participatory approaches for the conservation and sustainable use of natural resources' (UNCCD 1994, Art. 19). The 1998 Aarhus Convention specifically addresses public participation aiming to operationalize the provisions in various declarations and resolutions on sustainable development that emphasize the importance of participation to sustainable development, in part through ensuring access to environmental information (Art. 4) and ensuring public participation in environmental decision-making processes (Art. 6–8).

Environmental impact assessment

EIA is one of the most developed principles of sustainable development. The Espoo Convention on Environmental Impact Assessment in a Transboundary Context was adopted by UNECE governments in 1991 (Espoo Convention 1991). However, as it was adopted before 1992 and the momentum generated by the Earth Summit it does not reference sustainable development. However, the Protocol on Strategic Environmental Assessment to the Espoo Convention (Protocol on SEA), adopted in 2003, recognizes the importance of EIA to sustainable development (Protocol on SEA 2003, Preamble ¶ 2).

Sustainable development and national water law

While sustainable development is a principle developed in soft law materials at the international level and is emerging as a principle of international law, sustainable development must largely be accomplished at much smaller scales, particularly the national and local levels. Sustainable development, applied to water resources, means the sustainable management of water resources to ensure that people living within individual river basins or aquifer systems have access to improved water sources and adequate sanitation, water to support their livelihoods and water to support the ecosystems that depend on it. At the national level this means balancing, as at the international level, economic and environmental considerations and through legislation and regulation at the national level taking approaches that are likely to result in sustainable development of water resources, such as ecosystem-based approaches and IWRM. Schwarz identifies the need for a three-level process to translate the international principle of sustainable development at the domestic level into progress, including through localization, legalization and institutionalization. She describes the process as one of: adopting national-level policy objectives related to sustainable development, followed by passing laws and regulations to make it actionable; adjusting and modifying national-, state- and local-level laws and regulations, reviewing effectiveness, ensuring all new laws include sustainable development; and reorienting institutions towards integrated approaches and developing institutional capacity (Schwarz 2005, pp. 147–51).

The Global Water Partnership (GWP) defines IWRM as promoting 'the co-ordinated development and management of water, land, and related resources, by maximising economic and social welfare without compromising the sustainability of vital environmental systems' (Rogers and Hall 2003, front material).

IWRM has been identified as a key means of implementing sustainable development in the field of water since the Earth Summit in 1992. Chapter 18 of Agenda 21 focuses on the 'application of integrated approaches to the development, management and use of water resources' (UNCED 1992b, chapter 18, p. 196). Agenda 21 stresses the objective of IWRM should be undertaken at the basin or sub-basin level and should ensure the protection of water resources,

> taking into account the functioning of aquatic ecosystems and the perenniality of the resources, in order to satisfy and reconcile the needs for water in human activities. In developing and using water resources, priority has to be given to the satisfaction of basic needs and the safeguarding of ecosystems.
>
> *UNCED 1992b, ¶ 18.8*

The *JPOI* calls for states to 'promote integrated water-resources development', including by developing and implementing 'integrated river basin and watershed management strategies and plans for all major water bodies' (WSSD 2002b, ¶ 66(b)). The SDG on water calls for the implementation of IWRM at all levels by 2030 (UNGA 2015, Goal 6).

Meeting the SDG on water

The SDGs have generated significant political momentum towards sustainable development. In order to achieve the SDGs, however, lessons must be learned from both the strengths and weaknesses of the MDGs, in particular the failure to reach all of the MDGs. So why did the world fall short of their goals? Among reasons identified by scholars, include: they cover only part of the sustainable development agenda (Schmidt-Traub 2009, p. 77; Wisor 2012, p. 118; Harlin and Kjellén 2015, pp. 9, 12); they were adopted through a top-down process without public participation (Schmidt-Traub 2009, p. 78); there is a lack of data and use of imprecise metrics in part due to national assessment of what were originally designed to be global goals (Attaran 2005, pp. 955–6 and 959; Easterly 2009, p. 26; Wisor 2012, pp. 117–18; and Harlin and Kjellén 2015, p. 10). Yet the MDGs remain widely lauded for mobilizing the world towards the achievement of a series of time-bound goals that were accessible and easy to understand. The SDGs in the process of their adoption, arguably, did learn from these lessons: they remain simple to understand; they were developed through a long and much more participatory process that involved 83 national consultations, 11 thematic consultations, global citizen engagement through online platforms and intergovernmental negotiations; and an ongoing process led by the Inter-agency Expert Group on SDG Indicators (UN-Stats) to propose an indicator framework for the SDGs to the UN Statistical Commission in March 2016 that aims to address the challenges posed by MDG indicators.

While the SDGs do build on lessons learned from the MDGs, achieving the SDG on water and realizing the sustainable development of water resources will take a considerable effort on the part of the international community, national governments and local communities at the basin level. There are several reasons to be optimistic about sustainable development in the field of water resources at the international level. First, at the international soft law level, there is general coherence between the soft law outcomes on sustainable development and policy processes related to water management, including the promotion of integrated approaches, such as IWRM, participatory approaches and basin-level approaches. Second, sustainable development has been integrated into international water law, both in the 1997 UN Watercourses Convention and in State practice subsequent to the Earth Summit in 1992 through its regular inclusion in basin-level water treaties, making sustainable development of water resources a legally binding obligation for parties to those treaties and conventions. Third, the SDG on water, unlike the MDG on water, addresses water resources holistically, making the goal itself more coherent with international policy approaches to water resources management.

At the national level, the SDG on water should benefit from increased ownership by governments, civil society and other stakeholders, providing increased impetus to undertake the necessary transformations towards the sustainable development of water resources. While respecting national circumstances, steps taken at the national level should include the creation and adoption of national action plans on sustainable development and a process to amend or modify existing laws and regulations, and ensure that new ones on water resources integrate environmental considerations into development processes and devolve water management to the local or basin/sub-basin levels. Institutional structures should also be reformed to ensure integrated consideration of water resources across sectors. The level of effort required, and the resources needed in terms of financial and technical assistance, by each government will vary vastly depending on existing legal and regulatory frameworks and institutional capacities at the national, basin and local levels.

References

Aarhus Convention (1998) UNECE Convention on Access to Information, Public Participation in Decision-Making and Access to Justice in Environmental Matters. In UN, 2003. *UN Treaty Series* 2161. New York: UN.

ASEAN (1985) Agreement on the Conservation of Nature and Natural Resources. ECOLEX [online]. Available from: http://www.ecolex.org/ecolex/ledge/view/RecordDetails;DIDPFDSIjsessionid=AE86A31369FCF752ABB7AF209DA6C465?id=TRE-000820&index=treaties [accessed on 17 January 2016].

Attaran, A. (2005) An immeasurable crisis? A criticism of the millennium development goals and why they cannot be measured. *PLoS Medicine* 2(10), p. e318.

CBD (1992) Convention on Biodiversity. In UN, 2001. *UN Treaty Series* 1760. New York: UN.

CFA (2010) [online]. Available from: http://www.nilebasin.org/images/docs/CFA%20-%20English%20%20FrenchVersion.pdf [accessed on 17 January 2016].

Convention on the Protection of the Rhine (1999) ECOLEX [online]. Available from: http://www.ecolex.org/ecolex/ledge/view/RecordDetails;DIDPFDSIjsessionid=AE86A31369FCF752ABB7AF209DA6C465?id=TRE-001307&index=treaties [accessed on 17 January 2016].

Cordonier Segger, M. and Khalfan, A. (2004) *Sustainable Development Law: Principles, Practices and Prospects.* Oxford: Oxford University Press.

Danube River Convention (1994) ECOLEX [online]. Available from: http://www.ecolex.org/ecolex/ledge/view/RecordDetails;DIDPFDSIjsessionid=AE86A31369FCF752ABB7AF209DA6C465?id=TRE-001207&index=treaties [accessed on 17 January 2016].

Easterly, W. (2009) How the MDGs are unfair to Africa. *World Development* 37(1), pp. 26–35.

ENB (2013) Summary of the High-Level Consultation on Water in the Post-2015 Agenda. *Earth Negotiations Bulletin* [online]. Available from: http://www.iisd.ca/post2015/water/hlcw/html/crsvol208num8e.html [accessed 15 January 2016].

Espoo Convention (1991) Espoo Convention on Environmental Impact Assessment in a Transboundary Context. In UN, 1997. *UN Treaty Series* 1989. New York: UN.

Fitzmaurice, M., Maljean-Dubois, S. and Negri, S. (eds) (2014) *Environmental Protection and Sustainable Development from Rio to Rio+20.* Leiden: Martinus Nijhoff/Brill.

Framework Agreement on the Sava River Basin (2002) ECOLEX [online]. Available from: http://www.ecolex.org/ecolex/ledge/view/RecordDetails;DIDPFDSIjsessionid=AE86A31369FCF752ABB7AF209DA6C465?id=TRE-001824&index=treaties [accessed on 17 January 2016].

Handl, G. (1995) Sustainable development: general rules versus specific obligations. In W. Lang (ed.), *Sustainable Development and International Law*. London: Graham and Trotman.

Harlin, J., and Kjellén, M. (2015) Water and development: from MDGs towards SDGs. In A. Jägerskog, T.J. Clausen, T. Holmgren and K. Lexén (eds), *Water for Development: Charting a Water Wise Path*. Report No. 35. Stockholm: SIWI.

Hendry, S. (2014) *Frameworks for Water Law Reform*. Cambridge, UK: Cambridge University Press.

ICJ (1997a) *Gabčíkovo-Nagymaros Project (Hungary/Slovakia) Judgement*. ICJ Reports 1997. The Hague: ICJ.

ICJ (1997b) *Gabčíkovo-Nagymaros Project (Hungary/Slovakia) Judgement: Separate Opinion of Vice-President Weeramantry*. ICJ Reports 1997. The Hague: ICJ.

ILA (2012) Sofia Guiding Statements on the Judicial Elaboration of the 2002 New Delhi Declaration of Principles of International Law Related to Sustainable Development. Res. No. 7/2012. Sofia: ILA.

ILC (1988) *Report of the International Law Commission on the Work of Its Fortieth Session to the General Assembly*. 43 UN GAOR Supp. (No. 9), UN Doc. A/43/10. New York: UN.

International Agreement on the River Meuse (2002) ECOLEX [online]. Available from: http://www.ecolex.org/ecolex/ledge/view/RecordDetails;DIDPFDSIjsessionid=AE86A31369FCF752ABB7AF209DA6C465?id=TRE-001376&index=treaties [accessed on 17 January 2016].

ITTA (1994) ECOLEX [online]. Available from: http://www.ecolex.org/ecolex/ledge/view/RecordDetails;DIDPFDSIjsessionid=AE86A31369FCF752ABB7AF209DA6C465?id=TRE-001195&index=treaties [accessed on 17 January 2016].

IUCN (1980) *World Conservation Strategy: Living Resource Conservation for Sustainable Development*. Gland, Switzerland: IUCN.

Lake Ohrid Agreement (2004) ECOLEX [online]. Available from: http://www.ecolex.org/ecolex/ledge/view/RecordDetails;DIDPFDSIjsessionid=AE86A31369FCF752ABB7AF209DA6C465?id=TRE-151137&index=treaties [accessed on 17 January 2016].

Lake Tanganyika Convention (2003) ECOLEX [online]. Available from: http://www.ecolex.org/ecolex/ledge/view/RecordDetails;DIDPFDSIjsessionid=AE86A31369FCF752ABB7AF209DA6C465?id=TRE-001482&index=treaties [accessed on 17 January 2016].

Maljean-Dubois, S. and Negri, S. (2014) Introduction. In M. Fitzmaurice, S. Maljean-Dubois and S. Negri (eds), *Environmental Protection and Sustainable Development from Rio to Rio+20*. Leiden: Martinus Nijhoff/Brill.

Marrakesh Agreement (1994) Marrakesh Agreement to Establish the WTO. *International Legal Materials* XXXIII.

Mekong River Agreement (1995) ECOLEX [online]. Available from: http://www.ecolex.org/ecolex/ledge/view/RecordDetails;DIDPFDSIjsessionid=AE86A31369FCF752ABB7AF209DA6C465?id=TRE-001223&index=treaties [accessed on 17 January 2016].

Minamata Convention on Mercury (2013) In UN, UN Doc. CTC-XXVII-17. UN: New York.

NAFTA (1993) North American Free Trade Agreement. *International Legal Materials* XXXII.

Niger Water Charter (2008) ECOLEX [online]. Available from: http://www.ecolex.org/ecolex/ledge/view/RecordDetails;DIDPFDSIjsessionid=33E0D569EA7D763BF0F80FA686711157?id=TRE-146761&index=treaties [accessed on 17 January 2016].

OKACOM Agreement (1994) ECOLEX [online]. Available from: http://www.ecolex.org/ecolex/ledge/view/RecordDetails;DIDPFDSIjsessionid=AE86A31369FCF752ABB7AF209DA6C465?id=TRE-001851&index=treaties [accessed on 17 January 2016].

Protocol for the Protection and Sustainable Development of the Lake Victoria Basin (2003) ECOLEX [online]. Available from: http://www.ecolex.org/ecolex/ledge/view/RecordDetails;DIDPFDSIjsessionid=AE86A31369FCF752ABB7AF209DA6C465?id=TRE-159877&index=treaties [accessed on 17 January 2016].

Protocol on SEA (2003) Available from: http://www.unece.org/env/eia/about/sea_text.html [accessed 17 January 2016].

Revised SADC Protocol (2000) ECOLEX [online]. Available from: http://www.ecolex.org/ecolex/ledge/view/RecordDetails;DIDPFDSIjsessionid=AE86A31369FCF752ABB7AF209DA6C465?id=TRE-001360&index=treaties [accessed on 17 January 2016].

Rieu-Clarke, A. (2005) *International Law and Sustainable Development: Lessons from the Law of International Watercourses*. London: IWA.

Rio Apa Agreement (2006) ECOLEX [online]. Available from: http://www.ecolex.org/ecolex/ledge/view/RecordDetails;DIDPFDSIjsessionid=AE86A31369FCF752ABB7AF209DA6C465?id=TRE-151485&index=treaties [accessed on 17 January 2016].

Rogers, P. and Hall, A. (2003) *Effective Water Governance*. TEC Background Papers No. 7. Stockholm: GWP.

SADC Protocol (1995) ECOLEX [online]. Available from: http://www.ecolex.org/ecolex/ledge/view/RecordDetails;DIDPFDSIjsessionid=AE86A31369FCF752ABB7AF209DA6C465?id=TRE-001267&index=treaties [accessed on 17 January 2016].

Sands, P. (1995) International law in the field of sustainable development: emerging legal principles. In W. Lang (ed.), *Sustainable Development and International Law*. London: Graham and Trotman.

Schmidt-Traub, G. (2009) The MDGs and human rights-based approaches: moving towards a shared approach. *International Journal of Human Rights* 13(1), pp. 72–85.

Schwarz, P. (2005) Sustainable development in international law. *Non-State Actors and International Law* 5, pp. 127–52.

Tripartite Interim Agreement (2002) ECOLEX [online]. Available from: http://www.ecolex.org/ecolex/ledge/view/RecordDetails;DIDPFDSIjsessionid=AE86A31369FCF752ABB7AF209DA6C465?id=TRE-001811&index=treaties [accessed on 17 January 2016].

UN Watercourses Convention (1997) UN Convention on the Law of the Non-Navigational Uses of International Watercourses. In UNGA, 21 May, Res. 51/229. New York: UN.

UNCCD (1994) UN Convention to Combat Desertification. In UN, 1999. *UN Treaty Series* 1954. New York: UN.

UNECE Convention (1992) UNECE Convention on the Transboundary Effects of Industrial Accidents. In UN, 2002. *UN Treaty Series* 2105. New York: UN.

UNECE Water Convention (1992) UNECE Convention on the Protection and Use of Transboundary Watercourses and International Lakes. In UN, 2001. *UN Treaty Series* 1936. New York: UN.

UNCED (1992a) *Rio Declaration on Environment and Development*. UN Doc. A/CONF.151/26/Rev.1 (Vol. I), annex I. New York: UN.

UNCED (1992b) *Agenda 21*. UN Doc. A/CONF.151/26 (Vol. I), annex II. New York: UN.

UNCHE (1972) *Stockholm Declaration*. Stockholm, Sweden, 5–16 June 1972. UN Doc. A/CONF.48/14/Rev.1. New York: UN.

UNDG (2012) *The Millennium Development Goals Report 2012*. UN Report No. 12-24532. New York: UN.

UNFCCC (1992) UN Framework Convention on Climate Change. In UN, 2000. *UN Treaty Series* 1771. New York: UN.

UNGA (1982) 28 October. Res. 37/7. New York: UN.

UNGA (1983) 19 December. Res. 38/61. New York: UN.

UNGA (1997) 28 June. Res. S/19-2. New York: UN.

UNGA (2000) 18 September. Res. 55/2. New York: UN.

UNGA (2012) 11 September. Res. 66/288, annex. New York: UN.

UNGA (2015) 21 October. Res. 70/1. New York: UN.

Vienna Convention (1988) Vienna Convention for the Protection of the Ozone Layer. In UN, 1997. *UN Treaty Series* 1513. New York: UN.

Von Carlowitz, H.C. (1713) *Sylvicultura Oeconomica*. Liepzig: Johann Friedrich Braun.

Way, C. (2015) *The Millennium Development Goals Report 2015*. UN Report No. 15-04513. New York: UN.

WCED (1987) *Report of the WCED: Our Common Future*. UN Doc. A/42/427, annex. New York: UN.

Wisor, S. (2012) After the MDGs: citizen deliberation and the post-2015 development framework. *Ethics and International Affairs* 26(1), pp. 113–33.

WSSD (2002a) *Johannesburg Declaration on Sustainable Development*. UN Doc. A/CONF.199/20. New York: UN.

WSSD (2002b) *Johannesburg Plan of Implementation*. UN Doc. A/CONF.199/20, annex. New York: UN.

WWAP (2015) *The UN World Water Development Report 2015: Water for a Sustainable World*. Paris: UNESCO.

ZACPLAN Agreement (1987) ECOLEX [online] Available from: http://www.ecolex.org/ecolex/ledge/view/RecordDetails;DIDPFDSIjsessionid=AE86A31369FCF752ABB7AF209DA6C465?id=TRE-000971&index=treaties [accessed on 17 January 2016].

28

WATER SECURITY AS AN EVOLVING PARADIGM

Local, national, regional and global considerations

Bjørn-Oliver Magsig

Introduction

On Earth Day 2016, the UN Security Council held an Arria-formula meeting on water, peace and security (What's in Blue 2016). It was the first time Council members discussed water as a separate issue connected to peace and security. On the same day, the same UN body also released a press statement concerning the 'maritime tragedy in Mediterranean Sea' which resulted in the tragic and evitable deaths of almost 500 people when their overcrowded boat capsized on its passage from Libya to Europe (Security Council 2016). Acknowledging that the UN Security Council is the United Nations' most powerful body, with 'primary responsibility for the maintenance of international peace and security' (Article 24 of the UN Charter), those two apparently separate news items convey two important messages. First, water security has reached the highest political forum in international relations. The fact that many local actors, states and regional bodies perceive the water crisis as a threat to their respective security cannot be argued away. Second, the discussion about how water security should be defined, which dimensions it should entail and how this immense challenge should be addressed has only just begun. The debate is still being dominated by voices who base their arguments on a traditional understanding of security, which misses fundamental components – like the complex relations between environmental scarcity, conflict and migration (Selby and Hoffmann 2012).

As previous chapters have demonstrated, few challenges have the potential to create as much friction within and between states as the allocation and utilisation of freshwater resources. In its recently published report *High and Dry: Climate Change, Water, and the Economy*, the World Bank predicts that water scarcity will act as a risk multiplier, fuelling cycles of resource-driven conflict and migration, especially in already water-stressed regions which heavily depend on agriculture as an important economic factor (World Bank 2016, p. 19). The various simmering water-related conflicts around the world – like Egypt's difficult relationship with Ethiopia on the Nile (Abseno 2013; Zhang *et al.* 2015), the dispute over a more equitable distribution of water in the Middle East (Fröhlich 2012; Weinthal *et al.* 2015), the highly contentious region of Himalayan Asia (Wirsing *et al.* 2013; Magsig 2015b), or the intra-state conflict over water and coal in South Africa (Turton 2010) – bear witness to the fact that freshwater has acquired an independent status within the national and international security discourse (Wouters *et al.* 2009, p. 103). However, just as the concept of security has gone through a widening and deepening process, so has the perception of water security (Magsig 2014, p. 442). The argument that water security can be solely addressed as a nation's internal affair of securing its access to freshwater seems to be an untenable position in today's interrelated world. If this is the case, how should the water security debate be framed instead?

Water security: from buzz word to paradigm

In order to understand why the global water crisis – including its regional and local impacts – is often being phrased as a security concern, one has to comprehend the changing perception of security.

The securitisation of water

The origins of the term 'security' can be derived from the Latin phrase *sine cura*, meaning a state of living without care and concern (Warner and Meissner 2008, p. 254). Wolfers suggests that '[s]ecurity… in an objective sense, measures the absence of threats to acquired values, [and] in a subjective sense, [refers to] the absence of fear that such values might be attacked' (1962, p. 150). Hence, when taking a realist approach, objective security is achieved when the dangers posed by various risks, vulnerabilities and threats are being prevented or can be coped with; while from a social constructivist perspective, security is achieved once the perceptions of security risks, vulnerabilities and threats are overcome (Brauch 2011, p. 61). This implies that although objective factors are essential for the conceptualisation of security, they clearly are not sufficient, as the subjective factors play an important role in the perception of security. The perception, in turn, depends on the certain value given by society (*Wertidee*) and a universally applied normative concept (Brauch 2011). Within the discipline of security studies, traditional approaches focused primarily on military threats to the integrity of nation states (Brauch 2008, p. 28). This line of discourse changed with the end of the Cold War, when the distinction between internal and external security became increasingly blurred, and conceptual approaches needed to take into account the security-related issues arising out of new (or rather newly perceived) global threats – economic, social and environmental. In particular, the framing of environmental issues as security concerns appears to have gained currency rather quickly. In 1989, Eduard Shevardnadze, in his role as Soviet Foreign Minister, called for the establishment of an Environmental Security Council entrusted with issues of ecological security (Shevardnadze 1990); while politicians in the West argued along the same line. For instance, Johan Jørgen Holst, the then Norwegian Defence Minister, pointed out that 'environment degradation may be viewed as a contribution to armed conflict in the sense of exacerbating conflicts or adding new dimensions' (Holst 1989, p. 123).

Thus, in realising that the various new threats to both national and international security could not simply be addressed by looking through the military lens alone, the perception of security had changed considerably (Buzan 1991). The inclusion of non-military threats – the so called 'widening' process – was accompanied by efforts to also 'deepen' security studies. Here, the approach was to regard the individual, rather than the state, as the main referent object by introducing the concept of human security. In 1997, the UN Development Programme's Human Development Report observed that '[s]ecurity is increasingly interpreted as the security of people, not just territory; security of individuals, not just of nations; security through development, not through arms; security of all people everywhere – in their homes, in their jobs, in their streets, in their communities, and in the environment' (Brown 2005, p. 2).

One of the consequences of the awakened political interest in this concept was, again, an increase in research on environmental security, arguing that environmental degradation poses a threat to international security which justifies the re-evaluation of the traditional notion of security (Scholtz 2009, p. 138). Interestingly, it was not environmental research which initially drove the development forward. Peace research used the concept of environmental security to overcome the classic notion of a military-focused security of the individual state, while researchers of international development incorporated environmental factors in their analysis of violent conflicts in the developing world to argue for restructuring the global economy (Brock 1999, p. 37). This process led to defining the environmental security of a state as the 'absence of non-conventional threats against the environmental substratum essential to the well-being of its population and to the maintenance of its functional integrity' (Frédérick 1999, p. 100). Evidently, it reasserts the responsibilities of the state towards its citizens, which is the logical consequence of associating 'national security' with 'quality of life' – or human security (ibid., p. 94). This is why other scholars put the individual in the centre of security research, arguing that '[i]n the field of international security, a focus on the individual as the nexus of concern enables us to understand both the broad spectrum of threats, and their interlocking nature, in any given context' (McRae 2001, 21). While this concept does not break with the notion of national security, since 'the security of the state is necessary for human security', it nevertheless challenged the traditional thinking to become more human-centred (Von Tigerstrom 2007, p. 50).

From this new focus on human security, the 'essential freedoms' discourse evolved, primarily embraced by the United Nations, which placed the security paradigm within four fundamental freedoms: 1) the freedom from want; 2) the freedom from fear; 3) the freedom to live with human dignity and 4) the freedom from hazardous impact (United Nations 2004, p. 77). In his report 'In larger freedom', Kofi Annan, the former UN Secretary-General, identified the 'imperative of collective action' as essential for achieving the core purposes of the UN Charter (Annan 2005):

> In a world of interconnected threats and challenges, it is in each country's self-interest that all of them are addressed effectively. Hence, the cause of larger freedom can only be advanced by broad, deep and sustained global cooperation among states. Such cooperation is possible if every country's policies take into account not only the needs of its own citizens but also the needs of others. This kind of cooperation not only advances everyone's interests but also recognises our common humanity.
>
> *Annan 2005, p. 6*

Providing input to the UN Secretary-General's report, the High-Level Panel on Threats, Challenges and Change offered interesting insights on the security discourse. The Panel noted that,

> [t]he threats to peace and security in the twenty-first century include not just international war and conflict but... also include poverty, deadly infectious disease and environmental degradation since these can have equally catastrophic consequences. All of these threats can cause death or lessen life chances on a large scale. All of them can undermine states as the basic unit of the international system.
>
> *Ibid., p. 24*

The Panel endorsed the central role of the UN in the context of a more broadly defined concept of security by insisting:

that the Charter as a whole continues to provide a sound legal and policy basis for the organisation of collective security, enabling the Security Council to respond to threats to international peace and security, both old and new in a timely and effective manner.

United Nations 2004, p. 93

It recommended that the UN be provided with 'new expertise to deal with new threats – for example, the scientific advice necessary to address questions of environmental and biological security' (ibid., p. 91). While recent efforts of some nations have put the threats of climate change and water insecurity on the agenda of the UN Security Council (President of the Security Council 2011; What's in Blue 2016), other states remain very sceptical as to whether the Council is the right forum for discussing security issues linked to the environment or freshwater; only agreeing to debate them in informal Arria-formula meetings (Worsnip for Reuters 2011).

Despite international policy developments, most governments have adopted a security paradigm which is based on the flawed premise that all kinds of insecurity can be controlled by using military force or containment (Abbott *et al.* 2006, p. 28). This is why a new approach addressing the drivers of insecurity by 'curing the disease' rather than 'fighting the symptoms' is urgently needed (Brock 2011, p. 2). Recent research on collective and sustainable security is trying to pave the way towards a mutual understanding that security can no longer be regarded as a zero-sum game between states but instead has to be based on the concept of sustainable development in order to affectively address the root causes of insecurity and to ensure long-lasting peace and development (Scholtz 2009; Voigt 2009, p. 164). In combining sustainable security thinking with the more advanced concept of collective security, which is being perceived as one of the core purposes of the UN Charter (United Nations 2004, p. 93), a promising platform for discourse is finally emerging which is capable of facilitating a meaningful debate about how to address the various security issues the international community is facing. While states are the bedrock of the international system, and thus achieving collective security is impossible without being based on the various perceptions of states' securities, collective security is operating somewhat 'above and beyond' orthodox patterns of international relations – that is, by adding 'universal moral obligations' to the table of international negotiations (Orakhelashvili 2011, p. 6).

Defining water security

Against this backdrop, this chapter follows a rather broad understanding of security and regards a community to be 'water secure when it has sustainable access to freshwater of sufficient quantity and quality, or to the benefits derived therefrom; and the ability to minimise water-related risk and its various repercussions to an acceptable level – without compromising the supporting ecosystems' (Magsig 2015a, p. 31). This definition draws from both the widening and deepening process of the general security debate, while, at the same time, acknowledging the complexity of the global water crisis. Accordingly, the definition has several advantages over previous (mostly more restrictive) ones. First, by focusing on 'communities', it is scalable to the level one wants to look at water security – local, national, regional or even global. It also acknowledges the fact that in water resources management the overlapping of several levels of governance is the rule rather than the exception. Second, by including the 'benefits derived' from access to freshwater and the repercussions of water-related risks, the true complexity of the water crisis is being pulled into play. Not only are we looking at access to and threats from the resource water, but also the opportunities and issues linked (directly or indirectly) to it. Here, the concepts of virtual water and benefit sharing come to mind (Wouters and Moynihan 2013). Finally, by entailing undetermined parameters like 'sufficient quantity and quality of freshwater' and 'acceptable level of water-related risk and repercussions', the definition provides the respective community with considerable room to manoeuvre concerning the implementation of the concept of water security – geared to its own needs, capacities and preferences. The relative vagueness of the concept guarantees its resilience as well as global applicability while, at the same time, it avoids becoming arbitrary (Magsig 2015a, p. 32). Rather than being a somewhat constricting stipulation of the term 'water security', it aims at providing a platform for stimulating continuous discourse across all relevant disciplines and levels of governance.

Potential pitfalls of securitisation

Securitisation is a strategy for managing risk perceptions of stakeholders which aims at moving an issue to the top of the agenda in order to generate the political will needed to address it. Hence, in theory, a matter becomes a security issue when the securitising actor convinces the relevant audience that it poses an existential threat and can only be handled with exceptional effort (Buzan et al. 1998). While the language of 'water wars' is mainly driven by the media, many scholars and practitioners still ignore the complexity of the issues involved in addressing the water crisis as well (Chellaney 2011). This indicates that the desperately needed change in how water security is being perceived is very hard to achieve in practice. However, acknowledging that the water security paradigm does not automatically result in a more peaceful management of the resource does not mean it has to be regarded as a 'regrettable detour to a virtual blind-alley' (Mekonnen 2010, p. 421) either.

Certainly, there are risks involved in the securitisation of water. The main argument of opponents of the concept is that the security component may bring up discursive absolutes which are perceived to be non-negotiable between the conflicting parties, ultimately promoting disparities between riparians and enhancing 'nationalistic feelings' (Kibaroglu et al. 2008, p. 231). Egypt, for instance, appears to be following this path in the Nile Basin by slowing down the negotiation process of the Cooperative Framework Agreement (CFA) by arguing that the provision on water security (Article 14 of the CFA) should focus on historical water rights and uses – which have been established under a legal framework based on the colonial legacy (Wouters and Moynihan 2013, p. 342). It seems like Egypt is trying to make this legacy indelible by reinforcing its right to 75 per cent of the Nile's waters under the cover of water security. In doing so, it disregards Ethiopia's interests, which contributes nearly 85 per cent of the Nile waters (Mekonnen 2011, p. 361). While this is a prime example of how the paradigm can be misinterpreted, it should not be used as a thought-terminating cliché in order to disregard the water security concept in general. Not only is the bargaining tool employed by Egypt incompatible with the fundamental basis of international water law, it also is in conflict with the contemporary understanding of security, as the previous sections have shown. Egypt has applied various tactics not just to maintain the control over the freshwater resource, but also to be proactive in challenging the increase in power of its upstream riparians (Cascão 2008, p. 19). Hence, Egypt's attempt to derail the water security discussion was a mere continuation of its water diplomacy strategy. One has to be aware that, after all, water security can only be what states make of it (Julien 2012) – which becomes even more obvious in cases where a hydro-hegemon, like Egypt, is involved.

Another interesting example of how states use the water security paradigm to their national security advantage is the US Water Partnership, which aims at bringing together the private sector, NGOs, academia, civil society and the government to 'unite and mobilize "best of US" expertise, resources and ingenuity to address water challenges around the globe, with a special focus on developing countries where needs are greatest' (US Water Partnership 2012). According to the National Intelligence Council, the future competition for increasingly scarce freshwater resources will fuel instability in regions such as the Middle East and Asia, which are of particular importance to US national security (National Intelligence Council 2012). While a fully fledged war on water is considered unlikely in the near future, 'water in shared basins will increasingly be used as leverage; the use of water as a weapon or to further terrorist objectives also will become more likely beyond 10 years' (ibid., p. 3). It comes as no surprise that, by definition of the report, the authors adopt a narrow US security perspective.

> [M]any countries important to the United States will experience water problems – shortages, poor water quality, or floods – that will risk instability and state failure, increase regional tensions, and distract them from working with the United States on important US policy objectives.
>
> *Ibid., p. 1*

Hence, the Council does not try to hide the fact that one of the main reasons for the analysis and recommendations was to pave the way for the US to lead in water security issues around the globe and 'forestalling other actors from achieving the same influence at US expense' (Office of the Director of National Intelligence 2012). Since this approach strictly follows US national security interests, the outcome of the engagement can be either a push for change in transboundary water cooperation or the manifestation of the hegemonic structures in the respective basin – solely depending on US policy interests. While this is nothing new, since power games have been played under the guise of other concepts before, one has to closely watch the impact it has on water cooperation, as imported perceptions are not always compatible with the situations on the ground.

Water security and international law

The cross-cutting nature of water creates global interdependencies which make solutions to the water scarcity crisis highly complex, as water cooperation cannot be separated from global trends and drivers outside the 'water box'. Here, law can and should provide the normative content, as it: 1) defines and identifies the legal rights and obligations regarding the use of water and provides the prescriptive parameters for the management of the resource; 2) provides tools for ensuring the continuous integrity of the regime (including dispute prevention and settlement); and 3) allows for modifications of the existing regime, in order to be able to accommodate change (Wouters *et al.* 2009, p. 107).

Water security through a legal lens

In order to be able to analyse international legal regimes, the paradigm of water security (as introduced above) needs to be fleshed out further. Earlier work has developed an analytical framework for examining international law through a water security lens by focusing on issues of 1) availability; 2) access; 3) adaptability and 4) ambit (Magsig 2009). Issues of availability relate to concerns of water quality as well as quantity. This facet deals primarily with the actual management of the resource – including its control and sustainable protection. This includes the need to maintain the natural integrity of the freshwater resource by calling for an ecosystem approach (Moynihan 2016; McIntyre 2014). The element of access is central to the water security debate, as it deals with the issues revolving around the right to utilise a shared water resource. Given the complexity of cooperation over water resources, access covers a broad spectrum of concerns across the growing range and number of users and uses with regard to matters of (re)allocation. Here, the principle of equitable and reasonable utilisation, the cornerstone of international water law, is key to the process (Wouters *et al.* 2005). It determines the right of a state to use the waters of an international watercourse in two distinct ways: 1) by establishing the objective to be achieved, which then specifies the lawfulness of the new (or changed) utilisation of an international watercourse; and 2) by incorporating an operational function, since it requires all relevant factors and circumstances to be taken into account when determining what exactly qualifies as an equitable and reasonable use (Rieu-Clarke *et al.* 2012, p. 101). In order to support the obligation to weigh and balance all of the stakeholders' interests, dispute prevention and settlement mechanisms are of vital importance (Salman 2006). As, very often, the key factor of water cooperation is not absolute water scarcity, but rather the resilience of the institutions which govern the shared resource, the legal regime has to include flexibility and ensure adaptability to address changing conditions – while still providing for some level of predictability (Magsig 2014, p. 447). This element deals with the various uncertain variables – for example, impacts of global environmental change, migration patterns and economic development – which considerably influence cooperation over freshwater resources. If a treaty lacks flexible tools and water stress soars, disputes over the shared resource are likely to intensify in cases where one party to the agreement may find it difficult to reduce its consumption in order to comply with its legal obligations. In case the water stress causes asymmetric harm, the harmed state may be eager to terminate the agreement, while the co-riparian may find it beneficial to stick with it. In this regard, the International Court of Justice (ICJ) concluded in its Gabčíkovo-Nagymaros judgment that 'the stability of treaty relations requires that the plea of fundamental change of circumstances be applied only in exceptional cases' (ICJ 1997, p. 104). The ICJ further noted that new developments or changing conditions should be dealt with on the level of implementation of the treaty, not by simply terminating it (ibid., p. 112). However, several studies come to the conclusion that many states will have to renegotiate their basin treaties in order to avoid an increase in water insecurity (Goldenman 1990).

The final element is the concept of ambit, which describes and delimits the scope of water security – that is, the sphere of influence of the notion. In addition to the traditional (hydrological and geographical) meaning of scope, the approach here is to better reflect the common challenges of water insecurity. So far, one of the main weaknesses of water cooperation is the inability to link various influencing factors in a comprehensive manner. The extent of the breadth of objectives covered by a freshwater agreement ranges from merely quantitative agreements to much more sophisticated institutions which also govern aspects of water quality and emergency situations (see also Chapters 15 and 19, this volume). Evidently, the most effective management of transboundary watercourses, for the benefit of the whole basin, can only be achieved through a truly joint strategy involving all sectors and disciplines across borders (Magsig 2014, p. 450). In addition to the predominant perception of scope, the element of ambit also does justice to the fact that water security has to be seen as a collective security issue (Magsig 2009, p. 67). Owing to the interconnectedness of the globalised world and the role water plays in linking the various emerging crises, negative impacts may even be felt outside the basin. Thus, the times when water can solely be regarded as a national security issue are long gone, as one of our most fundamental common values is under threat – international peace and security. The linkages between different scales of cooperation over water (local, national, regional and global) are fluid; and international law has to act as an interface between those layers while illustrating ways towards truly regional solutions.

The general obligation to cooperate

International environmental governance in general, and transboundary water management in particular, have long been dominated by the either/or debate on sovereignty versus the joint management of natural resources. While most states have now accepted a more nuanced inter-pretation of sovereignty, the question about how sovereignty over freshwater resources should be interpreted today is still hotly debated. Critically, the notion of sovereignty carries with it a responsibility to cooperate. As indicated by Article 1 of the UN Charter: '[t]he purposes of the United Nations are:... (3) [t]o achieve international co-operation in solving international prob-lems of an economic, social, cultural, or humanitarian character'. This unspecified duty to coop-erate was partially clarified by the 1970 Declaration on Principles of International Law concerning Friendly Relations and Co-operation among States in accordance with the Charter of the United Nations, which stipulates that:

> states have the duty to co-operate with one another, irrespective of the differences in their political, economic and social systems, in the various spheres of international relations, in order to maintain international peace and security and to promote inter-national economic stability and progress, the general welfare of nations and interna-tional co-operation free from discrimination based on such differences.
>
> *See also Chapter 18, this volume*

While the Declaration does not constitute binding international law, it is being given considerable legal weight by its universal recognition as a standard of conduct and perception of it as an elabora-tion of principles of international law (Sands and Peel 2012, p. 11). In the MOX Case before the International Tribunal for the Law of the Sea, Judge Wolfrum argued that the duty to cooperate,

> balances the principle of sovereignty of states and thus ensures that community of interests are taken into account vis-à-vis individualistic state interests. It is the matter of prudence and caution as well in keeping with the overriding nature of the obligation to cooperate that the parties should engage therein.
>
> *The MOX Plant Case 2001*

From this follows that there is a need to ultimately arrive at a stage where the concept of state sovereignty is understood as one of 'cooperative sovereignty' (Perrez 2000). This necessity becomes particularly blatant when addressing the difficulty of managing common pool resources, where the collective action problem leads to unsatisfactory outcomes (Hardin 1968). Rather than treating sovereignty as a stumbling block in international negotiations – due to its apparent incompatibility with relinquishing freedoms and making concessions – acknowledging that the responsibility to cooperate is a key element of sovereignty itself seems to be a more promising strategy in addressing the tragedy of the commons (Schreuer 2002; Delbrück 2012). Hence, international law has to provide a path for moving from 'sovereignty as independence' to 'sovereignty as interdependence'.

In the arena of international water law, the general obligation to cooperate contains the pro-cedural duties of prior information and of prior consultation, which aim to operationalise the rather vague principle. Yet, it leaves a lot to be desired when it comes to fundamentally changing the way states perceive their sovereign rights over freshwater resources, as they still have much discretion regarding the particular means of cooperation. For instance, the setting up of joint institutions is not compulsory, which is a particular weakness of international water law.

Regional common concern for water security

In order to strengthen the general obligation to cooperate, international law has to be developed further. The urgency to act jointly on issues which necessitate cooperative action – like the management of freshwater resources – has led to the understanding of common security and revealed the limits of the current international legal regime. Earlier work has addressed the question as to how communality has been treated in international environmental law and what lessons can be learned for international water cooperation (Magsig 2015a, p. 123). While some approaches are too limited as they only apply to certain geographical areas beyond national jurisdictions and their resources (common area and common heritage), the notion of common concern appears to be the most promising in tackling issues of water security.

At the 1992 UN Conference on Environment and Development, a global framework for environmental responsibilities was designed which, for the first time, was based on a common concern, rather than the concept of good neighbourliness (Birnie *et al.* 2009, p. 128). This concern is based on the understanding that some kind of harm to the environment has the potential to adversely affect humanity as a whole; and, thus, mitigating those impacts can only be achieved effectively if the international community in its entirety is involved. Accepting this argument carries with it both a right and an obligation of the international community as a whole to have concern for the global environment (IUCN Commission on Environmental Law and International Council of Environmental Law 2010, p. 39). Both treaties negotiated in 1992 in Rio – the UN Framework Convention on Climate Change (UNFCCC) and the Convention on Biological Diversity (CBD) – follow this approach. While the inclusion of common concern in the respective preambles kick-started theoretical discussions, neither convention managed to overcome the state-centrism of the governance system.

The Draft International Covenant on Environment and Development (IUCN Commission on Environmental Law and International Council of Environmental Law 2010) tried to inspire the debate, with Article 3 stipulating that '[t]he global environment is a common concern of humanity and under the protection of the principles of international law, the dictates of the public conscience and the fundamental values of humanity'. According to the commentary, the concept should be interpreted as 'the basis upon which the international community at all levels can and must take joint and separate action to protect the environment' (ibid., p. 39). In urging that not only single issues, like climate change or the loss of biodiversity, should be treated as being of common concern to the international community, but also the environment as a whole, it constitutes a departure from previous approaches to common concern (Magsig 2015a, p. 133). Yet, due to the lack of legitimacy, the Covenant has lived in the shadows of international legal scholarship – despite it having breathed new life into the debate about communality in international law.

However, even in its embryonic stage, the concept of common concern is of particular interest to the advancement of international water law. Although its focus lies again on (common) benefits, it considers the benefits from common action rather than those derived from the mere exploitation of a resource (Brunnée 2007, p. 553). Moreover, it fixes its attention on what renders a concern as being common, rather than targeting one particular area or resource, and thus defuses discussions about common property and territorial sovereignty (Magsig 2015a, p. 133). One of its main advantages over other approaches to communality is that it triggers a shift from the orthodox reciprocity and material benefit sharing we often find in treaties of joint action in the long-term interest of the community. Still, the weaknesses of the notion are evident in both UNFCCC and CBD, as both fail in achieving strong legal impact with regard to the common concern at the global level. While it should be possible to construct an analogous mindset for freshwater cooperation to the loss of biodiversity or climatic changes, it seems illusive to agree on a perception of water security as a common concern of humankind (ibid., p. 206).

Here, scaling down to the regional layer is a more pragmatic middle road which can sufficiently accommodate the national interests within the global challenge. Since, contrary to a river basin, a region does not have to be a narrowly defined geographical area, framing water security as a regional common concern also opens up the enormous potential of including (non-state) actors and interests beyond the basin. By including non-riparian interests in the design and performance of international water cooperation, it may add a new dimension to international freshwater cooperation, which is still being perceived as a zero-sum game (Brunnée and Toope 1994, p. 71). Regional common concern provides a vehicle for inducting communality into already existing rules and principles of international water law in order to arrive at more resilient agreements. In acknowledging that a particular challenge must be perceived as a matter of common concern, states appreciate that transboundary water management can no longer be considered as a mere national issue and that a shift in the responsibilities of states from individual to concerted action is inevitable. For instance, regional common concern applied to the principle of equitable and reasonable use would strengthen cooperation by obliging states to regularly exchange data and information (see Article 9 of the UN Watercourses Convention) rather than merely for planned measures; establish joint institutions (see Article 8(2) of the UN Watercourses Convention); and provide technical assistance to less capable riparians. In short, it would put the principle of equitable and reasonable utilisation on a more common footing.

Concluding remarks

A contemporary understanding of water security, rooted in the notion of common concern, provides an impetus for riparian countries to rethink their existing approach to water diplomacy and base it on a more common strategy. Not only does such an approach provide promising avenues for strengthening cooperation over shared freshwater resources outside the traditional 'water box' – for instance, through human rights (see Chapter 8, this volume) and sustainable development goals (see Chapter 27, this volume) – it will ultimately lead to more peaceful relations. This can be achieved by strengthening two major elements of the securitisation of water – ethicisation and regionalisation. States will be very cautious in implementing an approach to water cooperation which comes from outside, as they are much more likely to develop their own concepts – based on their regional identity, the specific political and cultural environment and what they perceive as being just and equitable (Moynihan and Magsig 2014, p. 57). The water security paradigm pushes international law towards a regime which evokes shared responsibilities – and thus, is able to address common concerns like water insecurity more effectively by striving for truly joint and long-term regional water management.

References

Abbott, C., Rogers, P. and Sloboda, J. (2006) *Global Responses to Global Threats: Sustainable Security for the 21st Century*. Oxford: Oxford Research Group.

Abseno, M.M. (2013) The influence of the UN Watercourses Convention on the development of a treaty regime in the Nile River Basin. *Water International* 38(2), pp. 192–203.

Annan, K. (2005) In larger freedom: towards development, security and human rights for all. *Report of the Secretary-General, Fifty-ninth session, Agenda items 45 and 55*. New York: United Nations General Assembly.

Birnie, P., Boyle, A. and Redgwell, C. (2009) *International Law and the Environment*. Oxford: Oxford University Press.

Brauch, H.G. (2008) Introduction: globalization and environmental challenges: reconceptualizing security in the 21st century. In H.G. Brauch, Ú. Oswald Spring, C. Mesjasz, J. Grin, P. Dunay, N.C. Behera, B. Chourou, P. Kameri-Mbote and P.H. Liotta (eds), *Globalization and Environmental Challenges: Reconceptualizing Security in the 21st Century*. Berlin: Springer.

Brauch, H.G. (2011) Concepts of security threats, challenges, vulnerabilities and risks. In H.G. Brauch, Ú. Oswald Spring, C. Mesjasz, J. Grin, P. Dunay, N.C. Behera, B. Chourou, P. Kameri-Mbote and P.H. Liotta (eds), *Globalization and Environmental Challenges: Reconceptualizing Security in the 21st Century*. Berlin: Springer.

Brock, H. (2011) *Competition Over Resources: Drivers of Insecurity and the Global South*. London: Oxford Research Group.

Brock, L. (1999) Environmental conflict research: paradigms and perspectives. In A. Carius and K.M. Lietzmann (eds), *Environmental Change and Security: A European Perspective*. Berlin and New York: Springer.

Brown, O. (2005) The environment and our security: how our understanding of the links has changed. *International Conference on Environment, Peace and Dialogue Among Civilizations*. Tehran, Iran: International Institute for Sustainable Development.

Brunnée, J. (2007) Common areas, common heritage, and common concern. In D. Bodansky, J. Brunnée and E. Hey (eds), *The Oxford Handbook of International Environmental Law*. Oxford: Oxford University Press.

Brunnée, J. and Toope, S.J. (1994) Environmental security and freshwater resources: a case for international ecosystem law. *Yearbook of International Environmental Law* 5.

Buzan, B. (1991) *People, States and Fear: An Agenda for International Security Studies in the Post-Cold War Era*. London: Harvester Wheatsheaf.

Buzan, B., Wæver, O. and De Wilde, J. (1998) *Security: A New Framework for Analysis*. Boulder, CO: Lynne Rienner Publishers.

Cascão, A.E. (2008) Ethiopia: challenges to Egyptian hegemony in the Nile Basin. *Water Policy* 10(2), pp. 13–28.

Chellaney, B. (2011) *Water: Asia's New Battleground*. Washington, DC: Georgetown University Press.

Convention on Biological Diversity (1992) 5 June; entered into force 29 December 1993. 1760 UNTS 79; reprinted in 31 ILM 818 (1992).

Delbrück, J. (2012) The international obligation to cooperate: an empty shell or a hard law principle of international law? A critical look at a much debated paradigm of modern international law. In H.P. Hestermeyer, D. König, N. Matz-Lück, V. Röben, A. Seibert-Fohr, T.-P. Stoll and S. Vöneky (eds), *Coexistence, Cooperation and Solidarity*. Leiden: Martinus Nijhoff.

Frédérick, M. (1999) A realist's conceptual definition of environmental security. In D. Deudney and R.A. Matthew (eds), *Contested Grounds: Security and Conflict in the New Environmental Politics*. Albany: State University of New York Press.

Fröhlich, C.J. (2012) Security and discourse: the Israeli–Palestinian water conflict. *Conflict, Security and Development* 12(2), pp. 123–48.

Goldenman, G. (1990) Adapting to climate change: a study of international rivers and their legal arrangements. *Ecology Law Quarterly* 17(4), pp. 741–802.

Hardin, G. (1968) The tragedy of the commons. *Science* 162(3859), pp. 1243–8.

Holst, J.J. (1989) Security and the environment: a preliminary exploration. *Bulletin of Peace Proposals* 20(2), pp. 123–8.

IJC (1997) International Court of Justice, Gabčíkovo-Nagymaros Project (Hungary v Slovakia) Judgment, 25 September. ICJ Reports 1997.

IUCN Commission on Environmental Law and International Council of Environmental Law (2010) *Draft International Covenant on Environment and Development*. Gland, Switzerland: IUCN.

Julien, F. (2012) Hydropolitics is what societies make of it (or why we need a constructivist approach to the geopolitics of water). *International Journal of Sustainable Society* 4(1/2), pp. 45–71.

Kibaroglu, A., Brouma, A.D. and Erdem, M. (2008) Transboundary water issues in the Euphrates–Tigris River Basin: some methodological approaches and opportunities for cooperation. In L. Jansky, M. Nakayama and N.I. Pachova (eds), *International Water Security: Domestic Threats and Opportunities*. Tokyo and New York: United Nations University Press.

Magsig, B.-O. (2009) Introducing an analytical framework for water security: a platform for the refinement of international water law. *Journal of Water Law* 20(2/3), pp. 61–9.

Magsig, B.-O. (2014) Pushing the boundaries: rethinking international law in light of the common concern for water security. In V. Sancin and M. Kovič Dine (eds), *International Environmental Law: Contemporary Concerns and Challenges in 2014*. Ljubljana: GZ Založba.

Magsig, B.-O. (2015a) *International Water Law and the Quest for Common Security*. London: Routledge.

Magsig, B.-O. (2015b) Water security in Himalayan Asia: first stirrings of regional cooperation? *Water International* 40(2), pp. 342–53.

McIntyre, O. (2014) The protection of freshwater ecosystems revisited: towards a common understanding of the 'ecosystems approach' to the protection of transboundary water resources. *Review of European, Comparative and International Environmental Law* 23(1), pp. 88–95.

McRae, R.G. (2001) Human Security in a Globalized World. In R.G. McRae and D. Hubert (eds), *Human Security and the New Diplomacy: Protecting People, Promoting Peace*. Montreal: McGill-Queen's University Press.

Mekonnen, D.Z. (2010) The Nile Basin Cooperative Framework Agreement negotiations and the adoption of a 'water security' paradigm: flight into obscurity or a logical cul-de-sac? *European Journal of International Law* 21(2), pp. 421–40.

Mekonnen, D.Z. (2011) Between the Scylla of water security and Charybdis of benefit sharing: the Nile Basin Cooperative Framework Agreement – failed or just teetering on the brink?. *Göttingen Journal of International Law* 3(1), pp. 345–72.

Moynihan, R. (2016) International law on protection of transboundary freshwater ecosystems and biodiversity. In J. Razzaque and E. Morgera (eds), *Biodiversity and Nature Protection Law*. Cheltenham: Edward Elgar.

Moynihan, R. and Magsig, B.-O. (2014) The rising role of regional approaches in international water law: lessons from the UNECE water regime and Himalayan Asia for strengthening transboundary water cooperation. *Review of European Community and International Environmental Law* 23(1), pp. 43–58.

National Intelligence Council (2012) Global Water Security: Intelligence Community Assessment. Washington DC: Office of the Director of National Intelligence.

Office of the Director of National Intelligence (2012) ODNI News Release No. 4-12, ODNI releases assessment on global water security. 22 March. https://www.dni.gov/index.php/newsroom/press-releases/96-press-releases-2012/529-odni-releases-global-water-security-ica

Orakhelashvili, A. (2011) *Collective Security*. Oxford: Oxford University Press.

Perez, F.X. (2000) *Cooperative Sovereignty: From Independence to Interdependence in the Structure of International Environmental Law*. The Hague and Boston: Kluwer Law International.

Rieu-Clarke, A., Moynihan, R. and Magsig, B.-O. (2012) *UN Watercourses Convention: User's Guide*. Dundee: IHP-HELP Centre for Water Law, Policy and Science.

Salman, S.M.A. (2006) International water disputes: a new breed of claims, claimants, and settlement institutions. *Water International* 31(1), pp. 2–11.

Sands, P. and Peel, J. (2012) *Principles of International Environmental Law*. Cambridge: Cambridge University Press.

Scholtz, W. (2009) Collective (environmental) security: the yeast for the refinement of international law. *Yearbook of International Environmental Law* 19, pp. 135–62.

Schreuer, C. (2002) State sovereignty and the duty to cooperate - two incompatible notions?. In J. Delbrück (ed.), *International Law of Cooperation and State Sovereignty: Proceedings of an International Symposium of the Kiel Walther Schücking-Institute of International Law, May 23–26, 2001*. Berlin: Duncker & Humblot.

Security Council (2016) Security Council Press statement on maritime tragedy in Mediterranean Sea, 22 April. http://www.un.org/press/en/2016/sc12334.doc.htm

Selby, J. and Hoffmann, C. (2012) Water scarcity, conflict, and migration: a comparative analysis and reappraisal. *Environment and Planning C: Government and Policy* 30(6), pp. 997–1014.

Shevardnadze, E. (1990) Ecology and diplomacy. *Environmental Policy and Law* 20(1/2), pp. 20–4.

Statement by the President of the Security Council (2011) 20 July. UN Doc S/PRST/2011/15. Available from: http://www.securitycouncilreport.org/atf/cf/%7B65BFCF9B-6D27-4E9C-8CD3-CF6E4FF96FF9%7D/CC%20SPRST%202011%205.pdf

The MOX Plant Case (2001) (Ireland v United Kingdom), Provisional Measures, Order of 3 December, ITLOS, at Para 82. Available from: http://www.itlos.org/fileadmin/itlos/documents/cases/case_no_10/Order.03.12.01.E.pdf

Turton, A. (2010) The politics of water and mining in South Africa. In K. Wegerich and J. Warner (eds), *The Politics of Water: A Survey*. London: Routledge.

UN General Assembly (1970) Declaration of Principles of International Law Concerning Friendly Relations and Co-operation Among States in Accordance with the Charter of the United Nations, 24 October. UNGA Res 2625, 25 UN GAOR Supp 18, UN Doc A/5217 at 121; reprinted in 65 AJIL 243 (1971). Available from: http://www.un.org/ga/search/view_doc.asp?symbol=A/RES/2625(XXV)

UN Secretary-General (2008) Ban Ki-moon in *Time Magazine*, The right war, 17 April. http://www.time.com/time/specials/2007/article/0,28804,1730759_1731383_1731345,00.html

UN Security Council (2011) 6587th Meeting, 20 July. UN Doc SC/10332. Available from: http://www.un.org/News/Press/docs/2011/sc10332.doc.htm

United Nations (2004) *A More Secure World: Our Shared Responsibility – Report of the Secretary-General's High-Level Panel on Threats, Challenges and Change*. New York: United Nations.

United Nations Convention on the Law of the Non-Navigational Uses of International Watercourses (1997) UN Doc A/51/869, 21 May; in force 17 August 2014; reprinted in 36 ILM 700 (1997).

United Nations Framework Convention on Climate Change (1992) 9 May; entered into force 21 March 1994) 1771 UNTS 107; UN Doc A/AC.237/18 (Part II)/Add.1; reprinted in 31 ILM 849 (1992).

US Water Partnership (2012) Best of US solutions are available. 22 March. http://uswaterpartnership.org/

Voigt, C. (2009) Sustainable Security. *Yearbook of International Environmental Law* 19, pp. 163–96.

Von Tigerstrom, B. (2007) *Human Security and International Law*. Oxford: Hart.

Warner, J.F. and Meissner, R. (2008) The politics of security in the Okavango River Basin: from civil war to saving wetlands (1975–2002): a preliminary security impact assessment. In L. Jansky, M. Nakayama and N.I. Pachova (eds), *International Water Security: Domestic Threats and Opportunities*. Tokyo and New York: United Nations University Press.

Weinthal, E., Zawahri, N. and Sowers, J. (2015) Securitizing water, climate, and migration in Israel, Jordan, and Syria. *International Environmental Agreements: Politics, Law and Economics* 15(3), pp. 293–307.

What's in Blue (2013) Arria formula meeting on climate change. 14 February. http://www.whatsinblue. org/2013/02/arria-formula-meeting-on-climate-change.php

What's in Blue (2016) Arria formula meeting on water, peace and security. 21 April. http://www. whatsinblue.org/2016/04/arria-formula-meeting-on-water-peace-and-security.php

Wirsing, R.G., Stoll, D.C. and Jasparro, C. (2013) *International Conflict over Water Resources in Himalayan Asia*. Basingstoke: Palgrave Macmillan.

Wolfers, A. (1962) *Discord and Collaboration: Essays on International Politics*. Baltimore: Johns Hopkins Press.

World Bank (2016) *High and Dry: Climate Change, Water, and the Economy*. Washington DC: World Bank Group.

Worsnip, P. for Reuters (2011) West, Russia divided on UN council climate role. 20 July. http://www. reuters.com/article/2011/07/20/us-climate-un-idUSTRE76J7QY20110720

Wouters, P. and Moynihan, R. (2013) Water security: legal frameworks and the UN Watercourses Convention. In F.R. Loures and A. Rieu-Clarke (eds), *The UN Watercourses Convention in Force: Strengthening International Law for Transboundary Water Management*. London: Earthscan.

Wouters, P., Vinogradov, S. and Magsig, B.-O. (2009) Water security, hydrosolidarity and international law: a river runs through it... *Yearbook of International Environmental Law* 19(1), pp. 97–134.

Wouters, P., Vinogradov, S., Allan, A., Jones, P. and Rieu-Clarke, A. (2005) *Sharing Transboundary Waters: An Integrated Assessment of Equitable Entitlement: The Legal Assessment Model*. Paris: UNESCO.

Zhang, Y., Block, P., Hammond, M. and King, A. (2015) Ethiopia's grand renaissance dam: implications for downstream riparian countries. *Journal of Water Resources Planning and Management* 141(9), 05015002_1-05015002_10.

CONCLUSIONS

Andrew Allan, Sarah Hendry and Alistair Rieu-Clarke

As the international community works towards the 2030 sustainable development agenda, the need to strengthen laws and policy relating to water is likely to become critical. Water – as a key to sustainable development – will require effective frameworks to be in place to ensure that water and its multiple uses are allocated across a diverse range of stakeholders that operate at different spatial and multiple levels. In particular, law and policy reform will be a central component in the success of achieving Sustainable Development Goal 6.5, which calls upon states to implement integrated water resources management (IWRM) at all levels, including transboundary cooperation where appropriate (UN General Assembly 2015).

The task is a significant one. In assessing water laws and policy across OECD countries, a number of key challenges remain, including: multi-level governance gaps; lack of engagement with sub-national actors in the design of laws and policies; the need for horizontal coordination of water law and policy across governmental departments; weak and incoherent water information systems and databases for sharing water policy needs at basin, country and international levels; the paucity of water law and policy monitoring and evaluation mechanisms; capacity building; and the need for more open and inclusive water law and policy frameworks that engage with a wide range of stakeholders (OECD 2009). Within the context of transboundary waters, significant challenges have also been highlighted. UN-Water, for example, observe that:

> existing agreements are sometimes not sufficiently effective to promote integrated water resources management due to problems at the national and local levels such as inadequate water management structures and weak capacity in countries to implement the agreements as well as shortcomings in the agreements themselves (for example, inadequate integration of aspects such as the environment, the lack of enforcement mechanisms, limited – sectoral – scope and non-inclusion of important riparian States.
> *UN-Water 2008, p. 6*

This latter observation, together with OECD's emphasis on multi-level governance, recognises the need to consider both the national and international dimensions of water law and policy.

Some positive signs are evident. Within the transboundary context, two significant milestones have been reached in the last few years: first, with the entry into force of the 1997 UN Watercourses Convention and, second, with the opening of the 1992 UNECE Water Convention to all UN Member States. These two framework instruments, if promoted in a coordinated manner, have much to offer in strengthening legal frameworks relating to transboundary waters across the world. Significant activity has also taken place at the national level. For instance, a survey by UN-Water suggests that, by 2012, 80 per cent of countries had reformed their water laws in the last 20 years; although the report also recognised that 'translating policy and legal changes into implementation is a slow process' (UN-Water 2012, p. 17).

The contributions to this *Handbook* clearly demonstrate that the field of water law and policy is both vibrant and capable of making a significant contribution to the sustainable development agenda, while also responding to other global challenges, including food and energy security, biodiversity loss and ecosystem degradation, climate change and the promotion of peace and security. The influence of different global policy agendas on water also comes across clearly in all of the chapters. While extrapolating all the rich insights that have come out of the previous chapters is impossible in this Conclusion, an overview of some key messages are offered here.

Van Koppen's examination (Chapter 1) of the sometimes problematic relationship between customary frameworks for water use and formal permit systems addresses an issue that is rarely dealt with in legal texts. It is a critical issue for the majority of the world's water users, however, as government capacity to enable permit systems to percolate down to local and community levels is often severely curtailed. This results in inequitable situations for much of the world's poor, compounded by unequal power relations that have emanated from colonial periods. She focuses on practical solutions that recognise the difficulty in rolling out permit-based pro-grammes, suggesting that formal state recognition of customary systems would be a first step in ameliorating the position communities often find themselves in. She also identifies the international human rights framework as being a possible source of succour, linking rights to food, water, gender equity and to fair and equal treatment with water resources management more generally, and sees this as an important avenue for enforcing indigenous rights against potentially hostile central governments.

Continuing on the underlying theme of protecting the public interest, Scanlan (Chapter 2) addresses the fascinating developments in the application of the public trust doctrine. Over a long period of time, the doctrine has been used in the USA as an effective way of limiting private rights to natural resources, and the baton has been taken up in a number of other countries, notably India. The interplay between these two jurisdictions has created not only enhanced public protection of waters in India, but also is potentially now working the other way. Indian jurisprudence applying the principle has broadened its scope to apply to ecological protection and has linked it to the national Constitution directly, in a way that has not happened in the USA. It will be interesting to see how cases currently proceeding through the US court system fare when they attempt to extrapolate the public trust doctrine to apply to climate change. India's courts have taken up the doctrine in the absence of formal policy and legal direction from the central government. This will resonate strongly in other countries where political conflict has created mummified approaches to water resources management that resist change.

Kidd's (Chapter 3) analysis of the South African National Water Act has many useful insights for others. Many of us, in countries far away, have studied the South African reforms and com-mended the legislative integrity and cohesiveness of the law on the statute books. He reminds us that the implementation, and the law in practice, may be rather different. He analyses the legisla-tion in its historical context, and looks briefly at the IWRM framework before focusing on the rules for water allocation and water rights. One of the criticisms sometimes made of IWRM is that it diverts attention, and resources, away from more pressing matters – many countries might be better to prioritise water rights reform. South Africa, of course, did both, in an integrated way, whereby catchment management (should) provide the overarching context for more detailed reform. However, his analysis of the (ongoing) reform of water rights and the difficulties with reallocating water, will resonate in many countries at different stages of development. Meeting multiple needs where water and money are scarce, even where political priorities are clear, is one problem. Another is striking the right balance between moving too fast, and risking challenge to the law; or moving so slowly that nothing is achieved and political momentum is lost. South Africa water law reforms continue to present us with challenges, opportunities and cause for reflection.

Van Risjwick and Keessen (Chapter 4) demonstrate that the story of the EU's Water Framework Directive (WFD) is well known but not always well understood. As well as mandating river basin planning, the WFD also establishes an overall objective of 'good ecological status', which Member States should achieve progressively in all their surface waters. Van Risjwick and Keessen trace the implementation of the WFD, and the related Floods Directive, in terms of its integration and in the context of the development of the EU's wider environmental *acquis*. Many states around the world, meeting their Johannesburg commitments, have adopted IWRM-type approaches; fewer have explicitly mandated improvements to aquatic ecology as a goal, still less a binding objective. The EU is, of course, a special case, and the analysis recognises those specialities – economic, social and political. Nonetheless there are lessons that can be learned – including the advantages and disadvantages of a flexible and adaptive approach. For countries setting out on a similar path, of a high level and integrative water management instrument, the authors suggest that more attention should be paid to integrating climate change concerns, agricultural activities and the human rights to water and sanitation.

Hantke-Domas (Chapter 5) reflects on the utility of water markets, one of the most polarising areas of water management. In a pragmatic assessment of their effectiveness, he cautions against relying on them as the main tool for implementing IWRM, but he recognises their potential role in achieving economic efficiency. The problems of incorporating public interest and environmental considerations are comprehensively dealt with, as is the emerging interface between water markets and application of the human right to water. A novel analogue to the economic efficiency arguments for water markets is drawn with the Competition Law, drawing out competition-related aspects of water markets that should be considered by authorities. He cautions against the piecemeal implementation of water markets, where they are appropriate, in order to avoid potential future abuse of these markets by privileged water rights holders.

As we move towards a post-industrial society, Howarth (Chapter 6) traces the development of water quality law in one jurisdiction – England – and uses that story to demonstrate how water law adapts as we reconceptualise the underlying environmental problems. The four stages he identifies of regulating water pollution – prohibitive responses to industrial pollution; anticipatory landuse controls; strategic water quality management; and post-industrial environmental regulation – will each be applicable to different societies, at different times and different stages of development. Each will also relate to different modes of, and approaches to, water management. As England industrialised early and has moved beyond its industrial phase, it can be well placed to evidence that development and allow others to reflect on lessons learned, both positive and negative.

Hendry (Chapter 7) analyses the development of law and policy regarding contaminants of emerging concern. As she notes, these substances may be novel, but equally, they may have been discharged to the environment over many decades – yet the concerns over their effects are rising. These articles and substances can be categorised in many different ways, and this is itself a challenge for regulators: should the law focus on user groups, or product types, or the environmental media to which discharges are made? There are international frameworks, both policy (the work of UNEP and WHO) and law (the Chemicals Conventions and broad rules for waste management and air pollution), but most relevant controls, especially for water, exist at regional or national levels. The chapter looks at pharmaceuticals, recognising the role of agriculture as well as human use; at pesticides; and at cross-cutting issues such as endocrine disruptors, in the context of wastewater management. Although all developed jurisdictions are looking at the problem, and the science evidence base is emerging, progress is slow. The EU's comprehensive approach to chemicals is one way ahead, but, in the end, consumers, as well as policy-makers and regulators, need to be more involved in this debate.

Winkler (Chapter 8) has been at the forefront of the legal analysis of the human rights to water and sanitation, arguing consistently for several years that the right to water at least has emerged as a customary right in international law. Her chapter provides a state-of-the-art review but also points us towards the future, especially in relation to the human right to sanitation. We know that water is a cross-cutting issue, linked to many other rights and sustainable development goals; but the same is also true of sanitation, especially, rights to housing and an adequate standard of living; and goals related to health and well-being. As ever with social and economic rights, the poorest suffer most, so realising these rights also lifts people out of poverty. Examining water rights through the human rights prism can be uncomfortable, especially if water for subsistence farming is included – IWRM conceptions of stakeholder engagement and prioritising uses across a basin does not always fit with conceptions of absolute and indivisible human rights. Yet it is vital that we find ways forward that recognise unmet needs. Water services remains a key driver for global water policy, keeping it high on the agenda, and the human rights to water and sanitation are important tools to give health and dignity to the poorest.

Franceys and Hutchings (Chapter 9) also frame their chapter around human needs and development goals, while focusing more directly on the regulation and governance of water services, especially larger municipal systems with their infrastructure and financing requirements. Drawing on Franceys' lifetime experience of law and practice around the globe and based in a 'pro-poor' framework, the authors examine accountability, and the role of regulations and standards, before assessing the various contractual and ownership models that are typically used to deliver services, especially where use is being made of the private sector. The authors note the time-consuming and data-demanding nature of recent innovations in regulation, and suggest that, in many countries, the political context will remain a key driver for the economic regulation of services. This is an area where many countries still need to make progress, to deliver an adequate service effectively, and there are examples of good practice from the developing world. The final conclusion is simple – 'governance and regulation works'.

Allan (Chapter 10) analyses legal responses to the perennial challenge of flood management. Structuring the chapter around a framework of hazard, risk and vulnerability allows him to draw on cutting-edge science in flood management. As he notes, many dimensions of (national and regional) law and policy are relevant: disaster management and emergency response; tort; civil defence; water; urban planning; coastal zone management; and land use. Flooding is, at once, a natural event bringing benefits in terms of hydrological management and sediment distribution; a huge problem for civil society, especially in countries least well resourced to make comprehensive responses; and something likely to increase in the context of climate change. Drawing on examples from regional frameworks such as the EU Floods Directive, and national laws from a wide range of jurisdictions, Allan gives examples of good practice and ways in which the law can support, or hinder, both 'hard' defences and 'soft' or 'green' infrastructure, along with analysis of flood management for dams and the need for appropriate information for citizens.

One of the key sources of dispute among users is the allocation of water during periods of shortage. Tarlock (Chapter 11) examines the role that law has to play during droughts, recognising the significance of equitable apportionment among users as a way of spreading the burden and incentivising efficiency measures. He identifies the principle elements of modern water allocation and permit systems, and comprehensively assesses the preconditions that are needed from the legal perspective if equitable apportionment of the hardship caused by drought is to be achieved. The need for those managing water entitlements to be able to balance use rights against available resources requires those rights to be sufficiently flexible, a characteristic that will be of particular importance in the context of climate change and increasing inter-annual variability. Tarlock identifies best practice around the world, highlighting developments in Australia, as well as unpicking the problems in applying existing systems such as water markets and strict priority enforcement.

Akmouch and Clavreul (Chapter 12) reflect on the importance of stakeholder involvement in improving effectiveness, efficiency and inclusiveness with respect to water management. They identify the factors that are needed to ensure that engagement involves more than simply paying lip-service to stakeholders' needs. Building on a comprehensive study conducted by the OECD, they highlight six principles for engagement that are linked directly to criteria for good water governance. Using best practice examples from OECD state respondents, Akhmouch and Clavreul tease out the main benefits of stakeholder engagement over the long term, highlighting the utility in harmonising methodological approaches across countries in an effort to better understand what works and when. While they are realistic about the relative importance of stakeholder involvement in achieving water resource management goals, they identify four governance prerequisites essential for maximising the contribution from stakeholders: equity, transparency, accountability and trust.

Lessons that might be learned from the US experience of monitoring and enforcing water pollution provisions, principally as regards point sources, are examined in forensic detail by Paddock and Mulherin (Chapter 13). The effectiveness of the progressive development and elaboration of the Clean Water Act since 1972 is assessed against biophysical observations. Comprehensive monitoring and enforcement procedures are critical for the success of pollution control legislation, but these can be rendered impotent if capacity is not commensurate with need. Elaborate self-monitoring provisions are in place in order to optimise resource use, and data collection platforms must be capable of receiving, collating and processing all data inputs, while at the same time providing a publicly available database. A mixture of civil and criminal penalties are employed. Although the chapter focuses on the USA, transferable principles can be gleaned from this experience, most notably that clear discharge standards should be incorporated into permitting processes along with stringent monitoring and reporting procedures that can influence permit applicability. In addition, supplying regulatory authorities with a range of enforcement tools allows much greater flexibility in dealing with routine breaches and those that have a much bigger impact. The role of citizen enforcement complements the financial and technological resource optimisation evident in self-monitoring techniques.

Rieu-Clarke (Chapter 14) illustrates the importance of transboundary waters to almost all countries of the world; whether they be the 145 countries that directly share such waters, or those that are reliant on goods and services produced from the 263 international rivers and lakes, or over 275 transboundary aquifers. As maintained by Rieu-Clarke, the need to establish treaty arrangements over these transboundary resources is compelling, and that need is well recognised in the global and regional framework instruments that have been developed. However, the overview that Rieu-Clarke provides at the basin level demonstrates that much more work needs to be done at the basin and sub-basin level to establish cooperative arrangements that are capable of responding to contemporary challenges, such as climate change, biodiversity loss, population increase and unsustainable patterns of consumption. By breaking down treaty practice into regions, Rieu-Clarke offers some stark warnings, particularly for the fragmented treaty frameworks that are evident in Asia and South America. In these regions and beyond, much more needs to be done to raise awareness of the benefits of treaties in fostering cooperation between states.

McCaffrey (Chapter 15) provides a rich insight into the possible future direction of the law relating to transboundary waters by tracing its evolution. He goes back to 2450BC and the adoption of a treaty between two ancient Sumerian city states of Umma and Lagash that dealt with the diversion of waters from the Euphrates. It is clearly evident from McCaffrey's chapter that the law relating to transboundary waters has adapted to the changes in both the uses and scientific understanding of transboundary water systems. This is reflected in an early dominance for law to regulate issues relating to navigation, particularly in the nineteenth century, although McCaffrey ascertains growing attention to non-navigational uses in the twentieth century. After tracing different types of uses of transboundary waters, McCaffrey maintains that the current position on priorities is that: first, each international watercourse is unique, second, natural and human related conditions are not static and, third, no use enjoys inherent priority. Any conflicts over uses must therefore be weighed up on the basis of all relevant factors and circumstances in order to determine what is equitable and reasonable. In tracing the evolution of the law relating to transboundary waters, McCaffrey also highlights a move from surface water channels to systems of water; from piecemeal problem-solving to integrated management and development, from protection of fisheries to the protection of fish, and from no harm to equitable utilisation. These observations offer suggestions as to the areas in which the law is likely to develop.

While touched upon by McCaffrey and Rieu-Clarke, the contribution by Eckstein (Chapter 16) is solely dedicated to examining the law relating to transboundary aquifers. Eckstein recognises the critical link between water scarcity and groundwater. As the availability of surface water continues to become more unpredictable the importance of groundwater is clear. This is therefore an area that is likely to grow in stature in years to come, and international law, as Eckstein clearly demonstrates, has an important role to play. However, the present status of the law does not adequately reflect the critical strategic importance of transboundary aquifers to many states across the world. Examples of such cooperative arrangements exist, such as the first agreement exclusively dedicated to the management of a transboundary aquifer, the 1978 Genevese Convention. Other examples include cooperative arrangements relating to the Guarani Aquifer in South America, the Al-Sag/Al-Disi Aquifer between Jordan and Saudi Arabi, the Iullemeden and Taoudeni/Tanezrouft aquifer system in West Africa, the Hueco Bolson Aquifer between Mexico and the United States of America, and the Abbortsford–Sumas Aquifer between the US State of Washington and the Canadian province of British Columbia. Eckstein also points to the ongoing efforts taking place to develop the law relating to transboundary aquifers at the global level, namely through the Model Provisions on Transboundary Groundwaters adopted under the auspices of the UNECE Water Convention and the 2008 International Law Commission's Draft Articles on Transboundary Aquifers. The analysis of the aquifer-specific and general arrangements leads Eckstein to conclude that a number of important norms are emerging relating to the regular exchange of data and information, the monitoring and generation of supplemental data and information, the prior notification of planned activities, the creation of joint institutional mechanisms, and certain substantive obligations. Eckstein suggests that the unique characteristics and functioning of groundwater require careful consideration in the future development of laws relating to transboundary aquifers, and merit specific attention.

McIntyre (Chapter 17) recognises that while international water law as a discrete sub-topic of international law is relatively recent, there is already quite well-settled practice around three key rules: equitable and reasonable utilisation, the duty to prevent significant transboundary harm and the duty to cooperate in the management of shared waters – with the latter encompassing a suite of ancillary procedural obligations. However, McIntyre also notes that international water law does not operate in isolation, but is also influenced by developments in other fields such as international environmental law, international human rights law and international investment law. In his analysis of equitable and reasonable utilisation, McIntyre accepts that there is no hierarchy between uses, but he also suggests that certain considerations will usually be accorded more significance than others. The significance of vital human needs and linkages to the human right to water are highlighted in this regard; together with the growing emphasis placed on environmental issues. Environmental protection is also addressed through the *due diligence* obligation to prevent significant harm. While McIntyre accepts the criticism that the rule is open texture, he also explains how legal scholars and practitioners have developed a decent understanding of what is meant by 'harm', and also the extent of due diligence obligation.

Insights from Leb (Chapter 18) build upon those of McIntyre. Leb firmly establishes the duty to cooperate as a core principle of international water law, and as a key factor in the implementation of other principles of international water law. Leb also recognises the central importance of the duty to cooperate in the achievement of the objectives of sustainable development and ecosystem protection. While often criticised as too general, Leb explains the normative content of the duty to cooperation and emphasises that the importance of the regular exchange of data and information among riparian states – and the need for a framework by which states can notify and consult over planned measures – are seen as key aspects of the duty to cooperate. Leb maintains that these procedures, together with joint institutions, provide important foundations by which states can identify and realise mutual benefits.

Schmeier (Chapter 19) also emphasises the importance of joint institutional arrangements for the governance of transboundary waters. The chapter takes on the understudied question of what factors lead to the establishment of river basin organisations (RBOs), and concludes that two key factors are influential, namely the nature of the problem and the constellation of actors within a respective basin. Schmeier also explains diversity in design of RBOs, such as in their membership structure, the types of problems they address, their functions and their mandate. Ultimately, Schmeier concludes by asking the question whether RBOs matter. A key contribution of the chapter here is to first explore the meaning of effectiveness and then identify a series of exogenous and endogenous factors likely to influence effectiveness. These factors align closely to the reasons for establishing the RBOs in the first place, and their design features. While Schmeier concludes that effectiveness factors are becoming increasingly understood, new challenges, including climate change and the proliferation of large-infrastructure projects, are likely to put greater strain on RBOs.

The significance of Nikiforova's contribution (Chapter 20) is two-fold. First, it provides an analysis of the UNECE Water Convention's Implementation Committee. The committee itself has great potential significance given that the membership of the Convention comprises 41 parties across Europe, Central Asia and the Caucasus. Additionally, with the amendment to the UNECE Water Convention, which opens it up to non-UNECE states, the reach of the implementation may well extend. Second, the implementation committee can be seen as the first of its kind. While it picks up on the trend of establishing such committees under international treaty regimes, it is the first that falls under a framework instrument for transboundary waters. The committee therefore offers an interesting example that might be considered by other treaty regimes relating to transboundary waters. Given the growing challenges over transboundary waters, the implementation committee may provide an important non-adversarial means by which to avoid disputes escalating.

Sangbana (Chapter 21) picks up on a growing trend in international water law, namely the development of legal rules and mechanisms for integrating individuals and communities into the management and protection of transboundary freshwater. The roles that non-state actors play in transboundary water management are explored. Such roles include having observer status, whereby non-state actors can participate in the meetings of RBOs, and may also be permitted to submit documents and proposals to the main decision-making body. Sangbana suggests that ability to submit projects proved a useful means by which NGOs in the Danube could raise their concerns over certain hydropower developments. However, the limits of this right to participate are highlighted, namely that it is ultimately up to the decision-makers whether they take on board the advice. Sangbana also highlights the role of public consultations in allowing non-state actors a say in policy development processes; and the use of experts in the activities of basin organisations. Through the examination of a number of cases, Sangbana maintains that these roles by non-state actors are critical to the implementation of treaty arrangements, and they should therefore be strengthened in treaty regimes where they are lacking.

Closely associated with the difficulties of implementing transboundary treaty frameworks, Farnum, Hawkins and Tamarin (Chapter 22) provide important insights into how power influences international law. Key observations in this chapter are that political processes that create international law are largely shaped by power relations; legal norms and institutions often serve to reinforce existing power dynamics; and legal norms and institutions can be leveraged for or against extant hegemonic orders. The authors suggest that, in the context of transboundary waters, different 'pillars' of power exist, including geography, material, bargaining and ideational. These pillars, in turn, influence control over transboundary rivers, lakes or aquifers. International law is recognised as a significant source of 'soft power', which can influence state relations over transboundary waters. A further key insight from the chapter is to recognise the importance of accounting for non-state actors in analysing the relationship between power, politics and international water law.

A key insight underlying the chapter by Daza Vargas (Chapter 23) also relates to the role of non-state actors and, specifically, the influence of business. The chapter explores a relatively new area of international law – that is, international investment law. While new, Daza Vargas notes this is a rapidly developing area of law due to the ongoing effort by states to liberalise, promote and attract foreign investment; the active negotiation and conclusion of bilateral investment treaties; free trade agreements and other types of investment treaties; and a dynamic system of dispute settlement. Daza Vargas explores the key normative features of international investment regimes, and analyses key cases related to water. A central question, which has only just begun to be explored, is how to reconcile an adaptive approach to the management of water resources with the requirements of consistency and stability under international investment treaty arrangements. From her analsyis, Daza Vargas concludes that while tribunals deciding international investment disputes have been cautious in considering other international obligations, such as those relating to human rights or the environment, there is a growing trend to modify and adapt international investment treaties to secure the exercise of regulatory prerogatives that are for the public interest.

Patrick and Wenz (Chapter 24) address water governance and, specifically, decision-making in the context of a philosophical analysis. Starting with an overview of international norms within the UN Watercourses Convention, regional treaties and negotiation theory, they then identify a series of 'classical' philosophical (and some more modern) understandings of human behaviour, including religious beliefs, Kant's categorical imperative, utilitarianism, economic cost–benefit analysis and cultural relativism – and give examples of how water decision-making could be framed, supported, or applied, by each of these. Although the theories will be familiar to all the lawyers who have studied jurisprudence, it is rare, and refreshing, to see them given such an applied analysis. Often when we work with stakeholders, on large interdisciplinary projects perhaps, or in river basin management contexts (whether transboundary or not), we tend to assess stakeholder perceptions, wishes and views from a financial or economic perspective. It is helpful, in a complex and rapidly changing world, to remember that there are many subconscious and deeply rooted reasons for human behaviour, and reflecting on these in our negotiations may increase our chances of a successful outcome.

A key message from Sánchez Remírez, Pacheco Mollinedo and Aguilar Porras (Chapter 25) is the importance of incorporating an ecosystem approach into measures that are designed to address the impacts of climate change. The chapter recognises the centrality of adopting a holistic approach to the management of water systems, while also enhancing the essential role of ecosystems and the services they provide for human well-being. Three main pillars for enhancing the linkages between IWRM, adaptive governance and ecosystem-based approaches are offered by the authors, namely: the incorporation of multi-level governance arrangements; effectively addressing participation and engagement of public and stakeholders; and embedding flexibility in all laws, policies, management practices and institutional mechanisms. The chapter explores these three pillars in detail, and also examines them in the context of the Mauri and Desaguadero River Basin in Bolivia. The case study offers a valuable insight into how application of the pillars can strengthen climate change adaptation plans.

In offering a brief history of global water governance, Newton (Chapter 26) provides the basis by which to explore how the international community is likely to respond to global water challenges in future years. Newton traces activities related to water at the international level back to intergovernmental meetings in 1851. A web of intergovernmental organisations and professional associations then evolved during the twentieth century and, as Newton suggested, gained significant momentum up to the present day. Newton notes that, with water rising on the political agenda national and internationally, global water governance is also likely to develop and transform; and discussion on whether the United Nations should contain a more formal mechanism for water is beginning.

Schulz (Chapter 27) tackles a critical topic that will attract considerable attention over the forthcoming years as the international community works towards the 2030 Development Agenda – that is, the sustainable development goals (SDGs) and water. Her chapter traces the meaning of sustainable development and convincingly argues for water's centrality within the concept. A key insight that Schulz makes is to recognise the need to move away from seeing water as a cross-cutting issue, which has tended to lead to piecemeal approaches. Schulz also highlights how both the principle of sustainable development and its constituent parts have infiltrated national and international laws relating to water. Such insights offer an important foundation by which SDGs related to water can be advanced. In this regard, Schulz cautions that there is a risk that SDGs may be seen as a 'top-down' solution and there that is, therefore, still a need to secure ownership by governments, civil society and other stakeholders in order to secure their implementation.

Magsig (Chapter 28) sheds new light on the concept of water security, a complex and cross-cutting paradigm that is likely to continue to shape water law and policy at multiple levels in the years to come. While more work is needed to unpack the meaning of water security, and as Magsig maintains, shift away from more traditional notions of security, the fact that water security has been recognised by institutions such as the UN Security Council is significant. In analysing the evolution and differing interpretations of the concept of security, Magsig does much to analyse what such a contemporary understanding of water security means. A key insight here is the need to shift away from a focus on controlling security issues via military force or containment, and rather adopt preventative measures that address the underlying factors contributing to insecurity. In examining water security specifically, Magsig recognises the need to focus on communities at multiple levels – not only to look at access to and threats from water, but also opportunities to share in the benefits that water offers. Through an analysis of international water law, Magsig demonstrates how key components of the legal framework might be viewed through a water security lens, and also advocates the benefits of framing water security as a regional common concern. Magsig also stresses the importance of engaging with non-state actors and their interests when contextualising what regional common concern might mean in practice.

In addition to these insights drawn from the specific chapters, reflecting on the *Handbook* as a whole also allows us to draw some more general observations. There is a growing momentum around both water and its governance. This is reflected at every level, international, regional, national and local; in both law and policy; in the work of policy-makers, regulators, academics and practitioners; and in the debates over both water resource management and the delivery of water services.

The challenges around implementation are not new, but remain immense. Central to address-ing them is the recognition that, ultimately, individuals are responsible for implementing laws and policy, and the realisation that capacity building is critical to such endeavours. The imple-mentation of international and transboundary law, whether global or regional, requires political commitment, it must be based on sound science (including, where appropriate, local knowledge) and it also requires conducive and coherent policy and legislation at national level. This needs to reflect the commitments made to other states as well as provide solutions to domestic problems. Benefit sharing offers great potential in incentivising cooperation, but governance mechanisms remain central to ascertaining and allocating benefits among stakeholders at multiple levels. At national level, even where law reform is enacted, it needs resources and direction to ensure it is effective, and to ensure focus on state priorities and linkages with other relevant areas of domestic law. Consistency between policy objectives and legal frameworks is imper-ative, so policy-makers must focus on policy implementation in order to ensure that legal tools are identified and used appropriately. Procedural rules around transparency, notification or environmental assessment are good examples of cross-cutting areas which support imple-mentation of domestic and international rules.

Challenges of scale are writ large in water. A small catchment in a country such as Scotland is not really comparable to the Murray Darling in Australia, still less to the Mekong or the Nile. Scale also affects the development and implementation of law. In a very large basin, a system that creates a small group of key stakeholders is unlikely to 'drill down' to the users on the ground. Where there are multiple administrative agencies, still more where there are multiple sovereign states, it will be difficult to reach agreement and also to give effect to such agreements. Structures, as well as the content of policy, must reflect those scale issues. Evidence from flood management, especially as regards disaster risk management, suggests that multi-scale responses across sectors can be effectively coordinated even in countries where capacity is low.

IWRM can still be seen as a key driver. The paradigm has been criticised, and it is not a pana-cea; nor is it a script. There is no 'one right way' and the priorities within an IWRM approach will vary – whether that is reform to allocation rules, management of pollution, or delivery of basic services. But, in the end, it is nothing more than the conceptualisation of the idea that all water uses should be identified, all stakeholders included and holistic decisions made on the use of this scarce and precious resource. The gradual expansion of human rights jurisprudence and approaches is influencing the way that water use rights are developed and applied. Behind all this looms the threat and reality of climate change, a force that is imposing ever more urgency on decision-makers, especially as they seek to meet the SDGs in the knowledge that the uncertainty over resource availability is greater than ever.

Perhaps the final word should be on the roles of stakeholders – in the broadest definition, including the policy-makers, law-makers and regulators; the professionals, whether water managers, hydrologists, engineers, economists or lawyers; and the users and consumers – domestic, agricultural and industrial. Indeed, the current focus of large commercial and industrial users on water is good evidence both of the pressure on the resource and its true value. In addition to engaging stakeholders with respect to decisions regarding water, a more active role is being slowly elaborated, one that acknowledges the inevitable capacity limitations of regulatory and enforcement authorities, and encourages self-monitoring of use, and direct enforcement of legislation by individuals. To meet the basic water needs of citizens, supply food and energy for a growing population, enable improvements to living standards and still maintain the fragile ecosystems on which humans also depend will require joined-up thinking, excellent information appropriately conveyed, political will and the active involvement of all the players.

References

OECD (2009) *Water Governance in OECD Countries: A multi-level approach*. Paris: OECD.

UN General Assembly (2015) Transforming our world: the 2030 Agenda for Sustainable Development. A/RES70/1.

UN-Water (2008) *Transboundary Waters: Sharing Benefits, Sharing Responsibilities. Thematic Paper* [online]. Available from: http://www.unwater.org/downloads/UNW_TRANSBOUNDARY.pdf

UN-Water (2012) *Status Report on the Application of Integrated Approaches to Water Resources Management* [online]. Available from: http://www.un.org/waterforlifedecade/pdf/un_water_status_report_2012.pdf

INDEX

Milton Keynes UK
Ingram Content Group UK Ltd.
UKHW051852071024
449327UK00025B/1919

9 780367 231064